Microbiology of Green Fuels

I0027566

Editors

Abu Yousuf
Department of Chemical Engineering and Polymer Science
Shahjalal University of Science and Technology
Sylhet, Bangladesh

Elia Tomás-Pejó
Biotechnological Processes Unit
IMDEA Energy Institute
Madrid, Spain

CRC Press
Taylor & Francis Group
Boca Raton London New York

CRC Press is an imprint of the
Taylor & Francis Group, an **informa** business

A SCIENCE PUBLISHERS BOOK

Cover credit: Cover illustration reproduced by kind courtesy of the editors.

First edition published 2023
by CRC Press
6000 Broken Sound Parkway NW, Suite 300, Boca Raton, FL 33487-2742

and by CRC Press
4 Park Square, Milton Park, Abingdon, Oxon, OX14 4RN

© 2023 Taylor & Francis Group, LLC

CRC Press is an imprint of Taylor & Francis Group, LLC

Library of Congress Cataloging-in-Publication Data (applied for)

ISBN: 978-0-367-77391-5 (hbk)
ISBN: 978-0-367-77392-2 (pbk)
ISBN: 978-1-003-17115-7 (ebk)

DOI: 10.1201/9781003171157

Typeset in Times New Roman
by Radiant Productions

Dedication

To
The researchers, who want to make the Earth greener and cleaner
for the next generation.

—Abu Yousuf and Elia Tomás-Pejó

Preface

A key priority in today's society is the implementation of a sustainable bio-based economy. For such a goal, the production of renewable bioproducts like biofuels to replace fossil-derived compounds is crucial. In this context, the utilization of microorganisms for the production of biofuels from renewable resources is advantageous in terms of environmental sustainability and it is expected to play an important role in bioeconomy in the near future. In this sense, green fuel synthesis from agro-industrial organic wastes by microorganisms would boost circular economy. The success of the biotechnological biofuel production process requires, however, a conversion microorganism capable of both efficiently assimilating the major derived carbon sources and diverting their metabolites towards the specific fuel.

This book aims to show recent advances in the production of green fuels by means of microorganisms. Promising processes and microorganisms involved in the biofuel production have been discussed to give an in-depth overview of the state of the art with broad spectrum of industrial microbiology without focusing on one single group of microorganisms and biofuels. For the sustainability of the green fuel technologies, the book has also addressed techno-economic, strategic, and commercial interest in promoting green fuels. These facts make this book very valuable for biofuel companies and scientific community.

Chapter 1 provides an introduction to the four generations of biofuels, the European Union (EU) and USA policies related to their development and an overview of the historical and commercial aspects of bioethanol, biobutanol and biodiesel production as green fuels.

Chapter 2 focuses on the properties of different varieties of algae and critically analyses existing and potential routes for producing algae derived biofuels including biological, chemical, and thermochemical methods.

Chapter 3 concentrates on yeast-biorefineries which can use inexpensive agro-industrial waste to obtain two main products: bio-oils (feed and food), and the enzyme lipase, which is a high-value product that can modify the bio-oil to get more products.

Chapter 4 and 5 discuss bio-hydrogen production pathways, sources, most promising fermentation methods including dark fermentation and photo fermentation, and the factors affecting the bio-hydrogen production. Microbial and ecological aspects of the dark fermentation process are addressed, with a particular focus on the microbial physiology of the main microorganisms involved in H_2 production. They also provide an overview of energy and economic analysis along with the challenges and future aspects for bio-hydrogen.

Chapter 6 and 7 explain detail process design required for second generation bioethanol production based on two large microbial consortia, genetically modified *S. cerevisiae* yeast and *Clostridium* bacteria species and so-called non-conventional microorganisms for producing bioethanol from lignocellulosic materials.

Chapter 8 and 9 describe the productivity of a few microbial strains such as *Clostridium beijerinckii, Clostridium saccaroperbutylacetonicum,* and *Clostridium saccharoacetobutylicum* to increase the yield of biobutanol, and the application of genetically modified clostridial strains to improve butanol concentration and tolerance to inhibitors derived from lignocellulose. Chapter 9 also stresses over the process intensification as an engineering tool to decrease production costs.

Chapter 10, 11 and 12 analyse the use of oleaginous microorganisms as lipid source for biodiesel production. Particularly, the microorganism cultivation and harvesting, and the lipid extraction and conversion to biodiesel are described. The chapters give an in-depth account to achieve biodiesel from microalgae in an economically and environmentally sustainable manner.

Chapter 13 and 14 address the performance of anaerobic digestion by elucidating the advantages of organic matters degradation acceleration, biogas enhancement, and functional microbes involved and microbial community structures, followed by the biogas upgrading. Chapter 13 focuses on bioelectrochemical system as a potential option for biogas production. Chapter 14 gives an overview of strategies and operating conditions for different bacteria and archaea involved in anaerobic digestion of sewage sludge to produce biogas.

Chapter 15 comprehensively assesses the advantages and challenges of biofuels from 1st to 4th generation including plant-based and microbial oils-based feedstock. It also discusses the technological, economic and financial issues that must be approached by techno-economic analysis and its constrains.

Researchers and scientists with strong academic background and practical experiences have shared their thoughts and findings of their investigations in this book. We believe the book will enrich the foresight of current researchers and industrialists who are dedicatedly working on Green Fuels.

Sylhet, Bangladesh   Abu Yousuf
Madrid, Spain   Elia Tomás-Pejó

Acknowledgement

We are eternally grateful to all distinguished authors for their thoughtful contribution to the success of the book project. Their patience and diligence in revising the first draft of the chapters after adapting the reviewers' suggestions and comments are highly appreciated.

We would like to acknowledge the solicitous contributions of all the reviewers who spent their valuable time in constructive and professional manner to improve the quality of the book.

We are very thankful to staff at CRC PRESS, particularly Raju Primlani, who supported us tremendously throughout the project and for his great and encouraging mind.

Contents

CHAPTER 1

Biofuels

Introduction, Historical and Commercial Aspects

Maria G. Savvidou, Styliani Kalantzi and *Diomi Mamma**

1. Introduction

Climate change is a major concern in the world and various efforts are made towards its mitigation. The main driver of climate change is the greenhouse gas (GHG) effect. According to IEA's (International Energy Agency) Global Energy Review 2021, CO_2 emissions, which make up the vast majority of GHG, declined by 5.8% in 2020 globally (mainly due to SARS-Cov-2 pandemic). In 2009, there was also a decrease in GHG emissions (due to global financial crisis) but the 2020 decrease was five times higher. Global energy-related CO_2 emissions reached the highest level in the atmosphere in 2020 accounting for 412.5 ppb (parts per million). This level is approximately 50% higher than the corresponding level in the beginning of industrial revolution.[1] This dramatic change in the atmospheric CO_2 concentration is attributed to the dependence of our society on fossil fuels for energy as well as on deforestation. The share of fossil fuels in the energy consumed worldwide is 80% while 58% of the latter is consumed by the transport sector (Guo et al. 2015, Raud et al. 2019). Transport sector contributes approximately one quarter of all energy related GHG emissions (Darda et al. 2019). According to IPCC's (Intergovernmental Panel on Climate Change) Sixth Assessment Report (AR6) on the physical science basis of climate change report, issued on August 2021, *"Human activities have warmed the atmosphere, ocean and land. The likely range of total human-induced warming global surface temperature increase from 1850–1900 to 2010–2019 is 0.8°C to 1.3°C, with a best estimate of 1.07°C"*.[2]

Further, on October 2018 the IPCC issued a special report on the impacts of global warming (GW) of 1.5°C. GW is associated with serious negative impacts on the ecosystems as well as on human health and wellbeing. Limiting GW to 1.5°C compared to 2°C could secure a more sustainable and equitable society.[3] Biofuel is predicted to be the alternative energy for CO_2 mitigation in the

Biotechnology Laboratory, School of Chemical Engineering, National Technical University of Athens, 9 Iroon Polytechniou Str, Zografou Campus, 15700, Athens, Greece.
* Corresponding author: dmamma@chemeng.ntua.gr

[1] https://iea.blob.core.windows.net/assets/d0031107-401d-4a2f-a48b-9eed19457335/GlobalEnergyReview2021.pdf.
[2] https://www.ipcc.ch/report/ar6/wg1/downloads/report/IPCC_AR6_WGI_SPM.pdf.
[3] https://www.ipcc.ch/sr15/.

transport sector (Oh et al. 2018). Furthermore, biofuels can provide easily stored energy source available on demand, can help countries depending on oil or natural gas imports, increase energy security, stabilize energy markets and boost the development of a circular economy (Demibras 2009, Nanda et al. 2018, Darda et al. 2019, Callegari et al. 2020).

Biofuels can be grouped based on their chemical properties, nature, feedstock type, conversion processes and technological characteristics and thus various categories could be found in literature. Two commonly used categories are "*first-, second-, third- and fourth-generation*" and "*conventional and advanced*" biofuels (Acheampong et al. 2017, Alalwan et al. 2019, Raud et al. 2019, Fivga et al. 2019, Callegari et al. 2020, Lin and Lu 2021).

First-generation biofuels are produced from food crops (i.e., rice, wheat, barley, potato, corn, sugarcane, vegetable oil, etc.), while second-generation biofuels are derived from non-food crops (e.g., dedicated energy crops, agricultural residues, forest residues and other waste materials such as used cooking oil-UCO and municipal solid waste). Feedstock for the production of third-generation biofuels is the algal biomass, while production of fourth-generation biofuel is conducted using genetically modified microorganisms (i.e., microalgae, yeast, fungi and cyanobacteria). First-generation biofuels are also referred to as "*conventional biofuels*" while second-, third- and fourth-generation as "*advanced biofuels*" (Gaurav et al. 2017, Alalwan et al. 2019, Fivga et al. 2019).

Biofuels can be found in different states including liquid (bioethanol, biobutanol, biodiesel), gaseous (biogas) and solid (densified solid biofuel) (Costa et al. 2020). Liquid biofuels, mainly used in transport sector, are characterized either by their ability to blend with existing petroleum fuels, or by their ability to be used in existing internal combustion engines (Yusoff et al. 2015, Mahmudul et al. 2017). It should be noted though that there are several other fuels such as liquefied petroleum gas (LPG), compressed natural gas (CNG) and electricity (for electric vehicles) that can be used in the transport sector. Even though the above fuels have advantages compared to petroleum, their use requires modifications in vehicles and new fueling infrastructure in contrast to biofuels (Chang et al. 2017).

2. EU and USA Legislation for Biofuels

The UNFCCC's (United Nations Framework Convention on Climate Change) main objective is the "*stabilization of greenhouse gas concentrations in the atmosphere at a level that would prevent dangerous anthropogenic interference with the climate system*".[4] The Kyoto Protocol is an international agreement, emerged from the UNFCCC, that imposed binding limits on the emissions of GHGs. Developed countries were committed to a 5% reduction in GHG compared to 1990 levels over the five year period 2008–2012 (the first commitment period). EU under this protocol committed to reduce GHG emissions by 8%.[5]

European Parliament and Council adopted the biofuel Directive in 2003 (Directive 2003/30/EC).[6] The Directive aimed at promoting the use of biofuels or other renewable fuels to replace diesel or petrol for transport purposes in EU Member States. The target set by the directive was a 5.75% replacement of all transport fossil fuels with biofuels by 31 December 2010. However, the share of biofuels in the transport sector for the years 2005 and 2010 was far from the target set, accounting for 1.0% in 2005 and 4.1% in 2010.[7]

---

[4]   https://unfccc.int/files/essential_background/background_publications_htmlpdf/application/pdf/conveng.pdf.

[5]   https://unfccc.int/process-and-meetings/the-kyoto-protocol/what-is-the-kyoto-protocol/kyoto-protocol-targets-for-the-first-commitment-period.

[6]   https://eur-lex.europa.eu/legal-content/EN/TXT/PDF/?uri=CELEX:32003L0030&from=EN.

[7]   https://op.europa.eu/en/publication-detail/-/publication/87b16988-f740-11ea-991b-01aa75ed71a1/language-en.

In 2009, EU issued Directive 2009/28/EC, also known as Renewable Energy Directive (RED).[8] RED required 10% renewable energy usage, by 2020 in the transport sector. Biofuels' share in transport sector over the time period 2010 to 2017 ranged from 4.1 to 4.6%.[7] Furthermore, Directive 2009/30/EC,[9] the revised form of Fuel Quality Directive (FQD) 98/70/EC,[10] required a 6% reduction of the life cycle GHG emissions of transportation fuels, by December 2020.

The suggestions of the European Commission in 2012 (COM. 595),[11] concerning the GHG emissions associated with the indirect land use changes (ILUC), not taken into consideration in previous Directives, were incorporated in Directive 2015/1513/EC.[12] Under this Directive, the maximum share of conventional biofuels (biofuels produced from food crops), set by RED to 10% target, was capped at 7% by 2020. Furthermore, this Directive introduced a non-binding target of 0.5% use of advanced biofuels by 2020, in transport sector.

In 2016, based on the "Clean Energy for all Europeans" package, EC proposed the recast of the RED for which the expiration date was the year 2021. The new Directive that succeeded RED is known as RED II (Directive 2018/2001/EU) and entered into force in December 2018.[13] RED II spans the years up to 2030, with the most important goal being, that by 2030, at least 32% of Europe's energy will come from renewable sources. Concerning the fuels used in transport sector, RED II dictates that 14% of the energy consumed by 2030 must be renewable. Furthermore, 0.2% of transport energy by 2022, 1% by 2025 and at least 3.5% by 2030 must be supplied by advanced biofuels. Advanced biofuels will be counted double towards both the 3.5% and the 14% targets. On the other hand, the contribution of conventional biofuels will be 7%, equal to 2020 levels. RED II sets a series of sustainability and GHG emission criteria that transport biofuels must have in order to be counted towards the overall 14% target. The GHG savings threshold for transport biofuels was set at 65% after January 2021. Another important aspect of RED II is associated with the land-use, as it sets limits on the use of biofuels that are produced from feedstocks with high ILUC risk. For the implementation of the above, EC adopted the Delegated Regulation (EU) 2019/807,[14] which supplements RED II and defines the criteria for the determination of high ILUC risk feedstock as well as the certification of low ILUC risk biofuels.

In the USA, biofuel production is regulated by the Renewable Fuel Standard (RFS).[15] It was established under the Energy Policy Act of 2005[16] which amended the Clean Air Act (CAA)[17] and expanded under the Energy Independence and Security Act of 2007.[18]

RFS is implemented by Environmental Protection Agency (EPA) in consultation with USA Department of Agriculture and the Department of Energy. The objective of the RFS program is the reduction of the volume of petroleum-based fuels used in transportation, for heating or as jet fuel, by replacing a certain volume of the above with renewable fuel. The target for total renewable biofuels set by RFS for the year 2022 is 36 billion gallons, while the 21 billion gallons must be advanced biofuels. According to RFS, advanced biofuel can be produced from qualifying renewable biomass with the exception of corn starch and the threshold for lifecycle GHG reduction was set at 50%. Under the RFS, the GHG emissions must include direct and indirect emissions as well as land use

---

[8] https://eur-lex.europa.eu/legal-content/EN/TXT/PDF/?uri=CELEX:32009L0028&from=EN.

[9] https://eur-lex.europa.eu/LexUriServ/LexUriServ.do?uri=OJ:L:2009:140:0088:0113:EN:PDF.

[10] https://eur-lex.europa.eu/resource.html?uri=cellar:9cdbfc9b-d814-4e9e-b05d-49dbb7c97ba1.0008.02/DOC_1&format=PDF.

[11] https://ec.europa.eu/energy/sites/ener/files/com_2012_0595_en.pdf.

[12] https://eur-lex.europa.eu/legal-content/EN/TXT/PDF/?uri=CELEX:32015L1513&from=EN.

[13] https://eur-lex.europa.eu/legal-content/EN/TXT/PDF/?uri=CELEX:32018L2001&from=EN.

[14] https://eur-lex.europa.eu/legal-content/EN/TXT/PDF/?uri=CELEX:32019R0807&from=EN.

[15] https://www.epa.gov/renewable-fuel-standard-program/overview-renewable-fuel-standard.

[16] https://www.epa.gov/laws-regulations/summary-energy-policy-act.

[17] https://www.epa.gov/laws-regulations/summary-clean-air-act.

[18] https://www.epa.gov/laws-regulations/summary-energy-independence-and-security-act.

change. Every year, EPA sets the annual targets. The proposed volume requirements for the year 2020, for cellulosic biofuel, advanced biofuel, and total renewable fuel were 0.54 billion gallons, 5.04 billion gallons and 20.04 billion gallons, respectively.[19]

3. Generations of Biofuels

Biofuels can be classified into two general categories: primary and secondary biofuels. Materials such as fuelwood, wood chips and pellets, and organic materials are referred to as primary biofuels. Those materials are used for heating, cooking or electricity production without being processed (Guo et al. 2015, Rodionova et al. 2017, Raud et al. 2019).

Secondary biofuels are divided into four generations, e.g., first-generation, second-generation, third-generation and fourth-generation biofuels (Fig. 1). The characteristics of the biofuel itself may not change between these "generations", what is changed is the source from which the fuel is derived (Alaswad et al. 2015).

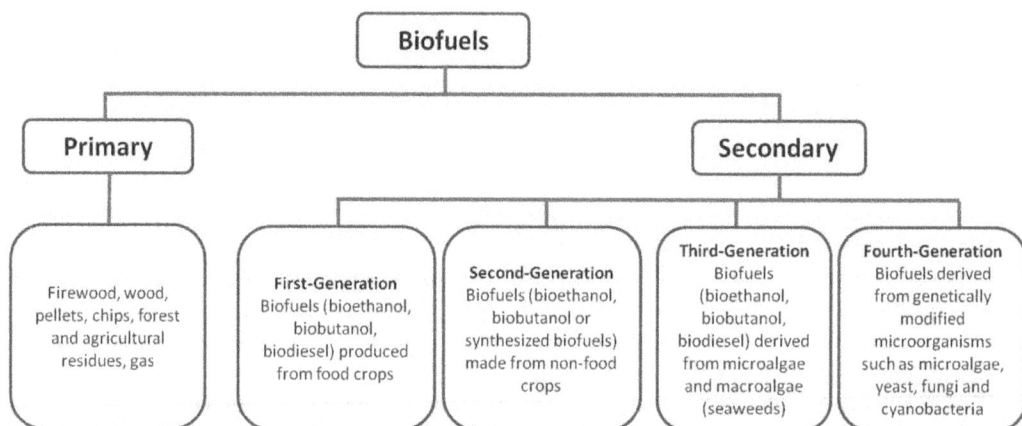

Fig. 1. Classification of biofuels (adapted from Raud et al. 2019 with modifications).

3.1 First-Generation Biofuels

First-generation biofuels represent a step towards energy independence and the abolition of fossil fuels as the sole energy source. The first-generation biofuels or conventional biofuels, referred to as fuels, are mainly produced from food crops such as rice, wheat, barley, potato, corn, sugarcane, vegetable oil, etc. The processes and technologies that convert food crops into biofuels are well established, making first-generation biofuels competitive with their fossil fuel counterparts (Raud et al. 2019, Lamichhane et al. 2021). On commercial level, the three main types of first-generation biofuels are biodiesel (Fatty Acid Methyl Esters, FAME or bio-esters), bioethanol and biogas (Naik et al. 2010). Biodiesel is produced from vegetable oils as well as residual oils and fats, through transesterification reaction (Rizwanul Fattah et al. 2020, Pasha et al. 2021). Biogas is produced by the anaerobic digestion (AD) (Olatunji et al. 2021). Bioethanol is produced from sugars by fermentation using the appropriate ethanologenic microorganism which converts simple sugars to ethanol. Starchy feedstocks, initially hydrolyzed by amylolytic enzymes to obtain simple sugars, are converted to ethanol by the appropriate ethanologenic microorganism (Bertrand et al. 2016). The world's largest producers of first-generation bioethanol are USA and Brazil using corn and sugarcane, respectively, as feedstock (Bertrand et al. 2016). Concerning the production costs,

---

[19] https://www.govinfo.gov/content/pkg/FR-2019-07-29/pdf/2019-15423.pdf.

according to Oliveira et al. (2019), historically, Brazil had lower production cost compared to USA, Europe, or China. Bioethanol production cost is now similar in both Brazil and USA. This could be attributed to the recent rapid increase in ethanol production in the USA combined with the decreased investment in ethanol production in Brazil.

The first-generation biofuels have been repeatedly criticized for the sustainability implications of their growing production scale. As stated by Nanda et al. (2018), those implications include competition for food, freshwater and land, land use changes, and deforestation.

Life Cycle Assessment (LCA) is a tool that is used widely to estimate global warming potential (GWP) as well as other environmental impacts of biofuels. Jeswani et al. (2020) demonstrated that first-generation biofuels can have lower GHG emissions than fossil fuels, provided that no land-use change (LUC) is involved.

3.2 Second-Generation Biofuels

The second-generation biofuels, also known as advanced biofuels, are fuels that can be produced from non-food crops, i.e., dedicated energy crops, agricultural and forest residues and other waste materials, such as used cooking oil (UCO) and municipal solid waste (Gaurav et al. 2017, Raud et al. 2019, Sharma et al. 2020). Second-generation biofuels exhibit advantages over their first-generation counterparts. Feedstocks used for second-generation biofuels production encounter the food *versus* fuel dilemma as they are not food crops and they do not require arable or fertile land for growing them. Second-generation feedstocks are characterized by high availability compared to food crops. Concerning their sustainability, implementation of second-generation biofuels has nearly no negative implications for the environment as they have greater potential than first-generation to reduce GHG emissions, provided there is no LUC (Nanda et al. 2018, Aron et al. 2020, Jeswani et al. 2020, Lamichhane et al. 2021, Lin and Lu 2021).

Lignocellulosic biomass (LB) represents the most abundant natural biomass with an approximate yield of about 200 billion tons per year (Kumar et al. 2020a). In recent years, LB has gained great interest as feedstock for energy production. LB is mainly composed of cellulose, hemicellulose and lignin. Cellulose is a linear homopolymer comprised of glucose units linked together by β-1,4-glycosidic bonds, with cellobiose residues as the repeating unit. Hemicellulose is a heteropolymer with side chains and is composed of pentose and hexose sugars. For different feedstock sources, the amount of sugar monomers can vary substantially. Lignin is an irregular and heterogeneous oxygenated polymer of p-propylphenol units, having as building blocks p-coumaryl, coniferyl, and sinapyl alcohols (Fig. 2) (Zhao et al. 2012, Tursi 2019).

Production of second-generation biofuels can be conducted *via* biochemical or thermochemical route. Through the controlled heating and/or oxidation of biomass, various intermediate energy carriers or heat could be generated. The thermochemical processes can convert biomass to fuel products *via* torrefaction (also called mild pyrolysis), pyrolysis, gasification, and hydrothermal liquefaction (HTL) (Fig. 3) (Damartzis and Zabaniotou 2011, Sikarwar et al. 2017, Gollakota et al. 2018, Ong et al. 2019, Okolie et al. 2021). During pyrolysis, biomass is thermally decomposed and as a result gaseous, liquid (pyrolysis oil) and solid intermediates are produced. Pyrolysis oil is a heterogeneous mixture of high alkalinity and oxygen content. This mixture can be upgraded to fuels or chemicals. The solid product (char) can be used as a fuel or soil amendment (Liu et al. 2020, Zadeh et al. 2020). Gasification is the partial oxidation of biomass resulting into the so called synthesis gas or syngas, which is rich in CO, H_2, CH_4, and CO_2. Syngas can be used in fueling internal combustion engines (Fiore et al. 2020). Furthermore, syngas can be fermented by specific acetogenic/anaerobic bacteria into bioethanol, biobutanol or platform chemicals (Sun et al. 2019, Ciliberti et al. 2020) or converted to liquid hydrocarbon fuels *via* the Fischer-Tropsch process (Sikarwar et al. 2017). HTL involves heating of the biomass, under pressure in the presence of water. Biomass at these conditions is subjected to a series of reactions including hydrolysis, dehydration,

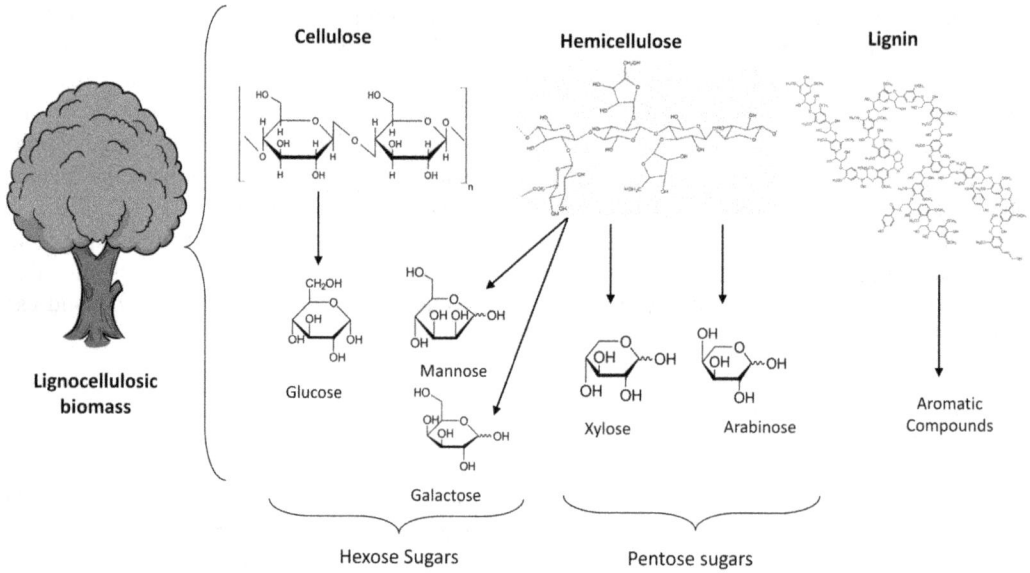

Fig. 2. The main components of LB.

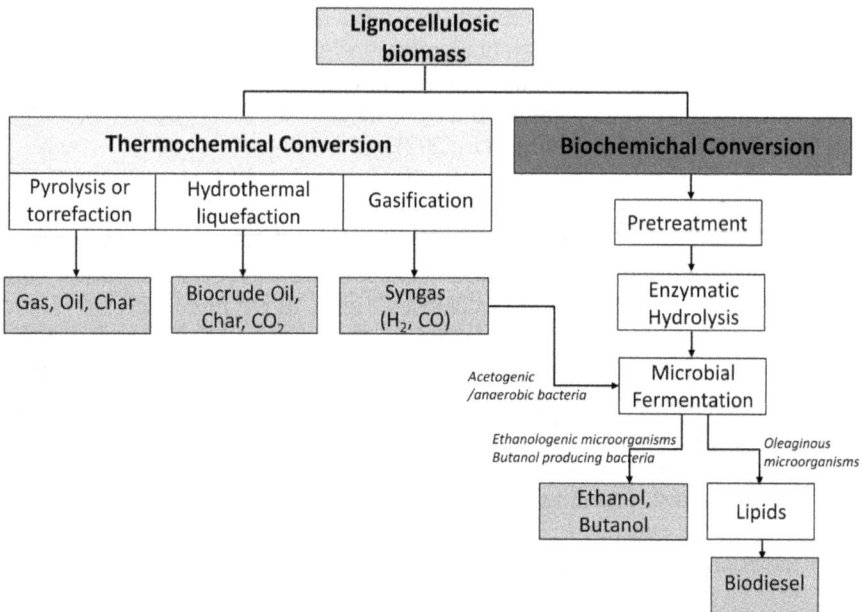

Fig. 3. LB conversion routes to biofuels.

decarboxylation, and condensation. HTL of biomass results in a mixture consisting of the biocrude oil which is water insoluble, an aqueous phase containing several organic compounds, a solid phase (biochar) and gases mainly in the form of CO_2. The main advantage of this technology is the use of wet biomass and thus the drying step, required for other processes, is omitted (Gollakota et al. 2018, Yang et al. 2020).

Biochemical conversion of LB is a multi-step process (Fig. 3), including (a) pretreatment to disrupt the recalcitrant structure of LB by removing lignin, degrade hemicellulose and alter the structure of cellulose, thus facilitating the accessibility of enzymes to cellulose, (b) enzymatic hydrolysis to generate sugars, using the appropriate cellulolytic and hemicellulolytic enzymes,

(c) fermentation of the resulted sugars, into liquid biofuels, such as bioethanol, bio-butanol and biodiesel. Oleaginous microorganisms, such as yeast and bacteria, can ferment hexose and pentose sugars present in LB hydrolysate and as a result accumulate lipids (mostly triglycerides, or TAGs) (*de novo* lipid accumulation). The latter are traditionally used as feedstock for lipid-based biofuels (Bhatia et al. 2017, Kumar et al. 2017, Birgen et al. 2019, Costa et al. 2020, Dey et al. 2020, Okolie et al. 2021, Sharma et al. 2020, Toor et al. 2020, Haq et al. 2021, Lamichhane et al. 2021, Machineni et al. 2020, Veza et al. 2021). In contrast to first-generation biofuels, large-scale production of second-generation biofuels still remains low as the production cost is high (Valdivia et al. 2016).

3.3 Third Generation Biofuels

Third-generation biofuels are those derived from algal biomass. Algae are a group of prokaryotic and eukaryotic organisms including single-celled genera as well as multicellular forms that can grow up to 60 m in length (Behera et al. 2015). Algae can be divided in two groups, microalgae and macroalgae. This distinction is based on their size (Fig. 4).

Fig. 4. (a) Cells of the microalgae *Nannochloropsis oceanica* and (b) *Macrocystis pyrifera* (Giant kelp) California, Channel Islands National Marine Sanctuary (USA National Oceanic and Atmospheric Administration; this photo is licensed under the Creative Commons Attribution 2.0 Generic license).

Algal biomass has several advantages as a feedstock for biofuel production. Microalgae can grow throughout the year and, having high growth rate compared to conventional crops, can provide more energy per unit of land area. Algae cultivation does not require arable land or fertilizers and microalgae can grow on water unsuitable for food crop production (e.g., waste, saline, brackish and non-potable water) (Benedetti et al. 2018, Shuba and Kifle 2018, Alalwana et al. 2019, Chowdhury and Loganathan 2019). Since autotrophic microalgae use CO_2 as a carbon source for their growth, their cultivation could help decrease the concentration CO_2 in the environment. As mentioned by Alalwana et al. (2019), for the production of 1 kg of algal biomass, approximately 1.8 kg of CO_2 must be consumed. Although algal biomass has advantages as a raw material for biofuel production, it also has disadvantages, mainly related to the high energy required for harvesting the microalgae, which represents 20–30% of the total production cost (Behera et al. 2015, Ananthi et al. 2021).

Microalgae can supply several different types of renewable biofuels. Biodiesel is derived from microalgal oil through acidic, alkaline, or enzyme-catalyzed reactions (Taparia et al. 2016, Dickinson et al. 2017). Biohydrogen could be produced photobiologically, due to inherent capacity of microalgae to split water into H_2 and O_2 using the solar energy. Furthermore, microalgae could be utilized as a feedstock for the production of bio-hydrogen through microbial dark fermentation (Khetkorn et al. 2017, Anwar et al. 2019). Finally, biomethane is produced by AD of algal biomass (Milledge et al. 2019) and bioethanol *via* fermentation of the carbohydrates (starch and cellulose) remained after lipids extraction (Debnath et al. 2021, Özçimen et al. 2020).

Microalgae can be processed through thermochemical, biochemical, chemical and direct combustion processes creating several end-use energy products (Raheem et al. 2018, Shuba et al.

2018, Khoo et al. 2019, Saad et al. 2019, Kumar et al. 2020b). Apart from biofuel production, microalgae are a source of bioactive compounds such as polyunsaturated fatty acids (PUFA), carotenoids, pigments, etc. with special interest in different industrial sectors such as food, feed, aquaculture cosmetics and health care industries (Suganya et al. 2016, Kumar et al. 2020b).

Concerning the sustainability of third-generation biofuels produced from microalgae, Aron et al. (2020) reported that they exhibit the lowest net GHG emissions compared to first- and second- generation biofuels. However, the high energy input required in several processing steps (e.g., harvesting of microalgae), mainly provided by fossil fuels, negatively contribute to GHG emissions.

Macroalgae (or seaweeds) are classified into three major groups: red (*Rhodophyta*), brown (*Phaeophyta*) and green (*Chlorophyta*). Macroalgae are photoautotrophic organisms and thus by consuming atmospheric CO_2 or HCO_3, they produce and store organic compounds (Bharathiraja et al. 2015, del Rio et al. 2020).

Concerning the carbohydrates present in seaweeds, those include cellulose, starch, sucrose, ulvan, carrageenan, agar, laminarin, mannitol, alginate, and fucoidan. Lignin, the major barrier in second-generation biofuels' production from LB, is nearly absent in seaweeds (Chen et al. 2015). Seaweeds are mainly used for the industrial production of hydrocolloids (also known as phycocolloids), i.e., alginate, carrageenan and agar (Biris-Dorhoi et al. 2020, del Rio et al. 2020). According to FAO, 40% of the total world hydrocolloids market is based on seaweeds.[20]

Even though the yield of hydrocolloid's extraction is generally high, there is a significant amount of solid residue that remains. For instance, as mentioned by del Rio et al. (2020), the solid residue remaining, after agar extraction, is approximately 30% of the total seaweed biomass used and this residue has a considerable amount of carbohydrates. This residual biomass could also be used as a feedstock for biofuel production. Furthermore, macroalgae's lipid content is usually low, compared to microalgae, and thus the production of biofuel from macroalgae is expected to depend on carbohydrates content rather than lipid content (Khoo et al. 2019).

Seaweeds have high water content (80–90%) which has a negative effect on several thermochemical processes that depend on dry biomass such as direct combustion, pyrolysis and gasification (Milledge and Harvey 2016). On the other hand, HTL for biocrude oil production (Djandja et al. 2020), fermentation for ethanol or butanol production, and AD for biogas production (McKennedy and Sherlock 2015, Maneein et al. 2018, del Rio et al. 2020) are among the processes appropriate for wet biomass.

Investigation of AD of macroalgae started in USA during the oil crisis of the 1970s as an alternative energy source and more recently as a solution in eutrophication problem caused by the proliferation of macroalgae in waterways (McKennedy and Sherlock 2015). Furthermore, apart from the whole seaweed, AD was performed in the residue after the removal of compounds of commercial interest, while the potential of digestate as a fertilizer has also been evaluated (Macura et al. 2019). Production of bioalcohols (bioethanol or biobutanol) from macroalgae is a multi-step process quite similar to that applied for bioalcohols production from LB. The major problem in bioethanol production is the inability of common ethanologenic microorganisms to metabolize the wide-range of sugars generated from the hydrolysis of carbohydrates present in seaweeds (Maneein et al. 2018, del Rio et al. 2020). Kawai and Murata (2016) summarized the developments in utilization and conversion of these sugars to bioethanol by different microorganisms. Biofuels production from macroalgae is not feasible from the economic point of view and thus the biorefinery concept for the simultaneous production of other commodities could be a solution (Khoo et al. 2019).

---

[20] http://www.fao.org/3/CA1121EN/ca1121en.pdf.

3.4 Fourth Generation Biofuels

The fourth-generation biofuels derive from genetically modified microorganisms such as microalgae, yeast, fungi and cyanobacteria. In contrast to the three other biofuel generations, no specific type of feedstock is characteristic of this generation. Synthetic biology approaches are the basis of fourth-generation biofuels production. Development of strains with their whole cellular networks optimized could simplify production stages. A widely accepted definition of synthetic biology is "*Synthetic biology is the design and construction of new biological parts, devices, and systems, and the re-design of existing, natural biological systems for useful purposes*" (Roberts et al. 2013). In this context, production of fourth-generation biofuels could be accomplished by developing photosynthetic microorganisms to produce photobiological solar fuels, by combining photovoltaics and microbial fuel production (electrobiofuels) or by developing synthetic cell factories or organelles (Acheampong et al. 2017, Aro et al. 2016, Alalwan et al. 2019, Malik et al. 2021, Mehmood et al. 2021).

Innovative research is directed towards photobiological solar fuels, which could be produced by photosynthetic organisms (single-celled algae or cyanobacteria), engineered if necessary.

Solar energy and water or water/CO_2 are the requirements for the production of photobiological solar fuels. The former raw material produces hydrogen-based fuels while the latter carbon-based ones (Aro et al. 2016, Chen et al. 2021). Production of biofuels directly from solar energy and Calvin–Benson cycle intermediates could be accomplished by introducing various fermentative metabolic pathways to cyanobacteria cells (Savakis and Hellingwer 2015, Liu et al. 2019, Xia et al. 2019).

Acheampong et al. (2017), concerning the maturity of the photobiological solar fuels technology, stated that "*[d]ue to proven fuel production potential of photobiological solar fuels, it is expected that during the upcoming decades, photobiological solar fuels will be arriving into the market*".

Microbial electrosynthesis (MES) is another innovative research area in the production of fourth-generation biofuel. MES relies on the ability of specific microorganisms to capture electrons from electrodes, incorporate them into their metabolism in order to convert CO_2 and excrete a reduced chemical as an electron sink (Prevoteau et al. 2020, Jourdin and Burdyny 2021). MES from CO_2 is carried out by homoacetogenic bacteria through the Wood-Ljungdahl pathway, which is considered the most energy efficient carbon fixation pathway (Ragsdale and Pierce 2008). Carbon is fixed in the form of several extracellular products, such as acetic acid, butyric acid, ethanol, and butanol, among others (Prevoteau et al. 2020).

In literature, under the term fourth-generation biofuel, mainly production of biofuel by genetically engineered algae is reported aiming at enhancing the quality and productivity of algae (microalgae and cyanobacteria). Homologous recombination, small interfering RNAs, clustered regularly interspaced short palindromic repeats/Cas (CRISPR/Cas), adaptive immune system and random insertion mutagenesis, are among the techniques that have been implemented to genetically modify microalgae and cyanobacteria (Gomaa et al. 2015, Abdulah et al. 2019, Brar et al. 2021, Godbole et al. 2021, Shokravi et al. 2021).

In general, cultivation of algae is conducted in both close and open systems. Both systems could be used for genetically engineered algae. The close cultivation system minimizes contamination and environmental exposure, while the capital and operating cost is higher compared to open system. Open cultivation systems exhibit higher risk of releasing the genetically engineered algae into the environment (Abdulah et al. 2019, Aron et al. 2020). Snow and Smith (2012) summarized the potential environmental problems caused by genetically engineered microalgae, including changes in the natural environment, toxicity, horizontal gene transfer, and competition with native species. In the study of Abdulah et al. (2019), the danger of using genetically engineered microalgae cultivated in an open pond system for the production of biofuels as well as the associated mitigation strategies were discussed. Apart from cultivation system (open or close system), another important issue is

Table 1. Advantages and disadvantages of the four generations of biofuels (Acheampong et al. 2017, Sikarwar et al. 2017, Abdullah et al. 2019, Chowdhury and Loganathan 2019).

Biofuel Generation	Advantages	Disadvantages
First	• Production and harvest of feedstock can be accomplished using already existing infrastructure and technology • The conversion to biofuels is easy • Commercial production/well established technologies • Regulations for first-generation biofuel production are clear	• Competition with foods crops • High cost of feedstock (food crops) • Rise of food prices • Arable land is required • Water and fertilizers are required • Reduction in net energy ratio
Second	• No food *vs* fuel conflict • Production and harvest of feedstock can be accomplished using already existing infrastructure and technology • Marginal land can be used for energy feedstock such Jatropha sp • Less water and no fertilizers required • Use of whole plant and/or residues results in more energy production per unit of land area • Regulations for second-generation biofuel production are clear	• Sophisticated processing technologies are required due to recalcitrant nature of LB • High production cost • No commercial production/immature technologies • Concerns about ecology preservation due to deforestation • Agricultural and forest residue use affects soil quality and contributes to soil erosion
Third	• No food *vs* fuel conflict • Non-arable land is required for algal cultivation • CO_2 fixation, waste water treatment, no fertilizers • Algal growth can be accomplished on water unsuitable for food production (e.g., waste, saline, brackish and non-potable water) • Algae is predicted to produce more energy per unit land area compared to conventional crops due to their high growth rate and short harvesting cycle • Apart from biofuels, algae provide several value-added products	• Insufficient biomass production for commercialization • Biomass contamination problem in open pond system • Requires new technologies in whole production line, i.e., from the production to processing into final biofuel product • High capital and production cost • No commercial production/immature technologies • Marine ecosystems could suffer eutrophication problems (e.g., algal blooms) • Marine cultivation is not yet subjected to specific regulations
Fourth	• No food *vs* fuel conflict • Non-arable land is required; growth can be accomplished on water unsuitable for food production (e.g., waste, saline, brackish and non-potable water) • High yield with high lipid containing algae; high production rate • Fourth-generation biofuel is considered as carbon negative	• Research is at initial stages • High capital and production cost • No commercial production/immature technologies • Leak of genetically modified microalgae could cause environmental and ecological problems • Marine cultivation is not yet subjected to specific regulations

how the byproducts and the residual water derived from the different stages of the production line will be handled. Both may contain plasmid or chromosomal DNA and simple disposal methods that do not include a step in which destruction of the genetic material will occur, involving the risk of lateral gene transfer.

Concerning the public acceptance of fourth-generation biofuels, especially those produced by genetically engineered algae across Europe, Villarreal et al. (2020) conducted a survey-based study. According to the survey's findings, the majority of respondents (a) believe that biofuels produced from genetically modified algae could provide benefits compared to other biofuels, (b) believe that closed production systems with high security standards should be applied to avoid unintentional impacts on humans and environment and (c) would choose to be final consumers of engineered algae biofuels, provided that there is clear evidence of their benefits and open communication of potential risks. However, a part of respondents were skeptical on the use of genetically modified microalgae, since the consequences of "genome editing" are still undefined.

4. Historical and Commercial Aspects of Selected Biofuels

4.1 Bioethanol

One of the widely used chemical compounds in the history of chemistry is ethyl alcohol, or ethanol. The production of alcoholic beverages *via* fermentation is one of the oldest procedures that mankind learned. Jars found in the Neolithic village Jiahu in northern China contained traces of a mixed fermented beverage. McGovern et al. (2004), after performing a chemical analysis of this residue, verified that the latter was a fermented drink made of grapes, hawthorn berries, honey, and rice. The date of its production was estimated in the range of 7000 to 6650 BC. Around 3400 BC, the Egyptians made different types of beer and wine.

Ethanol use as a fuel in engines dates back before the commercial production of gasoline. In 1826, the American inventor Samuel Morey designed an internal combustion engine for which he was granted with a patent (US Patent 4378 Issued April 1, 1826). This engine was intended to run a boat, using a mixture of ethanol and turpentine (refined from pine trees) as a fuel. Unfortunately, no investor could be found for further development of this engine. Thirty four years later, in 1860, the German engineer Nicholas August Otto created an internal combustion engine that could run on an ethanol-fuel blend. Even though he was denied a patent, Otto was funded by Eugen Langen, a sugar refining company owner who probably had associations with the European markets of ethanol (Songstad et al. 2009).

In 1896, Henry Ford in his tiny Detroit workshop created the Ford Quadricycle that could use pure ethanol as fuel. The Quadricycle reached 20 miles per hour, and due to its success, Ford founded the Ford Motor Company in 1903. In 1908, the famous Ford Model-T was produced, which was a flexible vehicle able of using ethanol or gasoline or a mixture of ethanol/gasoline as a fuel (Mussatto et al. 2010, Nanda et al. 2018). Henry Ford, in an interview in 1925 with The New York Times, argued on the production of bioethanol as well as on its use as a fuel, stating that ethanol is *"the fuel of the future"* (New York Times 1925). Earlier, in 1917, another great scientist and inventor, Alexander Graham Bell, in his National Geographic interview also spoke in favor of ethanol as a fuel (National Geographic 1917). After World War I, ethanol production became more expensive compared to the processing cost of petroleum-based fuel and as a result the demand for ethanol was reduced. Nevertheless, industries such as General Motors Corporation and DuPont were interested in using ethanol as antiknock agent (i.e., octane enhancer) as well as for possible replacement of petroleum-based fuels (Mussatto et al. 2010, Nanda et al. 2018). In 1923, leaded gasoline became commercially available, but scientists expressed concerns about public health due to lead poisoning perspective, underlining that there were alternatives (ethanol-gasoline blends) to the use of lead as antiknock agent (Kovarik 1998).

In the 1930s, gasoline-ethanol blends were popular in the Midwest States of USA since those states dominated the production of corn, which was used as a feedstock for ethanol production. From 1933 to 1939, various oil companies and the American Petroleum Institute claimed that tax incentives for ethanol would hurt the oil industry and that ethanol as a fuel was inferior to gasoline. During World War II, in USA, due to delays in rubber production, ethanol served as a feedstock for rubber products such as tires, raincoats, and engine gaskets for the war effort.

Interest in ethanol was revived worldwide in the 1970s due to Arab oil embargo and the concomitant oil crisis. Brazil created the National Alcohol Program (ProAlcool) in 1975. Production of ethanol in Brazil was conducted using sugarcane as feedstock. Sugarcane was selected due to problems in the international sugar market caused by overproduction of the commodity. In 1984, most new cars sold in Brazil were run by hydrated bioethanol as fuel. Furthermore, bioethanol and sugarcane production grew rapidly since 2000 due to rising oil prices and the availability of flex-fuel vehicles, which were supported by tax incentives (Mussato et al. 2010).

The USA, in 1970s, started the gradual removal of lead from gasoline (DiPardo 2000). Promotion of the production and use of bioethanol in USA was resumed in the late 1970s. One of the main reasons for that was the reinvigoration of the agricultural sector (Mussato et al. 2010).

In the EU, conventional bioethanol is produced from cereals and sugar beet derivatives. Wheat is mainly used in Germany and France, while corn is the feedstock mainly used in Central Europe (Hungary in particular).[21]

According to the data presented in IEA's (International Energy Agency) report on Renewables 2020, fuel ethanol production reached 115 billion L globally in 2019. However, in 2020 production decreased by 14.5% to 98 billion L due to SARS-Cov-2 crisis. Global ethanol production is projected to increase to 132 billion L by 2030. Production of ethanol in USA reached 59.5 billion L in 2019, which decreased by 12% to 52 billion L in 2020. Brazil produced 36 billion L of ethanol in 2019, while in 2020, bioethanol industry exhibited a 16.5% reduction in production. Ethanol production in China reached 3.9 billion L in 2019, and this level remained stable at around 4 billion L in 2020.[22] Together, USA and Brazil produce 84% of the world's ethanol, while EU is the third largest ethanol producer worldwide with 4.7 billion L.[23] At present, about 64% of ethanol is produced from corn, 26% from sugarcane, 3% from molasses, 3% from wheat, and the remainder from other grains, cassava or sugar beets.[24]

The three major ethanol producers, i.e., Brazil, USA and EU, use food-crops for its production and thus production suffers from sustainability implications (Nanda et al. 2018).

On the other hand, commercial size plants for the production of advanced bioethanol (second-generation bioethanol) have been constructed in several parts of the world. According to Padella et al. (2019), cellulosic ethanol production is significantly lower compared to the installed capacity. Bioethanol production process from advanced feedstocks exhibits high production cost while the technology is still in immature state. An overview of the global commercial scale cellulosic ethanol plants with capacity higher than 40.000 tons/year is presented in Table 2. It is evident from the data presented in the Table that the majority of plants appear to be in idle or on-hold state. The latter could be attributed to competitiveness issues between cellulosic and first-generation bioethanol (Padella et al. 2019).

---

[21] https://www.etipbioenergy.eu/value-chains/products-end-use/products/conventional-ethanol.
[22] https://iea.blob.core.windows.net/assets/1a24f1fe-c971-4c25-964a-57d0f31eb97b/Renewables_2020-PDF.pdf.
[23] https://afdc.energy.gov/data/.
[24] https://www.oecd-ilibrary.org/sites/3aeb7be3-en/index.html?itemId=/content/component/3aeb7be3-en.

Table 2. Commercial scale plants for bioethanol production from LB (Pandella et al. 2019, European Technology and Innovation Platform).[25]

Company	Country	Capacity Tons/Year	Status	Start-up Year
Abengoa Bioenergy Biomass of Kansas, LLC	USA	75.000	idle	2014
Beta Renewables (acquired by Versalis)	EU (Italy)	40.000	on hold	2013
DuPont Cellulosic Ethanol LLC (acquired by VERBIO)	USA	83.000	idle	2016
GranBio	Brazil	65.000	operational	2014
POET-DSM Advanced Biofuels	USA	75.000	idle	2014
Longlive Bio-technology Co. Ltd	China	60.000	idle	2012

Beta Renewables, a subsidiary of Chemtex and Grupo M&G, had developed the so called PROESA® technology for the production of cellulosic ethanol. This technology was applied in the world's first commercial-scale cellulosic ethanol plant that has been constructed in Crescentino, Italy and became operational in 2013. In 2017, the facility ceased production due to bankruptcy of the Grupo M&G, which was the parent company. The bio companies of Grupo M&G were acquired by Versalis, in 2018 (Padella et al. 2019).

In December 2015, Abengoa Bioenergy Biomass of Kansas, LLC ceased production at its plant, due to financial difficulties.[26] DuPont, in November 2018, after merging with Dow, idled the Iowa plant and sold it to Verbio North America Corp. The latter turned the plant into a renewable natural gas production facility. POET-DSM Advanced Biofuels, LLC, is based in South Dakota, USA and is a 50/50 joint venture between Royal DSM, Netherlands, and POET, LLC. The facility produced ethanol from corn crop residues. In November 2019, POET-DSM Advanced Biofuels, LLC announced that "*it will pause production of cellulosic biofuels at Project LIBERTY and shift to R&D focused on improving operational efficiency*".[27] Shandong Long live facility in China was established for the production of advanced bioethanol from corn stover and had a capacity of 60.000 tons per year. According to 2019 report on China's biofuel presented by United States Department of Agriculture (USDA)/Foreign Agricultural Service, the facility recently ceased operation.[28]

USDA/Foreign Agriculture Service reported in 2020 that there are three operational cellulosic ethanol plants of different capacity in Brazil, namely Bioflex/Granbio, Raizen-Costa Pinto Unit (capacity 36.000 tons/year) and Centro de Tecnologia Canavieira (CTC) demonstration plant (capacity 4.600 tons/year). Among the three facilities, the Raizen-Costa Pinto Unit is the only one producing at relatively large scale.[29]

4.2 Butanol

In 1862, Louis Pasteur reported on a microorganism that was able to grow in the absence of oxygen (anaerobic) and also to produce butyric acid and butanol. Pasteur gave the name *Vibrion butyrique* to that microorganism due to the fact that the major product was butyric acid (Pasteur 1862). Early in 20th century, the amount of natural rubber could not cover the demand and at that time, the idea

---

[25] https://www.etipbioenergy.eu/value-chains/products-end-use/products/cellulosic-ethanol.

[26] https://www.etipbioenergy.eu/value-chains/products-end-use/products/cellulosic-ethanol#ce1.

[27] https://poet.com/pr/epa-actions-trigger-project-liberty-shift.

[28] https://apps.fas.usda.gov/newgainapi/api/report/downloadreportbyfilename?filename=Biofuels%20Annual_Beijing_China%20-%20Peoples%20Republic%20of_8-9-2019.pdf.

[29] https://apps.fas.usda.gov/newgainapi/api/Report/DownloadReportByFileName?fileName=Biofuels%20Annual_Sao%20Paulo%20ATO_Brazil_08-03-2020.

of producing its synthetic counterpart emerged. Butanol was the starting material for butadiene production which in turn was the monomer used in synthetic rubber production. Butanol was not commercially available at that time and research efforts were directed towards its production (Jones and Woods 1986). Charles Weizmann isolated a bacterial culture (the main bacterial species of this culture was *Clostridium acetobutylicum*) with the ability of fermenting starchy materials to acetone and butanol (Sauer 2016). In 1915, Charles Weizmann was granted with a patent that claimed the acetone and butanol production by the isolated bacterial culture using starchy materials, with higher products' yield compared to other known cultures (Weizmann 1915). This innovative method of acetone/butanol production is also known as "the Weizmann process".

The demand for acetone, which was used in the production process of explosive cordite (an alternative to gunpowder) during World War I, especially in Britain, led to the implementation of the Weizmann process in industrial scale, in 1916 (Moon et al. 2016, Sauer 2016). After World War I, acetone production was sidelined while butanol became the chemical of interest, mainly due to increase in the production of cars. Butanol served as a precursor for the synthesis of butyl acetate which in turn was used for the production of quick drying lacquers applied in automobile manufacturing. During the start of World War II, Japan fueled airplanes with butanol produced from sugar plants. In the 1960s, the petrochemical route for butanol production emerged and the fermentation processes were abandoned. The reason was the availability of petroleum products in combination with the rising cost of molasses which were the feedstock used for ABE fermentation. The oil crisis of the 1970s revived the interest in butanol production, since it has superior fuel properties compared to bioethanol (Ndaba et al. 2015, Moon et al. 2016, Birgen et al. 2019, Vees et al. 2020, Veza et al. 2021).

Some of the major companies involved in butanol production are Celtic Renewables Ltd., Green Biologics Inc., Cobalt Biofuels, Butamax Advanced Biofuels, ButylFuel LLC and Gevo Inc.

Celtic Renewables Ltd. (based in UK), formed in 2012, uses a process based on a *Clostridium* sp. with the ability to convert both pentose and hexose sugars into acetone-butanol-ethanol and in thermally hydrolyzed whisky by-products as feedstock. Currently, the company is constructing a commercial scale processing plant in Grangemouth, Scotland.[30]

Green Biologics Inc. (based in UK) was founded in 2011. The company had developed a *Clostridium* strain that had the ability of converting a wide range of sugar into n-butanol through the ABE fermentation. Unfortunately, in 2019 the company ceased operation of its plant in Little Falls, USA.

Cobalt Biofuels (based in USA) was another company interested in biobutanol production from LB, but in 2015 the company was pivoted away from biofuels and turned into software development; also the company's name was changed to Cobalt Technologies.[31]

Butamax Advanced Biofuels LCC (based in USA) is a joint venture of BP and DuPont formed in 2009, aiming at the development and commercialization of bio-isobutanol as a next-generation renewable biofuel. It also should be noted that in 2018, the U.S. Environmental Protection Agency announced that Butamax has been granted registration of iso-butanol as a gasoline additive at up to 16%.[32]

Gevo Inc., focuses mainly on iso-butanol production applying a technology that has been developed by the same company, named GIFT® (Gevo Integrated Fermentation Technology), while

---

[30] https://www.celtic-renewables.com/process/.

[31] https://www.cobalttech.com/.

[32] https://www.govinfo.gov/content/pkg/FR-2018-03-29/pdf/2018-06119.pdf.

yellow dent corn #2 is used as a feedstock. The technology is based on a genetically modified yeast and on the continuous removal of iso-butanol from the fermentation broth.[33] The Gevo's facility in Luverne, Minessota, USA, according to the company's press release, re-started production operations of iso-butanol on Aug 04, 2021.[34]

4.3 Biodiesel

According to historical records, plant and seed oils have been used since 1500 BC. Oils and fats were not only utilized for light and heat purposes, fragrant oils were also employed as cosmetics, in religious events and medicine by the Ancient Egyptians. Furthermore, these oils have long been used in human diet. Whale oil was an important source of fuel in the 17th and 18th century in Europe and the USA (Balasubramanian et al. 2019).

The German engineer Rudolf Diesel was the inventor of the diesel engine, around 1890. In 1893, he was granted with a patent for this engine. Diesel engine could work with a variety of fuels including vegetable oil, and that was the major advantage of this engine compared to its petrol counterpart. However, in 1900, Nicolas Otto was the one who presented one of the new diesel engines at the Paris Exposition (Balasubramanian et al. 2019). This engine, upon request by the French Government, was fueled with peanut oil. The reason for that request was that French Government was exploiting the possibility of using oils derived from several kinds of nuts, widely available in its African colonies, ensuring that way energy autonomy in the colonies (Solomon and Krishna 2011). Implementation of diesel engine started 20 years after the expiration of Diesel's patent, i.e., in 1908. Rudolph Diesel was a pioneer on the exploitation of vegetable oils as fuels. He strongly believed that farmers could benefit from providing their own fuel. In 1912, a year before his death, he gave a speech arguing on the significance of vegetable oils as fuels, stating that "*the use of vegetable oils for engine fuels may seem insignificant today, but such oils may become, in the course of time, as important as petroleum and the coal-tar products of the present time*" (Knothe 2010). Fuels of vegetable origin became less important after Diesel's death considering that at the time various petroleum forms were available. As a consequence, the design of diesel engine was changed in order to be compatible with the properties of petroleum diesel fuel and thus vegetable oils due to their high viscosity could not be used anymore in those engines.[35] In the 1930s, lowering the viscosity of vegetable oils became a subject of interest. Different methods were investigated included pyrolysis, blending with solvents, and even emulsifying the fuel with water or alcohols; unfortunately, none of these solutions was suitable. In 1937, a patent for a "*procedure for the transformation of vegetable oils for their uses as fuels*", now called transesterification, was issued by Georges Chavanne, a professor of chemistry at the University of Brussels. In 1938, a passenger bus using palm oil ethyl ester as a fuel travelled between Brussels and Louvain (Knothe 2017). European countries like France, Belgium, and the UK expressed their interest in biodiesel production from oil crops. Those countries were not producers of oil crops, but the feedstocks were available in their colonies (Balasubramanian et al. 2019). During World War II, petroleum fuel supply faced serious restrictions. At that time, countries like Brazil, Argentina, China, India, and Japan turned to the use of vegetable oils, as fuel. However, the end of the War restored the petroleum fuel supply and once again vegetable oil fuel was sidelined (Knothe 2017).

The oil crisis of the 1970s revitalized the interest in vegetable oils as fuels. Currently, the term biodiesel includes traditional biodiesel (i.e., FAME) and hydrogenation derived renewable diesel

---

[33] https://gevo.com/products/isobutanol/.

[34] https://investors.gevo.com/news/gevos-luverne-facility-re-starts-production-operations.

[35] https://www.biodiesel.com/history-of-biodiesel-fuel/.

(HDRD) (previously known as hydrotreated or hydrogenated vegetable oil or HVO). HDRD is produced by subjecting fats or vegetable oils to a hydrotreating process. These fuels are also known as "renewable diesel fuels" and "green diesel" instead of "biodiesel" which is reserved for the fatty acid methyl esters (FAME) (Douvartzides et al. 2019).

According to the data presented in IEA's (International Energy Agency) report on Renewables 2020, global biodiesel and HDRD production was 48 billion L in 2019, while a decrease of 5% was recorded in 2020 mostly in European markets, mainly due to SARS-CoV-2 crisis. Global biodiesel production is projected to increase to 50 billion L by 2030 due to ongoing growth in Asian countries (such as Indonesia) and USA.[36]

Biodiesel was the first biofuel developed and used in the EU, adopted by the transportation sector in the 1990s. The EU is the world's largest biodiesel producer, accounting for approximately 32.3% of the global biodiesel production.[37] Biodiesel and HDRD production, in the EU, was 15.7 billion L in 2019 (approximately 12.8 billion L of FAME and 2.9 billion L of HDRD), while the production decreased to 13.6 billion L (11.9 billion L of FAME and 3.6 billion L of HDRD) in 2020, due to SARS-CoV-2 crisis. Projections over the period, 2023 to 2025, show that biodiesel production could reach 16.5 billion L in EU.[38] The annual capacity of facilities that produce biodiesel, in EU, range from 2.3 million L to 680 million L. Small sized plants are usually owned by farmers.

According to 2021 report on EU's biofuels presented by USDA/Foreign Agricultural Service, the majority of biodiesel in EU is produced by rapeseed oil (38% of total feedstock use in 2020), followed by used cooking oil (28%), palm oil (18%), animal fats (8%) and sunflower (2%). As evident, the share of palm oil in the total feedstock used for the production of biodiesel was high in 2020. RED II classifies biodiesel produced from palm oil as fuel with high indirect land-use change (ILUC)-risk. Germany is the major producer of biodiesel (FAME) in EU followed by France and Spain, while Netherlands produce the highest amount of HDRD followed by Italy and France.[39]

USA ranks second in biodiesel and HDRD production (approximately 18.1% of global production) with 8.4 billion L produced in 2019, while in 2020 production remained almost constant at around 8.2 billion L, not severely affected by SARS-CoV-2 crisis. Projections over the period 2023 to 2025 show an average annual biodiesel production of 14 billion L, mainly due to a fourfold increase in HDRD.[40] Major feedstocks used are soybean oil and UCOs.[41]

Indonesia ranks third in global biodiesel production, using palm oil as a feedstock. In 2019, the biodiesel production was 7.2 billion L, with a slight increase recorded in 2020 (7.9 billion L). Over the period 2023 to 2025, an increase of approximately 32% in biodiesel production is expected in Indonesia.[42]

Brazil produced 5.9 billion L of biodiesel, in 2019, from soybean oil, while production could reach 7 billion L by 2025.[43]

Replacement of petro-diesel by its bio-counterpart produced by oilseed plants is not a sustainable option for the future, mainly due to land use issues. To overcome this problem, current research is directed towards new feedstocks such as algae oil and dedicated energy crops that can be cultivated on land not suitable for food crops. Furthermore, research is also conducted for biodiesel production from LB using oleaginous microorganisms (Kumar et al. 2017).

[36] https://iea.blob.core.windows.net/assets/1a24f1fe-c971-4c25-964a-57d0f31eb97b/Renewables_2020-PDF.pdf.

[37] http://www.fao.org/3/cb5332en/Biofuels.pdf.

[38] https://iea.blob.core.windows.net/assets/1a24f1fe-c971-4c25-964a-57d0f31eb97b/Renewables_2020-PDF.pdf.

[39] https://apps.fas.usda.gov/newgainapi/api/Report/DownloadReportByFileName?fileName=Biofuels%20Annual_The%20 Hague_European%20Union_06-18-2021.pdf.

[40] https://iea.blob.core.windows.net/assets/1a24f1fe-c971-4c25-964a-57d0f31eb97b/Renewables_2020-PDF.pdf.

[41] http://www.fao.org/3/cb5332en/Biofuels.pdf.

[42] https://iea.blob.core.windows.net/assets/1a24f1fe-c971-4c25-964a-57d0f31eb97b/Renewables_2020-PDF.pdf.

[43] https://iea.blob.core.windows.net/assets/1a24f1fe-c971-4c25-964a-57d0f31eb97b/Renewables_2020-PDF.pdf.

5. Conclusions

Biofuels, a collective term used for liquid, gaseous and solid fuel sources, are primarily derived from biomass through many thermochemical and biochemical pathways. Biofuels can help in mitigating climate change caused by the increasing GHG emissions due to continuous use of fossil fuels, can help countries increase energy security, and stabilize energy markets and finally develop a new circular economy. Laws and regulations linked to the production and use of energy from renewable sources were legislated in several countries. Governments, through financial supports and tax incentives, promote the production and use of biofuels. Biofuels can be categorized into four generations. Depending on the feedstock used, biofuels are termed first-, second-, third-generation while production of fourth generation could be accomplished by developing designed microorganisms that could efficiently convert solar energy to fuel. Each generation of biofuel has advantages and limitations. First generation biofuels majorly come from food crops raising the food versus fuel dilemma, while second-, third-, and fourth-generation biofuels pose no threat to the food supply. Technology for the production of first-generation liquid biofuels, i.e., bioethanol and biodiesel, is mature and their production is the most cost-effective, resulting in commercial exploitation. USA is the world's largest producer of ethanol from corn, followed by Brazil where sugarcane is used as feedstock. The EU is the world's largest biodiesel producer, using rapeseed oil and UCOs as feedstocks.

On the other hand, second and third generation biofuels have high production costs, while the technology for its production is still immature. Second-generation biofuels have a greater potential than first-generation to reduce GHG emissions, provided that no LUC is involved, while third-generation biofuels from algae, at the present state of development, show the lowest net GHG emissions compared to first- and second-generation biofuels. Technology for the production of fourth generation biofuels is still in research stage. Overall, despite the potential of biofuel market, commercial scale production of advanced biofuels still requires technological development in order to achieve the coveted sustainability.

References

Abdulah, B., Muhammad, S.A.F.S., Shokravi, Z., Ismail, S., Kassim, K.A., Mahmood, A.N. and Aziz, M.M.A. 2019. Fourth generation biofuel: A review on risks and mitigation strategies. Renewable and Sustainable Energy Reviews 107: 37–50. https://doi.org/10.1016/j.rser.2019.02.018.

Acheampong, M., Ertem, F.C., Kappler, B. and Neubauer, P. 2017. In pursuit of Sustainable Development Goal (SDG) number 7: Will biofuels be reliable? Renewable and Sustainable Energy Reviews 75: 927–937. https://doi.org/10.1016/j.rser.2016.11.074.

Alalwan, H.A., Alminshid, A.H. and Aljaafari, H.A.S. 2019. Promising evolution of biofuel generations. Subject review. Renewable Energy Focus 28: 127–139. https://doi.org/10.1016/j.ref.2018.12.006.

Alaswad, A., Dassisti, M., Prescott, T. and Olabi, A.G. 2015. Technologies and developments of third generation biofuel production. Renewable and Sustainable Energy Reviews 51: 1446–1460. https://doi.org/10.1016/j.rser.2015.07.058.

Ananthi, V., Balaji, P., Sindhu, R., Kim, S.-H., Pugazhendhi, A. and Arun, A. 2021. A critical review on different harvesting techniques for algal based biodiesel production. Science of the Total Environment 780: 146467. https://doi.org/10.1016/j.scitotenv.2021.146467.

Anwar, M., Lou, S., Chen, L., Li, H. and Hu, Z. 2019. Recent advancement and strategy on bio-hydrogen production from photosynthetic microalgae. Bioresource Technology 292: 121972. https://doi.org/10.1016/j.biortech.2019.121972.

Aro, E.M. 2016. From first generation biofuels to advanced solar biofuels. Ambio 45: 24–31. https://doi.org/10.1007/s13280-015-0730-0.

Aron, N.S.M., Khoo, K.S., Chew, K.W., Show, P.L., Chen, W.-H. and Nguyen, T.H.P. 2020. Sustainability of the four generations of biofuels—A review. International Journal of Energy Research 44(12): 1–17. https://doi.org/10.1002/er.5557.

Balasubramanian, N. and Steward, K.F. 2019. Biodiesel: History of plant based oil usage and modern innovations. Substantia 3(2)Suppl. 1: 57–71. https://doi.org/10.13128/Substantia-281.

Behera, S., Singh, R., Arora, R., Sharma, K.N., Shukla, M. and Kumar, S. 2015. Scope of algae as third generation biofuels. Frontiers in Bioengineering and Biotechnology 2: 90. https://doi.org/10.3389/fbioe.2014.00090.

Benedetti, M., Vecchi, V., Barera, S. and Dall'Osto, L. 2018. Biomass from microalgae: The potential of domestication towards sustainable biofactories. Microbial Cell Factories 17: 173. https://doi.org/10.1186/s12934-018-1019-3.

Bertrand, E., Vandenberghe, L.P.S., Soccol, C.R., Sigoillot, J.-C. and Faulds, C. 2016. First generation bioethanol. pp. 175–212. *In*: Soccol, C., Brar, S., Faulds, C. and Ramos, L. (eds.). Green Fuels Technology. Green Energy and Technology. Springer International Publishing Switzerland. https://doi.org/10.1007/978-3-319-30205-8_8.

Bharathiraja, B., Chakravarthy, M., Kumar, R.R., Yogendran, D., Yuvaraj, D., Jayamuthunagai, J., Kumar, R.P. and Palani, S. 2015. Aquatic biomass (algae) as a future feedstock for bio-refineries: A review on cultivation, processing and products. Renewable and Sustainable Energy Reviews 47: 634–653. https://doi.org/10.1016/j.rser.2015.03.047.

Bhatia, S.K., Kim, S.-H., Yoon, J.-J. and Yang, Y.-H. 2017. Current status and strategies for second generation biofuel production using microbial systems. Energy Conversion and Management 148: 1142–1156. https://doi.org/10.1016/j.enconman.2017.06.073.

Birgen, C., Dürre, P., Preisig, H.A. and Wentzel, A. 2019. Butanol production from lignocellulosic biomass: Revisiting fermentation performance indicators with exploratory data analysis. Biotechnology for Biofuels 12: 167. https://doi.org/10.1186/s13068-019-1508-6.

Biris-Dorhoi, E.-S., Michiu, D., Pop, C.R., Rotar, A.M., Tofana, M., Pop, O.L., Socaci, S.A. and Farcas, A.C. 2020. Macroalgae—A sustainable source of chemical compounds with biological activities. Nutrients 12: 3085. https://doi.org/10.3390/nu12103085.

Brar, A., Kumar, M., Soni, T., Vivekanand, V. and Pareek, N. 2021. Insights into the genetic and metabolic engineering approaches to enhance the competence of microalgae as biofuel resource: A review. Bioresource Technology 339: 125597. https://doi.org/10.1016/j.biortech.2021.125597.

Callegari, A., Bolognesi, S., Cecconet, D. and Capodaglio, A.G. 2020. Production technologies, current role, and future prospects of biofuels feedstocks: A state-of-the-art review. Critical Reviews in Environmental Science and Technology 50(4): 384–436. https://doi.org/10.1080/10643389.2019.1629801.

Chang, W.-R., Hwang, J.-J. and Wu, W. 2017. Environmental impact and sustainability study on biofuels for transportation applications. Renewable and Sustainable Energy Reviews 67: 277–288. https://doi.org/10.1016/j.rser.2016.09.020.

Chen, H., Zhou, D., Luo, G., Zhang, S. and Chen, J. 2015. Macroalgae for biofuels production: Progress and perspectives. Renewable and Sustainable Energy Reviews 47: 427–437. https://doi.org/10.1016/j.rser.2015.03.086.

Chen, J., Li, Q., Wang, L., Fan, C. and Liu, H. 2021. Advances in whole-cell photobiological hydrogen production. Advanced NanoBiomed Research 1: 2000051. https://doi.org/10.1002/anbr.202000051.

Chowdhury, H. and Loganathan, B. 2019. Third-generation biofuels from microalgae: A review. Current Opinion in Green and Sustainable Chemistry 20: 39–44. https://doi.org/10.1016/j.cogsc.2019.09.003.

Ciliberti, C., Biundo, A., Albergo, R., Agrimi, G., Braccio, G., de Bari, I. and Pisano, I. 2020. Syngas derived from lignocellulosic biomass gasification as an alternative resource for innovative bioprocesses. Processes 8: 1567. https://doi.org/10.3390/pr8121567.

Costa, F.F., de Oliveira, D.T., Brito, Y.P., da Rocha Filho, G.N., Alvarado, C.G., Balum, A.M., Luque, R. and do Nascimento, L.A.S. 2020. Lignocellulosics to biofuels: An overview of recent and relevant advances. Current Opinion in Green and Sustainable Chemistry 24: 21–25. https://doi.org/10.1016/j.cogsc.2020.01.001.

Damartzis, T. and Zabaniotou, A. 2011. Thermochemical conversion of biomass to second generation biofuels through integrated process design—A review. Renewable and Sustainable Energy Reviews 15: 366–378. https://doi.org/10.1016/j.rser.2010.08.003.

Darda, S., Papalas, T. and Zabaniotou, A. 2019. Biofuels journey in Europe: Currently the way to low carbon economy sustainability is still a challenge. Journal of Cleaner Production 208: 575–588. https://doi.org/10.1016/j.jclepro.2018.10.147.

Debnath, C., Bandyopadhyay, T.K., Bhunia, B., Mishra, U., Narayanasamy, S. and Muthuraj, M. 2021. Microalgae: Sustainable resource of carbohydrates in third-generation biofuel production. Renewable and Sustainable Energy Reviews 150: 111464. https://doi.org/10.1016/j.rser.2021.111464.

del Rio, P.G., Gomes-Dias, J.S., Rocha, C.M.R., Romani, A., Garrote, G. and Domingues, L. 2020. Recent trends on seaweed fractionation for liquid biofuels production. Bioresource Technology 299: 122613. https://doi.org/10.1016/j.biortech.2019.122613.

Demibras, A. 2009. Political, economic and environmental impacts of biofuels: A review. Applied Energy 86: S108–S117. https://doi.org/10.1016/j.apenergy.2009.04.036.

Dey, P., Pal, P., Kevin, J.D. and Das, D.B. 2020. Lignocellulosic bioethanol production: Prospects of emerging membrane technologies to improve the process—a critical review. Reviews in Chemical Engineering 36(3): 333–367. https://doi.org/10.1515/revce-2018-0014.

Dickinson, S., Mientus, M., Frey, D., Amini-Hajibashi, A., Ozturk, S., Shaikh, F., Sengupta, D. and El-Halwagi, M.M. 2017. A review of biodiesel production from microalgae. Clean Technologies and Environmental Policy 19: 637–668. https://doi.org/10.1007/s10098-016-1309-6.

DiPardo. 2000. Outlook for Biomass Ethanol Production and Demand, United States Department of Energy.

Djandja, O.S., Wang, Z., Chen, L., Qin, L., Wang, F., Xu, Y. and Duan, P. 2020. Progress in hydrothermal liquefaction of algal biomass and hydrothermal upgrading of the subsequent crude bio-oil: A mini review. Energy & Fuels 34: 11723–11751. https://doi.org/10.1021/acs.energyfuels.0c01973.

Douvartzides, S.L., Charisiou, N.D., Papageridis, K.N. and Goula, M.A. 2019. Green diesel: Biomass feedstocks, production technologies, catalytic research, fuel properties and performance in compression ignition internal combustion engines. Energies 12: 809. https://doi.org/10.3390/en12050809.

Fiore, M., Magi, V. and Viggiano, A. 2020. Internal combustion engines powered by syngas: A review. Applied Energy 276: 115415. https://doi.org/10.1016/j.apenergy.2020.115415.

Fivga, A., Speranza, L.G., Branco, C.M., Ouadi, M. and Hornung, A. 2019. A review on the current state of the art for the production of advanced liquid biofuels. AIMS Energy 7(1): 46–76. https://doi.org/10.3934/energy.2019.1.46.

Ford Predicts Fuel from Vegetation, New York Times, Sept. 20, 1925, p. 24.

Gaurav, N., Sivasankari, S., Kiran, G.S., Ninawe, A. and Selvin, J. 2017. Utilization of bioresources for sustainable biofuels: A Review. Renewable and Sustainable Energy Reviews 73: 205–214. https://doi.org/10.1016/j.rser.2017.01.070.

Godbole, V., Pal, M.K. and Gautam, P. 2021. A critical perspective on the scope of interdisciplinary approaches used in fourth-generation biofuel production. Algal Research 58: 102436. https://doi.org/10.1016/j.algal.2021.102436.

Gollakota, A.R.K., Kishore, N. and Gu, S. 2018. A review on hydrothermal liquefaction of biomass. Renewable and Sustainable Energy Reviews 81: 1378–1392. https://doi.org/10.1016/j.rser.2017.05.178.

Gomaa, M.A., Al-Haj, L. and Abed, R.M.M. 2016. Metabolic engineering of Cyanobacteria and microalgae for enhanced production of biofuels and high-value products. Journal of Applied Microbiology 121: 919–931. https://doi.org/10.1111/jam.13232.

Guo, M., Song, W. and Buhain, J. 2015. Bioenergy and biofuels: History, status, and perspective. Renewable and Sustainable Energy Reviews 42: 712–725. https://doi.org/10.1016/j.rser.2014.10.013.

Haqul, I., Qaisar, K., Nawaz, A., Akram, F., Mukhtar, H., Zohu, X., Xu, Y., Mumtaz, M.W., Rashid, U., Ghani, W.A.W.A.K. and Choong, T.S.Y. 2021. Advances in valorization of lignocellulosic biomass towards energy generation. Catalysts 11: 309. https://doi.org/10.3390/catal11030309.

Jeswani, H.K., Chilvers, A. and Azapagic, A. 2020. Environmental sustainability of biofuels: A review. Proceedings of the Royal Society A 476: 20200351. https://doi.org/10.1098/rspa.2020.0351.

Jones, D.T. and Woods, D.R. 1986. Acetone-butanol fermentation revisited. Microbiology Reviews 50(4): 484–524. https://doi.org/10.1128/mr.50.4.484-524.1986.

Jourdin, L. and Burdyny, T. 2021. Microbial electrosynthesis: Where Do We Go from Here? Trends in Biotechnology 39(4): 359–369. https://doi.org/10.1016/j.tibtech.2020.10.014.

Kawai, S. and Murata, K. 2016. Biofuel production based on carbohydrates from both brown and red macroalgae: Recent developments in key biotechnologies. International Journal of Molecular Sciences 17: 145. https://doi.org/10.3390/ijms17020145.

Khetkorn, W., Rastogi, R.P., Incharoensakdi, A., Lindblad, P., Madamwar, D., Pandey, A. and Larroche, C. 2017. Microalgal hydrogen production—A review. Bioresource Technology 243: 1194–1206. https://doi.org/10.1016/j.biortech.2017.07.085.

Khoo, C.G., Dasan, Y.K., Lamb, M.K. and Lee, K.T. 2019. Algae biorefinery: Review on a broad spectrum of downstream processes and products. Bioresource Technology 292: 121964. https://doi.org/10.1016/j.biortech.2019.121964.

Knothe, G. 2010. 2—History of vegetable oil-based diesel fuels. pp. 5–19. In: Gerhard Knothe, Jürgen Krahl and Jon Van Gerpen (eds.). The Biodiesel Handbook (Second Edition), AOCS Press. https://doi.org/10.1016/B978-1-893997-62-2.50007-3.

Knothe, G. 2017. Georges Chavanne and the first biodiesel. INFORM International News on Fats, Oils, and Related Materials. (available at https://www.informmagazine-digital.org/informmagazine/july_august_2017/MobilePagedArticle.action?articleId=1127857#articleId1127857).

Kovarik, B. 1998. Henry Ford, Charles F. Kettering and the fuel of the future. Automotive History Review 32: 7–27. http://www.radford.edu/~wkovarik/papers/fuel.html.

Kumar, B., Bhardwaj, N., Agrawal, K., Chaturvedi, V. and Verma, P. 2020a. Current perspective on pretreatment technologies using lignocellulosic biomass: An emerging biorefinery concept. Fuel Processing Technology 199: 106244. https://doi.org/10.1016/j.fuproc.2019.106244.

Kumar, D., Singh, B. and Korstad, J. 2017. Utilization of lignocellulosic biomass by oleaginous yeast and bacteria for production of biodiesel and renewable diesel. Renewable and Sustainable Energy Reviews 73: 654–671. https://doi.org/10.1016/j.rser.2017.01.022.

Kumar, M., Sun, Y., Rathour, R., Pandey, A., Thakur, I.S. and Tsang, D.C.W. 2020b. Algae as potential feedstock for the production of biofuels and value-added products: Opportunities and challenges. Science of the Total Environment 716: 137116. https://doi.org/10.1016/j.scitotenv.2020.137116.

Lamichhane, G., Acharya, A., Poudel, D.K., Aryal, B., Gyawali, N., Niraula, P., Phuyal, S.R., Budhathoki, P., Bk, G. and Parajuli, N. 2021. Recent advances in bioethanol production from lignocellulosic biomass. International Journal of Green Energy 18(7): 731–744. https://doi.org/10.1080/15435075.2021.1880910.

Lin, C.-H. and Lu, C. 2021. Development perspectives of promising lignocellulose feedstocks for production of advanced generation biofuels: A review. Renewable and Sustainable Energy Reviews 136: 110445. https://doi.org/10.1016/j.rser.2020.110445.

Liu, G., Sheng, Y., Ager, J.W., Kraft, M. and Xu, R. 2019. Research advances towards large-scale solar hydrogen production from water. Energy Chem 1(2): 100014. https://doi.org/10.1016/j.enchem.2019.100014.

Liu, J., Hou, Q., Ju, M., Ji, P., Sun, Q. and Li, W. 2020. Biomass pyrolysis technology by catalytic fast pyrolysis, catalytic co-pyrolysis and microwave-assisted pyrolysis: A review. Catalysts 10: 742. https://doi.org/10.3390/catal10070742.

Liu, Y., Cruz-Morales, P., Zargar, A., Belcher, M.S., Pang, B., Englund, E., Dan, Q., Yin, K. and Keasling, J.D. 2021. Biofuels for a sustainable future. Cell 184(6): 1636–1647. https://doi.org/10.1016/j.cell.2021.01.052.

Ma, B.-L., Zheng, Z. and Ren, C. 2021. Chapter 6—Oat. pp. 222–248. *In*: Victor O. Sadras and Daniel F. Calderini (eds.). Crop Physiology Case Histories for Major Crops, Academic Press. https://doi.org/10.1016/B978-0-12-819194-1.00006-2.

Machineni, L. 2020. Lignocellulosic biofuel production: Review of alternatives. Biomass Conversion and Biorefinery 10: 779–791. https://doi.org/10.1007/s13399-019-00445-x.

Macura, B., Johannesdottir, S.L., Piniewski, M., Haddaway, N.R. and Kvarnström, E. 2019. Effectiveness of ecotechnologies for recovery of nitrogen and phosphorus from anaerobic digestate and effectiveness of the recovery products as fertilisers: A systematic review protocol. Environmental Evidence 8: 29. https://doi.org/10.1186/s13750-019-0173-3.

Mahmudul, H.M., Hagos, F.Y., Mamat, R., Adam, A.A., Ishak, W.F.W. and Alenezi, R. 2017. Production, characterization and performance of biodiesel as an alternative fuel in diesel engines—A review. Renewable and Sustainable Energy Reviews 72: 497–509. https://doi.org/10.1016/j.rser.2017.01.001.

Malik, S., Shahid, A., Liu, C.-G., Khan, A.Z., Nawaz, M.Z., Zhu, H. and Mehmood, M.A. 2021. Developing fourth-generation biofuels secreting microbial cell factories for enhanced productivity and efficient product recovery; A review. Fuel 298: 120858. https://doi.org/10.1016/j.fuel.2021.120858.

Maneein, S., Milledge, J.J., Nielsen, B.V. and Harvey, P.J. 2018. A review of seaweed pre-treatment methods for enhanced biofuel production by anaerobic digestion or fermentation. Fermentation 4: 100. https://doi.org/10.3390/fermentation4040100.

McGovern, P.E., Zhang, J., Tang, J., Zhang, Z., Hall, G.R., Moreau, R.A., Nuñez, A., Butrym, E.D., Richards, M.P., Wang, C.-S., Cheng, G., Zhao, Z. and Wang, C. 2004. Fermented beverages of pre- and proto-historic China. Proceedings of the National Academy of Sciences of the United States of America 101(51): 17593–17598. https://doi.org/10.1073/pnas.0407921102.

McKennedy, J. and Sherlock, O. 2015. Anaerobic digestion of marine macroalgae: A review. Renewable and Sustainable Energy Reviews 52: 1781–1790. https://doi.org/10.1016/j.rser.2015.07.101.

Mehmood, M.A., Shahid, A., Malik, S., Wang, N., Javed, M.R., Haider, M.N., Verma, P., Ashraf, M.U.F., Habib, N., Syafiuddin, A. and Boopathy, R. 2021. Advances in developing metabolically engineered microbial platforms to produce fourth-generation biofuels and high-value biochemicals. Bioresource Technology 337: 125510. https://doi.org/10.1016/j.biortech.2021.125510.

Milledge, J.J. and Harvey, P.J. 2016. Potential process 'hurdles' in the use of macroalgae as feedstock for biofuel production in the British Isles. Journal of Chemical Technology and Biotechnology 91: 2221–2234. https://doi.org/10.1002/jctb.5003.

Milledge, J.J., Nielsen, B.V., Maneein, S. and Harvey, P.J. 2019. A brief review of anaerobic digestion of algae for bioenergy. Energies 12: 1166. https://doi.org/10.3390/en12061166.

Moon, H.G., Jang, Y.-S., Cho, C., Lee, J., Binkley, R. and Lee, S.Y. 2016. One hundred years of clostridial butanol fermentation. FEMS Microbiology Letters 363: fnw001. https://doi.org/10.1093/femsle/fnw001.

Mussatto, S.I., Dragone, G., Guimarães, P.M.R., Silva, J.P.A., Carneiro, L.M., Roberto, I.C., Vicente, A., Domingues, L. and Teixeira, J.A. 2010. Technological trends, global market, and challenges of bio-ethanol production. Biotechnology Advances 28: 817–830. https://doi.org/10.1016/j.biotechadv.2010.07.001.

Naik, S.N., Goud, V.V., Rout, P.K. and Dalai, A.K. 2010. Production of first and second generation biofuels: A comprehensive review. Renewable and Sustainable Energy Reviews 14: 578–597. https://doi.org/10.1016/j.rser.2009.10.003.

Nanda, S., Rana, R., Sarangi, P.K., Dalai, A.K. and Kozinski, J.A. 2018. A broad introduction to first-, second-, and third-generation biofuels. pp. 1–25. *In*: Sarangi, P., Nanda, S. and Mohanty, P. (eds.). Recent Advancements in Biofuels and Bioenergy Utilization, Springer, Singapore. https://doi.org/10.1007/978-981-13-1307-3_1.

National Geographic, Alexander Graham Bell, Vol. 31, Feb. 1917, p. 131.

Ndaba, B., Chiyanzu, I. and Marx, S. 2015. n-Butanol derived from biochemical and chemical routes: A review. Biotechnology Reports 8: 1–9. https://doi.org/10.1016/j.btre.2015.08.001.

Oh, Y.-K., Hwang, K.-R., Kim, C., Kim, J.R. and Lee, J.-S. 2018. Recent developments and key barriers to advanced biofuels: A short review. Bioresource Technology 257: 320–333. https://doi.org/10.1016/j.biortech.2018.02.089.

Okolie, J.A., Mukherjee, A., Nanda, S., Dalai, A.K. and Kozinski, J.A. 2021. Next-generation biofuels and platform biochemicals from lignocellulosic biomass. International Journal of Energy Research 45: 14145–14169. https://doi.org/10.1002/er.6697.

Olatunji, K.O., Ahmed, N.A. and Ogunkunle, O. 2021. Optimization of biogas yield from lignocellulosic materials with different pretreatment methods: A review. Biotechnology for Biofuels 14: 159. https://doi.org/10.1186/s13068-021-02012-x.

Oliveira, C.C.N., Rochedo, P.R.R., Bhardwaj, P., Worrell, E. and Szklo, A. 2019. Bio-ethylene from sugarcane as a competitiveness strategy for the Brazilian chemical industry. Biofuels Bioproducts & Biorefining 14(2): 286–300. https://doi.org/10.1002/bbb.2069.

Ong, H.C., Chen, W.-H., Farooq, A., Gan, Y.Y., Lee, K.T. and Ashokkumar, V. 2019. Catalytic thermochemical conversion of biomass for biofuel production: A comprehensive review. Renewable and Sustainable Energy Reviews 113: 109266. https://doi.org/10.1016/j.rser.2019.109266.

Özçimen, D., Koçer, A.T., İnan, B. and Özer, T. 2020. Chapter 14—Bioethanol production from microalgae. pp. 373–389. *In*: Eduardo Jacob-Lopes, Mariana Manzoni Maroneze, Maria Isabel Queiroz and Leila Queiroz Zepka (eds.). Handbook of Microalgae-Based Processes and Products, Academic Press. https://doi.org/10.1016/B978-0-12-818536-0.00014-2.

Padella, M., O'Connell, A. and Prussi M. 2019. What is still limiting the deployment of cellulosic ethanol? Analysis of the current status of the sector. Applied Sciences 9: 4523. https://doi.org/10.3390/app9214523.

Pasha, M.K., Dai, L., Liu, D., Guo, M. and Du, W. 2021. An overview to process design, simulation and sustainability evaluation of biodiesel production. Biotechnology for Biofuels 14: 129. https://doi.org/10.1186/s13068-021-01977-z.

Pasteur, L. 1862. Quelques résultats nouveaux relatifs aux fermentations acétique et butyrique. Bulletin de la Société chimique de France, 52–53.

Prévoteau, A., Carvajal-Arroyo, J.M., Ganigué, R. and Rabaey, K. 2020. Microbial electrosynthesis from CO_2: Forever a promise? Current Opinion in Biotechnology 62: 48–57. https://doi.org/10.1016/j.copbio.2019.08.014.

Puricelli, S., Cardellini, G., Casadei, S., Faedo, D., van den Oever, A.E.M. and Grosso, M. 2021. A review on biofuels for light-duty vehicles in Europe. Renewable and Sustainable Energy Reviews 137: 110398. https://doi.org/10.1016/j.rser.2020.110398.

Ragsdale, S.W. and Pierce, E. 2008. Acetogenesis and the Wood–Ljungdahl pathway of CO_2 fixation. Biochimica et Biophysica Acta (BBA)—Proteins and Proteomics 1784: 1873–1898. https://doi.org/10.1016/j.bbapap.2008.08.012.

Raheem, A., Prinsen, P., Vuppaladadiyam, A.K., Zhao, M. and uque, R. 2018. A review on sustainable microalgae based biofuel and bioenergy production: Recent developments. Journal of Cleaner Production 181: 42–59. https://doi.org/10.1016/j.jclepro.2018.01.125.

Raud, M., Kikas, T., Sippula, O. and Shurpali, N.J. 2019. Potentials and challenges in lignocellulosic biofuel production technology. Renewable and Sustainable Energy Reviews 111: 44–56. https://doi.org/10.1016/j.rser.2019.05.020.

Rizwanul Fattah, I.M., Ong, H.C., Mahlia, T.M.I., Mofijur, M., Silitonga, A.S., Rahman, S.M.A. and Ahmad, A. 2020. State of the art of catalysts for biodiesel production. Frontiers in Energy Research 8: 101. https://doi.org/10.3389/fenrg.2020.00101.

Roberts, M.A.J., Cranenburgh, R.M., Stevens, M.P. and Oyston, P.C.F. 2013. Synthetic biology: Biology by design. Microbiology 159: 1219–1220. https://doi.org/10.1099/mic.0.069724-0.

Rodionova, M.V., Poudyal, R.S., Tiwari, I., Voloshin, R.A., Zharmukhamedov, S.K., Nam, H.G., Zayadan, B.K., Bruce, B.D., Hou, H.J.M. and Allakhverdiev, S.I. 2017. Biofuel production: Challenges and opportunities. International Journal of Hydrogen Energy 42(12): 8450–8461. https://doi.org/10.1016/j.ijhydene.2016.11.125.

Saad, M.G., Dosoky, N.S., Zoromba, M.S. and Shafik, H.M. 1920. Algal biofuels: Current status and key challenges. Energies 12: 1920. https://doi.org/10.3390/en12101920.

Sauer, M. 2016. Industrial production of acetone and butanol by fermentation-100 years later. FEMS Microbiology Letters 363: fnw134. https://doi.org/10.1093/femsle/fnw134.

Savakis, P. and Hellingwer, K.J. 2015. Engineering cyanobacteria for direct biofuel production from CO_2. Current Opinion in Biotechnology 33: 8–14. https://doi.org/10.1016/j.copbio.2014.09.007.

Sharma, B., Larroche, C. and Dussap, C.G. 2020. Comprehensive assessment of 2G bioethanol production. Bioresource Technology 313: 123630. https://doi.org/10.1016/j.biortech.2020.123630.

Shokravi, H., Shokravi, Z., Heidarrezaei, M., Ong, H.C., Koloor, S.S.R., Petrů, M., Lau, W.J. and Ismail, A.F. 2021. Fourth generation biofuel from genetically modified algal biomass: Challenges and future directions. Chemosphere 285: 131535. https://doi.org/10.1016/j.chemosphere.2021.131535.

Shuba, E.S. and Kifle, D. 2018. Microalgae to biofuels: 'Promising' alternative and renewable energy, review. Renewable and Sustainable Energy Reviews 81: 743–755. https://doi.org/10.1016/j.rser.2017.08.042.

Sikarwar, V.S., Zhao, M., Fennell, P.S., Shah, N. and Anthon, E.J. 2017. Progress in biofuel production from gasification. Progress in Energy and Combustion Science 61: 189–248. https://doi.org/10.1016/j.pecs.2017.04.001.

Snow, A.A. and Smith, V.H. 2012. Genetically engineered algae for biofuels: A key role for ecologists. Bioscience 62(8): 765–768. https://doi.org/10.1525/bio.2012.62.8.9.

Solomon, B.D. and Krishna, K. 2011. The coming sustainable energy transition: History, strategies, and outlook. Energy Policy 39: 7422–7431. https://doi.org/10.1016/j.enpol.2011.09.009.

Songstad, D.D., Lakshmanan, P., Chen, J., Gibbons, W., Hughes, S. and Nelson, R. 2009. Historical perspective of biofuels: Learning from the past to rediscover the future. *In Vitro* Cellular & Developmental Biology—Plant 45: 189–192. https://doi.org/10.1007/s11627-009-9218-6.

Suganya, T., Varman, M., Masjuki, H.H. and Renganathan, S. 2016. Macroalgae and microalgae as a potential source for commercial applications along with biofuels production: A biorefinery approach. Renewable and Sustainable Energy Reviews 55: 909–941. https://doi.org/10.1016/j.rser.2015.11.026.

Sun, X., Atiyeh, H.K., Huhnke, R.L. and Tanner, R.S. 2019. Syngas fermentation process development for production of biofuels and chemicals: A review. Bioresource Technology Reports 7: 100279. https://doi.org/10.1016/j.biteb.2019.100279.

Taparia, M.V.S.S., Mehrotra, R., Shukla, P. and Mehrotra, S. 2016. Developments and challenges in biodiesel production from microalgae: A review. Biotechnology and Applied Biochemistry 63(5): 715–726. https://doi.org/10.1002/bab.1412.

Toor, M., Kumar, S.S., Malyan, S.K., Bishnoi, N.R., Mathimani, T., Rajendran, K. and Pugazhendhi, A. 2020. An overview on bioethanol production from lignocellulosic feedstocks. Chemosphere 242: 125080. https://doi.org/10.1016/j.chemosphere.2019.125080.

Tursi, A. 2019. A review on biomass: Importance, chemistry, classification, and conversion. Biofuel Research Journal 22: 962–979. https://doi.org/10.18331/BRJ2019.6.2.3.

Valdivia, M., Galan, J.L., Laffarga, J. and Ramos, J.-L. 2016. Biofuels 2020: Biorefineries based on lignocellulosic materials. Microbial Biotechnology 9(5): 585–594. https://doi.org/10.1111/1751-7915.12387.

Vees, C.A., Neuendorf, C.S. and Pflügl, S. 2020. Towards continuous industrial bioprocessing with solventogenic and acetogenic clostridia: Challenges, progress and perspectives. Journal of Industrial Microbiology & Biotechnology 47: 753–787. https://doi.org/10.1007/s10295-020-02296-2.

Veza, I., Said, M.F.M. and Latiff, Z.A. 2021. Recent advances in butanol production by acetone-butanol-ethanol (ABE) fermentation. Biomass and Bioenergy 144: 105919. https://doi.org/10.1016/j.biombioe.2020.105919.

Villarreal, J.V., Burgués, C. and Rösch, C. 2020. Acceptability of genetically engineered algae biofuels in Europe: Opinions of experts and stakeholders. Biotechnology for Biofuels 13: 92. https://doi.org/10.1186/s13068-020-01730-y.

Weizmann, C. 1915. Improvement in the Bacterial Fermentation of Carbohydrates and in Bacterial Cultures for the Same. GB191504845A, March 29 1915.

Xia, P.-F., Ling, H., Foo, J.L. and Chang, M.W. 2019. Synthetic biology toolkits for metabolic engineering of cyanobacteria. Biotechnology Journal 14: 1800496. https://doi.org/10.1002/biot.201800496.

Yang, C., Wang, S., Yang, J., Xu, D., Li, Y., Li, J. and Zhang, Y. 2020. Hydrothermal liquefaction and gasification of biomass and model compounds: A review. Green Chemistry 22: 8210–8232. https://doi.org/10.1039/d0gc02802a.

Yusoff, M.N.A.M., Zulkifli, N.W.M., Masuma, B.M. and Masjukia, H.H. 2015. Feasibility of bioethanol and biobutanol as transportation fuel in spark-ignition engine: A review. RSC Advances 5: 100184–100211. https://doi.org/10.1039/C5RA12735A.

Zadeh, Z.E., Abdulkhani, A., Aboelazayem, O. and Saha, B. 2020. Recent Insights into lignocellulosic biomass pyrolysis: A critical review on pretreatment, characterization, and products upgrading. Processes 8: 799. https://doi.org/10.3390/pr8070799.

Zhao, X., Zhang, L. and Liu, D. 2012. Biomass recalcitrance. Part I: The chemical compositions and physical structures affecting the enzymatic hydrolysis of lignocellulose. Biofuels, Bioproducts and Biorefining 6: 465–482. https://doi.org/10.1002/bbb.1331.

CHAPTER 2

Green Fuels
Algae-based Sources and Production Routes

Jonathan S. Harris and *Anh N. Phan**

1. Introduction

Climate change and increasing environmental impact associated with the use of fossil fuels are driving governments, industries, and researchers to develop economical and sustainable alternatives. Although a number of approaches have been developed for producing renewable and sustainable energy, electricity storage methods remain limited in comparison to solid, liquid or gaseous fuels, with modern batteries having 15% of the energy capacity of petrol per cubic meter (Energy 2021). The demand for sustainable alternatives to fossil fuels thus continues to increase worldwide, increasing by 7% per annum (WBA 2019) despite the introduction of hybrids and electric cars.

Biofuels in particular are of key importance due to their high energy density, which makes them uniquely suited to applications that involve mobility or portability (Yu et al. 2020), particularly the transportation sector (Li et al. 2018). For stationary applications, electrical equivalents are widely available for most applications, and in many cases, electrically powered equipment is more reliable (Weber et al. 2019) due to fewer moving parts in the motor and reduced susceptibility to fouling. Biofuels can offer a more sustainable option for applications where direct connection to the electric grid is not feasible (Kargbo et al. 2021) or areas have unstable power grids. While biofuels are a more sustainable option than fossil fuels, there is still debate concerning the environmentally friendliness of biofuels, especially 1st generation biofuels. This is due to CO_2 emissions associated with land use changes, water consumption, electricity usage during production and other costs. Therefore, reducing or removing these requirements is an important goal along with making biofuels economically viable.

School of Engineering, Chemical Engineering, Newcastle University, Newcastle upon Tyne, NE1 7RU, United Kingdom.
* Corresponding author: anh.phan@newcastle.ac.uk

1.1 Biofuel Generations

Biofuels can be produced from carbohydrates, lignocellulosic materials (woody materials, waste, forest residues, etc.), triglyceride rich materials (Dahman et al. 2019) or from atmospheric CO_2 and water (Shokravi et al. 2019). Biofuels are differentiated into 4 generations depending on the feedstock used. 1st generation biofuels (commonly biodiesel and bioethanol) are produced from food grade sources (Santos et al. 2018), commonly soybean, corn, wheat, sugar cane and vegetable oils (Pratto et al. 2020) . The drawback of these is that they utilise food that is otherwise suitable for human consumption, increasing land usage and often prompting deforestation or other loss of biodiversity (Kargbo et al. 2021). The cost of the feedstock is high (up to 60% of the total operating costs (Jonker et al. 2019)).

2nd generation biofuels are produced from agricultural waste or other forms of waste, sawdust, wild grasses (Jindal and Jha 2016), corn stover (Jablonský et al. 2018), sewage or energy crops (Marin-Burgos and Clancy 2017). 2nd generation biofuels do not compete with agriculture and the feedstocks are 7–25% cheaper than 1st generation biofuels taking into account collection costs (Santos et al. 2018). However, 2nd generation feedstocks are generally more complex to valorise than 1st generation feedstocks, as the feedstocks (apart from waste cooking oils) contain mainly cellulose, lignin, water and inorganic content with small amounts of starch and triglycerides (Kumar et al. 2019). A well-developed example of 2nd generation biofuel is bio-ethanol from fermentation of lignocellulosic waste (Pratto et al. 2020). Many 2nd generation feedstocks are forms of waste, which means these sources are of limited availability. As a result, waste derived 2nd generation biofuels cannot be used exclusively and should be part of a diverse fuel solution. 3rd generation biofuels utilise aquatic plant life, generally algae, as the feedstock, due to the drastically reduced space requirements and growth periods (doubling biomass mass every 2–5 days, depending on strain) (Blifernez-Klassen et al. 2018). The primary challenges with producing 3rd generation biofuels are the cost of separating the algae from the water it grows in (< 1% algae by weight) (Fasaei et al. 2018) and common chemical approaches (i.e., transesterification of triglycerides to produce biodiesel or thermal gasification) are much less effective with high water content feedstocks (Im et al. 2014). Sustainably produced electricity can be used to produce 4th generation fuels from CO_2 and water in the air or CO_2 emissions from industrial plants (Abdullah et al. 2019). 4th generation biofuels rely on green energy being abundantly available and may not be economically viable (Shokravi et al. 2019) but has potential for increasing energy efficiency further. Thus, this paper will focus on 3rd generation biofuel production.

Green fuels, also known as biofuels, are fuels produced from biomass sources via biological, chemical or thermochemical processes. In theory, green fuels are carbon neutral due to the carbon content of these fuels originating from photosynthesis of CO_2 from the atmosphere into plant biomass (Yew et al. 2019). This should offset the CO_2 released during combustion of the fuel. However, this does not include CO_2 emissions from operating the biofuel plant or the electricity used in converting the biomass to fuel (Gnansounou and Kenthorai Raman 2016) or environmental impact from improper waste disposal, increased land and water usage (Somers and Quinn 2019). Depending upon types of feedstock and processes, biofuels result in significant decreases in CO_2 emissions compared to fossil fuels, e.g., 20–60% emission reduction for 1st generation whereas 70–90% emission reduction for 2nd generation biofuels (FAO 2008). 3rd and 4th generation biofuels are not widely used due to their relatively high production costs, so their carbon emissions have not been quantified at industrial scales. Most of these emissions originate from the use of unsustainably produced electricity in the biofuel production process and plant operation (Frieden et al. 2011); therefore the overall carbon balance of biofuels will likely improve as renewable energy sources become more widely used. Land use changes account for as much as 80% of the emissions attributable to 1st generation biofuels but are small for 2nd and negligible for 3rd and 4th generation biofuels (Singh et al. 2021). Algae can be cultivated in a wide variety of conditions either in ponds

Fig. 1. Illustration of main algae cultivation methods: raceway ponds (left) and photobioreactors (right).

or reactors (Fig. 1), requiring very little land area (15.5 tonnes biofuel $m^{-2}yr^{-1}$ for photobioreactor algae *vs* 0.004 tonnes biofuel $m^{-2}yr^{-1}$ for sugar cane) and have low nutrient requirements relative to terrestrial plants (Bošnjaković and Sinaga 2020). These factors make algae ideal for producing liquid fuels without resulting in additional CO_2 emissions from land use changes.

1.2 Biofuel Production Routes

Algae can be converted to biofuels via thermochemical, chemical, or biological methods in which a suitable method depends on the nature of algae such as the lipid/carbohydrate/protein content of the biomass and the remaining water content. These methods can be performed on the algae directly or on carbohydrates, proteins (Lupatini et al. 2017) or lipids extracted from the algae (Vasistha et al. 2021). Biological methods are generally performed after the cellulose of the algae cell wall is converted to sugars, as cellulose is resistant to biological degradation. Solvent extraction techniques can also be used to extract the cell contents, which can allow extraction and conversion to occur simultaneously. Biological methods such as fermentation (Vardon et al. 2012) and anaerobic digestion and chemical methods such as transesterification can then be used to convert a subset of the released cell contents.

Anaerobic digestion (AD) utilises microorganisms to convert the broken-open biomass into gaseous hydrocarbons (mainly methane), CO_2 and other impurities such as nitrogen/sulphur compounds (Rokicka et al. 2021) at a temperature range of 35–37°C in the absence of oxygen to minimise CO_2 production (Puligundla et al. 2019). The slow process for producing gaseous fuels (15–40 day residence time) and the requirement to separate the CO_2 and other impurities makes it less viable for producing biofuels compatible with current transportation infrastructure (Passos et al. 2014). Fermentation is another alternative biological method that uses yeast to convert carbohydrates to bioethanol (Puligundla et al. 2019). This process is slow compared to chemical or thermochemical methods (typically 4–5 days) (Hossain 2019) and the concentration of bioethanol is low (< 15% wt) (Puligundla et al. 2019) and therefore requires distillation to remove water, which is energy intensive.

Chemical processes, typically transesterification, are a commonly used method for producing biodiesel. Transesterification is the reaction between an alcohol, i.e., methanol, and glycerides (lipids) with/without a catalyst to produce fatty acid esters (biodiesel) and around 10% wt by-product glycerol. This method is very effective at converting lipids to fuel but has no effect on sugars or protein and is very sensitive to the presence of water when using alkaline as catalysts

(Thangaraj et al. 2018). Biodiesel is currently blended with petroleum diesel to meet the required properties of liquid fuels, e.g., 10% vol in the UK and EU. Research on converting biodiesel to long chain hydrocarbons is ongoing, though product yields are too low (21% wt yields of jet fuel fraction hydrocarbons after 2 hours), thus yet to be economically viable (Kaewmeesri et al. 2021).

Thermochemical methods (pyrolysis, gasification and hydrothermal liquefaction) utilise high temperature to decompose whole cells/biomass to liquid, solid and gaseous products (Kumar et al. 2019) in proportions dependent on the operating conditions. Gasification is performed at temperatures above 750°C in an oxidising environment (limited oxygen/air, steam, CO_2, or combination), to convert the biomass into syngas (containing mainly H_2, CO and small amount of CO_2 and other impurities). Drying is needed to reduce the moisture content to below 20% wt, which is highly energy intensive due to the high heat capacity of water. Pyrolysis is operated in an inert environment at temperatures below 750°C with or without a catalyst to convert biomass into liquid (bio-oil), gas and solid (char). The proportion of the three products depends upon operating conditions (such as heating rate, temperature, catalyst, feedstock size) and the nature of feedstock. The bio-oil is acidic (pH 4–6), thermally and chemically unstable, therefore spontaneously polymerising during storage (Adamakis et al. 2018). The bio-oil is a mixture of hundreds of organic components with an oxygen content of 15–30% and significant amounts of water (> 20% wt) (Marulanda et al. 2019); therefore, extensive downstream processing is required (Khoo et al. 2019) before using in internal combustion engines. As algae are generally rich in protein, nitrogenous products such as pyridines and indoles will be present (Vardon et al. 2012), which can potentially decompose to toxic gaseous hydrogen cyanide and NO_x and collect in the gaseous product (Wang and Brown 2013) or can be present in bio-oil fraction (need to be removed for using as fuels). Hydrothermal liquefaction (HTL) is performed with temperatures of 140–350°C in an aqueous environment pressurised (pressures up to 20 MPa) to ensure the water remains in liquid phase (Duan et al. 2018). Under these conditions, water can act as both a solvent and a catalyst for hydrolysis reactions. The aqueous conditions prevent decomposition of proteins to toxic nitrogenous products when using temperatures below 250°C (Chua and Schenk 2017) while reducing both lipids and carbohydrates to sugars, fatty acids and hydrocarbons (Ramli et al. 2020). Therefore, HTL is favourable for high water content feedstock such as algae.

Current research into biorefining algae to biofuels is extensive and diverse, and many potential methods are undergoing pilot scale testing worldwide (100–1000 L reactors) to test their economic viability and their environmental impact. This chapter critically analyses the methods to produce biofuels from micro or macroalgal biomass. Life cycle analysis (LCA) and Techno-economic analysis (TEA) have been systematically reviewed to compare these methods in terms of their economic and environmental benefits. Challenges and future improvement or research are also discussed.

2. Sources

3rd generation biofuels are defined as fuels produced from algal biomass, which have a very different growth pattern and cellular composition to terrestrial biomass sources. Although the 3rd generation biofuels also include other micro-organisms such as bacteria, yeast and fungi, these are generally used as an enzyme source to break down other biomass during biological processes such as fermentation or anaerobic digestion rather than as a feedstock (Wei et al. 2013a). In addition, no life cycle or techno-economic analyses were available for these feedstocks.

While any algae strain would be suitable for extraction of organics and biofuel production, including wild or a mixture of strains, certain algae strains have various advantages over others due to their different protein, carbohydrate and lipid content. In addition, these strains can also be induced to modify their composition through introducing environmental limitations such as low light intensity or low nutrient availability (Okoro et al. 2019, Choudhary et al. 2020).

Botryococcus braunii contains higher proportion of lipids (28–34% wt) (Tibbetts et al. 2015) than other strains and can generate hydrocarbons in the form of terpenes outside the cell, which results in a higher proportion of hydrocarbons when decomposed/extracted (15% wt without decomposition *vs* trace levels with other strains) (Ruangsomboon 2012). However, this strain has a slow growth rate (e.g., 0.16/day *vs* 0.39/day for Scenedesmus) (Goswami and Kalita 2011) and thus low productivity. HTL treatment of this strain resulted in a relatively high yield of a hydrophobic oil phase with low oxygen content (< 1% wt) compared to other strains and comparable to petroleum from crude oil. As this strain is lipid rich, lipid extraction techniques result in correspondingly high yields of oil, particularly supercritical CO_2 and ionic liquid extraction (Boni et al. 2018).

Scenedesmus strains are also lipid rich algae (Tang et al. 2019), which are studied for biodiesel production. However, they only grow well in warm waters (20–30°C) and high light intensity conditions (6000 lux to achieve maximum growth rate) (Blifernez-Klassen et al. 2018). This makes them particularly suitable for tropical locations such as the Caribbean, Brazil or Australia.

Chlorella is the most commonly used algae for laboratory testing of biofuel production methods, due to their consistency and tolerance of extremes of light and nutrient concentrations (Chen et al. 2020). Several *Chlorella* strains generate high lipid contents under stress, which means that higher lipid yields in *Chlorella* only occur under unfavourable growth conditions.

The most used *Nannochloropsis* strains are saltwater adapted, notable for their polyunsaturated fatty acid content (4–18% wt) (Dourou et al. 2018) and decent growth rate (0.59/day) (Manisali et al. 2019). Therefore, *Nannochloropsis* strains are one of the most common choices for microalgae cultivation and biofuel production (Chua and Schenk 2017). *Nannochloropsis* strains are also suitable for bio-organic production as some proteins and polyunsaturated fatty acids present are of significant value, e.g., eicosapentaenoic acid (EPA) (Chua and Schenk 2017).

Spirulina strains are adaptable and include both salt and freshwater variants, which are particularly high in protein, amino acids (up to 57% wt) (Tibbetts et al. 2015) and a range of valuable vitamins and dietary minerals. The high protein content means the algae has a high nitrogen content, which mandates more upgrading of any bio-oils downstream to be usable as fuels. *Spirulina* strains generate low yields of bio-oils compared to other strains (36% yield at 260°C) (Tang et al. 2016). However, this means HTL oil and aqueous phases are rich in various nitrogenous organics such as pyrrolidine derivatives, pyrazine and amides/amines that are valuable by-products for drug production and pesticides once separated out.

Dunaliella tertiolecta is a saltwater alga that has been used commercially as a source of pigments, due to its high β-carotene content (> 10% wt dry basis). β-carotene is unstable at most temperatures or under high light conditions. As a result, this alga is more generally used as a source of pharmaceutical feedstocks. Nonetheless, HTL can be used to generate moderate yields of bio-oils (up to 55% in 380°C, 225 bar) (Shahi et al. 2020). Other less widely known algae strains are being tested for biofuel or organic chemical production, such as *Chlamydomonas reinhardtii* and *Phaeodactylum tricornutum*.

While 3rd generation biofuels are typically produced from microalgae due to their rapid growth rate (around 10x faster than terrestrial plants), macroalgae such as the seaweed *Sargassum* can also be used to produce 3rd generation biofuels (Suutari et al. 2015). The larger size of macroalgae means dewatering can be performed with simpler methods, such as passing through mesh as opposed to centrifugation (Chen et al. 2015). The macroalgae species studied for biofuel production in terms of sequential extractive transesterification and fermentation include *Ulva intestinalis*, *Sargassum* and *Cladophora glomerata*, though many varieties are used depending on which species are easily available in the location.

Ulva macroalgae, a green macroalgae also known as sea lettuce, is particularly notable for its extremely rapid growth rate (more than double that of typical macroalgae) typically dominating in nature. In addition, this strain also absorbs all available nutrients 50–150% more rapidly than other macroalgae (Mourad and El-Azim 2019). Typically, *Ulva* strains contain 5–7% wt lipids and 25–30% wt carbohydrates (Osman et al. 2020), making them more suitable for fermentation based processes or HTL than lipid focused transesterification.

Sargassum species are brown macroalgae that can be found in tropical or temperate seawater, particularly in shallow areas and coral reefs (Borines et al. 2013). *Sargassum* strains are typically low in lipids (0.5–4% wt) and rich in carbohydrates (25–68% wt), with the remainder comprising inorganics and proteins (Pirian et al. 2020). In particular, many *Sargassum* strains are relatively rich in sulphated carbohydrates (1–2% wt), a potentially valuable by-product but also having a negative environmental impact if stored dry as it emits hydrogen sulphate under these conditions (Milledge and Harvey 2016).

In general, microalgae contain a higher concentration of lipids than macroalgae but lower levels of carbohydrates, due to physiological differences in nutrient diffusion rates caused by decreased surface area: volume ratios in larger organisms (Fares et al. 2020). This difference in composition makes macroalgae better suited to fermentation/hydrolysis-based approaches whereas microalgae are suited to extractive transesterification or *in situ* transesterification. Table 1 illustrates typical compositions of algae.

Table 1. Typical composition of commonly used algae strains (Blifernez-Klassen et al. 2018, Tibbetts et al. 2015, Jatmiko et al. 2019, Salosso 2019).

Strain	Carbohydrates (% wt)	Lipids (% wt)	Protein (% wt)	Other (% wt)
Botryococcus	25	29	26	20
Chlorella	26	20	41	13
Nannochloropsis	27	30	41	2
Spirulina	28	8	57	7
Scenedesmus	43	16	28	14
Ulva	58	0.2	14	27
Sargassum	60	1.7	10	28
Cladophora	63	2.6	16	18

3. State-of-the Art Technologies for Fuel Production

Due to fuels comparable to petroleum being most desirable, as these are compatible with current infrastructure, methods that produce gaseous or oxygen rich fuels are less desirable. In addition, algae retain significant amounts of water even after most drying processes (up to 10% wt), so methods that tolerate water are better suited to produce algal fuels. With these requirements, the most appropriate methods for algal fuels are hydrothermal liquefaction, hydrolysis/fermentation and potentially transesterification.

3.1 Thermal Process

HTL is operated at a temperature range of 140°C–350°C and pressures up to 20 MPa (Duan et al. 2018). Under these temperatures and pressures, the water acts as a solvent and catalyst due to reduced hydrogen bonding and increased concentration of H_3O^+ oxonium ions (Xu et al. 2019), thereby decomposing biomass at lower temperatures than other thermochemical processes (220°C *vs* > 250°C for pyrolysis). HTL can decompose lipids and carbohydrates to fatty acids, glycerol and monosaccharides with minimal decomposition of proteins when operated ≤ 220°C. This is beneficial as nitrogen rich proteins could potentially produce environmentally damaging or toxic nitrogenous by-products such as pyrazine or hydrogen cyanide (Jazrawi et al. 2015). HTL produces two liquid phases: an oil phase and aqueous phase. The oil phase has an oxygen content of 5–12% wt and high energy content (33–38 MJ/kg compared to 46.4 MJ/kg of petrol) (Tang et al. 2019) compared to other thermochemical techniques, thereby reducing the need for deoxygenation of the product bio-oil. The aqueous phase generally contains the products of carbohydrate and protein hydrolysis, generally sugars and amino acids with some glycerol and fatty acids when operated at low temperatures (< 200°C) (Xu and Savage 2015), which is suitable as a nutrient source for fermentation, anaerobic digestion or algae growth. At temperatures above 220°C, however, the sugars and amino acids decompose further to various ecotoxic/toxic chemicals, which would require remediation before the aqueous phase could be valorised to extract additional products or used for algae growth nutrients. A key finding for HTL is that deoxygenation and denitrogenation reactions occur above 250°C (Jazrawi et al. 2015), which would improve the energy density of the oil phase and permit its use as a "drop-in" fuel (E4tech 2017).

3.2 Biochemical Processes

Fermentation is the most common biological method of producing liquid fuels, which uses certain microorganisms to convert sugars (glucose, fructose, galactose) to ethanol (also known as bioethanol), biohydrogen and CO_2. This reaction occurs under aerobic conditions to maximise ethanol yield (Zhu et al. 2013). As glucose groups are primarily found in carbohydrates, including cellulose, cellulosic biomass provides the highest ethanol yields, with any non-glucose groups being either unaffected by the fermentation microorganisms or consumed by the microorganisms to replicate.

Fermentation rate depends on the temperature, pH, growth rate and genetic stability of the microorganism and the tolerance of the microorganism to osmosis, alcohol and inhibitors. *Saccharomyces cerevisiae* is a particularly widely used microorganism for fermentation, as it not only produces high yields of bioethanol (up to 44% wt) but also has a resistance to bioethanol and inhibitor components compared to most other microorganisms and tolerates pH down to 4.0 (Kabir et al. 2019), which can be present following acid hydrolysis of lignocellulosic biomass. Other strains such as *Zymomonas* produce bioethanol around 10% faster than *Saccharomyces* (maximum of 0.89 g/L/hr) (Ma'As et al. 2020), but generally cannot tolerate acetic acids or acid hydrolysis products that may arise during fermentation (Wei et al. 2013b). Thermophilic bacteria can also be used to allow operation at higher temperatures (up to 55°C), with corresponding benefits from increased solubilisation of biomass and resistance to contamination from wild bacteria (Beri et al. 2020). These thermophilic cultures are, however, less resistant to bioethanol concentration without mutations (< 10 g/l) and produce acetic/lactic acids as by-products (at around 15% selectivity in addition to ethanol), lowering the overall selectivity to ethanol compared to typical strains such as *Saccharomyces* (Lynd et al. 2017).

As most fermentation microorganisms cannot consume cellulose or other glucose polymers directly, saccharification or hydrolysis is generally performed first as a pre-treatment step before fermentation (Pratto et al. 2020). Hydrolysis can be performed chemically, thermally or enzymatically. Chemical hydrolysis utilises a cellulose disrupting solvent such as hexane or chloroform, followed by a catalyst for decomposing cellulose such as H_2SO_4 and HCl. Similarly, enzymatic hydrolysis is the use of cellulase enzymes to break the cell wall, while thermal hydrolysis utilises temperatures to induce decomposition. Work on combined enzymatic saccharification/fermentation has been performed, which provides fermentation rates close to fermentation of sugars, reducing the overall time required to achieve maximum conversion (Chohan et al. 2020).

As the fermentation process produces ethanol below 15% wt, distillation is necessary to remove the excess water so that the resulting bioethanol meets the standards to use as fuel. Generally, industrial bioethanol plants aim for 99.6% recovery of ethanol (Kumar et al. 2021), with a typical energy cost of 14.1 MJ/kg bioethanol recovered, accounting for around 60% of the total energy requirements (Chen et al. 2018).

Anaerobic digestion (AD) is the decomposition of biomass by bacteria to biogas and solid residues in the absence of air. The biogas is generally composed of 50–70% vol methane and 30–45% vol CO_2 (Sialve et al. 2009). Hydrogen sulphide and hydrogen are common by-products, at yields of < 3.5% vol and < 2% vol, respectively (Passos et al. 2014). This method works very well for water rich algae and can be operated in a single stage with relatively flexible conditions to valorise most common components present in the biomass. Some dewatering is still required to concentrate the algae broth to 5–10% wt algae, otherwise the reaction rate is slow. Therefore, product recovery becomes too costly due to the low concentrations and the bacteria culture is vulnerable to washout with the products (Ward et al. 2014). Cellulose, lignin and polyphenols require 3–5 times longer than carbohydrates to be broken down by AD, while sulphur rich algae species reduce gas evolution rate as sulphur concentrations above 25 mg/litre poison the fermentation bacterial culture (Chen et al. 2008). Without pre-treatment, algae cells can remain intact during anaerobic digestion for 64 days as a result (Solé-Bundó et al. 2019). The enzymatic approach is widely used at lab scale to enhance digestion rates with minimal additional energy expenditure (Timira et al. 2021), with lipase and cellulase increasing biogas yields by > 20% wt with an equivalent reaction time (Nitsos et al. 2020). AD of lipids produces higher methane yield than proteins and carbohydrates, but can inhibit bacterial activity through their intermediate products, i.e., long chain fatty acids. Overall, lipids reduce overall biogas conversion and reaction rate, which means lipid removal can be beneficial. However, their removal is only energy positive with lipid content above 40% wt (Ward et al. 2014).

The waste products of AD can be used as nutrients for algae cultivation, with the waste being rich in many algae compatible nutrients, nitrogen and phosphorous, with many other trace elements being beneficial. In contrast to the use of HTL waste as algae growth media (Chen et al. 2020), the waste from AD is non-toxic and can be recycled freely without dilution.

However, it is more complex to store biogas during transport and it has a lower energy density than liquid fuels. Converting biogas into methanol via intermediate processes such as steam reforming of biogas and hydrogenation can be performed but currently produces low yields (around 8% wt) (Giuliano et al. 2020).

Solid state fermentation of algae has the advantages of easier recovery of liquids and generates less wastewater. However, this process requires very low water content feedstock, therefore energy intensive drying step of the algae is required (Lizardi-Jiménez and Hernandez-Martinez 2017). Studies on solid state fermentation of algae successfully produce lipids or fatty acids, with yields of around 40 mg of lipids per gram of substrate (Cheirsilp and Kitcha 2015). The collected lipids are then converted to biodiesel; therefore, the solid state fermentation needs to be integrated with downstream processes. Solid state fermentation is highly effective for producing complex organics such as proteases due to simplified product recovery and improved selectivity.

Transesterification is a commonly used method for converting lipid sources (glycerides) with a short chain alcohol, mainly methanol to fatty acid esters (known as biodiesel) with or without catalysts and glycerol as a by-product. Alkaline catalysed transesterification, which is commercially used, produces high selectivity of fatty acid esters (> 90%) and high yield (90% wt) over 1–2 hour reaction time at a temperature range of 40–65°C (Dong et al. 2013). The molar ratio of methanol to triglycerides is typically 6:1 to achieve high yield of fatty acid esters over 1–2 hours, which is higher than stoichiometric ratio (3:1). Basic, acidic catalysts or enzymes can be used to initiate transesterification, with different advantages and disadvantages. Acidic catalysts are unaffected by the presence of water and free fatty acids (FFA), but at 4000 times slower rate than alkaline (Thangaraj et al. 2018). Basic catalysts are vulnerable to water and/or acids (water and FFA need to be kept below 0.05% wt and 0.5% wt, respectively) as these interact with the catalyst to produce soaps, which complicate production separation, lowering the yield of the products. The soap formation also consumes the catalyst, reducing reaction rates and requiring more catalyst to complete the reaction (Nagappan et al. 2019). The water present can also hydrolyse glycerides to release more FFA, which catalyses further saponification. For acidic catalysts, the FFA can react with alcohol to form biodiesel, which is not the case for basic catalysts. Despite this advantage for impure oils and fats, acid catalysed esterification is not widely used due to its slow reaction rate. Pre-treatment of wet or FFA rich oils with an acidic catalyst followed by base catalysed transesterification is a viable option (a two-stage process). Research into combined acid-base catalysts remains a topic of significant interest, and has proven effective for converting contaminated oils to biodiesel in a single step, as the acidic active sites convert any FFA or water present into fatty acid esters while the basic groups have a faster reaction rate for triglyceride conversion (Cornu et al. 2017). However, these catalysts are still vulnerable to water poisoning and the combined acid-base catalysts required (such as heteropolyacids (HPA)) are more expensive than common acid or base catalysts (£325/kg for HPA *vs* £90/ton NaOH) (Dong et al. 2013).

Acid catalysed transesterification of microalgae was performed at 300°C with the addition of a 36.4% wt sulphuric acid: 73.6% wt methanol solvent, to produce 85% wt biodiesel yields after 2 hours (Torres et al. 2017). Co-solvents such as hexane can be used to reduce the amount of methanol required for complete transesterification from 8:1 to 6:1 with wet algal biomass (Park et al. 2017). The use of hexane co-solvent allows > 90% conversion within 3 hours with 18% sulphuric acid catalyst regardless of water content. Enzymes are becoming more attractive as these have reaction rates comparable to basic catalysts and avoid saponification. Work into making these enzymes re-usable through immobilisation continues with whole cell biocatalysts immobilised in synthetic supports. However, to date, enzyme catalysed transesterification is not widely used commercially due to the high cost of the enzymes (Afifah et al. 2019).

Biodiesel and bioethanol currently require blending with petrol before use in unmodified engines (1:9 ratio by volume, known as B10 or E10, respectively) (Kargbo et al. 2021), which is a limitation with regards to its sustainability. Modified engines can utilise pure bioethanol or biodiesel, though these modifications are currently relatively uncommon. Extensive research at laboratory and pilot scale is underway to convert fatty acid esters to hydrocarbons (Sun et al. 2018).

4. Process Intensification and Integration

Process intensification is generally defined as "a set of innovative principles related to equipment and process design, which bring significant (> a factor of 2) improvements in process efficiency, fixed or operating expenses, improvements in quality and/or reductions in waste" (Reay et al. 2013). Process intensification techniques can be applied for producing biofuels, e.g., *in situ* transesterification, where lipid extraction and transesterification are performed simultaneously (Ehimen et al. 2010, Li et al. 2011). The disadvantage of *in situ* transesterification is that a high quantity of methanol is required (up to a 300:1 methanol: biomass ratio) to achieve complete conversion to bio-diesel within 4 hours (Velasquez et al. 2012). Therefore, recovery system of methanol and large reactor/ separation systems (increasing capital costs) are needed. Reactor design is another approach to intensify a process by reducing size of reactors and downstream processes via enhancing mass and heat transfer. Intensified reactors can be used for transesterification, obtaining 99% yields of fatty acid esters with 5–20 minute reaction time compared to 1–2 hours to obtain the same yields in a standard reaction vessel (Patle et al. 2021). These intensified reactors can be used in series to convert by-product glycerol without separation. Application of microwaves or ultrasonication (instead of conventional heating) during *in situ* transesterification can be used to extract lipids without a co-solvent, with low fatty acid ester yields (35–60% wt) (Luo et al. 2014) compared to a traditional two-stage transesterification (extraction-reaction: 90% wt yields of fatty acid esters).

For HTL, a two-stage HTL has been proposed to address the issue with formation of toxic by-products: the addition of a 140–180°C pre-treatment stage (Jazrawi et al. 2015) before dewatering and HTL at $\geq$ 200°C. The low temperatures of the first stage can induce hydrolysis to sugar and amino acids without producing toxic by-products (Sereewatthanawut et al. 2008). This added stage reduces the energy efficiency overall but increases bio-oil yields from 25 to 28% wt (Usami et al. 2020) while decreasing the nitrogen content of the product from 7.7 to 4.2% wt (Prapaiwatcharapan et al. 2015).

For biochemical methods, process intensification primarily focuses on reactor designs with superior mixing characteristics (Noorman et al. 2018), as mass transfer of nutrients for yeast/bacteria growth is the rate limiting step. Converting biochemical batch operated processes to continuous operation permits a 10–100 fold reduction in reactor size from enhanced mass transfer rates, and intensification of oxygen and nutrient transfer increases reaction rate by a factor of 3.6 (Noorman et al. 2018). Saccharification and fermentation can be performed in one intensified reactor with an overall relative increase of 20–25% for ethanol yield compared to a conventional stirred tank with the same residence time due to the uniform mixing (Jiang et al. 2020).

To accurately quantify the environmental impact of a process, all stages of the route from the initial feedstock to disposal of the final product must be included. Life cycle analysis (LCA) of fuels or electricity (Choudhary et al. 2020) require quantification of the environmental cost of:

- The end use of the fuel by consumers
- Distributing and dispensing the fuels (e.g., piping and pumping costs)
- Fuel production from the feedstock
- Transportation of the feedstock to fuel production
- Feedstock production/gathering (including emissions from land use changes)

In addition to these, all sources of greenhouse gases must also be considered (Singh and Olsen 2011), such as:

- Evaporation or leakage of produced fuels
- Venting or flaring of by-product gas mixtures
- Changes in CO_2 content of soils from land use changes

However, many studies do not include the end use of the fuel, disposal/handling of by-products or cultivation of the feedstock (Kargbo et al. 2021). For algae or seaweed feedstock, the cultivation credit for each should be consistent (growth of 1 ton of algae sequesters 1.8 tonnes of CO_2). All carbon present in the fuel is converted back to CO_2 when it is used, which partially offsets algae cultivation (Nawkarkar et al. 2019). The proportion of sequestered CO_2 in the fuel also depends primarily on fuel yield.

The values in Table 2 can be compared to conventional jet fuel (emissions of 3.16 $kgCO_{2\,eq}$/kg fuel) and conventional diesel (2.66 $kgCO_{2\,eq}$/kg fuel). Comparison of different approaches can be difficult, as each source uses a different method and very few include all sources of environmental impact. In general, thermochemical methods result in moderate to high environmental impact, with HTL providing the least impact and pyrolysis the most impact (Bennion et al. 2015). The reason why pyrolysis has a greater impact than gasification is because the product requires stabilisation after production to prevent polymerisation. In the LCA studies, this was performed using 4 stage extraction followed by hydrogenation, which would add significantly to the electricity requirements of the process (Bennion et al. 2015). If electricity required comes from renewable sources such as wind and solar, the environmental impact can be significantly reduced.

Beyond the requirement for pyrolysis oil stabilisation, the primary reason for the difference in environmental impact is the drying step, which accounts for around 30% (Fasaei et al. 2018) of the total energy use. If environmentally neutral electricity is used, the emissions for each method

Table 2. Life cycle analysis comparison for 3rd generation biofuels via numerous techniques.

Technique	Feedstock	Product	Emissions by Product Yield ($kgCO_{2eq}$/kg fuel)	References
Gasification	Microalgae	Syngas	1.05–4.29	(Azadi et al. 2015)
Gasification using solar energy	Microalgae	Syngas	0.6–0.88	(Azadi et al. 2015)
HTL	Microalgae cultivated in wastewater effluent	Bio-jet	0.82	(Fortier et al. 2014)
HTL	Microalgae cultivated in refinery effluent	Bio-jet	2.02	(Fortier et al. 2014)
HTL	Microalgae	Bio-oil	0.36	(Frank et al. 2013)
HTL	Generic microalgae	Bio-oil	1.38	(Bennion et al. 2015)
Pyrolysis	Generic microalgae	Bio-oil	6.4	(Bennion et al. 2015)
Basic transesterification	Generic microalgae	Biodiesel	0.14	(Sander and Murthy 2010)
Basic transesterification	Microalgae	Biodiesel	0.27	(Frank et al. 2013)
Basic transesterification	Chlorella vulgaris	Biodiesel	0.57	(Adesanya et al. 2014)
In situ transesterification (acid catalyst)	Generic microalgae	Biodiesel	0.81	(Brentner et al. 2011)
Fermentation	Seaweed	Bioethanol	0.96	(Alvarado-Morales et al. 2013)
Fermentation	Cyanobacteria	Bioethanol	0.35–0.49	(Luo et al. 2010)
Fermentation	Microalgae + food waste	Biohydrogen	1.88	(Sun et al. 2019)

are decreased by as much as 40% due to electricity derived emission sources (e.g., lighting, heating, pumping and product refining). HTL does not require drying beyond a 95% wt water algae slurry, while gasification and pyrolysis can only tolerate < 20% wt moisture. Transesterification results in relatively low environmental impact with all feedstocks (Table 2). However, *in situ* transesterification has significantly greater environmental impact, likely from the drastic increase in methanol requirements and required reaction time.

Fermentation also demonstrates relatively low environmental impact values with marine biomass (Table 2) despite the requirement for pre-treatment with hydrolysis. However, for seaweed, a higher environmental impact could be due to the large size of biomass particles (Alvarado-Morales et al. 2013) which require crushing or other mechanical methods for breaking down the large structures. LCA studies on AD of algae or seaweeds are very limited due to algae requiring additional treatment to decompose the cellulose cell wall before anaerobic digestion can obtain significant yields and the lack of a useful liquid fuel product (Solé-Bundó et al. 2019).

Fig. 2 below shows a comparison based on predictions for a specific case study, which allows direct comparison of the different techniques and their impact on greenhouse gas emissions. The boundary conditions for both Table 2 and Fig. 2 are defined to include all emissions from algae feedstock growth to its use in an engine, assuming complete combustion. The CO_2 emission ratios between techniques are generally comparable to those found in Table 2 above. However, HTL and especially transesterification appear to have significantly higher emissions than expected, which is due to the higher emissions associated with electricity used in the comparison study. HTL uses significant energy to heat the large volume of water, which results in high energy usage and thus

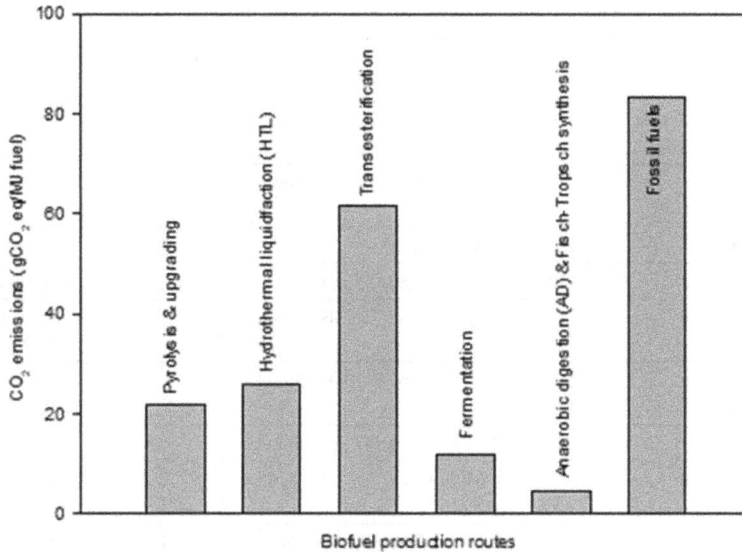

Fig. 2. Direct comparison of predicted greenhouse gas emissions for biofuel production techniques (Bošnjaković and Sinaga 2020, Naeini et al. 2020).

higher emissions. However, this study does not optimise waste heat recovery, which can drastically reduce energy requirements for HTL between batches and does not consider intensification techniques. Likewise, transesterification has particularly high emissions due to extensive dewatering requirements (Fu et al. 2013), which can potentially be reduced if *in situ* transesterification or acid-base catalysis becomes economically feasible.

The second key metric for determining the most effective biofuel production route is from the Techno-economic analysis (TEA). An important part of TEA studies is sensitivity analysis (Barlow et al. 2016), which determines where small changes in certain variables have significant impact on the results. This determines the variables that are most important to know accurately to obtain a good simulation, as well as the strengths and weaknesses of the simulation.

Table 3 is a comparison of TEA performed on potential 3rd generation biofuels with the cost of the product fuel to break even over predicted plant lifespan per gallon of gasoline equivalent used as the metric for comparison of the potential processes.

Among the techniques producing liquid fuels, HTL at large scales produces fuel for the lowest price due to lenient requirements for dewatering, permitting operation with water contents up to 95–98% wt (Gollakota et al. 2018). The dewatering accounts for 33% of the total capital costs (Batan et al. 2016) and 20–30% of the operating costs (Fasaei et al. 2018) when the target moisture content is 5% wt, which increases if lower water contents are required. This expense is minimised with HTL, which makes fuel production through HTL significantly cheaper than most alternatives. AD shows comparable fuel costs (£0.73/litre) to HTL (£0.54–1.03/litre), though this assumes that the productivity of 30 g/m²day¹ can be met, with actual measured algae growth rates under the modelled conditions being 20–25 g algae/m²day¹ (Zamalloa et al. 2011). Even with a lower productivity, AD produces fuel at only £1.42/litre due to dewatering only being required to a 1% wt algae slurry before digestion can begin (Ward et al. 2014). The downside is that this produces biogas, which is largely incompatible with current engines, particularly in the transport sector (Yu et al. 2020). Conversion of biogas to liquid biofuels can be performed using Fischer-Tropsch synthesis. However, this will significantly increase the fuel cost. Cars can be adapted to use natural gas/biogas directly (Yu et al. 2020); however, uptake of these designs is limited and storing the fuel requires pressurisation to contain enough fuel for an acceptable vehicle range.

Table 3. Techno-economic analysis for fuels from various approaches and sources.

Technique	Plant Size (MT/year)	Feedstock	Product	Production Cost (£/litre)	References
Gasification & Fischer-Tropsch synthesis/Transesterification	100	Microalgae	Biodiesel	1.65	(Taylor et al. 2013)
Gasification & Fischer-Tropsch synthesis	100	Microalgae	Biodiesel	2.56	(Taylor et al. 2013)
Transesterification	100	Microalgae	Biodiesel	2.45	(Taylor et al. 2013)
In situ transesterification	105	Microalgae	Biodiesel	0.86	(Nagarajan et al. 2013)
Pyrolysis and upgrading (Thermal drying)	2000	Microalgae	Biodiesel	1.70	(Thilakaratne et al. 2014)
Pyrolysis and upgrading (Mechanical drying)	2000	Microalgae	Biodiesel	1.47	(Thilakaratne et al. 2014)
HTL	2000	Microalgae	Biodiesel	0.54	(Ou et al. 2015)
HTL (algae from biofilm reactor)	100	Microalgae	Biodiesel	2.49	(Barlow et al. 2016)
Lipid extraction	3000	Microalgae	Lipids	1.78	(Davis et al. 2011)
Lipid extraction+ HTL	3000	Microalgae	Biodiesel	2.05	(Davis et al. 2011)
Solvent extraction	2000	Microalgae	Biodiesel	2.64	(Sun et al. 2011)
HTL-Upgrading	100	Sewage Sludge	Biodiesel	1.03	(Snowden-Swan et al. 2016)
AD (productivity: 20g /m²day¹ dry algae)	73	Microalgae	Biogas	1.42	(Zamalloa et al. 2011)
AD (productivity: 30g /m²day¹ dry algae)	73	Microalgae	Biogas	0.73	(Zamalloa et al. 2011)

Conventional transesterification requires a low water content ($< 0.5\%$ wt) to prevent saponification. Despite the cheap catalysts and mild process conditions, biodiesels from algae are relatively costly due to water removal costs (£2.5/litre vs £0.5–0.9/litre for other biodiesels). In addition, most solvents for lipid extraction, such as hexane, butanol or chloroform, are hydrophobic (as lipids are generally hydrophobic), which makes lipid extraction from wet algae broths extremely challenging due to the extraction solvents being isolated from the cells by the formation of hydrophobic-hydrophilic phase boundaries (Jeevan Kumar et al. 2017). *In situ* transesterification, however, is significantly cheaper to operate (£0.86/litre), as the methanol/sulphuric acid solvent used for extraction is hydrophilic (Velasquez et al. 2012) and the cultivated algae broth often does not need to be extensively dewatered due to the acid catalysts not catalysing soap production (Kim et al. 2017). Despite the additional costs of the bulk methanol and longer reaction time from *in situ* transesterification, the fuel price remains comparable to thermochemical and biological methods (Davis et al. 2011).

5. Conclusions and Future Development

Each method of producing biofuel has challenges associated with its use, which hinder its usage as part of a green energy solution. HTL produces liquid fuels relatively quickly and can be energy efficient when proper energy recycling is done, but downstream processing to refine the mixed HTL products remains the main challenge. HTL decomposition of proteins can produce many toxic nitrogenous compounds such as pyrazines and imidazole, while producing little by way of energy dense fuel compounds. Operating HTL at a temperature suitable to hydrolyse the sugars and lipids without breaking down the proteins is widely studied, but the suitable temperature/pressure range is small and is not well defined to date. These by-products could also be recycled as nutrients, but this still requires further conversion to less toxic forms.

Transesterification is a well-known method for producing biofuel, but the current alkaline catalysed transesterification is vulnerable to wet or acidic feedstocks. More energy efficient methods of drying algae would address this issue. Most methanol used in transesterification is produced from fossil fuels. A version of transesterification that can tolerate high water content and methanol derived from sustainable and green source is of interest. Combined acid-base catalysts that tolerate water or extractive transesterification that can separate the water from the biomass and catalyst would be effective solutions to this challenge. Currently, combined acid-base catalysts remain expensive whereas extractive transesterification reduces overall reaction rate and has not been studied enough to determine if it is an effective method at large scale.

Fermentation is a highly effective and selective method for producing biofuels, but like other biological methods, has a slow reaction rate. Although the reaction rate can be improved using intensification techniques, it remains slower than thermochemical or chemical methods. AD also has a slow reaction rate and thus is operated at large scales for biogas production. The main challenge with AD is the requirement to break open the cell wall before digestion and inability to valorise cellulose without prior decomposition. Many methods for combining AD and cellulose hydrolysis have been discovered such as combined alkali/electrochemical hydrolysis to improve the rate, but more work on how these could best be utilised is required. AD could also be improved through improvement of nutrient mass transfer rates through process intensification or more energy efficient methods of converting biogas into liquid biofuels.

Despite the many challenges that utilising algal biomass poses for each potential biofuel production route, algal biofuels are rapidly becoming more feasible due to rising fossil fuel costs and improved biofuel production methods (Bošnjaković and Sinaga 2020, Somers and Quinn 2019). Algal biorefineries are already in use for extraction of valuable organics from algae (Harvey and Ben-Amotz 2020, Marzorati et al. 2020), but producing biofuels is not yet viewed as economically viable (Hannon et al. 2010), though the cost of producing algal biofuels is continuously dropping for all potential processes due to ongoing research. Of the current methods for producing biofuels, HTL, *in situ* transesterification and AD are thus far the more economically viable methods as quantified by TEA studies due to the relatively simple process mechanism and relatively little downstream processing requirements. These techniques also do not require extensive algae dewatering.

In terms of environmental impact, all methods reduce significant CO_2 emissions compared to conventional fuels. Anaerobic digestion also only produces gaseous fuels, so this would need to be paired with a gaseous to liquid fuel conversion process to produce valuable liquid fuels. Thermochemical methods such as HTL, pyrolysis or gasification release 2–20% of the feedstock mass in the form of CO_2 during processing in the gaseous product stream, but the environmental impact of them depends on the process tolerance of water. HTL is, as a result, the most appropriate thermochemical method, though the high heating requirement for raising the algae broth to HTL conditions can make HTL relatively poor, both economically and environmentally, if the heat cannot be effectively recycled between batches. To improve HTL, methods of optimising conversion, such as catalysts, reactor designs and better process modelling, to reduce the required temperature without reducing conversion, would have the greatest effect on the economic and environmental impact of the process.

Overall, HTL, transesterification and AD appear to be the best options for producing algal biofuels, though each has benefits and disadvantages. In addition, as the proportion of electricity from sustainable and renewable sources improves, the environmental impact of biofuel production will decrease with time.

References

Abdullah, B., Syed Muhammad, S.A.F.A., Shokravi, Z., Ismail, S., Kassim, K.A., Mahmood, A.N. and Aziz, M.M.A. 2019. Fourth generation biofuel: A review on risks and mitigation strategies. Renew. Sust. Energ. Rev. 107: 37–50.

Adamakis, I.-D., Lazaridis, P.A., Terzopoulou, E., Torofias, S., Valari, M., Kalaitzi, P., Rousonikolos, V., Gkoutzikostas, D., Zouboulis, A., Zalidis, G. and Triantafyllidis, K.S. 2018. Cultivation, characterization, and properties of Chlorella vulgaris microalgae with different lipid contents and effect on fast pyrolysis oil composition. Environ. Sci. Pollut. Res. 25: 23018–23032.

Adesanya, V.O., Cadena, E., Scott, S.A. and Smith, A.G. 2014. Life cycle assessment on microalgal biodiesel production using a hybrid cultivation system. Bioresour. Technol. 163: 343–355.

Afifah, A.N., Syahrullail, S., Wan Azlee, N.I., Che Sidik, N.A., Yahya, W.J. and Abd Rahim, E. 2019. Biolubricant production from palm stearin through enzymatic transesterification method. Biochem. Eng. J. 148: 178–184.

Alvarado-Morales, M., Boldrin, A., Karakashev, D.B., Holdt, S.L., Angelidaki, I. and Astrup, T. 2013. Life cycle assessment of biofuel production from brown seaweed in Nordic conditions. Bioresour. Technol. 129: 92–99.

Azadi, P., Brownbridge, G., Mosbach, S., Inderwildi, O. and Kraft, M. 2015. Simulation and life cycle assessment of algae gasification process in dual fluidized bed gasifiers. Green Chem. 17: 1793–1801.

Barlow, J., Sims, R.C. and Quinn, J.C. 2016. Techno-economic and life-cycle assessment of an attached growth algal biorefinery. Bioresour. Technol. 220: 360–368.

Batan, L.Y., Graff, G.D. and Bradley, T.H. 2016. Techno-economic and Monte Carlo probabilistic analysis of microalgae biofuel production system. Bioresour. Technol. 219: 45–52.

Bennion, E.P., Ginosar, D.M., Moses, J., Agblevor, F. and Quinn, J.C. 2015. Lifecycle assessment of microalgae to biofuel: Comparison of thermochemical processing pathways. Appl. Energy 154: 1062–1071.

Beri, D., York, W.S., Lynd, L.R., Peña, M.J. and Herring, C.D. 2020. Development of a thermophilic coculture for corn fiber conversion to ethanol. Nat. Commun. 11: 1937.

Blifernez-Klassen, O., Chaudhari, S., Klassen, V., Wördenweber, R., Steffens, T., Cholewa, D., Niehaus, K., Kalinowski, J. and Kruse, O. 2018. Metabolic survey of Botryococcus braunii: Impact of the physiological state on product formation. PloS One 13: e0198976–e0198976.

Boni, J., Aida, S. and Kalsum, L. 2018. Lipid extraction method from microalgae Botryococcus braunii as raw material to make biodiesel with soxhlet xxtraction. J. Phys.: Conference Series 1095: 012004.

Borines, M.G., De Leon, R.L. and Cuello, J.L. 2013. Bioethanol production from the macroalgae Sargassum spp. Bioresour. Technol. 138: 22–29.

Bošnjaković, M. and Sinaga, N. 2020. The perspective of large-scale production of algae biodiesel. Appl. Sci. 10: 8181.

Brentner, L.B., Eckelman, M.J. and Zimmerman, J.B. 2011. Combinatorial life cycle assessment to inform process design of industrial production of algal biodiesel. Environ. Sci. Technol. 45: 7060–7067.

Cheirsilp, B. and Kitcha, S. 2015. Solid state fermentation by cellulolytic oleaginous fungi for direct conversion of lignocellulosic biomass into lipids: Fed-batch and repeated-batch fermentations. Ind. Crops Prod. 66: 73–80.

Chen, H., Cai, D., Chen, C., Wang, J., Qin, P. and Tan, T. 2018. Novel distillation process for effective and stable separation of high-concentration acetone–butanol–ethanol mixture from fermentation–pervaporation integration process. Biotechnol. Biofuels 11: 286.

Chen, H., Zhou, D., Luo, G., Zhang, S. and Chen, J. 2015. Macroalgae for biofuels production: Progress and perspectives. Renew. Sust. Energy Rev. 47: 427–437.

Chen, P.H., Venegas Jimenez, J.L., Rowland, S.M., Quinn, J.C. and Laurens, L.M.L. 2020. Nutrient recycle from algae hydrothermal liquefaction aqueous phase through a novel selective remediation approach. Algal Res. 46: 101776.

Chen, Y., Cheng, J.J. and Creamer, K.S. 2008. Inhibition of anaerobic digestion process: A review. Bioresour. Technol. 99: 4044–4064.

Chohan, N.A., Aruwajoye, G.S., Sewsynker-Sukai, Y. and Gueguim Kana, E.B. 2020. Valorisation of potato peel wastes for bioethanol production using simultaneous saccharification and fermentation: Process optimization and kinetic assessment. Renew. Energy 146: 1031–1040.

Choudhary, P., Assemany, P.P., Naaz, F., Bhattacharya, A., Castro, J.D.S., Couto, E.D.A.D.C., Calijuri, M.L., Pant, K.K. and Malik, A. 2020. A review of biochemical and thermochemical energy conversion routes of wastewater grown algal biomass. Sci. Total Environ. 726: 137961.

Chua, E.T. and Schenk, P.M. 2017. A biorefinery for Nannochloropsis: Induction, harvesting, and extraction of EPA-rich oil and high-value protein. Bioresour. Technol. 244: 1416–1424.

Cornu, D., Lin, L., Daou, M.M., Jaber, M., Krafft, J.-M., Herledan, V., Laugel, G., Millot, Y. and Lauron-Pernot, H. 2017. Influence of acid–base properties of Mg-based catalysts on transesterification: Role of magnesium silicate hydrate formation. Catal. Sci. Technol. 7: 1701–1712.

Dahman, Y., Syed, K., Begum, S., Roy, P. and Mohtasebi, B. 2019. Biofuels: Their characteristics and analysis. pp. 277–235. *In*: Biomass, Biopolymer-Based Materials, and Bioenergy. Elsevier.

Davis, R., Aden, A. and Pienkos, P.T. 2011. Techno-economic analysis of autotrophic microalgae for fuel production. Appl. Energy 88: 3524–3531.

Dong, T., Wang, J., Miao, C., Zheng, Y. and Chen, S. 2013. Two-step *in situ* biodiesel production from microalgae with high free fatty acid content. Bioresour. Technol. 136: 8–15.

Dourou, M., Tsolcha, O.N., Tekerlekopoulou, A.G., Bokas, D. and Aggelis, G. 2018. Fish farm effluents are suitable growth media for Nannochloropsis gaditana, a polyunsaturated fatty acid producing microalga. Eng. Life Sci. 18: 851–860.

Duan, P., Yang, S.-K., Xu, Y.-P., Wang, F., Zhao, D., Weng, Y. and Shi, X.-L. 2018. Integration of hydrothermal liquefaction and supercritical water gasification for improvement of energy recovery from algal biomass. Energy 155: 734–745.

E4tech. 2017. Advanced drop-in biofuels: UK production capacity outlook to 2030. Final Report. E4tech (UK) Ltd.

Ehimen, E.A., Sun, Z.F. and Carrington, C.G. 2010. Variables affecting the *in situ* transesterification of microalgae lipids. Fuel 89: 677–684.

Energy, U.S.D.O. 2021. Fuel properties comparison. Alternative Fuels Data Center.

FAO. 2008. The State of Food and Agriculture. Rome.

Fares, A.L.B., Calvão, L.B., Torres, N.R., Gurgel, E.S.C. and Michelan, T.S. 2020. Environmental factors affect macrophyte diversity on Amazonian aquatic ecosystems inserted in an anthropogenic landscape. Ecol. Indic. 113: 106231.

Fasaei, F., Bitter, J.H., Slegers, P.M. and Van Boxtel, T. 2018. Techno-economic evaluation of microalgae harvesting and dewatering systems. Algal Res. 31: 347–362.

Fortier, M.-O. P., Roberts, G.W., Stagg-Williams, S.M. and Sturm, B.S.M. 2014. Life cycle assessment of bio-jet fuel from hydrothermal liquefaction of microalgae. Appl. Energy 122: 73–82.

Frank, E.D., Elgowainy, A., Han, J. and Wang, Z. 2013. Life cycle comparison of hydrothermal liquefaction and lipid extraction pathways to renewable diesel from algae. Mitig. Adapt. Strateg. Glob. Chang. 18: 137–158.

Frieden, D., Peña, N., Bird, D.N., Schwaiger, H. and Canella, L. 2011. Emission balances of first- and second-generation biofuels: Case studies from Africa, Mexico and Indonesia. Centre for International Forestry Research (CIFOR), pp. 48–49.

Fu, X., Li, D., Chen, J., Zhang, Y., Huang, W., Zhu, Y., Yang, J. and Zhang, C. 2013. A microalgae residue based carbon solid acid catalyst for biodiesel production. Bioresour. Technol. 146: 767–770.

Giuliano, A., Catizzone, E., Freda, C. and Cornacchia, G. 2020. Valorization of OFMSW digestate-derived syngas toward methanol, hydrogen, or electricity: Process simulation and carbon footprint calculation. Processes 8: 526.

Gnansounou, E. and Kenthorai Raman, J. 2016. Life cycle assessment of algae biodiesel and its co-products. Appl. Energy 161: 300–308.

Gollakota, A.R.K., Kishore, N. and Gu, S. 2018. A review on hydrothermal liquefaction of biomass. Renew. Sust. Energ. Rev. 81: 1378–1392.

Goswami, R.D. and Kalita, M. 2011. Scenedesmus dimorphus and Scenedesmus quadricauda: Two potent indigenous microalgae strains for biomass production and CO2 mitigation—A study on their growth behavior and lipid productivity under different concentration of urea as nitrogen source. J. Algal Biomass Util. 2: 2–4.

Hannon, M., Gimpel, J., Tran, M., Rasala, B. and Mayfield, S. 2010. Biofuels from algae: Challenges and potential. Biofuels 1: 763–784.

Harvey, P.J. and Ben-Amotz, A. 2020. Towards a sustainable Dunaliella salina microalgal biorefinery for 9-cis β-carotene production. Algal Res. 50: 102002.

Hossain, S.M.Z. 2019. Biochemical conversion of microalgae biomass into biofuel. Chem. Eng. Technol. 42: 2594–2607.

Im, H., Lee, H., Park, M.S., Yang, J.-W. and Lee, J.W. 2014. Concurrent extraction and reaction for the production of biodiesel from wet microalgae. Bioresour. Technol. 152: 534–537.

Jablonský, M., Škulcová, A., Malvis, A. and Šima, J. 2018. Extraction of value-added components from food industry based and agro-forest biowastes by deep eutectic solvents. J. Biotechnol. 282: 46–66.

Jatmiko, T., Prasetyo, D., Poeloengasih, C. and Khasanah, Y. 2019. Nutritional Evaluation of *Ulva* sp. from Sepanjang Coast, Gunungkidul, Indonesia. IOP Conf. Ser.: Earth Environ. Sci. 251: 012011.

Jazrawi, C., Biller, P., He, Y., Montoya, A., Ross, A.B., Maschmeyer, T. and Haynes, B.S. 2015. Two-stage hydrothermal liquefaction of a high-protein microalga. Algal Res. 8: 15–22.

Jeevan Kumar, S.P., Vijay Kumar, G., Dash, A., Scholz, P. and Banerjee, R. 2017. Sustainable green solvents and techniques for lipid extraction from microalgae: A review. Algal Res. 21: 138–147.

Jiang, F., Peng, Z., Li, H., Li, J. and Wang, S. 2020. Effect of hydraulic retention time on anaerobic baffled reactor operation: Enhanced biohydrogen production and enrichment of hydrogen-producing acetogens. Processes 8: 339.

Jindal, M.K. and Jha, M.K. 2016. Effect of process parameters on hydrothermal liquefaction of waste furniture sawdust for bio-oil production. RSC Adv. 6: 41772–41780.

Jonker, J.G.G., Junginger, M., Posada, J., Ioiart, C.S., Faaij, A.P.C. and Van Der Hilst, F. 2019. Economic performance and GHG emission intensity of sugarcane- and eucalyptus-derived biofuels and biobased chemicals in Brazil. Biofuel, Bioprod. Biorefin. 13: 950–977.

Kabir, F., Gulfraz, M., Raja, G.K., Inam-Ul-Haq, M., Batool, I., Awais, M., Habiba, U. and Gul, H. 2019. Comparative study on the usability of lignocellulosic and algal biomass for production of alcoholic fuels. Bioresources 14: 8135–8154.

Kaewmeesri, R., Nonkumwong, J., Kiatkittipong, W., Laosiripojana, N. and Faungnawakij, K. 2021. Deoxygenations of palm oil-derived methyl esters over mono- and bimetallic NiCo catalysts. J. Environ. Chem. Eng. 9: 105128.

Kargbo, H., Harris, J.S. and Phan, A.N. 2021. "Drop-in" fuel production from biomass: Critical review on techno-economic feasibility and sustainability. Renew. Sust. Energ. Rev. 135: 110168.

Khoo, C.G., Dasan, Y.K., Lam, M.K. and Lee, K.T. 2019. Algae biorefinery: Review on a broad spectrum of downstream processes and products. Bioresour. Technol. 292: 121964.

Kim, B., Chang, Y.K. and Lee, J.W. 2017. Efficient solvothermal wet *in situ* transesterification of Nannochloropsis gaditana for biodiesel production. Bioprocess Biosyst. Eng. 40: 723–730.

Kumar, B.R., Mathimani, T., Sudhakar, M.P., Rajendran, K., Nizami, A.-S., Brindhadevi, K. and Pugazhendhi, A. 2021. A state of the art review on the cultivation of algae for energy and other valuable products: Application, challenges, and opportunities. Renew. Sust. Energ. Rev. 138: 110649.

Kumar, G., Dharmaraja, J., Arvindnarayan, S., Shoban, S., Bakonyi, P., Saratale, G.D., Nemestóthy, N., Bélafi–Bakó, K., Yoon, J.J. and Kim, S.H. 2019. A comprehensive review on thermochemical, biological, biochemical and hybrid conversion methods of bio-derived lignocellulosic molecules into renewable fuels. Fuel 251: 352–367.

Li, M., Xu, J., Xie, H. and Wang, Y. 2018. Transport biofuels technological paradigm based conversion approaches towards a bio-electric energy framework. Energy Convers. Manag. 172: 554–566.

Li, Y., Lian, S., Tong, D., Song, R., Yang, W., Fan, Y., Qing, R. and Hu, C. 2011. One-step production of biodiesel from Nannochloropsis sp. on solid base Mg–Zr catalyst. Appl. Energy 88: 3313–3317.

Lizardi-Jimenez, M.A. and Hernandez-Martinez, R. 2017. Solid state fermentation (SSF): Diversity of applications to valorize waste and biomass. Biotech. 7: 44.

Luo, D., Hu, Z., Choi, D.G., Thomas, V.M., Realff, M.J. and Chance, R.R. 2010. Life cycle energy and greenhouse gas emissions for an ethanol production process based on blue-green algae. Environ. Sci. Technol 44: 8670–8677.

Luo, J., Fang, Z. and Smith, R.L. 2014. Ultrasound-enhanced conversion of biomass to biofuels. Prog. Energy Combust. Sci. 41: 56–93.

Lupatini, A.L., Colla, L.M., Canan, C. and Colla, E. 2017. Potential application of microalga Spirulina platensis as a protein source. J. Sci. Food Agric. 97: 724–732.

Lynd, L.R., Liang, X., Biddy, M.J., Allee, A., Cai, H., Foust, T., Himmel, M.E., Laser, M.S., Wang, M. and Wyman, C.E. 2017. Cellulosic ethanol: status and innovation. Curr. Opin. Biotechnol. 45: 202–211.

Ma'as, M.F., Ghazali, H.M. and Chieng, S. 2020. Bioethanol production from Brewer's rice by *Saccharomyces cerevisiae* and *Zymomonas mobilis*: Evaluation of process kinetics and performance. Energy Sources A: Recovery Util. Environ. Eff., pp. 1–14.

Manisali, A.Y., Sunol, A.K. and Philippidis, G.P. 2019. Effect of macronutrients on phospholipid production by the microalga Nannochloropsis oculata in a photobioreactor. Algal Res. 41: 101514.

Marin-Burgos, V. and Clancy, J.S. 2017. Understanding the expansion of energy crops beyond the global biofuel boom: evidence from oil palm expansion in Colombia. Energy Sustain. Soc. 7: 21.

Marulanda, V.A., Gutierrez, C.D.B. and Alzate, C.A.C. 2019. Thermochemical, biological, biochemical, and hybrid conversion methods of bio-derived molecules into renewable fuels. pp. 59–81. *In*: Advanced Bioprocessing for Alternative Fuels, Biobased Chemicals, and Bioproducts. Woodhead Publishing Series in Energy.

Marzorati, S., Schievano, A., Idà, A. and Verotta, L. 2020. Carotenoids, chlorophylls and phycocyanin from Spirulina: Supercritical CO_2 and water extraction methods for added value products cascade. Green Chem. 22: 187–196.

Milledge, J.J. and Harvey, P.J. 2016. Golden tides: Problem or golden opportunity? The valorisation of sargassum from beach inundations. J. Mar. Sci. Eng. 4: 60.

Mourad, F. and El-Azim, H. 2019. Use of green alga *Ulva lactuca* (L.) as an indicator to heavy metal pollution at intertidal waters in Suez Gulf, Aqaba Gulf and Suez Canal, Egypt. Egypt J. Aquat. Biol. Fish 23: 437–449.

Naeini, M.A., Zandieh, M., Najafi, S.E. and Sajadi, S. 2020. Analyzing the development of the third-generation biodiesel production from microalgae by a novel hybrid decision-making method: The case of Iran. Energy 195: 116895.

Nagappan, S., Devendran, S., Tsai, P.-C., Dinakaran, S., Dahms, H.-U. and Ponnusamy, V.K. 2019. Passive cell disruption lipid extraction methods of microalgae for biofuel production—A review. Fuel 252: 699–709.

Nagarajan, S., Chou, S.K., Cao, S., Wu, C. and Zhou, Z. 2013. An updated comprehensive techno-economic analysis of algae biodiesel. Bioresour. Technol. 145: 150–156.

Nawkarkar, P., Singh, A.K., Abdin, M.Z. and Kumar, S. 2019. Life cycle assessment of Chlorella species producing biodiesel and remediating wastewater. J. Biosci. 44: 89.

Nitsos, C., Filali, R., Taidi, B. and Lemaire, J. 2020. Current and novel approaches to downstream processing of microalgae: A review. Biotechnol. Adv. 45: 107650.

Noorman, H., Van Winden, W., Heijnen, J. and Van Der Lans, R. 2018. Intensified fermentation processes and equipment. pp. 1–41. *In*: Intensification of Biobased Processes. Green Chemistry Series.

Okoro, V., Azimov, U., Munoz, J., Hernandez, H.H. and Phan, A.N. 2019. Microalgae cultivation and harvesting: Growth performance and use of flocculants—A review. Renew. Sust. Energ. Rev. 115: 109364.

Osman, M.E.H., Abo-Shady, A.M., Elshobary, M.E., Abd El-Ghafar, M.O. and Abomohra, A.E.-F. 2020. Screening of seaweeds for sustainable biofuel recovery through sequential biodiesel and bioethanol production. Environ. Sci. Pollut. Res. 27: 32481–32493.

Ou, L., Thilakaratne, R., Brown, R.C. and Wright, M.M. 2015. Techno-economic analysis of transportation fuels from defatted microalgae via hydrothermal liquefaction and hydroprocessing. Biomass Bioenerg. 72: 45–54.

Park, J., Kim, B., Chang, Y.K. and Lee, J.W. 2017. Wet *in situ* transesterification of microalgae using ethyl acetate as a co-solvent and reactant. Bioresour. Technol. 230: 8–14.

Passos, F., Uggetti, E., Carrère, H. and Ferrer, I. 2014. Pretreatment of microalgae to improve biogas production: A review. Bioresour. Technol. 172: 403–412.

Patle, D.S., Pandey, A., Srivastava, S., Sawarkar, A.N. and Kumar, S. 2021. Ultrasound-intensified biodiesel production from algal biomass: A review. Environ. Chem. Lett. 19: 209–229.

Pirian, K., Jeliani, Z.Z., Arman, M., Sohrabipour, J. and Yousefzadi, M. 2020. Proximate analysis of selected macroalgal species from the persian gulf as a nutritional resource. Trop. Life Sci. Res. 31: 1–17.

Prapaiwatcharapan, K., Sunphorka, S., Kuchonthara, P., Kangvansaichol, K. and Hinchiranan, N. 2015. Single- and two-step hydrothermal liquefaction of microalgae in a semi-continuous reactor: Effect of the operating parameters. Bioresour. Technol. 191: 426–432.

Pratto, B., Dos Santos-Rocha, M.S.R., Longati, A.A., De Sousa Júnior, R. and Cruz, A.J.G. 2020. Experimental optimization and techno-economic analysis of bioethanol production by simultaneous saccharification and fermentation process using sugarcane straw. Bioresour. Technol. 297: 122494.

Puligundla, P., Smogrovicova, D., Mok, C. and Obulam, V.S.R. 2019. A review of recent advances in high gravity ethanol fermentation. Renew. Energy 133: 1366–1379.

Ramli, R.N., Lee, C.K. and Kassim, M.A. 2020. Extraction and characterization of starch from microalgae and comparison with commercial corn starch. IOP Conf. Ser.: Mater. Sci. Eng. 716: 012012.

Reay, D., Ramshaw, C. and Harvey, A. 2013. Process intensification: Engineering for efficiency, sustainability and flexibility. Butterworth-Heinemann.

Rokicka, M., Zieliński, M., Dudek, M. and Dębowski, M. 2021. Effects of ultrasonic and microwave pretreatment on lipid extraction of microalgae and methane production from the residual extracted biomass. BioEnergy Res. 14: 752–760.

Ruangsomboon, S. 2012. Effect of light, nutrient, cultivation time and salinity on lipid production of newly isolated strain of the green microalga, Botryococcus braunii KMITL 2. Bioresour. Technol. 109: 261–265.

Salosso, Y. 2019. Nutrient and alginate content of macroalgae Sargassum sp. from Kupang Bay waters, East Nusa Tenggara, Indonesia. Aquac. Aquar. Conserv. Legis. 12: 2130–2136.

Sander, K. and Murthy, G.S. 2010. Life cycle analysis of algae biodiesel. Int. J. Life Cycle Assess. 15: 704–714.

Santos, C.I., Silva, C.C., Mussatto, S.I., Osseweijer, P., Van Der Wielen, L.A.M. and Posada, J.A. 2018. Integrated 1st and 2nd generation sugarcane bio-refinery for jet fuel production in Brazil: Techno-economic and greenhouse gas emissions assessment. Renew. Energy 129: 733–747.

Sereewatthanawut, I., Prapintip, S., Watchiraruji, K., Goto, M., Sasaki, M. and Shotipruk, A. 2008. Extraction of protein and amino acids from deoiled rice bran by subcritical water hydrolysis. Bioresour. Technol. 99: 555–61.

Shahi, T., Beheshti, B., Zenouzi, A. and Almasi, M. 2020. Bio-oil production from residual biomass of microalgae after lipid extraction: The case of Dunaliella Sp. Biocatal. Agric. Biotechnol. 23: 101494.

Shokravi, Z., Shokravi, H., Aziz, M. and Shokravi, H. 2019. The fourth-generation biofuel: A systematic review on nearly two decades of research from 2008 to 2019. pp. 213–251. *In*: Fossil Free Fuels. CRC Press.

Sialve, B., Bernet, N. and Bernard, O. 2009. Anaerobic digestion of microalgae as a necessary step to make microalgal biodiesel sustainable. Biotechnol. Adv. 27: 409–16.

Singh, A. and Olsen, S.I. 2011. A critical review of biochemical conversion, sustainability and life cycle assessment of algal biofuels. Appl. Energy 88: 3548–3555.

Singh, D., Sharma, D., Soni, S.L., Inda, C.S., Sharma, S., Sharma, P.K. and Jhalani, A. 2021. A comprehensive review of physicochemical properties, production process, performance and emissions characteristics of 2nd generation biodiesel feedstock: Jatropha curcas. Fuel 285: 119110.

Snowden-Swan, L.J., Zhu, Y., Jones, S.B., Elliott, D.C., Schmidt, A.J., Hallen, R.T., Billing, J.M., Hart, T.R., Fox, S.P. and Maupin, G.D. 2016. Hydrothermal liquefaction and upgrading of municipal wastewater treatment plant Sludge: A preliminary techno-economic analysis, Rev. 1. Pacific Northwest National Lab.(PNNL), Richland, WA (United States).

Solé-Bundó, M., Passos, F., Romero-Güiza, M.S., Ferrer, I. and Astals, S. 2019. Co-digestion strategies to enhance microalgae anaerobic digestion: A review. Renew. Sust. Energ. Rev. 112: 471–482.

Somers, M.D. and Quinn, J.C. 2019. Sustainability of carbon delivery to an algal biorefinery: A techno-economic and life-cycle assessment. J. CO_2 Util. 30: 193–204.

Sun, A., Davis, R., Starbuck, M., Ben-Amotz, A., Pate, R. and Pienkos, P.T. 2011. Comparative cost analysis of algal oil production for biofuels. Energy 36: 5169–5179.

Sun, P., Liu, S., Zhou, Y., Zhang, S. and Yao, Z. 2018. Production of renewable light olefins from fatty acid methyl esters by hydroprocessing and sequential steam cracking. ACS Sustain. Chem. Eng. 6: 13579–13587.

Suutari, M., Leskinen, E., Fagerstedt, K., Kuparinen, J., Kuuppo, P. and Blomster, J. 2015. Macroalgae in biofuel production. Phycological. Res. 63: 1–18.

Tang, S., Shi, Z., Tang, X. and Yang, X. 2019. Hydrotreatment of biocrudes derived from hydrothermal liquefaction and lipid extraction of the high-lipid Scenedesmus. Green Chem. 21: 3413–3423.

Tang, X., Zhang, C., Li, Z. and Yang, X. 2016. Element and chemical compounds transfer in bio-crude from hydrothermal liquefaction of microalgae. Bioresour. Technol. 202: 8–14.

Taylor, B., Xiao, N., Sikorski, J., Yong, M., Harris, T., Helme, T., Smallbone, A., Bhave, A. and Kraft, M. 2013. Techno-economic assessment of carbon-negative algal biodiesel for transport solutions. Appl. Energy 106: 262–274.

Thangaraj, B., Solomon, P.R., Muniyandi, B., Ranganathan, S. and Lin, L. 2018. Catalysis in biodiesel production—A review. Clean Energy 3: 2–23.

Thilakaratne, R., Wright, M.M. and Brown, R.C. 2014. A techno-economic analysis of microalgae remnant catalytic pyrolysis and upgrading to fuels. Fuel 128: 104–112.

Tibbetts, S.M., Milley, J.E. and Lall, S.P. 2015. Chemical composition and nutritional properties of freshwater and marine microalgal biomass cultured in photobioreactors. J. Appl. Phycol. 27: 1109–1119.

Timira, V., Meki, K., Li, Z., Lin, H., Xu, M. and Pramod, S.N. 2021. A comprehensive review on the application of novel disruption techniques for proteins release from microalgae. Crit. Rev. Food Sci. Nutr. 1–17.

Torres, S., Acien, G., García-Cuadra, F. and Navia, R. 2017. Direct transesterification of microalgae biomass and biodiesel refining with vacuum distillation. Algal Res. 28: 30–38.

Usami, R., Fujii, K. and Fushimi, C. 2020. Improvement of bio-oil and nitrogen recovery from microalgae using two-stage hydrothermal liquefaction with solid carbon and HCL acid catalysis. ACS Omega 5: 6684–6696.

Vardon, D.R., Sharma, B.K., Blazina, G.V., Rajagopalan, K. and Strathmann, T.J. 2012. Thermochemical conversion of raw and defatted algal biomass via hydrothermal liquefaction and slow pyrolysis. Bioresour. Technol. 109: 178–187.

Vasistha, S., Khanra, A., Clifford, M. and Rai, M.P. 2021. Current advances in microalgae harvesting and lipid extraction processes for improved biodiesel production: A review. Renew. Sust. Energ. Rev. 137: 110498.

Velasquez, S., Lee, J. and Harvey, A. 2012. Alkaline in situ transesterification of Chlorella vulgaris. Fuel 94: 544–550.

Wang, K. and Brown, R.C. 2013. Catalytic pyrolysis of microalgae for production of aromatics and ammonia. Green Chem. 15: 675–681.

Ward, A.J., Lewis, D.M. and Green, F.B. 2014. Anaerobic digestion of algae biomass: A review. Algal Res. 5: 204–214.

WBA. 2019. Global Bioenergy Statistics 2019.

Weber, De A.B.N., Da Rocha, B.P., Smith Schneider, P., Daemme, L.C. and De Arruda Penteado Neto, R. 2019. Energy and emission impacts of liquid fueled engines compared to electric motors for small size motorcycles based on the Brazilian scenario. Energy 168: 70–79.

Wei, N., Quarterman, J. and Jin, Y.-S. 2013b. Marine macroalgae: An untapped resource for producing fuels and chemicals. Trends Biotechnol. 31: 70–77.

Wei, Q., Wang, H., Chen, Z., Lv, Z., Xie, Y. and Lu, F. 2013a. Profiling of dynamic changes in the microbial community during the soyce sauce fermentation process. Appl. Microbiol. Biotechnol. 97: 9111–9119.

Xu, D., Guo, S., Liu, L., Lin, G., Wu, Z., Guo, Y. and Wang, S. 2019. Heterogeneous catalytic effects on the characteristics of water-soluble and water-insoluble biocrudes in chlorella hydrothermal liquefaction. Appl. Energy 243: 165–174.

Xu, D. and Savage, P.E. 2015. Effect of reaction time and algae loading on water-soluble and insoluble biocrude fractions from hydrothermal liquefaction of algae. Algal Res. 12: 60–67.

Yew, G.Y., Lee, S.Y., Show, P.L., Tao, Y., Law, C.L., Nguyen, T.T.C. and Chang, J.-S. 2019. Recent advances in algae biodiesel production: From upstream cultivation to downstream processing. Bioresour. Technol. Rep. 7: 100227.

Yu, X., Sandhu, N.S., Yang, Z. and Zheng, M. 2020. Suitability of energy sources for automotive application—A review. Appl. Energy 271: 115169.

Zamalloa, C., Vulsteke, E., Albrecht, J. and Verstraete, W. 2011. The techno-economic potential of renewable energy through the anaerobic digestion of microalgae. Bioresour. Technol. 102: 1149–1158.

Zhu, G., Xiao, Z., Zhu, X., Yi, F. and Wan, X. 2013. Reducing sugars production from sugarcane bagasse wastes by hydrolysis in sub-critical water. Clean Technol. Environ. Policy 15: 55–61.

CHAPTER 3

Agro-Industrial Wastes to Sustainable Bio-Oil Fuels, Enzymes and Biobased Chemicals in Yeast-Biorefineries

R. Cosío-Cuadros,[1] Gema Núñez-López,[1] Martha F. Martín del Campo,[2] Jorge A. Rodríguez,[1] Juan C. Mateos-Díaz[1] and Georgina Sandoval[1,]*

1. Introduction

1.1 Biorefineries

In the context of circular economy, recent studies are focused on the utilization of lower-cost and nonedible feedstocks for biofuels and other chemicals production, particularly through biorefinery, which according to the International Energy Agency (IEA) Bioenergy Task 42, is "the sustainable processing of biomass into a spectrum of bio-based products (food, feed, chemicals, materials) and energy (fuels, power, heat)". Regarding its processes, it includes biochemical, thermochemical, chemical, and mechanical processes (IEA 2012).

A classification of biorefinery feedstock includes energy crops, agricultural waste, forestry waste, and industrial and municipal wastes (Maity 2015). Agricultural residues include stalks, leaves, husks, cobs, hulls, and bagasse that represent leftovers of the process and have no commercial application (Casas-Godoy et al. 2020).

Among the most interesting but also challenging biobased products, we found biofuels. To improve the economic viability of the biofuels production process, a biorefinery scheme can co-produce chemicals, feed, and food to generate added value.

1.2 Bio-oils

As mentioned before, biofuels' economic production can be improved in a biorefinery scheme, and this is especially interesting for oil-based biofuels (i.e., biodiesel, HVO, etc.). Indeed, with the

[1] Biotecnología Industrial, Centro de Investigación y Asistencia en Tecnología y Diseño del Estado de Jalisco A.C., Guadalajara, Jalisco, México.
[2] Departamento de Fundamentos del Conocimiento, Centro Universitario del Norte, Universidad de Guadalajara, Colotlán, Jalisco, México.
* Corresponding author: gsandoval@confluencia.net/gsandoval@ciatej.mx

development of new fuels from fats and oils, the reduction of land available for food crops and the increase in the price of vegetable oils, microbial bio-oils have regained interest.

Microbial bio-oils (also called single-cell oils, SCO) have emerged as alternative raw materials for biofuels, feed and food. Microbial oils are defined as the oils produced by oleaginous microorganisms, which are microorganisms able to accumulate more than 20% of their dry cell weight as lipids bodies (droplets inside the cells) (Ratledge 1991). Microbial bio-oils have many advantages compared to vegetable oils, such as a short life cycle, less labor required and contrary to plants, yeasts are not impacted by the place of growth, season and climate (Masri et al. 2019, Niehus et al. 2018a).

Lipids from yeasts are mainly triacylglycerols (TAG), which have very similar chemical composition to lipids obtained from plant oilseeds. Furthermore, yeast bio-oils have the advantage that can be produced using a wide range of nutrient sources, including industrial wastes, which could reduce the production costs of bio-oils.

On the other hand, fatty acid alkyl esters (biodiesel) and biolubricants are currently produced at a large scale by alkali-catalyzed transesterification of vegetable oil and waste fats with short-chain alcohols (Leung et al. 2010). However, this chemical technology has several negative aspects (e.g., difficulties in removing catalysts from the final product, it does not work with oils/fats containing a high fatty acid or water content, among others). For this reason, enzymes as biocatalysts for biodiesel (Sandoval et al. 2017) and biolubricants (Fernandes et al. 2021) are gaining interest, and lipases produced from yeast in a biorefinery scheme are included in this chapter.

1.3 Non-conventional Yeasts

Non-conventional yeasts, also called 'non-*Saccharomyces*' yeasts, can hydrolyze and catabolize a variety of substrates different from sugars. Recent advances in the understanding of its catabolic pathways and in metabolic engineering led to further optimize cell factories to utilize alternative waste feedstocks (Do et al. 2019). Despite the recent advances to engineer *S. cerevisiae* to produce large quantities of fatty acids, oleaginous yeast continues to outperform engineered *S. cerevisiae* (Spagnuolo et al. 2019). Recently, the most intensively studied oleaginous yeast have been *Cutaneotrichosporon oleaginosus* (20% of publications), *Rhodotorula toruloides* (19%) and *Yarrowia lipolytica* (19%) (Abeln and Chuck 2021).

Y. lipolytica is one of the most studied non-conventional yeasts and it is also 'oleaginous' as it accumulates bio-oils. Various reports on biobased chemicals and enzymes production by *Y. lipolytica* appeared recently (de Souza et al. 2019, Fickers et al. 2020, López-Pérez and Viniegra-González 2016, Madzak 2021, Pereira et al. 2019, Posso Mendoza et al. 2020, Sales et al. 2020, Yan et al. 2018). In addition, modern genetic tools are available to engineer this yeast (Wheeldon and Blenner 2021).

R. toruloides (previously known as *Rhodosporidium toruloides*) is a red yeast. Therefore, it naturally produces carotenoids responsible for its red colour and it accumulates other lipids. In addition, it produces enzymes interesting for pharma and chemical industries such as L-phenylalanine ammonia-lyase and D-amino acid oxidase. *R. toruloides* can naturally grow on a wide range of carbon sources including agro-industrial wastes as it presents good tolerance to inhibitors found in unrefined substrates. Synthetic biology and metabolic engineering tools for this yeast are also available (Park et al. 2018).

C. oleaginosus (formerly known as *Cryptococcus curvatus*) also stands out as an oleaginous yeast, able to accumulate up to 60% of lipids in its cells. It grows rapidly on several agro-industrial wastes, such as biodiesel-derived glycerol and lignocellulosic hydrolysates, being very resistant to hydrolysate byproducts. In addition, its genetic accessibility makes *C. oleaginosus* a promising yeast for biorefineries (Bracharz et al. 2017, Di Fidio et al. 2021).

The yeast *Pichia pastoris* also outstands by its utilization of non-conventional substrates such as methanol and glycerol efficiently; both carbon sources (which can come from byproduct wastes) are commonly used for constitutive recombinant protein production in this yeast (Do et al. 2019, Valero 2018).

1.4 Enzymes

Biocatalysis is nowadays an important part of bioeconomy for greener manufacturing process, as enzymes are biodegradable, require mild conditions to work and are highly specific, carrying out complex reactions in a simple way. Enzymes produced by yeast have many applications as biocatalyst for biobased chemicals for pharmaceutical, cosmetic, food and bulk chemicals industry (De Regil and Sandoval 2013). This chapter focuses on lipases as one of the most applied enzymes in organic chemistry and particularly for biobased oleochemicals.

1.4.1 Lipases

Lipases are carboxyl ester hydrolases (EC 3.1.1.3) that hydrolyze triglycerides in an aqueous medium, but in organic media, they perform synthesis reactions (Casas-Godoy et al. 2018). Lipases simultaneously perform the triglycerides (TAG) transesterification and free fatty acids (FFA) esterification, overcoming the chemical synthesis drawbacks and allowing to obtain higher quality products, such as biodiesel, bio-oil fuels and biobased lubricants (Fernandes et al. 2021, Sandoval et al. 2017, Vargas et al. 2018).

Oleaginous yeast can also produce lipases. For instance, *Y. lipolytica* has 16 lipases in its genome, Lip2 (YLL2) being extracellular (Fickers et al. 2011). YLL2 showed wide applications in organic synthesis and bioremediation (Cancino et al. 2008, Fraga et al. 2018, Guieysse et al. 2004, Posso Mendoza et al. 2020, Sales et al. 2020). YLL2 can be also produced using agro-industrial wastes as described later.

1.4.1.1 Solid State Fermentation

The high cost of commercial lipases biocatalysts limits their use in oils/fats transformation in processes of low-cost products such as biofuels. Solid state fermentation (SSF) is a low-cost process where microorganisms grow close to the natural environment on moist solids, inert supports or insoluble substrates with a very low content or absence of free water (Abdul Manan and Webb 2017). SSF has been employed as an economic process for lipases production through the use of agro-industrial wastes as substrates.

Moreover, fermented solids (FS) may be directly employed as biocatalysts allowing the production of more cost-competitive bio-oil fuels and biobased chemicals (Aguieiras et al. 2019). This strategy may use inexpensive agro-industrial wastes as solid support/substrates for microorganism growth and lipase production, avoiding expensive steps of enzyme purification and immobilization (Pereira et al. 2019).

1.4.1.2 Recombinant Enzymes

The use of recombinant DNA technology to produce recombinant enzymes in a host microorganism has several advantages such as repeatability between lots, scalability, purification facilitated, economic production of large quantities of enzyme and the possibility of producing engineered tailor-made enzymes (Valero 2018).

In this chapter, an overview of the production of the recombinant lipase from *Thermomyces lanuginosus* (TLLr) by growing *P. pastoris* in suitable SSF supports, achieving a solid biocatalyst for the synthesis of ethyl esters, is presented. This example gives an additional contribution to the yeast-biorefineries concept using agro-industrial wastes and bio-oils for sustainable biofuels, enzymes, and biobased chemicals production.

Additional examples of enzymes produced in *P. pastoris* using agro-industrial wastes are also presented.

1.5 Other Biobased Chemicals

Simultaneous production of bio-oils or enzymes with other biobased chemicals is also possible. Although studied for a while now, production of single-cell protein (SCP) regained interest because of the need of non-animal proteins and proteins with a low environmental footprint (Jach et al. 2022).

Organic acids and polyols are other valuable chemicals that can also be produced by non-conventional yeasts using agro-wastes as described later.

A representation of yeast biorefinery and products described in this chapter is given in Fig. 1. According to a recent review (Abeln and Chuck 2021), oleaginous yeasts have been primarily investigated for the production of biofuels (77%), food/supplements (24%), oleochemicals (19%) animal feed (3%), and as carbon sources mainly saccharides (60%), but also agro-industrial wastes such as hydrolysates (26%), and glycerol (19%).

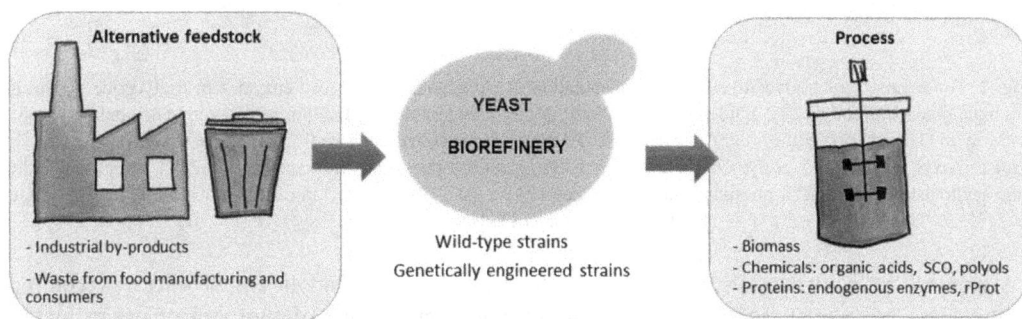

Fig. 1. Yeast biorefinery representation. SCO: single-cell oils (bio-oils), rProt: recombinant proteins. Adapted from Do et al. (2019) under CC BY 4.0 license.

2. Bio-oils and Biobased Chemicals' Production in Yeast Biorefineries

2.1 Yeast Bio-oils from Agro-industrial Wastes

In oleaginous yeasts, accumulation of bio-oils (mainly TAG) can occur by two mechanisms. *De novo* synthesis occurs under nitrogen depleting and excess of carbon, leading to a metabolic switch favouring the lipogenic phase (Fig. 2), while *ex novo* synthesis occurs by a direct incorporation into the cell of hydrophobic substrates such as fatty acids (FA), TAG and sterols, which are either used for energy or as storage lipids (Papanikolaou and Aggelis 2011).

Y. lipolytica, *C. oleaginosus* and *R. toruloides* have been described as the most efficient yeasts in terms of yield of bio-oil (Abeln and Chuck 2021). Table 1 presents the agro-industrial wastes that have been used as carbon sources to produce bio-oils by the oleaginous yeasts focused on in this chapter. As has been observed, the metabolic versatility of these yeast allows a wide repertoire of agro-industrial wastes to be harnessed. Yields depend on raw material and can be improved by optimization of fermentation conditions, metabolic engineering and synthetic biology.

Y. lipolytica can metabolize hydrophobic and acidic substrates such as hydrocarbons (Das and Chandran 2011) and volatile fatty acids (Llamas et al. 2020b). In addition, metabolic engineering can also allow the consumption of agro-industrial wastes not naturally consumed by the yeast. For

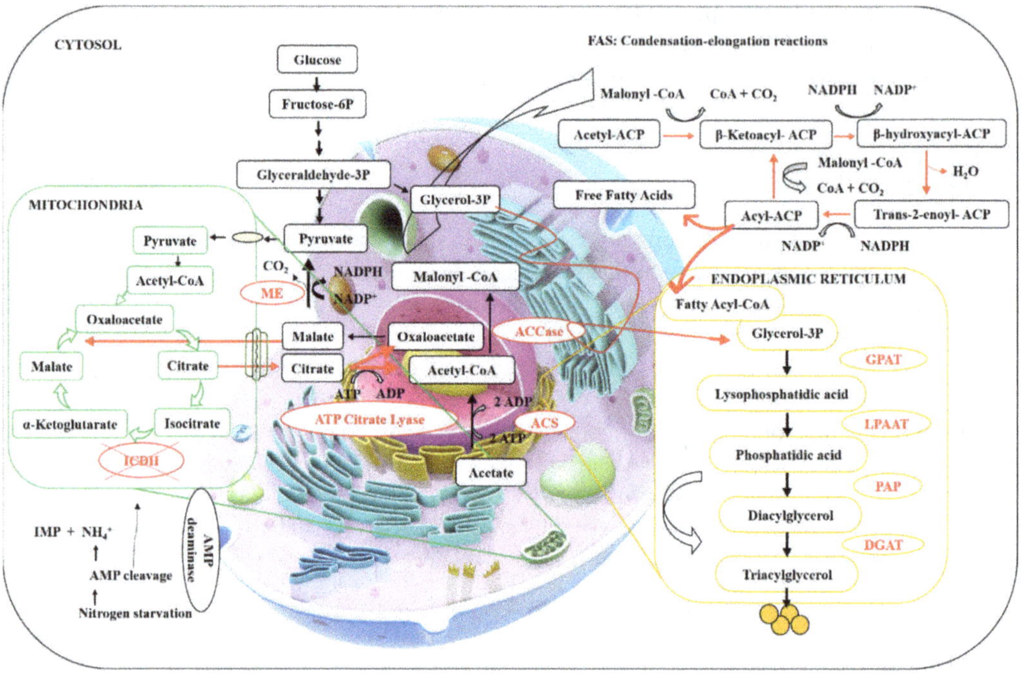

Fig. 2. Biochemistry of triglycerides (TAG) accumulation in oleaginous yeasts. Key enzymes in triglyceride synthesis are highlighted with red circles. Mitochondrial pathway in green, endoplasmic reticulum pathway in yellow and cytosolic pathway in black. Abbreviations: fatty acid synthase (FAS), isocitrate dehydrogenase (ICDH), malic enzyme (ME), acetyl-CoA carboxylase (ACCase), acetyl-CoA synthetase (ACS), glycerol-3-phosphate acyltransferase (GPAT), lysophosphatidic acid acyltransferase (LPAAT), phosphatidate phosphatase (PAP), diacylglycerol acyltransferase (DGAT). From Caporusso et al. (2021a) under CC BY 4.0 license.

instance, xylose is not naturally taken by *Y. lipolytica*; therefore, metabolic engineering was required to use agave bagasse hydrolysate as substrate. Introduction of genes of phosphoketolase pathways into an already "obese" (overproducer of bio-oils) strain allowed the consumption of xylose and other sugars from agave bagasse hydrolysate as the sole carbon source, reaching a production of 16.5 g/L of bio-oil (Niehus et al. 2018b). Regarding tolerance to inhibitors in lignocellulosic hydrolysates, (Konzock et al. 2021), identified in a synthetic hydrolysate that formic acid, furfural, and coniferyl aldehyde were the major inhibitors for *Y. lipolytica* growth.

In Table 1 it can be seen that *C. oleaginosus* converted various lignocellulosic hydrolysates without the need of detoxification. *C. oleaginosus* also became a model oleaginous yeast for metabolic engineering and synthetic biology approaches (Bracharz et al. 2017, Di Fidio et al. 2021, Pham et al. 2021) to increase lipid productivity even at a comparable level of vegetable oil plants (Masri et al. 2019). Pham et al. (2021) reported a constraint-based metabolic model containing 1553 reactions involving 1373 metabolites in 11 compartments, aiming at massive lipid accumulation. This model suggests that ATP-citrate lyase is a possible target in *C. oleaginosus* to further improve lipid production.

Regarding culture conditions, fed-batch increases productivity and some additives (surfactants in particular) also increased the yield of *R. toruloides* lipids (Xu et al. 2016). In addition to bio-oils (Table 1), *R. toruloides* was also engineered to produce special high-value lipids such as carotenoids, omega-3 for food industry and very long-chain fatty acids (VLCFA) that are important renewable feedstocks in plastic, cosmetics, nylon, and lubricant industries (Park et al. 2018).

Indeed, depending on its composition and FA profile, bio-oils can be used for biodiesel industry, food and biopolymers (Vasconcelos et al. 2019). A comparison of yeast bio-oil FA profile shows

Table 1. Bio-oils produced by oleaginous yeasts with agro-industrial wastes as C source.

Carbon Source	Yeast (W/D/E)*	Bio-oil Titre (g/L)	Reference
*Crude glycerol***	*Y. lipolytica (W)*	13.6	(Sara et al. 2016)
Crude glycerol	*R. toruloides (W)*	12.2	(Yang et al. 2014)
Crude glycerol	*C. oleaginosus (W)*	4.2	(Gong et al. 2016)
Crude glycerol plus methanol	*C. oleaginosus (W)*	20.8	(Chen et al. 2018)
Lignocellulosic hydrolysates			
Agave bagasse	*Y. lipolytica (E)*	16.5	(Niehus et al. 2018b)
Sugarcane bagasse	*Y. lipolytica (E)*	6.7	(Tsigie et al. 2011)
Potato peel	*R. toruloides (W)*	26.7	(Carmona-Cabello et al. 2021)
Corn stover	*R. toruloides (D)*	31.9	(Fei et al. 2016)
Corn stover	*R. toruloides (D)*	23.3	(Sànchez I Nogué et al. 2018)
Corn stover plus crude glycerol	*C. oleaginosus (W)*	21.7	(Gong et al. 2016)
Cardboard/paper	*C. oleaginosus (W)*	9.1	(Zhou et al. 2017)
Sorghum stalk	*C. oleaginosus (W)*	13.1	(Lee et al. 2017)
Switchgrass	*C. oleaginosus (W)*	12.3	(Lee et al. 2017)
Microalgae	*C. oleaginosus (W)*	30.6	(Meo et al. 2017)
Cardoon stalks	*C. oleaginosus (W)*	7.5	(Caporusso et al. 2021b)
Beech wood	*C. oleaginosus (W)*	21.9	(Siebenhaller et al. 2018)
Giant reed	*C. oleaginosus (W)*	28.4	(Di Fidio et al. 2019)
Other C sources			
Acetic acid	*Y. lipolytica (E)*	115***	(Xu et al. 2017)
Acetic acid and glycerol	*Y. lipolytica (W)*	12.4	(Fontanille et al. 2012)
Acetic acid	*R. toruloides (D)*	4.4	(Huang et al. 2016)
Acetic acid	*C. oleaginosus (W)*	6.5	(Liu et al. 2017)
Food waste	*R. toruloides (D)*	6.4	(Ma et al. 2018)
Food waste	*R. toruloides (W)*	7.3	(Zeng et al. 2017)
Volatile fatty acids	*Y. lipolytica (W)*	3.5	(Pereira et al. 2021)
Volatile fatty acids	*Y. lipolytica (W)*	8.0 (%w/w)	(Llamas et al. 2020a)
Volatile fatty acids	*R. toruloides (W)*	5.0 (%w/w)	(Llamas et al. 2020a)
Volatile fatty acids	*C. oleaginosus (W)*	7.4	(Liu et al. 2017)

* W: wild type. D: diploid, hyphal conjugate. E: engineered.
** "Crude glycerol" refers to glycerol which is a biodiesel or soap by-product.
*** Semicontinuous system.

an elevated monounsaturated FA content, comparable or superior to vegetable oils (Vasconcelos et al. 2019). Also, some industrial companies (Dupont, DSM, Cargill, Nestlé among others) already launched the production and commercialization of engineered yeast bio-oils, mainly omega-3 for food and feed applications (Caporusso et al. 2021a).

Besides, other valuable SCO-derivatives can also be synthesized using yeast lipases. For instance, bioactive amides using *C. rugosa* lipase (El-Baz et al. 2021b) and glucose esters were catalyzed by *C. antactica* B lipase (El-Baz et al. 2021a).

In addition to substrate and yeast strain, other parameters affecting bio-oil production in oleaginous yeasts (C/N ratio, temperature, pH, additives, reactor conditions and oxygenation) have been recently reviewed by Caporusso et al. (2021a). However, given the variety of substrates and processes presented here, it is evident that individual process optimization is required in each case.

2.2 Polyols and Organic Acids in Yeast Biorefineries

Glycerol is a versatile carbon source for yeasts and is also a byproduct of biodiesel and soap industries that, besides bio-oils (Table 1), can generate other valuable chemicals such as polyols and organic acids (citric acid mainly). Some commercial applications of polyols and organic acids from *Y. lipolytica* are: citric acid production by ADM (USA) company and erythritol production by Baolingbao Biology Co. (China) (Madzak 2021).

Depending on fermentation conditions and with crude glycerol as substrate, *Y. lipolytica* strains can produce either the polyols mannitol and erythritol (33.6 g/L) with constant nitrogen supplementation, or citric acid (39 g/L) in batch (Papanikolaou et al. 2017). A more recent study on new wild-type strains of *Y. lipolytica* that efficiently grow on biodiesel-derived crude glycerol wastes has shown levels of polyols of 52 g/L (in flask) and citric acid of 102 g/L in bioreactor (Papanikolaou et al. 2020).

Rakicka-Pustułka et al. (2021) investigated polyol (erythritol, mannitol and arabitol) production by *Yarrowia* clade using crude glycerol from the biodiesel (obtaining 59.8–62.7 g/dm^3) and with soap-derived glycerol (76.8–79.5 g/dm^3). Jagtap et al. (2021) engineered a *Y. lipolytica* strain to increase the erythritol production by increasing the expression of native glycerol kinase, and transketolase. This strain increased crude glycerol utilization by 2.5-fold and was able to produce 58.8 g/L erythritol in fed-batch fermentation.

Partly deproteinized whey was another waste-derived substrate used for sustainable citric acid production by a cold-adapted *Y. lipolytica* strain giving 33 g/L of citric acid in non-sterile medium (Arslan et al. 2016). As for bio-oils, productivity of citric acid by *Y. lipolytica* was dependent on strain type, carbon source, C/N ratio, as well as physicochemical conditions (pH, temperature, oxygen transfer rate, etc.) (Carsanba et al. 2019).

The development of industrial production platforms for other organic acids and biomolecules for organic synthesis also has great interest. Indeed, oxo- and hydroxy-carboxylic acids are of special interest in organic synthesis (Aurich et al. 2012).

For instance, Fina et al. (2021) developed engineered *P. pastoris* strains to produce the commodity chemical 3-hydroxypropionic acid (3-HP), which is a chemical platform that can be converted into acrylic acid and to other alternatives to petroleum-based products. The best strain was cultured on glycerol in fed-batch mode, achieving a final concentration of 3-HP of 24.75 g/L.

(2R,3S)-isocitric acid has long been used only as a specific biochemical reagent. As chiral intermediate of the tricarboxylic acid cycle, (2R,3S)-isocitric acid and its derivatives are promising compound with many applications (Aurich et al. 2012). There is an increasing evidence that it can also be used for prevention and treatment of some diseases such as blood clots, some forms of Parkinson's disease, to ameliorate impaired spatial memory and for obtaining HIV/AIDS protease inhibitors (Aurich et al. 2012, Kamzolova and Morgunov 2019, Morgunov et al. 2020). The production of isocitric acid with *Y. lipolytica* using biodiesel waste glycerol has been reported by Morgunov et al. (2020) with titres of 58 to 90 h/L, yield of 40% and 3:1 isocitric/citric ratio. A recent review of the production of this interesting organic acid by natural, mutant, and recombinant strains of *Y. lipolytica* has also appeared (Kamzolova and Morgunov 2019).

2.3 SCP from Oleaginous Yeasts in Waste Biorefineries

The protein in microorganism's biomass is also called bioprotein, protein biomass, or single-cell protein (SCP), though filamentous yeast, algae and fungi may be multicellular (Jach et al. 2022). It is humankind's challenge to meet the need of protein products which SCP cultivated on agro-industrial wastes can address (Gervasi et al. 2018). Therefore, there is a need to obtain higher efficiencies of SCP production and balanced aminoacid and nutrients in SCP (Jach et al. 2022).

Composition of some oleaginous yeast have been already investigated. In the case of *R. toruloides*, Shen et al. (2017) found that under nitrogen limitation, the cellular lipid content decreased but the carbohydrate and protein contents increased, while under carbon limitation, the cellular lipid, protein, and carbohydrate contents remained relatively constant (Shen et al. 2017).

Y. lipolytica is considered GRAS by Food and Drug Administration (FDA), and therefore as safe for humans and animals (Groenewald et al. 2014). Dried heat-killed *Y. lipolytica* biomass cultured on biofuel waste was used as feed additive since 2010 and in general, it has benefits in animal feed (Guardiola et al. 2021). *Y. lipolytica* has been studied as feed in calves, piglets and turkeys (Jach et al. 2022). An already commercial application is the use of *Y. lipolytica* biomass as fodder yeast for farm and pet animals by Skotan SA (Poland), who also uses this yeast in prebiotic/probiotic applications (Madzak 2021).

Y. lipolytica SCP has been also used or studied as feed in aquaculture. It was tested on shrimp and salmon and found to support an increase in fish weight (Jones et al. 2020). *Y. lipolytica* was also applied as feed in Atlantic salmon and Pacific red snapper (Jach et al. 2022). Dupont and AquaChile also commercialized an EPA-rich *Y. lipolytica* (Tocher et al. 2019).

For humans, the European Food Safety Authority (EFSA) has recently authorized *Y. lipolytica* biomass as a novel food in dietary supplements for the general population above 3 years of age as safe and nutritionally advantageous, with maximum daily use levels of 3 g/day for children from 3 years up to 10 years of age, and 6 g/day thereafter (EFSA et al. 2019). Also, in 2020, *Y. lipolytica* biomass enriched with selenium was included as food supplement in the list of authorized novel foods under Commission Regulation (EU) 2020/1999 (2002/46/WE) (EFSA et al. 2020).

Y. lipolytica biomass contains 29–65% of protein, depending on the carbon source and culture conditions. In crude glycerol, Rakicka-Pustułka et al. (2021) obtained 30%, Juszczyk et al. (2013) 45%, while in rye and oat wastes Drzymała et al. (2020) got 30–44.5%, and in palm oil mill effluent 20.2% (Louhasakul et al. 2019). In the case of crude glycerol, Yan et al. (2018) showed that the trace compounds that could be present are not harmful; on the contrary, they could also contribute as N and C sources assimilable by the yeast.

Regarding essential amino acids profile in *Y. lipolytica* SCP, it agrees with FAO standards for fodder yeast (Jach and Serefko 2018) and most amino acids exceed FAO requirements (Table 2).

Table 2. Mean contents of essential amino acids (mg/g) in yeast SCP and other protein sources, compared with FAO requirements for adults. Data from Jach et al. (2022).

Amino Acid	FAO Requirements	*Y. lipolytica*	*S. cerevisiae*	*C. utilis*	Egg	Cow Milk
Arginine	–	48	46.5	32	11.5	33
Histidine	15	26	23.5	16	4	37
Isoleucine	30	44	37	48	68	40
Leucine	59	68	63	71	90	88
Lysine	45	70	65	51	63	78
Cysteine	–	11	9	24	24	9
Methionine	–	12	14	15.5	32	29
Sulphur AA	22	23	23	39.5	56	38
Phenylalanine	–	40	33	41	63	47
Tryptophan	–	47	9	39	16	ND
Tyrosine	–	66	26	20	195	16
Aromatic AA	38	153	68	100	98.5	63
Threonine	23	48	48	41	50	48.7
Valine	39	53	53	55	74	47.9

SCP from *Y. lipolytica* contains high level of digestible ether extract. This yeast is also rich in minerals (such as calcium, chromium, cooper, iron, iodine, magnesium, manganese, phosphorus, potassium, selenium, sulphur, and zinc), and B-complex vitamins (Table 3). Furthermore, *Y. lipolytica* also have in its cell wall immunomodulator β and α-D-glucans with low molecular weight (Angulo et al. 2021). Therefore, SCP from *Y. lipolytica* biorefineries is a very promising source of feed, food, and nutraceuticals.

Table 3. Dry matter, ether extract and microutrients in yeast biomass. Data from Czech et al. (2016) and Michalik et al. (2013).

Component	*Y. lipolytica*	*S. cerevisiae*	Units
Dry matter	952	925	g/kg DW
Ether extract	200	7	g/kg DW
Ash	36	62	g/kg DW
Sodium	16	8	g/kg DW
Potassium	22	17	g/kg DW
Sulphur	5	4	g/kg DW
Calcium	4	2	g/kg DW
Phosphorus	4	10	g/kg DW
Magnesium	2	1	g/kg DW
Manganese	15	4	mg/kg DW
Zinc	70	61	mg/kg DW
Iron	110	101	mg/kg DW
Vitamin E	7	45	mg/kg DW
Vitamin B1	98	119	mg/kg DW
Vitamin B2	16	6	mg/kg DW
Vitamin B6	28	26	mg/kg DW
Vitamin B12	56	4	μg/kg DW

3. Enzymes Produced in Yeast Biorefineries

Enzymes are one of the most value-added products that can be produced in a biorefinery. Therefore, besides the examples of biobased chemicals given above, this section describes: (i) an example of coproduction of lipase and SCP in *Y. lipolytica* in media supplemented with agro-industrial wastes; (ii) the production of the recombinant enzymes in *P. pastoris* using agro-industrial wastes; (iii) the example of recombinant lipase from *T. lanuginosus* (TLLr) expressed in the yeast *P. pastoris* and produced by SSF using agro-industrial wastes. To achieve a circular economy, the application of this solid as biocatalyst to obtain biodiesel is also presented.

3.1 Lipase and SCP Coproduction in Y. lipolytica

As mentioned before, YLL2 extracellular lipase from *Y. lipolytica* has wide applications in organic synthesis and bioremediation (Cancino et al. 2008, Fraga et al. 2018, Guieysse et al. 2004, Posso Mendoza et al. 2020, Sales et al. 2020). In the context of biorefinery, *Y. lipolytica* can also be

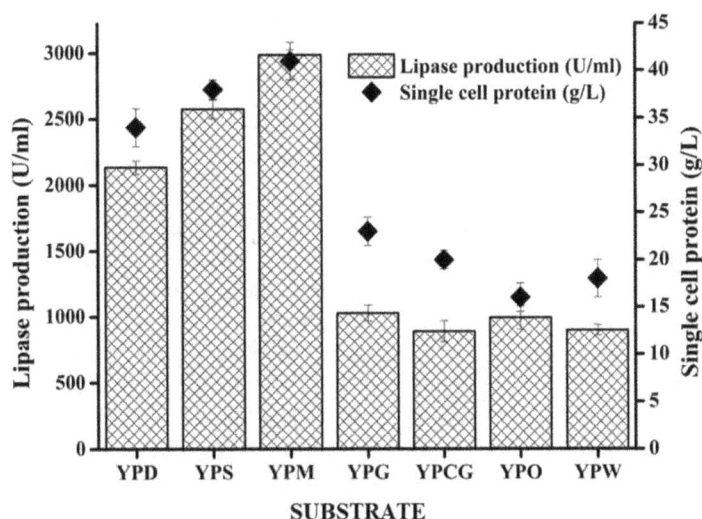

Fig. 3. Lipase activity and SCP obtained with an engineered *Y. lipolytica* lipase and different rich media (YDP) modified with agro-industrial wastes as carbon source. D: dextrose, S: sucrose, M: molasses, G: pure glycerol, CG: crude glycerol, O: olive oil, W: waste cooking oil. Single cell protein is expressed as DCW. Lipase activity was measured by titration of TG. From Yan et al. (2018) under CC 4.0 license.

engineered to overproduce YLL2 and at the same time SCP. Fig. 3 presents lipase activity (titration assay) and SCP expressed as DCW obtained with an engineered *Y. lipolytica* and rich media (YDP) modified with usual (destrose, sucrose, pure glycerol, olive oils) and agro-industrial wastes (molasses, crude glycerol, waste cooking oils) as carbon source. Up to 3000 U/mL of lipase activity and 40 g/L of SCP can be obtained, respectively, using sugarcane molasses as substrate at low scale. Scaling up to 10-L fermentation yielded 151.2 g/L of SCP, and an enzyme production of 16420 U/mL (4.13 g/L) after 89 h. Both YLL2 and SCP were tested as feed for the fish *Cynoglossus semilaevis* Günther, observing after 32 days that average weights were significantly higher compared to fish fed with the control diet (Yan et al. 2018).

3.2 *P. pastoris as Host for Enzymes in Yeast Biorefinery*

As mentioned in the introduction, the yeast *P. pastoris* can efficiently grow on methanol and glycerol, which can come from byproduct wastes contributing to the sustainability of recombinant protein production in this yeast (Do et al. 2019, Valero 2018).

Indeed, *P. pastoris* can efficiently utilize crude biodiesel-derived glycerol containing NaOH or KOH for α-amylase production, obtaining 3.5 U/ml within 30 h (Anastácio et al. 2014). Equally, with crude glycerol as carbon source, thermostable β-mannanase was produced in *P. pastoris* reaching 2385 U/mL (Luo et al. 2018).

The combination of crude glycerol and methanol to produce recombinant bovine chymosin in fed-batch with a methanol-inducible strain (AOX1 promoter) was studied by Noseda et al. (2016), giving a coagulant activity of 192 IMCU/mL after 120 h of induction.

The promoter PGK can also be used to culture *P. pastoris* strains, using crude glycerol derived from biodiesel for the production of the widely applied lipase B from *Candida antarctica* (rCALB). The rCALB obtained in fed-batch was compared with the commercially available CALB (Novozymes) and it had similar properties of stability and triglyceride selectivity (Robert et al. 2017).

3.3 Solid Ferment Biocatalyst for Biodiesel Production TTL in P. pastoris by SSF

3.3.1 Agro-industrial wastes used in solid-state fermentation (SSF) to obtain enzymes

In most cases, agro-industrial wastes have little or no use in the process that generated them. Thus, several investigations around the world are focused on the use of agro-industrial wastes as raw materials to obtain high-value products, under a biorefinery concept. Agro-industrial wastes are all those materials with organic content that are generated in large amounts from agricultural activities (Beltrán-Ramírez et al. 2019). They can be liquid or solid, of animal or vegetable origin, and according to Galanakis (2012) and Ravindran et al. (2018), they can be divided into seven groups, depending on their origin: cereals, roots and tubers, fruits and vegetables, meat products, fish and shellfish, dairy products, and oilseed plants. Among the most common highly available residues, there are cane molasses and bagasse, oilseed pastes, corn liquor, and beer residues. Additionally, other abundant agricultural crop residues obtained from cereals (e.g., sorghum, rice, corn, wheat, and barley) also include straw, stems, leaves, bark, seeds, among others (Salihu et al. 2012).

All the aforementioned residues can be used as supports or substrates to produce interesting enzymes by SSF. This fermentation technique gained important attention during the last three decades because it allows obtaining high enzymatic titres at low costs (Hidayatullah et al. 2018). It is also an interesting alternative to revalue oleaginous residues as substrates for lipases production, which can be easily employed to achieve heterogeneous lipase-based biocatalysts for the synthesis of biofuels and biobased chemicals (Aguieiras et al. 2019, Collaço et al. 2021, Ojeda-Hernández et al. 2018).

Obtaining lipases by SSF has been reported under different process conditions, using diverse agro-industrial residues as substrate, e.g., husks (wheat, rice, soybeans, barley), oilseeds (soybeans, olive, castor, jatropha), bagasse (sugarcane) and known microorganisms (Godoy et al. 2011, Ilmi et al. 2017, Ruiz Flores 2014).

3.3.2 Yeast lipases produced by SSF

Among the various groups of microorganisms used in SSF to produce lipases, filamentous fungi are the most reported, due to their ability to penetrate and naturally grow on solid substrates. Other microorganisms, such as bacteria (Kumar Sahoo et al. 2018), archaea (Martin del Campo et al. 2015), and yeasts (de Freitas et al. 2021, Sales et al. 2020), have been used to produce lipases in SSF to a lesser extent. Previous research reported yeasts from *Candida* and *Yarrowia* genera as interesting native lipases producers in SSF (Ramos-Sánchez et al. 2015). Farias et al. (2014) reported *Y. lipolytica* yeast able to produce 102 ± 6 U/g and 139 ± 4 U/g of lipase in SSF, using cottonseed and soy cakes, respectively, without any type of supplementation, which minimized lipase production costs.

Likewise, Lopes et al. (2016) showed that *Y. lipolytica* yeast (IMUFRJ 50682) reached a lipase activity of 486 U/g after 96 h of fermentation in SSF, using oil mill and wheat bran wastes as substrates. More recently, Sales et al. (2020) demonstrated the impact of watermelon rind supplementation on the *Y. lipolytica* (IMUFRJ 50682) lipase production. Watermelon rind is a cutin-rich waste, its supplementation led to an increase in the porosity of the SSF support, allowing to obtain a 46% improvement in lipase productivity.

3.3.3 Biodiesel and biolubricants synthesized by FS with lipase activity

The dried FS obtained after lipase production in SSF can be directly employed as a heterogeneous biocatalyst with lipase activity for the synthesis of diverse interesting products (Aguieiras et al. 2017). This simple strategy allows reducing or avoiding costly extraction processes, purification, and immobilization of the enzyme (Borges et al. 2021, Ojeda-Hernández et al. 2018).

Ojeda-Hernández et al. (2018) described a comprehensive SSF method to produce bio-oil fuels using FS with the native lipase activity from *Rhizomucor miehei*. This dried FS was successfully employed as a solid biocatalyst for the production of ethyl esters (biodiesel) with a final conversion > 90% at 24 h, similar to that obtained using a commercial immobilized lipase. Aguieiras et al. (2019) also reported the production of FS from filamentous fungi by SSF, using babassu cake and cottonseed flour as supports. These FS were employed for the enzymatic synthesis of alkyl esters from macauba, soybean, and palm oils, to be used as biofuel (biodiesel) and biolubricant (Aguieiras et al. 2019). Besides, various native yeasts products (proteins, ethanol, and secondary metabolites) were obtained by SSF using *P. pastoris, Kluyveromyces marxianus, S. cerevisiae*, and *Y. lipolytica* (López-Pérez and Viniegra-González 2016, Šelo et al. 2021). *Y. lipolytica* is one of the employed yeasts to obtain FS used for the synthesis of different esters, taking advantage of the use of agro-industrial wastes (de Souza et al. 2019, Pereira et al. 2019).

3.3.4 Recombinant T. lanuginosus lipase produced by P. pastoris in SSF

As previously overviewed, SSF is an interesting fermentation tool to obtain native yeast lipases. Nonetheless, one of the little-explored fields in the SSF is obtaining recombinant lipases from heterologous yeasts expression systems such as the widely used *P. pastoris*. Thus, an example of the TLLr production, by growing *P. pastoris* in suitable SSF supports to achieve lipase based heterogeneous biocatalysts for the synthesis of biodiesel, is reported here.

TLL is an industrially important lipase that still is subject to studies and applications (Bohr et al. 2019). For economically pertinent enzymatic processes, immobilization is almost always necessary (Khan 2021). In the case of SSF, the enzyme is auto-immobilized in agro-industrial wastes, contributing both to the environmental and economical sustainability (Sales et al. 2020).

The choice of the support to produce high titre of lipase activity in SSF is one of the most important parameters since it determines the humidity content, heat-mass transfer, and usefulness of the dried solid ferment as a heterogeneous biocatalyst, during the fermentation and synthesis processes.

Fig. 4 shows the kinetics of the *T. lanuginosus* recombinant lipase production by *P. pastoris* in SSF, using different supports. For the initial screening, Erlenmeyer flasks were employed to

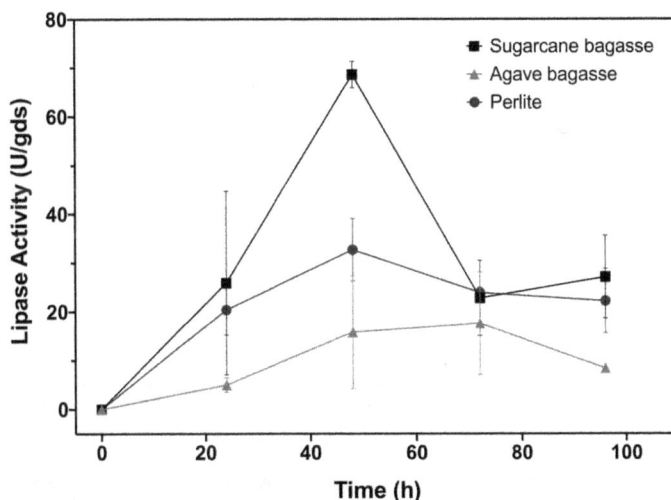

Fig. 4. Kinetics of the *T. lanuginosus* recombinant lipase production by *P. pastoris* in SSF, using different supports (■ – Sugar cane bagasse, ▲ – Agave bagasse and ● – Perlite) incubated at 30°C with saturated humidity for 96 h. Erlenmeyer flasks were used for the SSF. Culture protocol as in (Ojeda-Hernández et al. 2018). The kinetics of the p-nitrophenolate (pNP) release during the reaction at 37°C was monitored, taking readings every 30 s for 15 min at 410 nm.

select the best SSF support. For all supports, a progressive increase in lipase activity was observed, reaching its maximum activity at 48 h (for Perlite and sugarcane bagasse) and 72 h (for agave bagasse); subsequently, it is possible to see a sharp fall in enzyme activity. This decrease in the activity of lipase may probably be due to the presence of proteases secreted by the microorganism in the culture medium, which could be acting on the lipases after a shortage of nutrients in the later time of fermentation.

Some authors demonstrated that after the maximum enzyme activity was reached by SSF, subsequently, a sharp drop in enzyme activity could be seen (Sales et al. 2020). The authors attribute the decrease in lipase amount to proteolysis. In the same interval, an increase in proteolytic activity was observed and, probably, proteases were capable of hydrolyzing the enzyme, decreasing their activity.

TLLr presented its highest activity (68.6 U/g) when using sugarcane bagasse as support during SSF, being 2 and 4 times higher than that obtained with perlite (32.7 U/g) and agave bagasse (17.6 U/g), respectively (Fig. 4). Thus, sugarcane bagasse was employed as support for TLLr production.

Another factor that significantly influences lipase production during SSF is temperature and humidity control, which may be modulated by the type of SSF bioreactor employed. SSF bioreactors are classified into diverse categories, depending on their mode of employment (Arora et al. 2018). Static flasks are one of the different laboratory-scale SSF bioreactors, offering simplicity (Vaseghi et al. 2012) and effective temperature and humidity control during fermentation (Londoño-Hernandez et al. 2020).

As shown in Fig. 5, the kinetics of the *T. lanuginosus* recombinant lipase production in 50 mL conical centrifuge tubes as bioreactors, reached the highest lipase activity (264.3 U/g) after 48 h of fermentation. As expected, the activity was 3.8 times higher than that obtained using Erlenmeyer flasks. It is worth noting that the reduction in the fermentation volume allowed better humidity and temperature control, achieving solid ferments with greater heterogeneity throughout the fermentation, and thus higher lipase activity. Expression of *T. lanuginosus* lipase in *P. pastoris*

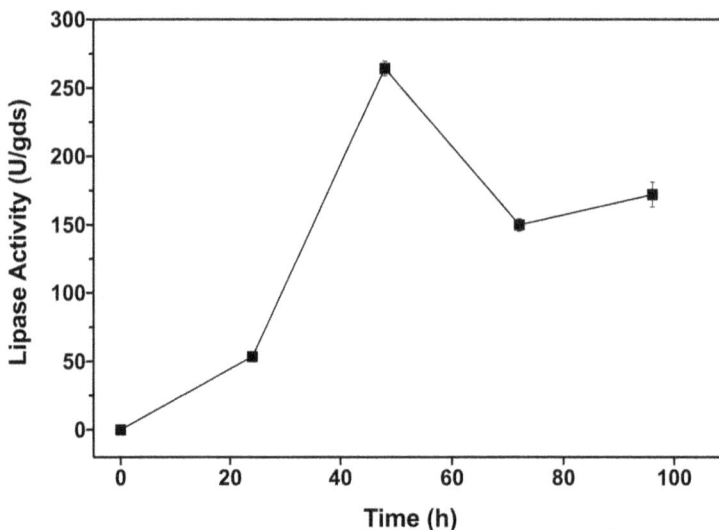

Fig. 5. Kinetics of the *T. lanuginosus* recombinant lipase production by *P. pastoris* in SSF, using 50 mL conical centrifuge tubes as bioreactors, incubated at 30°C with saturated humidity for 96 h. Sugar cane bagasse was employed as support. Culture protocol as reported by Ojeda-Hernández et al. (2018). The kinetics of the p-nitrophenolate (pNP) release during the reaction at 37°C was monitored taking readings every 30 s for 15 min at 410 nm.

was previously reported by Yan et al. (2014) and also obtained a solid biocatalyst; however, instead of using SSF, whole cells of *P. pastoris* were prepared. This whole-cell biocatalyst (WCB) had an activity of 0.73 U/mg of dried WCB (Yan et al. 2014).

3.3.5 Ethyl esters (biodiesel) produced by TLLr fermented solids

A solid biocatalyst of TLLr (TLF) was dried after culturing *P. pastoris* in SSF and employed for the ethyl oleate synthesis. To determine the synthesis capacity, the esterification of oleic acid with ethanol (96%) was performed using different isooctane contents (Cosio-Cuadros 2022). For comparative purposes, the commercial lipase immobilized on immobead 150 from *T. lanuginosus* and the obtained TLF were employed under the same conditions, using in both cases 5 U of lipase per mL. After 24 h, ethyl ester yields obtained with the TLF were 99.7, 51.4, and 32.4% at 1/20, 1/10, and 1/5 isooctane dilutions, respectively (Fig. 6A). However, in Fig. 6B when the synthesis of ethyl oleate with the commercial TLL as biocatalyst was followed, the final conversion reached was 100, 99.1, and 81% after 24 h for the 1/20, 1/10, and 1/5 isooctane dilutions, respectively. Solvents are extensively used during lipase-catalyzed esterification because they increase the solubility of substrates and products and decrease diffusional limitations, favouring reaction rates and final conversions. The best esterification conversion obtained with TLF is similar to that reported by de Souza et al. (2019) (90% in 24 h), for ethyl octanoate and decanoate, using a FS from *Y. lipolytica* growth on soybean meal (de Souza et al. 2019).

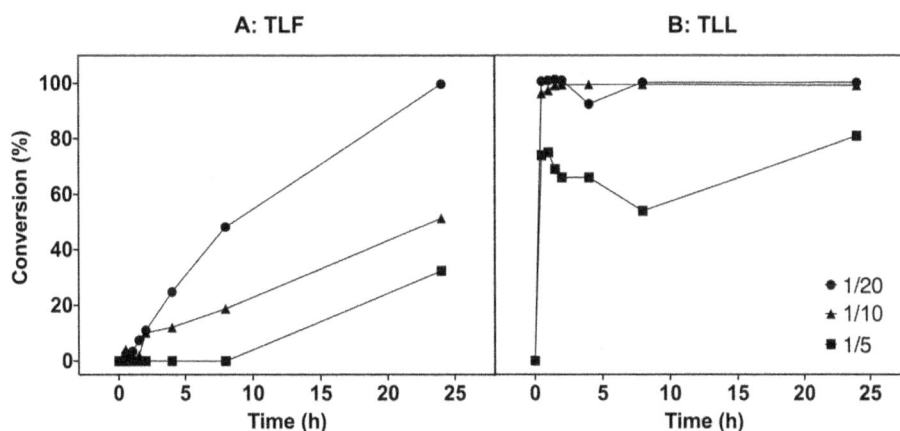

Fig. 6. Kinetics of ethyl oleate (biodiesel) synthesis using different solvent contents. TLF and TLL were employed as biocatalysts and oleic acid and 96% ethanol as substrates (unpublished data). A spectrophotometric method reported by Armendáriz-Ruiz et al. (2015) was employed to quantify the concentration of the remaining oleic acid during synthesis reactions. Symbols represent isooctane dilutions.

4. Concluding Remarks

This chapter collects some interesting examples and will serve as proof of concept of different yeast applications in biorefineries. Optimizing cultures with wild type oleaginous yeasts and engineering strains allowed to reduce the costs associated with the production of enzymes and lipid-derived compounds, such as biofuels or chemicals, as well as the co-production of organic acids and SCP. A variety of agro-industrial residues can be used by oleaginous yeasts to serve as a feasible fermentation-based biorefinery.

Examples of non-saccharomyces yeasts biorefineries are summarized here: (i) a *Y. lipolytica* biorefinery based on crude glycerol can generate a variety of products: bio-oil, lipase, single-cell protein, polyols and organic acids; (ii) lignocellulosic hydrolysates, that in the context of agro-industrial wastes biorefinery, can be efficiently converted by *R. toruloides* and *C. oleaginosus*; (iii) the TLL solid biocatalyst produced by SSF using a recombinant *P. pastoris* strain and sugar cane bagasse as support, that can be directly employed as biocatalyst for biodiesel production.

The first example of biorefinery illustrates how multiple products can be obtained with a single feedstock (crude glycerol) and a yeast biorefinery with *Y. lipolytica*, which has become a model for genetic engineering of yeast and for the isolation of new species of this interesting clade. The second case in point open possibilities of economically-viable and sustainable production of bio-oils and biochemicals with *R. toruloides* and *C. oleaginosus*.

The third approach offers the advantage of using cheap FS biocatalysts based on yeast biorefineries and agro-industrial wastes, warding off any extraction, purification, and immobilization steps. Solid biocatalysts thus produced have good potential to reduce the enzymatic process costs for bio-oil fuels and biobased chemicals production.

Acknowledgements

Author Ricardo Cosío-Cuadros was supported by a scholarship from the *Consejo Nacional de Ciencia y Tecnología* (CONACyT Mexico), number 588511. Authors are grateful to CONACyT and Energy Ministry (SENER Mexico), for grant number FSE-250014.

References

Abdul Manan, M. and Webb, C. 2017. Modern microbial solid state fermentation technology for future biorefineries for the production of added-value products. Biofuel Research Journal 4(4): 730–40.

Abeln, F. and Chuck, C.J. 2021. The history, state of the art and future prospects for oleaginous yeast research. Microbial Cell Factories 20(1): 221.

Aguieiras, E.C., Cavalcanti-Oliveira, E.D., De Castro, A.M., Langone, M.a.P. and Freire, D.M.G. 2017. Simultaneous enzymatic transesterification and esterification of an acid oil using fermented solid as biocatalyst. Journal of the American Oil Chemists' Society 94(4): 551–58.

Aguieiras, E.C., De Barros, D.S., Fernandez-Lafuente, R. and Freire, D.M. 2019. Production of lipases in cottonseed meal and application of the fermented solid as biocatalyst in esterification and transesterification reactions. Renewable Energy 130: 574–81.

Anastácio, G.S., Santos, K.O., Suarez, P.a.Z., Torres, F.a.G., De Marco, J.L. and Parachin, N.S. 2014. Utilization of glycerin byproduct derived from soybean oil biodiesel as a carbon source for heterologous protein production in pichia pastoris. Bioresource Technology 152: 505–10.

Angulo, M., Reyes-Becerril, M. and Angulo, C. 2021. Yarrowia lipolytica n6-glucan protects goat leukocytes against *Escherichia coli* by enhancing phagocytosis and immune signaling pathway genes. Microbial Pathogenesis 150: 104735.

Annisa, A.N. and Widayat, W. 2018. A review of bio-lubricant production from vegetable oils using esterification transesterification process. *In* MATEC Web of Conferences, 06007: EDP Sciences.

Armendáriz-Ruiz, M., Mateos-Díaz, E., Rodríguez-González, J.A., María Camacho-Ruiz, R., Gutiérrez-Mora, A., Sandoval-Fabian, G. et al. 2015. Carica papaya by-products as new biocatalysts for the synthesis of oleic acid esters. Biocatalysis and Biotransformation 33(4): 216–23.

Arora, S., Rani, R. and Ghosh, S. 2018. Bioreactors in solid state fermentation technology: Design, applications and engineering aspects. Journal of Biotechnology 269: 16–34.

Arslan, N.P., Aydogan, M.N. and Taskin, M. 2016. Citric acid production from partly deproteinized whey under non-sterile culture conditions using immobilized cells of lactose—positive and cold-adapted *Yarrowia lipolytica* b9. Journal of Biotechnology 231: 32–39.

Aurich, A., Specht, R., Müller, R.A., Stottmeister, U., Yovkova, V., Otto, C. et al. 2012. Microbiologically produced carboxylic acids used as building blocks in organic synthesis. pp. 391–423. *In*: Wang, X., Chen, J. and Quinn, P. (eds.). Reprogramming Microbial Metabolic Pathways. Dordrecht: Springer Netherlands.

Beltrán-Ramírez, F., Orona-Tamayo, D., Cornejo-Corona, I., González-Cervantes, J.L.N., De Jesús Esparza-Claudio, J. and Quintana-Rodríguez, E. 2019. Agro-industrial waste revalorization: The growing biorefinery. Biomass for Bioenergy-recent Trends and Future Challenges.

Bohr, S.S.R., Lund, P.M., Kallenbach, A.S., Pinholt, H., Thomsen, J., Iversen, L. et al. 2019. Direct observation of thermomyces lanuginosus lipase diffusional states by single particle tracking and their remodeling by mutations and inhibition. Scientific Reports 9(1): 16169.

Borges, J.P., Quilles Junior, J.C., Ohe, T.H.K., Ferrarezi, A.L., Nunes, C., Boscolo, M. et al. 2021. Free and substrate-immobilised lipases from *fusarium verticillioides* p24 as a biocatalyst for hydrolysis and transesterification reactions. Appl. Biochem. Biotechnol. 193(1): 33–51.

Bracharz, F., Beukhout, T., Mehlmer, N. and Brück, T. 2017. Opportunities and challenges in the development of *cutaneotrichosporon oleaginosus* atcc 20509 as a new cell factory for custom tailored microbial oils. Microbial Cell Factories 16(1): 178.

Cancino, M., Bauchart, P., Sandoval, G., Nicaud, J.-M., André, I., Dossat, V. et al. 2008. A variant of yarrowia lipolytica lipase with improved activity and enantioselectivity for resolution of 2-bromo-arylacetic acid esters. Tetrahedron: Asymmetry 19(13): 1608–12.

Caporusso, A., Capece, A. and De Bari, I. 2021a. Oleaginous yeasts as cell factories for the sustainable production of microbial lipids by the valorization of agri-food wastes. Fermentation 7(2): 50.

Caporusso, A., De Bari, I., Valerio, V., Albergo, R. and Liuzzi, F. 2021b. Conversion of cardoon crop residues into single cell oils by lipomyces tetrasporus and cutaneotrichosporon curvatus: Process optimizations to overcome the microbial inhibition of lignocellulosic hydrolysates. Industrial Crops and Products 159: 113030.

Carmona-Cabello, M., García, I.L., Papadaki, A., Tsouko, E., Koutinas, A. and Dorado, M.P. 2021. Biodiesel production using microbial lipids derived from food waste discarded by catering services. Bioresource Technology 323: 124597.

Carsanba, E., Papanikolaou, S., Fickers, P. and Erten, H. 2019. Screening various yarrowia lipolytica strains for citric acid production. Yeast 36(5): 319–27.

Casas-Godoy, L., Barrera-Martínez, I., Ayala-Mendivil, N., Aguilar-Juárez, O., Arellano-García, L., Reyes, A.L. et al. 2020. Chapter 4—Biofuels. pp. 125–70. *In*: Galanakis, C.M. (ed.). Biobased Products and Industries. Elsevier.

Casas-Godoy, L., Gasteazoro, F., Duquesne, S., Bordes, F., Marty, A. and Sandoval, G. 2018. Lipases: An overview. pp. 3–38. *In*: Sandoval, G. (ed.). Lipases and Phospholipases: Methods and Protocols. New York, NY: Springer New York.

Collaço, A.C.A., Aguieiras, E.C., Cavalcanti, E.D. and Freire, D.M. 2021. Development of an integrated process involving palm industry co-products for monoglyceride/diglyceride emulsifier synthesis: Use of palm cake and fiber for lipase production and palm fatty-acid distillate as raw material. LWT 135: 110039.

Cosío-Cuadros, R. 2022. Producción de Lipasas Fúngicas por Fermentación en Medio Sólido para la Síntesis de Ésteres Alquílicos [PhD Thesis]. Centro de Investigación y Asistencia en Tecnología y Diseño del Estado de Jalisco.

Czech, A., Smolczyk, A., Ognik, K. and Kiesz, M. 2016. Nutritional value of yarrowia lipolytica yeast and its effect on growth performance indicators n piglets. Annals of Animal Science 16(4): 1091–100.

Chen, J., Zhang, X., Tyagi, R.D. and Drogui, P. 2018. Utilization of methanol in crude glycerol to assist lipid production in non-sterilized fermentation from trichosporon oleaginosus. Bioresource Technology 253: 8–15.

Das, N. and Chandran, P. 2011. Microbial degradation of petroleum hydrocarbon contaminants: An overview. Biotechnology Research International 2011: 941810.

De Freitas, M.F.M., Cavalcante, L.S., Gudina, E.J., Silverio, S.C., Rodrigues, S., Rodrigues, L.R. et al. 2021. Sustainable lipase production by *diutina rugosa* nrrl y-95 through a combined use of agro-industrial residues as feedstock. Appl. Biochem. Biotechnol. 193(2): 589–605.

De Regil, R. and Sandoval, G. 2013. Biocatalysis for biobased chemicals. Biomolecules 3(4): 812–47.

De Souza, C.E.C., Ribeiro, B.D. and Coelho, M.a.Z. 2019. Characterization and application of yarrowia lipolytica lipase obtained by solid-state fermentation in the synthesis of different esters used in the food industry. Applied Biochemistry and Biotechnology 189(3): 933–59.

Di Fidio, N., Liuzzi, F., Mastrolitti, S., Albergo, R. and De Bari, I. 2019. Single cell oil production from undetoxified *arundo donax* l. Hydrolysate by *cutaneotrichosporon curvatus*. Journal of Microbiology and Biotechnology 29(2): 256–67.

Di Fidio, N., Minonne, F., Antonetti, C. and Raspolli Galletti, A.M. 2021. Cutaneotrichosporon oleaginosus: A versatile whole-cell biocatalyst for the production of single-cell oil from agro-industrial wastes. Catalysts 11(11): 1291.

Do, D.T.H., Theron, C.W. and Fickers, P. 2019. Organic wastes as feedstocks for non-conventional yeast-based bioprocesses. Microorganisms 7(8): 229.

Drzymała, K., Mirończuk, A.M., Pietrzak, W. and Dobrowolski, A. 2020. Rye and oat agricultural wastes as substrate candidates for biomass production of the non-conventional yeast yarrowia lipolytica. Sustainability 12(18): 7704.

Efsa, D. Turck, Castenmiller, J., De Henauw, S., Hirsch-Ernst, K.-I. Kearney, J. et al. 2019. Efsa panel on nutrition, novel foods food allergens. Safety of yarrowia lipolytica yeast biomass as a novel food pursuant to regulation (eu) 2015/2283. EFSA Journal 17(2): e05594.

Efsa, D. Turck, Castenmiller, J., De Henauw, S., Hirsch-Ernst, K.I., Kearney, J. et al. 2020. Efsa panel on nutrition, novel foods, food allergens. Safety of selenium-enriched biomass of yarrowia lipolytica as a novel food pursuant to regulation (eu) 2015/2283. EFSA Journal 18(1): e05992.

El-Baz, H.A., Elazzazy, A.M., Saleh, T.S., Dourou, M., Mahyoub, J.A., Baeshen, M.N. et al. 2021a. Enzymatic synthesis of glucose fatty acid esters using scos as acyl group-donors and their biological activities. Applied Sciences 11(6): 2700.

El-Baz, H.A., Elazzazy, A.M., Saleh, T.S., Dritsas, P., Mahyoub, J.A., Baeshen, M.N. et al. 2021b. Single cell oil (sco)–based bioactive compounds: I—enzymatic synthesis of fatty acid amides using scos as acyl group donors and their biological activities. Applied Biochemistry and Biotechnology 193(3): 822–45.

Fei, Q., O'brien, M., Nelson, R., Chen, X., Lowell, A. and Dowe, N. 2016. Enhanced lipid production by rhodosporidium toruloides using different fed-batch feeding strategies with lignocellulosic hydrolysate as the sole carbon source. Biotechnology for Biofuels 9(1): 130.

Fernandes, K.V., Cavalcanti, E.D., Cipolatti, E.P., Aguieiras, E.C., Pinto, M.C., Tavares, F.A. et al. 2021. Enzymatic synthesis of biolubricants from by-product of soybean oil processing catalyzed by different biocatalysts of candida rugosa lipase. Catalysis Today 362: 122–29.

Fickers, P., Cheng, H. and Sze Ki Lin, C. 2020. Sugar alcohols and organic acids synthesis in yarrowia lipolytica: Where are we? Microorganisms 8(4): 574.

Fickers, P., Marty, A. and Nicaud, J.M. 2011. The lipases from yarrowia lipolytica: Genetics, production, regulation, biochemical characterization and biotechnological applications. Biotechnology Advances 29(6): 632–44.

Fina, A., Brêda, G.C., Pérez-Trujillo, M., Freire, D.M.G., Almeida, R.V., Albiol, J. et al. 2021. Benchmarking recombinant pichia pastoris for 3-hydroxypropionic acid production from glycerol. Microbial Biotechnology 14(4): 1671–82.

Fontanille, P., Kumar, V., Christophe, G., Nouaille, R. and Larroche, C. 2012. Bioconversion of volatile fatty acids into lipids by the oleaginous yeast yarrowia lipolytica. Bioresource Technology 114: 443–49.

Fraga, J.L., Penha, A.C.B., Da A., Pereira, S., Silva, K.A., Akil, E., Torres, A.G. et al. 2018. Use of yarrowia lipolytica lipase immobilized in cell debris for the production of lipolyzed milk fat (lmf). International Journal of Molecular Sciences 19(11): 3413.

Galanakis, C.M. 2012. Recovery of high added-value components from food wastes: Conventional, emerging technologies and commercialized applications. Trends in Food Science & Technology 26(2): 68–87.

Gervasi, T., Pellizzeri, V., Calabrese, G., Di Bella, G., Cicero, N. and Dugo, G. 2018. Production of single cell protein (scp) from food and agricultural waste by using saccharomyces cerevisiae. Natural Product Research 32(6): 648–53.

Godoy, M.G., Gutarra, M.L., Castro, A.M., Machado, O.L. and Freire, D.M. 2011. Adding value to a toxic residue from the biodiesel industry: Production of two distinct pool of lipases from *penicillium simplicissimum* in castor bean waste. J. Ind. Microbiol. Biotechnol. 38(8): 945–53.

Gong, Z., Zhou, W., Shen, H., Zhao, Z.K., Yang, Z., Yan, J. et al. 2016. Co-utilization of corn stover hydrolysates and biodiesel-derived glycerol by cryptococcus curvatus for lipid production. Bioresource Technology 219: 552–58.

Groenewald, M., Boekhout, T., Neuvéglise, C., Gaillardin, C., Van Dijck, P.W.M. and Wyss, M. 2014. Y*arrowia lipolytica*: Safety assessment of an oleaginous yeast with a great industrial potential. Critical Reviews in Microbiology 40(3): 187–206.

Guardiola, F.A., Esteban, M.Á. and Angulo, C. 2021. *Yarrowia lipolytica*, health benefits for animals. Applied Microbiology and Biotechnology 105(20): 7577–92.

Guieysse, D., Sandoval, G., Faure, L., Nicaud, J.-M., Monsan, P. and Marty, A. 2004. New efficient lipase from yarrowia lipolytica for the resolution of 2-bromo-arylacetic acid esters. *Tetrahedron: Asymmetry* 15(22): 3539–43.

Hidayatullah, I., Arbianti, R., Utami, T.S., Suci, M., Sahlan, M., Wijanarko, A., Gozan, M. and Hermansyah, H. 2018. Techno-economic analysis of lipase enzyme production from agro-industry waste with solid state fermentation method. IOP Conference Series: Materials Science and Engineering.

Huang, X.-F., Liu, J.-N., Lu, L.-J., Peng, K.-M., Yang, G.-X. and Liu, J. 2016. Culture strategies for lipid production using acetic acid as sole carbon source by rhodosporidium toruloides. Bioresource Technology 206: 141–49.

Iea. 2012. Bio-based chemicals: Value added products from biorefineries. https://www.ieabioenergy.com/blog/publications/bio-based-chemicals-value-added-products-from-biorefineries/ (accessed November 2021).

Ilmi, M., Hidayat, C., Hastuti, P., Heeres, H.J. and Van Der Maarel, M.J.E.C. 2017. Utilisation of jatropha press cake as substrate in biomass and lipase production from aspergillus niger 65i6 and rhizomucor miehei cbs 360.62. Biocatalysis and Agricultural Biotechnology 9: 103–07.

Jach, M.E. and Serefko, A. 2018. Chapter 9—Nutritional yeast biomass: Characterization and application. pp. 237–70. *In*: Holban, A.M. and Grumezescu, A.M. (eds.). Diet, Microbiome and Health. Academic Press.

Jach, M.E., Serefko, A., Ziaja, M. and Kieliszek, M. 2022. Yeast protein as an easily accessible food source. Metabolites 12(1): 63.

Jagtap, S.S., Bedekar, A.A., Singh, V., Jin, Y.-S. and Rao, C.V. 2021. Metabolic engineering of the oleaginous yeast yarrowia lipolytica po1f for production of erythritol from glycerol. Biotechnology for Biofuels 14(1): 188.

Jones, S.W., Karpol, A., Friedman, S., Maru, B.T. and Tracy, B.P. 2020. Recent advances in single cell protein use as a feed ingredient in aquaculture. Current Opinion in Biotechnology 61: 189–97.

Juszczyk, P., Tomaszewska, L., Kita, A. and Rymowicz, W. 2013. Biomass production by novel strains of yarrowia lipolytica using raw glycerol, derived from biodiesel production. Bioresource Technology 137: 124–31.

Kamzolova, S.V. and Morgunov, I.G. 2019. Microbial production of (2r,3s)-isocitric acid: State of the arts and prospects. Applied Microbiology and Biotechnology 103(23): 9321–33.

Khan, M.R. 2021. Immobilized enzymes: A comprehensive review. Bulletin of the National Research Centre 45(1): 207.

Konzock, O., Zaghen, S. and Norbeck, J. 2021. Tolerance of yarrowia lipolytica to inhibitors commonly found in lignocellulosic hydrolysates. BMC Microbiology 21(1): 77.

Kumar Sahoo, R., Kumar, M., Mohanty, S., Sawyer, M., Rahman, P., Sukla, L.B. et al. 2018. Statistical optimization for lipase production from solid waste of vegetable oil industry. Prep. Biochem. Biotechnol. 48(4): 321–26.

Lee, J., Vadlani, P. and Min, D. 2017. Sustainable production of microbial lipids from lignocellulosic biomass using oleaginous yeast cultures. Journal of Sustainable Bioenergy Systems 7: 36–50.

Leung, D.Y., Wu, X. and Leung, M. 2010. A review on biodiesel production using catalyzed transesterification. Applied Energy 87(4): 1083–1095.

Liu, J.N., Huang, X., Chen, R., Yuan, M. and Liu, J. 2017. Efficient bioconversion of high-content volatile fatty acids into microbial lipids by cryptococcus curvatus atcc 20509. Bioresource Technology 239: 394–401.

Londoño-Hernandez, L., Ruiz, H.A., Toro, C.R., Ascacio-Valdes, A., Rodriguez-Herrera, R., Aguilera-Carbo, A. et al. 2020. Advantages and progress innovations of solid-state fermentation to produce industrial enzymes. pp. 87–113. *In*: Microbial Enzymes: Roles and Applications in Industries. Springer, Singapore.

López-Pérez, M. and Viniegra-González, G. 2016. Production of protein and metabolites by yeast grown in solid state fermentation: Present status and perspectives. Journal of Chemical Technology & Biotechnology 91(5): 1224–31.

Louhasakul, Y., Cheirsilp, B., Maneerat, S. and Prasertsan, P. 2019. Potential use of flocculating oleaginous yeasts for bioconversion of industrial wastes into biodiesel feedstocks. Renewable Energy 136: 1311–19.

Luo, Z., Miao, J., Luo, W., Li, G., Du, Y. and Yu, X. 2018. Crude glycerol from biodiesel as a carbon source for production of a recombinant highly thermostable β-mannanase by pichia pastoris. Biotechnology Letters 40(1): 135–41.

Llamas, M., Dourou, M., González-Fernández, C., Aggelis, G. and Tomás-Pejó, E. 2020a. Screening of oleaginous yeasts for lipid production using volatile fatty acids as substrate. Biomass and Bioenergy 138: 105553.

Llamas, M., Tomás-Pejó, E. and González-Fernández, C. 2020b. Volatile fatty acids from organic wastes as novel low-cost carbon source for *yarrowia lipolytica*. New Biotechnology 56: 123–29.

Ma, X., Gao, Z., Gao, M., Ma, Y., Ma, H., Zhang, M. et al. 2018. Microbial lipid production from food waste saccharified liquid and the effects of compositions. Energy Conversion and Management 172: 306–15.

Madzak, C. 2021. *Yarrowia lipolytica* strains and their biotechnological applications: How natural biodiversity and metabolic engineering could contribute to cell factories improvement. Journal of Fungi 7(7): 548.

Maity, S.K. 2015. Opportunities, recent trends and challenges of integrated biorefinery: Part I. Renewable and Sustainable Energy Reviews 43: 1427–45.

Martin Del Campo, M., R.M. Camacho, J.C. Mateos-Díaz, M. Müller-Santos, J. Córdova and J.A. Rodríguez. 2015. Solid-state fermentation as a potential technique for esterase/lipase production by halophilic archaea. Extremophiles 19(6): 1121–32.

Masri, M.A., Garbe, D., Mehlmer, N. and Brück, T.B. 2019. A sustainable, high-performance process for the economic production of waste-free microbial oils that can replace plant-based equivalents. Energy & Environmental Science 12(9): 2717–32.

Meo, A., Priebe, X.L. and Weuster-Botz, D. 2017. Lipid production with trichosporon oleaginosus in a membrane bioreactor using microalgae hydrolysate. Journal of Biotechnology 241: 1–10.

Michalik, B., Biel, W., Lubowicki, R. and Jacyno, E. 2013. Chemical composition and biological value of proteins of the yeast <i>yarrowia lipolytica</i> growing on industrial glycerol. Canadian Journal of Animal Science 94(1): 99–104, 6.

Morgunov, I.G., Kamzolova, S.V., Karpukhina, O.V., Bokieva, S.B., Lunina, J.N. and Inozemtsev, A.N. 2020. Microbiological production of isocitric acid from biodiesel waste and its effect on spatial memory. Microorganisms 8(4): 462.

Niehus, X., Casas-Godoy, L., Rodríguez-Valadez, F.J. and Sandoval, G. 2018a. Evaluation of *yarrowia lipolytica* oil for biodiesel production: Land use oil yield, carbon, and energy balance. Journal of Lipids 2018: 6393749.

Niehus, X., Crutz-Le Coq, A.-M., Sandoval, G., Nicaud, J.-M. and Ledesma-Amaro, R. 2018b. Engineering *yarrowia lipolytica* to enhance lipid production from lignocellulosic materials. Biotechnology for Biofuels 11(1): 11.

Noseda, D.G., Recúpero, M., Blasco, M., Bozzo, J. and Galvagno, M.Á. 2016. Production in stirred-tank bioreactor of recombinant bovine chymosin b by a high-level expression transformant clone of pichia pastoris. Protein Expression and Purification 123: 112–21.

Ojeda-Hernández, D.D., Cosío-Cuadros, R., Sandoval, G., Rodríguez-González, J.A. and Mateos-Díaz, J.C. 2018. Solid-state fermentation as an economic production method of lipases. pp. 217–28. *In*: Sandoval, G. (ed.). Lipases and Phospholipases: Methods and Protocols. New York, NY: Springer New York.

Papanikolaou, S. and Aggelis, G. 2011. Lipids of oleaginous yeasts. Part I: Biochemistry of single cell oil production. European Journal of Lipid Science and Technology 113(8): 1031–51.

Papanikolaou, S., Diamantopoulou, P., Blanchard, F., Lambrinea, E., Chevalot, I., Stoforos, N.G. et al. 2020. Physiological characterization of a novel wild-type yarrowia lipolytica strain grown on glycerol: Effects of cultivation conditions and mode on polyols and citric acid production. Applied Sciences 10(20): 7373.

Papanikolaou, S., Kampisopoulou, E., Blanchard, F., Rondags, E., Gardeli, C., Koutinas, A.A. et al. 2017. Production of secondary metabolites through glycerol fermentation under carbon-excess conditions by the yeasts yarrowia lipolytica and rhodosporidium toruloides. European Journal of Lipid Science and Technology 119(9): 1600507.

Park, Y.-K., Nicaud, J.-M. and Ledesma-Amaro, R. 2018. The engineering potential of *rhodosporidium toruloides* as a workhorse for biotechnological applications. Trends in Biotechnology 36(3): 304–17.

Pereira, A.D.S., Fontes-Sant'ana, G.C. and Amaral, P.F. 2019. Mango agro-industrial wastes for lipase production from yarrowia lipolytica and the potential of the fermented solid as a biocatalyst. Food and Bioproducts Processing 115: 68–77.

Pereira, A.S., Miranda, S.M., Lopes, M. and Belo, I. 2021. Factors affecting microbial lipids production by yarrowia lipolytica strains from volatile fatty acids: Effect of co-substrates, operation mode and oxygen. Journal of Biotechnology 331: 37–47.

Pham, N., Reijnders, M., Suarez-Diez, M., Nijsse, B., Springer, J., Eggink, G. et al. 2021. Genome-scale metabolic modeling underscores the potential of cutaneotrichosporon oleaginosus atcc 20509 as a cell factory for biofuel production. Biotechnology for Biofuels 14(1): 2.

Posso Mendoza, H., Pérez Salinas, R., Tarón Dunoyer, A., Tatis, C.C., Morgado-Gamero, W.B., Castillo Ramírez, M. et al. 2020. Evaluation of enzymatic extract with lipase activity of yarrowia lipolytica. An application of data mining for the food industry wastewater treatment. Cham: Springer International Publishing, pp. 304–13.

Rakicka-Pustułka, M., Miedzianka, J., Jama, D., Kawalec, S., Liman, K., Janek, T. et al. 2021. High value-added products derived from crude glycerol via microbial fermentation using yarrowia clade yeast. Microbial Cell Factories 20(1): 195.

Ramos-Sánchez, L.B., Cujilema-Quitio, M.C., Julian-Ricardo, M.C., Cordova, J. and Fickers, P. 2015. Fungal lipase production by solid-state fermentation. Journal of Bioprocessing & Biotechniques 5(2): 1.

Ratledge, C. 1991. Microorganisms for lipids. *Acta Biotechnologica* 11(5): 429–38.

Ravindran, R., Hassan, S.S., Williams, G.A. and Jaiswal, A.K. 2018. A review on bioconversion of agro-industrial wastes to industrially important enzymes. Bioengineering 5(4): 93.

Robert, J.M., Lattari, F.S., Machado, A.C., De Castro, A.M., Almeida, R.V., Torres, F.a.G. et al. 2017. Production of recombinant lipase b from candida antarctica in pichia pastoris under control of the promoter pgk using crude glycerol from biodiesel production as carbon source. Biochemical Engineering Journal 118: 123–31.

Rosa, D.R., Cammarota, M.C. and Freire, D.M.G. 2006. Production and utilization of a novel solid enzymatic preparation produced by penicillium restrictum in activated sludge systems treating wastewater with high levels of oil and grease. Environmental Engineering Science 23(5): 814–23.

Ruiz Flores, L.a.F.C., Guillermo Camacho Ruiz, M. Ángeles Amaya Delgado, Lorena Jiménez Ocampo, Rafael, Rodriguez Gonzzález, Jorge A. 2014. Aprovechamiento de la pasta de jatropha para la producción de lipasas de hongos filamentosos endógenos. Paper presentat at the Memorias del XXXV Encuentro Nacional de la AMIDIQ in Puerto Vallarta, Jalisco, México.

Sales, J.C.S., De Castro, A.M., Ribeiro, B.D. and Coelho, M.A.Z. 2020. Supplementation of watermelon peels as an enhancer of lipase and esterase production by yarrowia lipolytica in solid-state fermentation and their potential use as biocatalysts in poly (ethylene terephthalate)(pet) depolymerization reactions. Biocatalysis and Biotransformation 38(6): 457–68.

Salihu, A., Alam, M.Z., Abdulkarim, M.I. and Salleh, H.M. 2012. Lipase production: An insight in the utilization of renewable agricultural residues. Resources, Conservation and Recycling 58: 36–44.

Sànchez I. Nogué, V., Black, B.A., Kruger, J.S., Singer, C.A., Ramirez, K.J., Reed, M.L. et al. 2018. Integrated diesel production from lignocellulosic sugars via oleaginous yeast. Green Chemistry 20(18): 4349–65.

Sandoval, G., Casas-Godoy, L., Bonet-Ragel, K., Rodrigues, J., Ferreira-Dias, S. and Valero, F. 2017. Enzyme-catalyzed production of biodiesel as alternative to chemical-catalyzed processes: Advantages and constraints. Current Biochemical Engineering 4(2): 109–41.

Sara, M., Brar, S.K. and Blais, J.F. 2016. Lipid production by yarrowia lipolytica grown on biodiesel-derived crude glycerol: Optimization of growth parameters and their effects on the fermentation efficiency. RSC Advances 6(93): 90547–58.

Šelo, G., Planinić, M., Tišma, M., Tomas, S., Koceva Komlenić, D. and Bucić-Kojić, A. 2021. A comprehensive review on valorization of agro-food industrial residues by solid-state fermentation. Foods 10(5): 927.

Shen, H., Zhang, X., Gong, Z., Wang, Y., Yu, X., Yang, X. et al. 2017. Compositional profiles of rhodosporidium toruloides cells under nutrient limitation. Applied Microbiology and Biotechnology 101(9): 3801–09.

Siebenhaller, S., Kirchhoff, J., Kirschhöfer, F., Brenner-Weiß, G., Muhle-Goll, C., Luy, B. et al. 2018. Integrated process for the enzymatic production of fatty acid sugar esters completely based on lignocellulosic substrates. Frontiers in Chemistry 6.

Spagnuolo, M., Yaguchi, A. and Blenner, M. 2019. Oleaginous yeast for biofuel and oleochemical production. Current Opinion in Biotechnology 57: 73–81.

Tocher, D.R., Betancor, M.B., Sprague, M., Olsen, R.E. and Napier, J.A. 2019. Omega-3 long-chain polyunsaturated fatty acids, epa and dha: Bridging the gap between supply and demand. Nutrients 11(1): 89.

Tsigie, Y.A., Wang, C.-Y., Truong, C.-T. and Ju, Y.-H. 2011. Lipid production from yarrowia lipolytica po1g grown in sugarcane bagasse hydrolysate. Bioresource Technology 102(19): 9216–22.

Valero, F. 2018. Recent advances in pichia pastoris as host for heterologous expression system for lipases: A review. pp. 205–16. *In*: Sandoval, G. (ed.). Lipases and Phospholipases: Methods and Protocols. New York, NY: Springer New York.

Vargas, M., Niehus, X., Casas-Godoy, L. and Sandoval, G. 2018. Lipases as biocatalyst for biodiesel production. pp. 377–90. *In*: Sandoval, G. (ed.). Lipases and Phospholipases: Methods and Protocols. New York, NY: Springer New York.

Vasconcelos, B., J.C. Teixeira, G. Dragone and J.A. Teixeira. 2019. Oleaginous yeasts for sustainable lipid production—from biodiesel to surf boards, a wide range of "green" applications. Applied Microbiology and Biotechnology 103(9): 3651–67.

Vaseghi, Z., Najafpour, G.D., Mohseni, S., Mahjoub, S. and Hosseinpour, M.N. 2012. Lipase production in tray-bioreactor via solid state fermentation under desired growth conditions. Iranica Journal of Energy and Environment 3(1): 59–65.

Wheeldon, I. and Blenner, M. 2021. *Yarrowia lipolytica*: Methods and Protocols Methods in Molecular Biology. New York, NY: Humana.

Xu, J., Du, W., Zhao, X. and Liu, D. 2016. Renewable microbial lipid production from oleaginous yeast: Some surfactants greatly improved lipid production of rhodosporidium toruloides. World Journal of Microbiology and Biotechnology 32(7): 107.

Xu, J., Liu, N., Qiao, K., Vogg, S. and Stephanopoulos, G. 2017. Application of metabolic controls for the maximization of lipid production in semicontinuous fermentation. Proceedings of the National Academy of Sciences 114(27): E5308–E16.

Yan, J., Han, B., Gui, X., Wang, G., Xu, L., Yan, Y. et al. 2018. Engineering *yarrowia lipolytica* to simultaneously produce lipase and single cell protein from agro-industrial wastes for feed. Scientific Reports 8(1): 758.

Yan, J., Zheng, X. and Li, S. 2014. A novel and robust recombinant *pichia pastoris* yeast whole cell biocatalyst with intracellular overexpression of a thermomyces lanuginosus lipase: Preparation, characterization and application in biodiesel production. Bioresource Technology 151: 43–48.

Yang, X., Jin, G., Gong, Z., Shen, H., Bai, F. and Zhao, Z.K. 2014. Recycling biodiesel-derived glycerol by the oleaginous yeast rhodosporidium toruloides y4 through the two-stage lipid production process. Biochemical Engineering Journal 91: 86–91.

Zeng, Y., Bian, D., Xie, Y., Jiang, X., Li, X., Li, P. et al. 2017. Utilization of food waste hydrolysate for microbial lipid and protein production by rhodosporidium toruloides y2. Journal of Chemical Technology & Biotechnology 92(3): 666–73.

Zhou, W., Gong, Z., Zhang, L., Liu, Y., Yan, J. and Zhao, M. 2017. Feasibility of lipid production from waste paper by the oleaginous yeast *cryptococcus curvatus*. Bioresources 12(3): 5249–63.

CHAPTER 4

Microbial and Ecological Aspects in Biohydrogen Production by Dark Fermentation

*Braga Nan, L., Noguer, M., Dauptain, K. and Trably, E.**

1. Introduction

Developing clean and renewable technologies for energy production is nowadays at an urgent stage due to the continuously growing energy demand and the environmental issues associated to fossil fuels' uses (Zhang et al. 2020). As emerging and promising alternative to fossil fuels, hydrogen is an interesting energy carrier, because it presents the highest energy yield per mass unit (122 kJ/g) when compared to other fuels as crude oil (44 kJ/g) or methane (54 kJ/g). H_2 is also a carbon-free combustion fuel that can be used in the transportation sector, avoiding diffuse emissions of greenhouse gases (Akhlaghi and Najafpour-Darzi 2020). However, H_2 is currently produced mainly from fossil fuels, either by natural gas steam reforming or coal gasification. These technologies release large amounts of fossil carbon dioxide into the environment (around 10 kg_{CO2} per kg of H_2 generated) (Akhlaghi and Najafpour-Darzi 2020). As emerging alternatives, H_2 can be produced from renewable energies and sources, such as by electrolysis from water splitting with renewable electricity, or by thermochemical processes such as gasification or pyrolysis of biomass. In these cases, the produced H_2 is called "green H_2" (Nikolaidis and Poullikkas 2017). Hydrogen can also be biologically produced by using low-cost organic materials, such as agricultural or food waste (FW), at ambient pressures and temperatures, making these biological processes the most environmental-friendly (Ghimire et al. 2015). The H_2 issued from biological technologies is named biohydrogen (bioH_2).

Biohydrogen can be generated by light-dependent processes involving photosynthetic microorganisms such as microalgae, cyanobacteria, and purple non-sulphur bacteria (PNSB) or by light-independent processes implying strict and facultative anaerobic microorganisms or electroactive bacteria. Photosynthetic hydrogen-generating microorganisms can be either autotrophic or heterotrophic. They generate H_2 from incomplete photosynthesis in a physiological process named biophotolysis, or from organic compounds by photo-fermentation (Sivaramakrishnan et al.

INRAE, Univ Montpellier, LBE, 102 avenue des Etangs, 11100, Narbonne, France.
* Corresponding author: eric.trably@inrae.fr

2021). These light-dependent processes present the advantage to generate only simple products such as H_2, CO_2 and O_2. In counterpart, the main drawbacks of these technologies rely on the low light conversion efficiency, the complexity of reactor design and operation, and the low hydrogen production rates (Hallenbeck et al. 2012).

Dark fermentation (DF) and microbial electrolysis are light-independent processes. Microbial electrolysis principles are similar to water electrolysis, although the energy requirements are 10 times lower since most of the energy comes from the oxidation of organic molecules. As main advantage of microbial electrolysis, different organic substrates can be used and can be almost completely converted into H_2 (Sivaramakrishnan et al. 2021). Even though bioelectrochemical processes show the highest H_2 conversion yields, their performances are highly dependent on the organic molecules used at the anode, the material used for abiotic H_2 evolution at the cathode, the type of membrane and the process design (Fu et al. 2021). In contrast, dark fermentation (DF) presents the simplest reactor design and operation which reduce drastically the costs of production in regard to the other processes. DF is carried out by strict or facultative anaerobic bacteria, that release H_2 during the fermentation of organic substrates, and more specifically carbohydrates (Monlau et al. 2012). As main advantage, DF contributes to waste treatment and valorisation because carbohydrate-rich waste can be used as feedstock. Although H_2 yields are constrained due to thermodynamic limitations, DF exhibits the highest hydrogen production rates among all the biological processes. Given these characteristics and the fact that the fermentation processes are already used at industrial scale, DF is one of the most promising biological process to be rapidly upscaled to produce biohydrogen at large scale (Dauptain et al. 2020). For that, economic aspects are also a major concern and to have a more profitable process, production costs (2.5 \$/kgH$_2$) need to be reduced to be competitive with regard to fossil fuel-based technologies (< 1 \$/kgH$_2$) (Dincer and Acar 2015).

2. Microbial Physiology of Key Microbial Species in Dark Fermentation

2.1 Enzymes Involved in H$_2$ Evolution

2.1.1 Hydrogenases

In dark fermentation, H_2 evolution is mainly based on the action of one type of metalloenzymes named hydrogenases, that have the ability to reduce protons into H_2 (Hallenbeck and Benemann 2002). Hydrogenases are key players of the H_2 metabolism and are widely distributed in all the three domains of life (Vignais and Billoud 2007). They are classified into three groups, according to the metal content of their active sites (Fig. 1): (i) the [Ni-Fe] hydrogenases, (ii) the [Fe-Fe] hydrogenases and (iii) the [Fe] hydrogenases.

The [Ni-Fe] and the [Fe-Fe] hydrogenases catalyse the reaction: $2H^+ + 2e^- \leftrightarrow H_2$, while the [Fe] hydrogenases are principally present in methanogens and have different catalytic properties for H_2 fixation than the other hydrogenases.

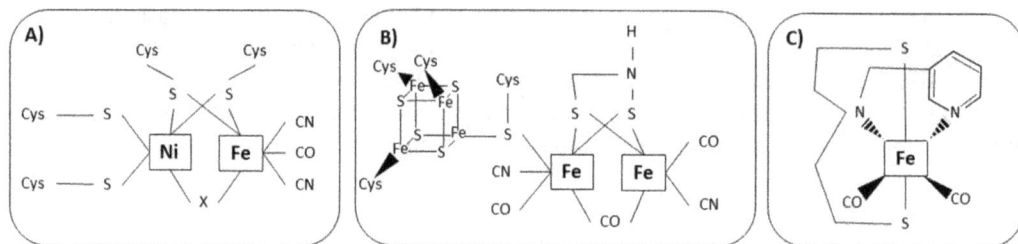

Fig. 1. Configuration of the active site of [Ni-Fe] (X = O^{2-}, OH, OH$_2$, SO, and in the reduced form X = H) (A), [Fe-Fe] (B) and [Fe] (C) hydrogenases. Adapted from (Kim and Kim 2011).

2.1.2 The [Ni-Fe] hydrogenases

[Ni-Fe] hydrogenases have been found in both Archaea and Bacteria and constitute the most studied and largest group of hydrogenases (Cadoux and Milton 2020). Several functions have been described in the literature for the [Ni-Fe] hydrogenases, such as (i) H_2 uptake, mostly observed in hydrogenotrophic methanogens, sulphate-reducing, Fe^{3+}-reducing and denitrifying bacteria, which use the electrons coming from the oxidation of H_2 to produce ATP (Vignais and Billoud 2007); (ii) H_2 evolution, present in fermentative bacteria as *Escherichia coli*; and (iii) H_2 detection in the local environment of the cell, which triggers a cascade of reactions implicated in the activation of genes coding for H_2-uptaking hydrogenases (Kim and Kim 2011). This type of hydrogenases is present in *Bradyrhizobium japonicum, Rhodobacter eutropha, Rhodobacter capsulatus,* and *Rhodopseudomonas palustris* (Vignais and Billoud 2007).

The [Ni-Fe] hydrogenases are formed by two subunits: an α-subunit, which contains the bimetallic active site and a β-subunit, which contains Fe-S centres. The [Ni-Fe] active site is linked to the protein by a S-bound of four cysteines residues. The Fe atom of the active site is also linked to 3 non-protein ligands: 1 CO and 2 CN (Fig. 1a). The principal function of the β-subunit is to transfer electrons between the active site and the electron donor/acceptor binding site (Vignais 2008). Because of their structure, [Ni-Fe] hydrogenases are less sensitive to O_2 and CO than [Fe-Fe] hydrogenases (Shafaat et al. 2013).

2.1.3 The [Fe-Fe] hydrogenases

The [Fe-Fe] hydrogenases are commonly found in anaerobic prokaryotes, such as *Clostridium* sp. (Vignais and Billoud 2007). They are the only hydrogenases found in eukaryotes. These enzymes act as redox balance controllers of the cell. Their main function is to reduce the excess of electrons into H_2, as produced by the metabolism in fermentative anaerobic bacteria, as *Clostridium* sp., or by the chloroplasts in some green microalgae (Hallenbeck and Benemann 2002, Vignais and Billoud 2007). In the cells, the electrons generated by the cell metabolism are carried by ferredoxins (Fd_{ox}) or adenine dinucleotides ($NAD(P)^+$) on their reduced form, i.e., Fd_{red} and $NAD(P)H^+$. However, to ensure equilibrium of the cell, the reduced carriers need to be re-oxidised by transferring the electrons to another molecule. The [Fe-Fe] hydrogenases are involved in the re-oxidation of these carriers by using the electrons to reduce protons into H_2.

Many of the [Fe-Fe] hydrogenases are monomeric proteins consisting of one subunit containing the active site, also called the H cluster. More rarely, dimeric, trimeric and tetrameric [Fe-Fe] hydrogenases have also been identified (Vignais and Billoud 2007). The active site of the enzyme is composed by a binuclear [Fe-Fe] centre bound to a [4Fe–4S] cluster by a bridging cysteine and is attached to the protein through four cysteine ligands. One CN and two CO ligands are bound to both Fe atoms and two bridging sulphur atoms are linked to the Fe atoms (Fig. 1b) (Gao et al. 2020). [Fe-Fe] hydrogenases are extremely sensitive and irreversibly inactivated by O_2 traces (Koo et al. 2016).

2.1.4 The [Fe] hydrogenases

[Fe] Hydrogenases, also called Hmd enzymes (H_2-forming methylenetetrahydromethanopterin dehydrogenase), are structurally different from other hydrogenases as they do not contain Fe-S centres. They present an active centre composed of only one metallic Fe atom. Their catalytic properties are different from the [Ni-Fe] and [Fe-Fe] hydrogenases since the Fe in their active site is not redox-active. Thus, they do not carry out the reversible reaction $2H^+ + 2e^- \leftrightarrow H_2$. These hydrogenases are implicated in the fixation of H_2 for the reduction of CO_2 into CH_4. They are mainly found in hydrogenotrophic methanogens, such as *Methanothermobacter marburgensis* (Dey et al. 2013, Heinekey 2009).

This enzyme is expressed under nickel-limiting growth conditions and in absence of light. Hmd is composed by two identical subunits and contains two Fe atoms per homodimer. This enzyme is implicated in the reduction of the methenyltetrahydromethanopterin (methenyl-H_4MPT^+) intermediary to methylene-H_4MPT and H^+ in the CH_4 production pathway (Dey et al. 2013, Vignais and Billoud 2007).

2.2 *Major H_2 Related Metabolic Pathways*

Two main H_2-producing pathways have been described: (i) the pyruvate-ferredoxin-oxydoreductase (PFOR) pathway and (ii) the pyruvate-formate-lyase (PFL) pathway (Fig. 2). Both pathways present similar initial reactions with the production of pyruvate from glucose, through the Entner-Doudoroff pathway also called glycolysis (Angenent et al. 2004). Table 1 summarizes the H_2 generating reactions, the related metabolic pathways and an example of a microorganism performing this pathway.

In the PFOR pathway, pyruvate is transformed to acetyl-CoA and CO_2, generating a reduced ferredoxin (Fd_{red}), which is further oxidised by a ferredoxin-dependent [Fe-Fe] hydrogenase enzyme to produce H_2 (Tapia-Venegas et al. 2015). The acetyl-CoA, resulting from this reaction, is then oxidised to acetate to form one ATP. When H_2 pressure is lower than 60 Pa, additional H_2 production can occur through re-oxidation of NADH produced during the glycolysis by a NADH-dependent [Fe-Fe] hydrogenase or a NADH-ferredoxin-dependent [Fe-Fe] hydrogenase (shown in Fig. 2 as alternative hydrogenases) (Angenent et al. 2004, Hallenbeck et al. 2012). Therefore, if all the NADH is oxidized by this reaction, a maximum theoretical hydrogen yield of 4 $mol_{H2}/mol_{glucose}$ (Equation 1) could be achieved. However, at higher H_2 partial pressures, NADH is regenerated through the reduction of acetyl-CoA to other metabolites, such as butyrate, reducing the H_2 yield to 2 $mol_{H2}/mol_{glucose}$ (Equation 2) (Dessì et al. 2018b). Hence, the most common products during H_2-producing fermentation are acetate and butyrate (Hawkes et al. 2007).

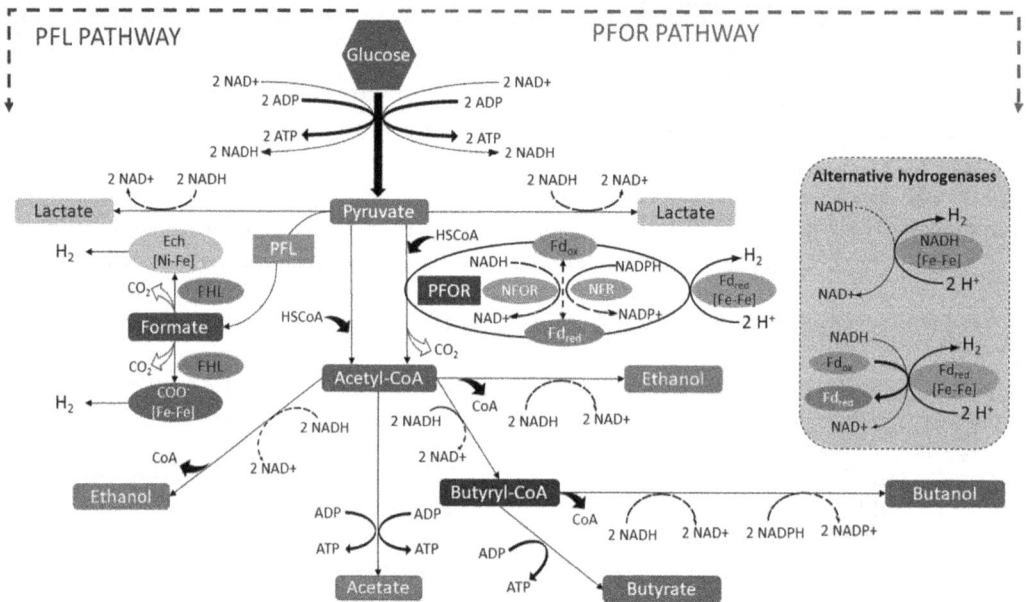

Fig. 2. Principal bioH_2 producing pathways from glucose. References: PFL: pyruvate-formate-lyase, FHL: formate-hydrogen-lyase, PFOR: pyruvate-ferredoxin-oxydoreductase, NFOR: NADH-ferredoxin oxydoreductase, NFR: NADH-ferredoxin reductase. Adapted from Ramírez-Morales et al. (2015) and Tapia-Venegas et al. (2015).

Table 1. Main H_2 producing reactions from glucose (modified from Castelló et al. (2020), Madigan et al. (2015)).

Type of Fermentation	Equation	Microorganisms	Metabolic Pathway	ΔG° (Kj/mol)	N° Eq.
Acetate fermentation	$C_6H_{12}O_6 + 2\,H_2O$ $\rightarrow 2\,CH_3COOH$ $+ 2\,CO_2 + 4\,H_2$	*Clostridium* sp.	PFOR	–215	1
Butyrate fermentation	$C_6H_{12}O_6 \rightarrow CH_3CH_2CH_2COOH$ $+ 2\,CO_2 + 2\,H_2$	*Clostridium* sp.	PFOR	–264	2
Acetate ethanol fermentation	$C_6H_{12}O_6 + 3\,H_2O$ $\rightarrow CH_3CH_2OH$ $+ CH_3COOH$ $+ 2\,CO_2 + 2\,H_2$	*Enterobacter* sp.	PFL	–255	3
Assumed mixed fermentation in mixed cultures	$4\,C_6H_{12}O_6 + 2\,H_2O$ $\rightarrow 3\,CH_3CH_2CH_2COOH$ $+ 2\,CH_3COOH + 6\,CO_2 + 10\,H_2$	Mixed culture	Mixed	–252	4

In the PFL pathway, pyruvate is transformed to acetyl-CoA and formate. Then, formate is converted to H_2 and CO_2 by an enzymatic complex, composed of the enzyme formate-hydrogen lyase (FHL) and either a formate-dependent [Ni-Fe] hydrogenase (so-called Ech hydrogenase) or a formate-dependent [Fe-Fe] hydrogenase, depending on the microorganism involved (Hallenbeck et al. 2012, Tapia-Venegas et al. 2015) (Fig. 2). The acetyl-CoA is further converted into acetate, butyrate, ethanol or butanediol depending on the type of microorganisms and the environmental conditions that can affect the metabolic fluxes (Li and Fang 2007). The microorganisms producing H_2 by this pathway, principally facultative anaerobic bacteria, cannot transform NADH to H_2 and therefore they are limited to a maximum yield of 2 $mol_{H2}/mol_{glucose}$ (Equation 3) (Ghimire et al. 2015, Hallenbeck et al. 2012).

Generally, in DF, the H_2 yields vary between 0.5–2.5 $mol_{H2}/mol_{glucose}$, which is lower than the theoretical maximum. One of the reasons of such lower H_2 yields is that low H_2 partial pressures are required (< 60 Pa) to reach the maximum yields, which is a difficult condition to achieve in efficient hydrogenogenic reactors (Castelló et al. 2020). Moreover, dark fermentation reactors are usually inoculated with mixed cultures which have the advantages of metabolic flexibility and robustness, but the disadvantage of favouring the presence of undesirable bacteria, such as non-H_2 producing or H_2-consuming bacteria. These bacteria outcompete the hydrogen-producing bacteria and/or consume the released H_2, hindering the global H_2 production (Tapia-Venegas et al. 2015).

2.3 Main H_2 Producing Bacteria (also called HPB)

Several fermentative anaerobic bacteria and archaea are able to produce H_2. These microorganisms can be obligate or facultative anaerobes (Łukajtis et al. 2018). Obligate anaerobes cannot survive if oxygen is present in the media, while facultative anaerobes will perform aerobic respiration if oxygen is available or will perform fermentation if oxygen is absent (Madigan et al. 2015).

Among the obligate anaerobic microorganisms able to produce H_2, the members of the genus *Clostridium* (phylum Firmicutes) are considered to be the most efficient hydrogen producers, as they are generally associated with high hydrogen yields and productivities (Palomo-Briones et al. 2017). Etchebehere et al. (2016) studied the microbial communities of 20 hydrogen-producing reactors at laboratory scale by sequencing 16S rRNA genes. These reactors presented different configurations (fixed–bed reactors (FBR), expanded granular sludge bed (EGSB), up-flow anaerobic sludge bed (UASB), sequencing batch reactors (SBR), and continuous stirred tank reactors (CSTR)) and were inoculated with different sources of inocula (indigenous, anaerobic sludge or compost), which were pre-treated with different methods (none, heat, aeration, etc.). Moreover, the substrates (simple and complex substrates) were variable between the reactors, while the applied operation parameters (hydraulic retention time (HRT), pH, temperature, and organic loading rate (OLR)) were also variable. Interestingly, the authors found a positive relationship between the composition in *Clostridium* species and the hydrogen productivity/yield.

One important characteristic of *Clostridium* species is their ability to produce spores, which allows the bacteria to survive difficult conditions, such as low or high pH, extreme temperature, dryness, or toxic chemical exposure. As a result of this feature, they can be selected and separated from the non-spore forming bacteria by inoculum pre-treatment (Bundhoo et al. 2015). Besides, they have the ability to ferment wide variety of carbohydrates and produce H_2 more efficiently than facultative bacteria (Łukajtis et al. 2018). Many *Clostridium* species are able to efficiently produce H_2. Illustratively, *Clostridium pasteurianum*, *Clostridium saccharobutylicum*, *Clostridium butyricum*, *Clostridium beijerinckii* are the hydrogen producers that are the most investigated. As already mentioned, these microorganisms perform the PFOR pathway producing 4 mol of H_2 by mol of consumed hexose, which is the maximal theoretical H_2 yield possible by dark fermentation. Nevertheless, due to the versatile metabolism of the *Clostridium* genus, the type of substrate and the operating conditions determine which metabolites are coproduced with H_2 (Castelló et al. 2020). Some detrimental metabolic reactions can also occur with some members of the *Clostridium* genus. In particular, at low pH and high carbohydrate concentrations, *C. acetobutylicum* can switch from hydrogen-producing butyrate and acetate pathways to acetone and butanol production (Xue and Cheng 2019). This reaction is named solventogenesis and is detrimental to H_2 production. Similarly, some *Clostridium* species, eg. *C. aceticum*, *C. ljungdahlii* and *C. carboxidivorans* can consume hydrogen and fix CO_2 to produce acetate by homoacetogenesis (Drake 1991, Lazaro et al. 2014).

Several non-spore-forming obligate anaerobic bacteria have also been reported to efficiently produce H_2 (Cabrol et al. 2017). Within the Firmicutes phylum, *Ethanoligenens harbinense* were found to be dominant in H_2-producing reactors with ethanol-type fermentation at acidic pH (< 3.0–4.5) (Mota et al. 2018, Ren et al. 2007). *Ethanoligenens harbinense* produces H_2 along with acetate, ethanol and CO_2 by the PFOR pathway (Li et al. 2019). *Acetanaerobacterium elongatum* (belonging to the Firmicutes phylum) were also reported to produce H_2, acetate, ethanol, and CO_2 (Chen and Dong 2004). Besides, they have been related to the production of H_2 when treating different substrates as food waste or silage (Kim et al. 2011, Li et al. 2012). The members of the *Megasphaera* genus have been attributed to be one of the major contributors to H_2 production in reactors treating vinasses from sugarcane industry, cheese whey, and solid food-waste (Castelló et al. 2009, Ferraz Júnior et al. 2015, Moreno-Andrade et al. 2015). *Megasphaera* sp. produces H_2, ethanol, acetate, and CO_2 from glucose (Ren et al. 2007). Some species of this genus, such as *Megasphaera elsdenii*, have the ability to produce H_2 and butyrate from lactate (Buitrón et al. 2020, Ohnishi et al. 2022). The bacteria belonging to the *Megasphaera* genus produce H_2 by the PFOR pathway (Ohnishi et al. 2022). Some species of the *Selenomonas* genus are often detected during DF, while some are related to H_2 production (from glucose or lactate) (Luo et al. 2008, Mariakakis et al. 2011, Rosa et al. 2014), others are identified as deleterious to H_2 production, due to their ability to produce propionate and lactate during DF (Kim et al. 2008, Sivagurunathan et al. 2014). It has been reported that *Selenomonas* species, such as *Selenomonas acidaminovorans* and *Selenomonas ruminatus*, produce lactate, acetate, propionate, H_2 and CO_2. Besides, it has been observed that in the presence of H_2-consuming microorganisms, these species increase their H_2 production (Guangsheng et al. 1992, Scheifinger et al. 1975). Both genera (*Megasphaera* and *Selenomonas*) belong to the *Veillonellaceae* family within the Firmicutes phylum, which were associated with H_2-producing reactors but with low productivities (Etchebehere et al. 2016).

Some facultative bacteria also produce H_2. *Bacillus* species (belonging to the *Bacillaceae* family and the Firmicutes phylum), such as *Bacillus cereus*, *Bacillus amyloliquefaciens*, *Bacillus coagulans*, *Bacillus cloacae*, *Bacillus macerans*, *Bacillus licheniformis* and *Bacillus polymyxa* carry-out mixed acid fermentation when growing without oxygen producing H_2 and organic acids

(Kotay and Das 2007, Łukajtis et al. 2018, Patel et al. 2014, Song et al. 2013). Similarly, the bacteria belonging to the *Enterobacteriaceae* family (belonging to the phylum Proteobacteria), such as *Enterobacter* sp., *Klebsiella* sp., *Shigella* sp., *Citrobacter* sp., among others, perform mixed-acid fermentations generating H_2, CO_2, ethanol, succinic, lactic, and acetic acid. Some are also able to perform butanediol fermentation producing butanediol, ethanol, H_2 and CO_2 (Madigan et al. 2015). The bacteria belonging to the *Enterobacteriaceae* family as well as some *Bacillus* species produce H_2 by the PFL pathway (Cabrol et al. 2017). The advantage of the facultative bacteria over the obligate anaerobes is their resistance to oxygen and their hydrolytic capabilities which could be required when working with complex substrates (Łukajtis et al. 2018, Patel et al. 2014). For instance, Patel et al. (2014) conducted a batch study with *Enterobacter cloacae*, *Klebsiella* sp., *Citrobacter* sp. and *Bacillus cereus* strains, by combining these strains in co-cultures pairwise. Interestingly, when *B. cereus* and *E. aerogenes* were co-cultivated, a production up to 3 $mol_{H2}/mol_{glucose}$ was observed likely by combining different metabolisms and the ability to consume traces of oxygen in the bulk. Bacteria from the genus *Prevotella* (belonging to the Bacteroidetes phylum) were also detected in several H_2 producing systems and identified as H_2-producing bacteria (Cieciura-Włoch et al. 2020, Detman et al. 2021a, b, Mariakakis et al. 2011).

Extreme thermophilic and hyperthermophilic hydrogen-producing-microorganisms, such as *Thermoanaerobacter* sp., *Thermotoga* sp. and *Pyrococcus* sp., have the highest reported H_2 yields (Verhaart et al. 2010). These microorganisms grow at high temperatures (> 60°C), which are thermodynamically more favourable for H_2 production, because sugar conversion to acetate becomes more energetically favourable as the temperature increases, which results in higher H_2 yields (Stams 1994). For instance, *Thermotoga* sp. (belonging to the Thermotogae phylum) and *Thermoanaerobacter* sp. (from the Firmicutes phylum) recycle almost all their reducing equivalents (NADH and Fd_{red}) generated during the glycolysis by forming H_2, achieving H_2 yields up to 3–4 $mol_{H2}/mol_{glucose}$ (Verhaart et al. 2010). The Archaea species *Pyrococcus furiosus* (Phylum Euryarchaeota) held one of the highest H_2 yields reported (3.8 $mol_{H2}/mol_{glucose}$) (Chou et al. 2007). This hyperthermophilic facultative microorganism growth optimally at a temperature of 98–100°C on sugars and small peptides, produced H_2, acetate (if growing from sugars), H_2S (if elemental sulphur is present), alanine and ethanol (Fiala and Stetter 1986, Schicho et al. 1993). *Pyrococcus furiosus* has two cytoplasmic hydrogenases and a membrane-bound hydrogenase complex, which accepts electrons from ferredoxin directly and it was shown to have a very high hydrogen evolving activity compared with other hydrogenases (Silva et al. 2000). Their cytoplasmic hydrogenases normally reduce sulphur but can switch to hydrogen production when sulphur is not available (Silva et al. 2000). At high H_2 partial pressures (> 2.2 kPa), part of the reducing equivalents is recycled by forming alanine and ethanol. Therefore, the production of H_2S, ethanol and alanine hinders the H_2 yield of *Pyrococcus furiosus* (Chou et al. 2007, Verhaart et al. 2010). Even though hyperthermophilic microorganisms exhibit the highest yields in dark fermentation, their volumetric productivities are lower than the ones reported for mesophilic microorganisms (Pawar et al. 2013). Besides, hyperthermophilic temperatures could cause energy loss (Dessì et al. 2018b). Perera et al. (2010) reported that high operation temperatures are detrimental to the overall energy gain when considering the H_2 produced. One possibility to face this challenge could be to treat high temperature effluents (often industrially produced) on site to avoid energy losses (Dessì et al. 2018b).

3. Ecology of Hydrogen-Producing Microbial Communities

3.1 Importance of the Microbial Diversity in Hydrogen Production

Dark fermentation process is easier to be scaled up with mixed cultures than pure cultures as sterile conditions are not necessary (Castelló et al. 2020). The microbial diversity in mixed cultures provides metabolic flexibility, robustness, function redundancies and facilitates the degradation and conversion into H_2 of complex substrates. In particular, the presence of microorganisms from *Bacillus* genus could help fermentative H_2-producing bacteria (HPB) by releasing hydrolytic enzymes that degrade complex substrates such as lignin, starch, or cellulose, increasing the amount of more readily degradable carbohydrates (Liu and Wang 2012, Mugnai et al. 2021). Anaerobic facultative bacteria can also consume oxygen traces, and support the growth of strict anaerobic HPB, such as *Clostridium* sp., by generating highly anaerobic local environments.

In counterpart, microbial diversity could also disfavour the hydrogen production process. One of the main disadvantages of using mixed cultures is the prevalence of microorganisms that outcompete the HPB for the organic substrate or directly consume the produced H_2, both resulting in a decrease of the global H_2 yield (Cabrol et al. 2017).

3.1.1 H_2-consuming microorganisms

Three groups of hydrogen-consuming microorganisms are commonly found in hydrogenogenic anaerobic reactors: hydrogenotrophic methanogenic archaea, homoacetogenic bacteria and sulphate-reducing bacteria (Castelló et al. 2020).

Sulphate-reducing bacteria (SRB) are extremely competitive in anaerobic environment and consume H_2 through a reaction that is more energetically favourable than homoacetogenesis and methanogenesis (Table 2). This reaction can even occur at very low H_2 concentrations ($\sim 10^{-6}$ atm). SRB growth can be limited by controlling the sulphate concentration in the media as well as the pH and the hydraulic retention time (HRT) (Guo et al. 2010b). In particular, pH lower than 6 have been reported to be effective for limiting the growth of SRB (Ghimire et al. 2015).

Table 2. H_2-consuming reactions (modified from Saady (2013)).

Microorganisms	Equation	$\Delta G°$ (Kj/mol)	N° Eq.
Sulphate reducing bacteria	$SO_4^{2-} + 4\,H_2 \rightarrow H_2S + 4\,H_2O$	−152.5	5
Hydrogenotrophic methanogenic archaea	$4\,H_2 + CO_2 \rightarrow CH_4 + 2\,H_2O$	−135	6
Homoacetogenic bacteria	$4\,H_2 + 2\,CO_2 \rightarrow CH_3COOH + 2\,H_2O$	−104.5	7

Hydrogenotrophic methanogenic archaea produce CH_4 using H_2 and CO_2. H_2 consumption by these microorganisms during dark fermentation has been widely reported in the literature (Carrillo-Reyes et al. 2012, Castelló et al. 2009, Etchebehere et al. 2016). Several strategies have been proposed to eliminate hydrogenotrophic methanogens from the microbial communities, such as the application of a heat-shock treatment to the inoculum. This pre-treatment is the most used and has been reported to be the most effective method for avoiding H_2 consumption by methanogenesis (Carrillo-Reyes et al. 2014). Another strategy consists in modifying the operating parameters of the fermenter. In particular, the use of a pH lower than 6 is strongly unfavourable for the growth of methanogens (Greses et al. 2021). Shortening the hydraulic retention time to a value lower than 24 h is also recommended to wash out slowly growing methanogenic archaea from the system (Akhlaghi and Najafpour-Darzi 2020). However, these strategies are not always affective, as methanogenic archaea were still detected despite heat-shock pre-treatment of the inoculum (Carrillo-Reyes et al. 2014), the use of a short HRT (< 24 h) (Mariakakis et al. 2011) and acidic pH (< 6) (Castelló et al. 2009).

Homoacetogenic bacteria found in hydrogenogenic reactors are mostly affiliated to the species *Acetobacterium woodii*, *Moorella thermoacetica*, or also concern a wide range of *Clostridium* species such as *Clostridium aceticum*, *Clostridium thermoautotrophicum*, *Clostridium thermoaceticum*, *Clostridium stercorarium*, and *Clostridium ljungdahlii* (Schuchmann and Müller 2016). Within the *Clostridium* genus, some homoacetogenic bacteria can perform both H_2 production or H_2 uptake due to their bidirectional hydrogenase enzymes (Westerholm et al. 2019). Huang et al. (2010)

observed that the abundance of *C. lundense, C. peptidivorans* and *C. vincentii* increased when H_2 production decreased, and they assumed that these microorganisms could consume H_2 via reverse hydrogenases. Since homoacetogenic bacteria are phylogenetically widespread bacteria, their identification and monitoring in hydrogenogenic reactors by 16S rRNA gene sequencing is not fully appropriate (Cabrol et al. 2017). As they accumulate acetate, a metabolite common to other reactions occurring in DF, and more precisely in hydrogenogenic reactions, their activity cannot be monitored through the accumulation of metabolites (see Equations 1, 3 and 4). Homoacetogenesis appears to be independent of the temperature, pH, microbial inoculum and is even enhanced in high cell density reactors where interspecies hydrogen transfer mechanisms are favoured (Saady 2013, Dinamarca et al. 2011). To counteract this effect, enhancing the gas-liquid mass transfer of dark fermentation reactors, using strategies such as gas recirculation or increasing the mixing of the reactor, can be effective to decrease the occurrence of homoacetogenesis (Buitrón et al. 2020, Montiel Corona and Razo-Flores 2018). Using an adapted C/N/P ratio of 100/0.5/0.3 has also been reported effective to prevent homoacetogenesis in anaerobic fluidized bed reactors (Carosia et al. 2021).

3.1.2 *The ambiguous role of lactic acid bacteria*

Due to their ability to produce lactic acid from carbohydrates without producing H_2 (Equations 25, 26 and 27), lactic acid bacteria (LAB) are commonly considered as competitors of HPB. In DF mixed cultures, the main LAB belong to the genera *Lactobacillus, Sporolactobacillus* or *Streptococcus*. Several species from the Bacillales order such as *Paenibacillus* sp. or *Bacillus* sp. were also reported as lactate producers (Chatellard et al. 2016). However, LAB are not only detrimental to H_2 production but their presence is also associated to either high or low hydrogen yields (Etchebehere et al. 2016). Although the accumulation of lactic acid is often considered as an indicator of fermentation failure, some studies suggested that the presence of LAB could have a positive effect on hydrogen production. Indeed, LAB could enhance H_2 yields through oxygen consumption as they are facultative anaerobic bacteria, creating an anaerobic environment beneficial to HPB. In addition, these bacteria produce enzymes that hydrolyse complex compounds, transforming them into simpler substrates that HPB can subsequently consume (Chang et al. 2008a, Li et al. 2011). Moreover, lactic acid can also be a substrate for hydrogen production (García-Depraect et al. 2021). In particular, Table 3 summarises the microorganisms involved in hydrogen production from lactate, the associated metabolic pathways and the related hydrogen yield.

The acrylate and the PFOR pathways were reported as the main H_2 producing pathways from lactate or lactate and acetate, respectively (García-Depraect et al. 2021). The need of acetate as supplementary electron acceptor is linked to a lower activity of the hydrogenase during lactate fermentation than glucose fermentation. This fact results in a lower hydrogen yield when HPB are growing on lactate using the PFOR pathway ($0.1–0.6$ $mol_{H2}/mol_{Lactate}$) (Diez-Gonzalez et al. 1995). Detman et al. (2019) showed that the bacteria which are able to convert lactate and acetate to butyrate, use electron-transferring flavoprotein complexes, which are very specific for lactate oxidation and butyrate formation. As shown in Table 3, some *Clostridium* sp. are able to consume acetate and lactate generated by LAB, producing butyrate and hydrogen. The association between LAB and lactate consumers that produce hydrogen could explain the ambiguous role of LAB in DF.

Nonetheless, LAB are commonly reported as responsible for unstable process operations and low hydrogen production rates (Dauptain et al. 2020, Palomo-Briones et al. 2017). This can be attributed to the ability of LAB to produce bacteriocins (bio-inhibitors), especially in presence of *Lactococcus* sp. (Castelló et al. 2020). Etchebehere et al. (2016) suggested that several *Clostridium* species could be inhibited by bacteriocins, which could explain the negative role of LAB. The accumulation of lactic acid due to the development of LAB could also contribute to generate unfavourable conditions for HPB through a pH decrease of the culture medium (below 5–6) (Jo et al. 2007, Kawagoshi et al. 2005).

Table 3. Empirical equations, hydrogen yields ($mol_{H2}/mol_{glucose}$), and microorganisms involved during hydrogen production from lactate.

Microorganism Involved or Mixed Culture	Reaction	Yield H_2	N° Eq.	Ref.
Clostridium neopropionicum	$Lactate \rightarrow H_2 + CO_2 + 0.5\ Butyrate$	1	8	(Tholozan et al. 1992)
Lactobacillus bifermentans	$Lactate \rightarrow 0.5\ Acetate + 0.5\ EtOH + H_2 + CO_2$	1	9	(Kandler et al. 1983)
soil culture	$Lactate \rightarrow 0.5\ Acetate + 0.35\ Butyrate + 0.1\ H_2 + 0.55\ CO_2$	0.1	10	(Bhat and Barker 1947)
Butyribacterium methylotrophicum	$Lactate \rightarrow 0.47\ Acetate + 0.37\ Butyrate + 0.005\ H_2 + 0.49\ CO_2$	0.005	11	(Shen et al. 1996)
Clostridium termitidis + *Clostridium beijerinckii*	$Lactate \rightarrow 2H_2 + CO_2 + Acetate$	2	12	(Flores et al. 2017)
Megasphaera elsdenii	$Lactate \rightarrow 0.8\ H_2 + 1.16\ CO_2 + 0.12\ Acetate + 0.26\ Butyrate$	0.8	13	(Prabhu et al. 2012)
Clostridium acetobutylicum P262	$Lactate + 0.4\ Acetate \rightarrow 0.6\ H_2 + CO_2 + 0.7\ Butyrate$	0.6	14	(Diez-Gonzalez et al. 1995)
Clostridium diolis JPCC H-3	$Lactate + 0.5\ Acetate \rightarrow 0.5\ H_2 + CO_2 + 0.75\ Butyrate$	0.5	15	(Matsumoto and Nishimura 2007)
Clostridium tyrobutyricum	$Lactate + 0.37\ Acetate \rightarrow 0.53\ H_2 + 0.97\ CO_2 + 0.63\ Butyrate$	0.53	16	(Bryant and Burkey 1956)
Clostridium beijerinckii	$Lactate + 0.48\ Acetate \rightarrow 0.59\ H_2 + CO_2 + 0.65\ Butyrate$	0.59	17	(Bhat and Barker 1947)
Clostridium sp. BPY5	$Lactate + 0.27\ Acetate \rightarrow 0.47\ H_2 + 0.52\ CO_2 + 0.66\ Butyrate$	0.47	18	(Tao 2016)
Clostridium tyrobutyricum	$Lactate + 0.27\ Acetate \rightarrow 0.09\ H_2 + 0.36\ CO_2 + 0.65\ Butyrate$	0.09	19	(Wu et al. 2012)
	$Lactate + 0.28\ Acetate \rightarrow 0.16\ H_2 + 0.43\ CO_2 + 0.69\ Butyrate$	0.16	20	
	$Lactate + 0.42\ Acetate \rightarrow 0.47\ H_2 + 0.7\ CO_2 + 0.71\ Butyrate$	0.47	21	

Table 3 contd. ...

...Table 3 contd.

Microorganism Involved or Mixed Culture	Reaction	Yield H_2	N° Eq.	Ref.
	Lactate + 0.38 *Acetate* $\rightarrow$ 0.45 H_2 + 0.67 CO_2 + 0.68 *Butyrate*	0.45	22	
	Lactate + 0.16 *Acetate* $\rightarrow$ 0.21 H_2 + 0.51 CO_2 + 0.66 *Butyrate*	0.21	23	
Acidogenic anaerobic reactor sludge	*Lactate* + 0.28 *Acetate* $\rightarrow$ 0.39 H_2 + 0.67 *Butyrate*	0.39	24	(Blanco et al. 2019)

3.1.3 Role of non-HPB: substrate competitors and hydrolytic bacteria

Some bacteria such as formate-producing bacteria may also outcompete the HPB for the substrate. Palomo-Briones et al. (2017) identified members of the *Enterobacteriaceae* family as formate producers. Mugnai et al. (2021) identified some genera such as *Romboustia, Caproiciproducens, Lachnoclostridium, Ruminiclostridium* and *Clostridium* as able to produce and consume formate to produce H_2 and CO_2, due to the presence of an enzyme formate C-acetyltransferase.

As shown in Table 4, several substrates are converted to metabolites that cannot be further transformed into hydrogen, such as propionate, succinate, ethanol, or butanol. Bacteria from the genera *Megasphaera, Propionibacterium, Clostridium, Selenomonas, Propionispira* or *Schwartzia* were previously identified as propionate producers (Cabrol et al. 2017). Propionate can be generated in pathways related to Equation 28 or Equation 29. Ethanol can be produced by bacteria belonging to the Enterobacterales order through the ethanol pathway according to Equantion 26 or Equation 30 (without hydrogen production) or the acetate-ethanol pathway with a yield of 2 $mol_{H2}/mol_{glucose}$ (Equation 3) (Zhou et al. 2018). Acetone, ethanol and butanol (ABE fermentation) pathways are also favoured after a metabolic stress resulting from a sudden pH or temperature decrease as well as the accumulation of volatile fatty acids (Dessì et al. 2018a, Van Ginkel and Logan 2005b).

In addition, biomass hydrolysis is often the limiting step of the dark fermentation process when dealing with complex substrates. Hydrolysis results from the activity of extracellular hydrolytic enzymes such as amylase, cellulase or xylanase (Cabrol et al. 2017, Mugnai et al. 2021). Various bacteria genera have been reported to release hydrolytic enzymes, such as *Bacillus* sp., *Proteus mirabilis*, members of the *Ruminococcaceae* family or some *Clostridium* sp. (*C. acetobutylicum,*

Table 4. Substrate competition equations: Non-hydrogen producing pathways.

Metabolites	Equation	Equation Number	Reference
Lactic acid and acetic acid	$C_6H_{12}O_6 \rightarrow CH_3CHOHCOOH + 1.5\ CH_3COOH$	25	(Wang et al. 2020)
Lactic acid and ethanol	$C_6H_{12}O_6 \rightarrow CH_3CHOHCOOH + CH_3CH_2OH + CO_2$	26	(Ferreira et al. 2018)
Lactic acid	$C_6H_{12}O_6 \rightarrow 2\ CH_3CHOHCOOH$	27	(Wang et al. 2020)
Propionic acid	$C_6H_{12}O_6 + 2\ H_2 \rightarrow 2\ CH_3CH_2COOH + 2\ H_2O$	28	(Ferreira et al. 2018)
Propionic and acetic acid	$1.5\ C_6H_{12}O_6 \rightarrow 2\ CH_3CH_2COOH + CH_3COOH + CO_2$	29	(Wang et al. 2020)
Ethanol	$C_6H_{12}O_6 + H_2O \rightarrow CH_3CH_2OH + 2\ CO_2$	30	(Wang et al. 2020)
Butanol	$C_6H_{12}O_6 \rightarrow CH_3CH_2CH_2OH + 2\ CO_2 + H_2O$	31	(Wang et al. 2020)

C. cellulosi or *C. stercorarium*). Hung et al. (2011) reported that *Bifidobacterium* sp. could also contribute to the biomass breakdown. Those bacteria assist hydrogen producers by converting the complex organic matter into monomers that can be further used by HPB. However, the exact role and interactions existing among these microorganisms and HPB in dark fermentation still need to be clarified as their exact functions are not fully understood (Rafrafi et al. 2013).

3.1.4 Indigenous versus exogenous bacteria

Indigenous bacteria, which are naturally living on the substrate, in contrast to exogenous bacteria that are added through an external inoculum, are particularly efficient to produce H_2 (Dauptain et al. 2020). Several studies showed the importance of indigenous bacteria on hydrogen yields in comparison to the addition of external inoculum (Dauptain et al. 2020, Favaro et al. 2013, François et al. 2021). Dauptain et al. (2020) reported that indigenous bacteria were as efficient as pre-treated exogenous bacterial inoculum to reach high hydrogen yields from seven organic substrates. Illustratively, uninoculated batch tests led to a hydrogen yield of 47 ± 10 mLH$_2$/ gVS (presenting a relative abundance of 62% Clostridiales and 35% Enterobacterales) while 60 ± 3 mL H$_2$/gVS (presenting a relative abundance of 63% Clostridiales and 31% Enterobacteriales) were observed with a pre-treated inoculum using sorghum. The authors attributed these results to the selection of similar bacterial communities, mainly composed of members from the Clostridiales and Enterobacterales orders. Interestingly, similar metabolic pathways were observed supporting the importance of indigenous bacteria (butyrate, acetate and H_2 as main metabolites). Consistently, François et al. (2021) reported an H_2 yield of ~2 mol$_{H2}$/mol$_{sugar}$ from Chardonnay grape must deposits with indigenous bacteria (with or without a heat treatment) and with exogenous heat-treated bacteria originated from a wastewater treatment plant.

Some authors tried to determine the interactions existing between indigenous and exogenous bacteria. Favaro et al. (2013) noticed a higher hydrogen yield (70 mLH$_2$/gVS) when a pre-treated inoculum was added, with regard to only indigenous bacteria (42 mLH$_2$/gVS) using the Organic Fraction of Municipal Solid Waste (OFMSW) as substrate. This result might be explained by the high abundance in LAB among the indigenous bacteria. Meanwhile, these authors showed also a lower H_2 production from sterilized OFMSW, which probably indicated positive interactions between indigenous and exogenous bacteria.

3.1.5 Microbial dynamics and their influence on process stability

Microbial composition can considerably differ along the dark fermentation process, leading to changes in metabolic pathways, especially during the different hydrogen production stages in batch tests (starting phase, production stage and steady state). To better determine the dynamics of the metabolic pathways and linking them to hydrogen production, some authors analysed the microbial communities at different times of the batch experiments. During co-fermentation of wastewaters (20%) and tequila vinasses (80%), García-Depraect et al. (2019) observed a high diversity microbial community at the beginning of the operation, which shifted towards a *Lactobacillus*-dominated community (80%) associated with the production of lactate. Meanwhile, *Clostridium* sp. gradually became the most dominant species, and was associated with lactate and acetate consumption to produce butyrate and hydrogen. At the end of the hydrogen production stage, some non-HPB as *Blautia* sp. (homoacetogenic bacteria) were also detected. In another study, Mugnai et al. (2021) observed lactate, formate and acetate production and further consumption during dark fermentation of olive-mill waste. During the initial stage of H_2 production, bacterial communities were dominated by *Bacillus* and *Clostridium* genera. *Clostridium* sp. proportions rapidly increased and were concomitant with butyrate and H_2 accumulation. When hydrogen production ceased, other bacterial communities emerged such as *Lysinbacillus* sp. or *Ruminoclostridium* species. Therefore, monitoring the microbial community dynamics together with the accumulation of metabolites is important to help in DF process operation by predicting and preventing process failures.

4. Strategies to Improve H_2 Production by Dark Fermentation

In Fig. 3, the strategies used to improve dark fermentation are shown. These strategies could be applied before the process start (selecting a suitable inoculum, engineering synthetic microbial communities, applying different pre-treatments to increase substrate accessibility for the HPB, etc.) or/and modifying some operational parameters during the process (temperature, pH, etc.). The strategies used to improve dark fermentation will be presented in the next two sections (3 and 4).

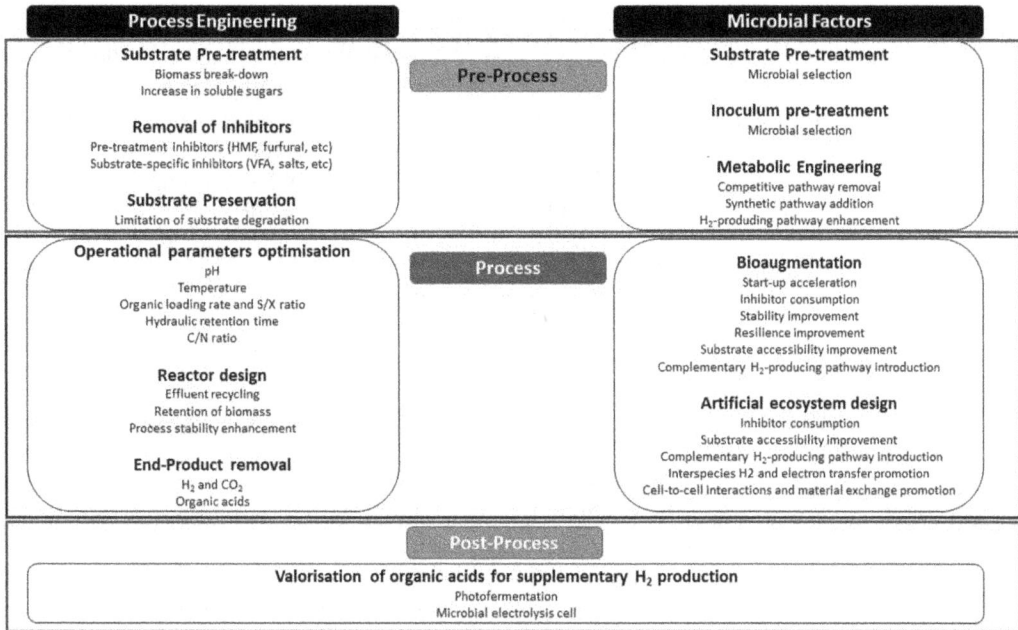

Fig. 3. Summary of the different methods that can be applied in order to improve H_2 production.

4.1 *Different Approaches for Selecting HPB*

4.1.1 *Inoculum pre-treatment*

The addition of an external inoculum is frequently mentioned in literature as favourable to support efficient H_2 production in DF. In particular, the inoculum source can significantly impact the H_2 yields of simple or complex organic matter. Pecorini et al. (2019) showed a large range of performances from 29 to 90 mLH_2/gVS from food waste, depending on the origin of the inoculum. Indeed, in mixed microbial cultures, many microorganisms can consume H_2, while many others can outcompete HPB for organic substrates. Fortunately, most of the H_2 producing bacteria (HPB) such as members of the *Clostridium* genus are able to sporulate, and fresh inoculum is often heat treated to select these microorganisms. Most of the non-sporulating bacteria will not survive to the extreme conditions of such treatment. The most used thermal treatments consist of applying temperatures from 65°C to 100°C for 15 min to 2 h (Rafieenia et al. 2018). Other inoculum treatment methods have been proposed to eliminate non-H_2-producing bacteria, such as aeration, the addition of chemical inhibitors as 2-bromoethansulphonate (BES) or chloroform, application of alkaline or acid environments, and the use of microwaves or sonication methods (Rafieenia et al. 2018). Many studies have investigated the impact of inoculum treatment on H_2 yields using simple sugars (mainly glucose) as substrate. As an example, Chang et al. (2011) performed DF experiments using glucose as substrate, in batch reactors set at 35°C, under stirring (150 rpm) and with a starting pH of 7. The authors showed a higher yield with acid treatment (1.51 $mol_{H2}/mol_{glucose}$) than with heat treatment (0.9 $mol_{H2}/mol_{glucose}$) or when using raw waste activated sludge as inoculum (0.38 $mol_{H2}/mol_{glucose}$).

4.1.2 Inoculum storage

Once an efficient microbial community is selected, the operators could be interested in storing the inoculum for further reuse or propagation to other systems. However, the method and time of storage could affect the functionality of the ecosystem, and some authors investigated the influence of inoculum storage on H_2 yields. In particular, Dauptain et al. (2021) investigated the effects of two storage modes such as freezing and freeze-drying for a short (1 week) and long period (1.5 months). Stored inocula were afterwards evaluated for their ability to produce H_2 from glucose and complex organic substrates as FW and OFMSW. A variable impact of the storage time and the storage mode was observed on glucose, and the changes were attributed to bacterial shifts from *Clostridium* sp. or *Escherichia-Shigella* to *Raoultella* sp. In contrast, for complex substrates (FW and OFMSW), no statistical differences on the hydrogen yields for the three inocula were observed. These results were attributed to a high and stable abundance in Clostridiales or Enterobacteriales, that led to similar metabolic pathways. Similarly, García-Depraect et al. (2020) reported a stable H_2 yield with an inoculum stored at 4°C (up to 730 days) and attributed the results to a stable abundance in Clostridiales between 190 and 730 days of storage.

4.1.3 Biomass pre-treatment

Biomass pre-treatments constitute a crucial step to break down biomass structure, to increase biodegradable compound accessibility and to assist the hydrolysis of cellulose and hemicelluloses into soluble sugars, which can be directly converted to H_2 (Rafieenia et al. 2018). Consequently, using biomass pre-treatments is frequently reported in the literature to further improve the H_2 yields in DF processes. Pre-treatment methods are generally categorized according to the nature of the disruption force, i.e., mechanical (as milling), physical (as microwaves, sonication or steam explosion), chemical (acid or alkaline), thermal or biological (Bundhoo et al. 2015, Rafieenia et al. 2018). Several criteria, such as pre-treatment duration, energy input, concentration in chemicals, temperature, the solid content and the power or frequency applied can considerably influence the overall substrate biodegradability and therefore, H_2 production.

During thermal pre-treatments, the biomass is generally heated to temperatures between 70°C and 200°C (Parthiba Karthikeyan et al. 2018, Rafieenia et al. 2018). In sonication pre-treatments, waves frequency range between 20 kHz and 10 MHz (ultrasound area). This pre-treatment consists in the propagation of ultrasound waves from a probe, leading to cavitation effects (Rafieenia et al. 2018). In contrast, when applying microwave pre-treatments, electromagnetic radiations with frequencies ranging from 300 MHz to 300 GHz are used (Aguilar-Reynosa et al. 2017). For chemical pre-treatments, an acid (HCl, H_2SO_4) or a base (NaOH, $Ca(OH)_2$) is directly added into the bulk in order to adjust the pH of the media to 3–4 or 10–12, respectively (Wong et al. 2014). During biological pre-treatments, enzymatic cocktails are generally added (Sambusiti et al. 2015). The effectiveness and applicability of a pre-treatment depend on the specific properties of the substrate, the operational conditions, the additional costs at large scale as well as the energy requirements.

Chemical and thermo-chemical pre-treatments seem to be the most efficient methods to improve the H_2 production. Rafieenia et al. (2018) reported increases in H_2 yield from 9%, when an alkaline pre-treatment was applied to FW (NaOH, pH = 11 during 24 h) up to 6550% when algal biomass was pre-treated with hydrochloric acid (200 mL/L) at 121°C for 20 minutes (Elbeshbishy et al. 2017, Rorke and Gueguim Kana 2016). However, using a pre-treatment can also impact the microbial metabolic pathways and the microbial communities. In Kim et al. (2014), food-waste was pre-treated by HCl at room temperature for 12 h at pH from 1 to 4 and batch experiments were performed at 35°C, pH 8, under stirring (100 rpm) but with no external inoculum addition. The authors observed an increase in H_2 production from 54 (raw) to 158 mLH_2/gVS (treated at pH = 2) due to a bacterial metabolic shift from *Lactobacillus* (80%) towards *Clostridium* sp. (90%). Interestingly, Dauptain et al. (2020) observed similar final bacterial compositions (mainly Clostridiales and Enterobacteriales) and H_2 yields for different organic waste (i.e., corn silage, sorghum, OFMSW and FW) after applying thermal pre-treatments to them (90°C for 15 min).

However, in some cases, biomass pre-treatments and especially thermo-chemical pre-treatments can lead to the production of inhibitory compounds such as 5-HMF (5-(hydroxymethyl) furfural), furfural or phenolic compounds. Monlau et al. (2014) observed a metabolic shift after thermochemical pre-treatment of sunflower stalks towards ethanol and lactate production together with a reduction of H_2 yield, likely due to the presence of furfural, 5-HMF and phenol compounds after pre-treatment. In addition, after chemical pre-treatment, the pH must be adjusted between 5 and 6, which is known to be the optimal pH to produce H_2 (Hawkes et al. 2007). However, acid and base addition could contribute to increase the total ionic strength of the solution and subsequently cause process inhibition and a possible decrease in the H_2 yield (Paillet et al. 2020).

4.2 Impact of the Operating Parameters on Bacterial Selection

Process instability is often caused by changes in biotic or abiotic factors that lower the H_2 production performances. Castelló et al. (2020) reported that a low microbial diversity was probably the cause of process instability due to the lack of functional redundancy within the microbial community. Indeed, the recovery of a function in a disturbed community is highly dependent of the functional redundancy. The diversity can vary according to the reactor design (CSTR, UASB, among others) and operation (Etchebehere et al. 2016). Consequently, changes in microbial communities due to perturbation in the operating conditions can deeply affect the H_2 production. While a high abundance in members from the *Clostridium* genus is generally associated with good DF performances, a shift towards low H_2 producers, such as *Megasphaera* sp., or H_2 consumers, such as homoacetogens and methanogens, could lead to instability (Etchebehere et al. 2016). Several strategies have been developed to drive the fermentation towards H_2 production. In particular, the modification of the fermentation operational parameters towards the selection of HPB aims to control the microbial ecosystem and improve its resilience and resistance. Among the operating conditions, pH, HRT, substrate concentration, and temperature were reported to have significant effect on microbial communities and metabolic pathways, as detailed below.

4.2.1 pH

pH significantly impacts microbial community, related metabolic pathways, and the enzymatic activities. For instance, the optimal pH for methanogenesis is between 6.5 to 8.5, while for lactic fermentation the optimal pH is 5.5–6.5. Most of the studies reviewed by Elbeshbishy et al. (2017) have reported that the optimal pH for H_2 production range between 5 and 6 in continuous and batch reactors, which coincide with the optimal pH range for some HPB, e.g., *Clostridium* sp. (Wong et al. 2014).

Out of this range, other metabolic reactions can occur, diverting the electrons flow from H_2 production to other metabolic end-products. Illustratively, at acidic pH (< 4), the accumulation of acids and alcohols could induce stress to the microorganisms as they can penetrate the cellular membrane entering the microbial cell and interfering with regular metabolic activities (Elbeshbishy et al. 2017). This has been observed to be detrimental to H_2 production (Lu et al. 2011) and substrate degradation (Fang and Liu 2002). However, no clear trend between pH and specific metabolic pathways has been established (Łukajtis et al. 2018). For instance, De Gioannis et al. (2014) studied the impact of pH from 5.5 to 8.5 on dark fermentation of cheese whey. Here, the optimal pH was strongly dependent on substrate characteristics and inoculum source. The authors concluded that no clear link exists between pH changes and metabolic pathways shifts due to the multiple pathways that can overlap during the fermentation process. The metabolic patterns observed at different pH are mostly due to the optimal growth of specific microorganisms within the microbial community. Neutral and alkaline pH (≥ 7) can induce a decrease in H_2 production by promoting methanogenesis or by shifting the metabolic activity towards acetate, propionate, ethanol or formate pathways (Temudo et al. 2008). However, some exceptions are found in the literature. For example, Lee et al. (2002) reported an optimal H_2 production for a *Clostridium*-related species at pH 9. Moreover, *Clostridium quinii* and *Clostridium interstinale* were found dominant at pH 7.5–8 (Temudo et al. 2008). On the other hand, Mota et al. (2018) observed an optimal H_2 production at pH 3 with a microbial community dominated by *Ethanoligenens* sp. and *Clostridium* sp. Whilst, the HPB *Klebsiella oxytoca* optimal pH and H_2 producing activity was reported to be at neutral pH (6–7) (Temudo et al. 2008).

4.2.2 HRT

In reactors operated in continuous mode, the Hydraulic Retention Time (HRT) specifically selects the active microorganisms having a sufficiently high growth rate to withstand the dilution caused by the liquid flow. Łukajtis et al. (2018) showed that the optimal HRT depends mainly on the substrate and its biodegradability but ranged between 0.5 h to 12 h. Using simple sugars as substrate (glucose), the H_2 yield of an anaerobic fluidized bed reactor, operated at 30°C, without pH control was increased from 1.41 to 2.49 mol H_2/mol glucose, by shortening the HRT from 8 h to 2 h (de Amorim et al. 2009). Similarly, for more complex sugars as lactose and sucrose, the lowest used HRT in the operations (6 h and 12 h, respectively) led to the best hydrogen productions rates (2.0 LH_2/Ld and 4.4 LH_2/Ld, respectively) (Palomo-Briones et al. 2017, Salem et al. 2018). Optimisation of the HRT could be directed to the selection of HPB over methanogens, because their generation time is shorter than the generation time of methanogens (Table 5). Hence, using a short HRT will favour the washing out of methanogens. For some other complex substrates, such as melon and watermelon residues, long HRT (> 20 days) has been used successfully with no methanogenic activity by maintaining a low pH (5.5) (Greses et al. 2021).

Palomo-Briones et al. (2017) studied the impact of a range of HRT, from 6 to 24 h, on H_2 production and community dynamics, using lactose as a substrate. A maximum H_2 yield of 0.86 mol_{H2}/$mol_{glucose}$ was observed at 6 h HRT, in link with the growth of a community dominated by *Clostridiaceae, Lachnospiraceae* and *Enterobacteriaceae*. A longer HRT favoured the emergence of lactate-producing bacteria, especially *Sporolactobacillaceae* and *Streptococcaceae*.

Table 5. Generation time and activity of different microorganisms occurring in DF reactors.

Microorganism	Generation Time (h)	Activity	Reference
Clostridium butyricum	2.1	H_2 producer	(Abbad-Andaloussi et al. 1995)
Clostridium tyrobutyricum	< 3.5	H_2 producer	(Linger et al. 2020)
Clostridium pasteurianum	1.4	H_2 producer	(Mallette et al. 1974)
Methanococcus maripaludis	8.3	Methanogenic hydrogenotrophic	(Costa et al. 2013)
Methanosarcina sp.	9–24	Methanogenic hydrogenotrophic	(Ferguson and Mah 1983)
Clostridium carboxidivorans	6.9–11.5	Homoacetogenic	(Lanzillo et al. 2020)
Clostridium ljungdahlii	4.6	Homoacetogenic	(Whitham et al. 2015)
Lactobacillus plantarum	2.7–3.4	Lactate producer	(Sedewitz et al. 1984)

Similarly, during fermentation of condensed molasses, the H_2 production rate increased from 152.5 $mmolH_2/Ld$ to 390 $mmolH_2/Ld$, while the HRT was shortened from 12 to 3 h (Chang et al. 2008b). Ueno et al. (2006) reported that the optimal H_2 production in their experiments was observed when *Thermoanaerobacterium thermosaccharolyticum* was the dominant HPB. These authors have employed an HRT < 1 day, and an artificial garbage slurry composed by dog food diluted in water and milled paper as substrate.

4.2.3 *Substrate concentration and organic loading rate (OLR)*

Substrate concentration also exerts a strong selective pressure on microbial populations. Indeed, too high substrate concentrations can lead to the accumulation of acids, that can induce cellular inhibition and/or microbial community shifts and reduce the H_2 production (Łukajtis et al. 2018). Kyazze et al. (2006) studied the influence of different sucrose concentrations (10 to 50 g/L) on H_2 production in a continuous operation. They reached stable H_2 production at 10, 20 and 40 g/L while at 50 g/L production the system was instable. Overall, the H_2 yield decreased with the increasing sucrose concentration from 1.7 mol_{H2}/mol_{hexose} at 10 g/L to 0.8 mol_{H2}/mol_{hexose} at 50 g/L. It was suggested that concentrations higher than 20 gCOD/L could hinder the H_2 production in batch reactors (Elbeshbishy et al. 2017). In addition, unsterile substrates can also carry some detrimental indigenous populations that can outcompete the HPB. As an illustration, during DF of real food waste, concentrations higher than 15 gVS/L were unfavourable to H_2 production and induced lactic acid fermentation due to the high amount of LAB introduced with the substrate (Pu et al. 2019).

In continuous reactor, the substrate concentration is not the sole factor to be considered but also the Organic Loading Rate (OLR). The OLR (g/Ld) is the amount of organic matter per unit of reactor volume provided in a given period of time. Equation 32 shows the OLR formula where C_{feed} represents the concentration of the feed solution (g/L), Q represents the rate of feed addition in the reactor (L/d), $V_{reactor}$ represents the effective volume of reactor (L) and *HRT* is the hydraulic retention time applied to the reactor (d).

$$Organic\ Loading\ Rate\ (OLR) = \frac{c_{feed} \times Q}{V_{reactor}} = \frac{c_{feed}}{HRT} \tag{32}$$

The optimal range of OLR for H_2 production is between 18–180 gCOD/Ld, depending on the substrate used (monosaccharides, sucrose, food waste), the reactor configuration (anaerobic fluidized bed reactor (AFBR), up-flow anaerobic sludge blanket (UASB), up-flow anaerobic fixed-bed reactor (UAFBR), anaerobic membrane reactor (AMB)), the temperature (from 23°C to 60°C) and the microbial inocula (Elbeshbishy et al. 2017). It has been proposed that reactors with biomass retention systems could handle higher OLR (Van Ginkel and Logan 2005a). For instance, Hafez et al. (2010) using a CSTR reactor coupled with a gravity settler, which helped to avoid the biomass washout of the reactors, allowed to maintain a high H_2 yield (2.8 $mol_{H2}/mol_{glucose}$) while increasing the OLR from 6.5 $gCOD_{glucose}/Ld$ to 103 $gCOD_{glucose}/Ld$. The highest H_2 yield and H_2 production rate were observed at an OLR of 103 $gCOD_{glucose}/Ld$. Nevertheless, higher OLR (from 154 $gCOD_{glucose}/Ld$ to 206 $gCOD_{glucose}/Ld$) led to a drop of the H_2 yield (from 2.8 to 1.2 and 1.1 $mol_{H2}/mol_{glucose}$, respectively). OLR shock can cause microbial community shifts as reported with multiple substrates and reactor configurations, and the OLR impact mainly depends on the microbial composition. For instance, Macías-muro et al. (2021) observed a shift from *Clostridium* sp. to *Sporolactobacillus* sp. in a CSTR reactor fed with vinasses at pH 5.5 and 6 h HRT when the OLR was increased from 80 to 160 gCOD/Ld. This change negatively affected the H_2 production rate. In contrast, Veeravalli et al. (2017) reported H_2 production rate improvement in UASB reactors (37°C, pH = 5, HRT = 24 h) fed with glucose, when the OLR was increased from 8.6 to 12.8 gCOD/Ld. This improvement was attributed to the replacement of *Synergistaceae* and *Propionibacteriaceae* by *Clostridiaceae* and *Ruminococcaceae*.

4.2.4 Temperature

Fermentation can be conducted at different temperatures: psychrophilic (0–25 °C), mesophilic (25–45°C), thermophilic (45–65°C), extreme thermophilic (65–80°C) and hyper thermophilic (> 80°C) temperature. Most of the studies dealing with DF were carried out under mesophilic conditions, because most HPB in seed sludge are mesophilic (Wong et al. 2014). Nevertheless, thermophilic temperatures can lead to higher H_2 yields, and in particularly hyperthermophilic temperatures, in which H_2 yields close to the theoretical maximum of 4 $mol_{H2}/mol_{glucose}$ can be reached (Chou et al. 2007). Fermentation under psychrophilic conditions can be carried out by specialist bacteria, where growth rate and activity is the highest at temperatures between 0 and 25°C. Alvarado-Cuevas et al. (2015) isolated microorganisms from Antarctica that were able to produce H_2 at psychrophilic conditions. These authors have reported that the maximum H_2 production rate (16.7 mLH_2/Lh) was achieved by a strain related to *Janthinobacterium agaricidamnosum* (98% identity), while the highest H_2 yield (1.57 $mol_{H2}/mol_{glucose}$) was attained by a strain related to *Polaromonas jejuensis* (99% identity). Performing dark fermentation at low temperature was shown to be an interesting solution to increase the net energy gains in dark fermentation (Rodríguez-Valderrama et al. 2019).

Similarly to pH, HRT, substrate concentration and OLR, different microorganisms are selected according to the operating temperature, a fact that can affect the metabolic pattern in DF. For example, in a semi-continuous acidogenic reactor treating OFMSW, concentrations of acetic acid were higher in thermophilic conditions than in mesophilic conditions (Valdez-Vazquez et al. 2005). Lazaro et al. (2014) showed that during DF of sugarcane vinasses, *Clostridium* sp. dominated in mesophilic conditions, while *Thermoanaerobacter* sp. dominated in thermophilic conditions with similar highest H_2 yields (2.23 and 2.31 $mol_{H2}/gCOD_{molasses}$, respectively), although with a different substrate concentration (7 $gCOD_{molasses}/L$ and 2 $gCOD_{molasses}/L$). Sivagurunathan et al. (2014) showed that increasing the temperature from 37°C to 45°C permitted to overcome propionate inhibition and increased the H_2 production rate and yield from 8.58 to 13.6 LH_2/Ld and from 1.04 to 1.68 mol_{H2}/mol_{hexose}, respectively. The authors reported that the H_2 production improvement was due to a shift from *Selenomonas lacticifex* and *Bifidobacterium catenulatum*, that are propionate producers, to a dominance of *Clostridium butyricum, Clostridium perfringens, Clostridium acetobutylicum* and *Ethanoligenens harbinense*.

4.3 Management of End-Product Inhibition

During DF, the operational parameters are selected to drive the microbial community towards efficient H_2 production. However, H_2 production can be inhibited by the accumulation of co-products. In particular, H_2, CO_2 and liquid metabolites are detrimental to hydrogen yields and productivities, at enzymatic, cellular and population levels.

4.3.1 Inhibition by organic acids and ethanol

Acetate and butyrate accumulation can deeply affect cellular homeostasis of the microorganisms. Indeed, the undissociated form of these acids can freely permeate through the cell membrane. Acetate and butyrate easily dissociate in the cytosol, which has a neutral pH, as their pKa are 4.76 and 4.82, respectively. The released protons induce an internal pH decrease, which disturbs cellular homeostasis and affects the enzymatic activities. To counteract this effect, the cell actively extrudes these protons by using ATP-consuming cellular pumps, increasing the maintenance energy, and the amount of energy available for biomass growth is reduced (Herrero et al. 1985). As a consequence, hydrogen production is also reduced (Van Ginkel and Logan 2005b). Furthermore, the dissociative forms of acetate and butyrate increase the osmotic pressure leading to growth inhibition (Van Niel et al. 2003).

The effect of lactate on H_2 production by dark fermentation and on associated microbial communities has been poorly studied. Several authors reported an inhibitory effect of lactate at concentration as high as 55 mM and 300 mM for starch and pre-fermented food waste, respectively (Baghchehsaraee et al. 2009, Noblecourt et al. 2018). Nevertheless, 44 mM L-Lactate induced a 35% decrease in hydrogen production during dark fermentation of food waste (Noblecourt et al. 2018). As both D-Lactate and L-Lactate are naturally produced by LAB (Chang et al. 1999), the inhibitory effect of enantiomeric lactate remains unclear.

An inhibitory effect of 10 mM of ethanol on hydrogen production was also reported by Wang et al. (2008), in mesophilic batch tests using glucose as substrate. Nevertheless, its inhibitory effects were lower than acetate, butyrate, and propionate.

The sensitivity to acids accumulation can vary among the species and strains of microorganisms. For butyrate, the minimum inhibitory concentration measured were 1.2, 2.0, 2.3 and 2.5 g/L for *Clostridium perfringens ATCC 12915*, *Enterococcus faecalis ATCC 29212*, *Escherichia coli ATCC 25922* and *Escherichia coli F18*, respectively (Kovanda 2019). The different inhibitory concentrations among these microorganism might be explained by differences in membrane permeation and pH homeostasis (Slonczewski et al. 2009). Intrinsic properties of the species can therefore be responsible of populational shifts occurring during fermentation, that can further affect the H_2 yield. Indeed, butyrate inhibition was reported to be associated with a decrease in *Clostridia* sp. growth and an increase in non-hydrogen-producing bacteria, such as *Pseudomonas* sp., *Klebsiella* sp., *Acinetobacter* sp. and *Bacillus* sp. during fermentation of glucose at pH 5.5–7 (Chen et al. 2021). A decrease in ethanol production was also observed.

4.3.2 Inhibition by H_2

Hydrogen production through electron transfer from NADH or Fd_{red} to protons depends on the relative redox potential of the redox couple, and can be limited by thermodynamics (Angenent et al. 2004). Overall, a high H_2 partial pressure disfavours H_2 production and favours alternative metabolic pathways to regenerate reducing equivalents. Moreover, H_2 partial pressure at which the H_2 production reaction is thermodynamically feasible depends on the temperature, the cellular internal pH and the NAD+/NADH ratio (Bastidas-Oyanedel et al. 2012). The partial pressure of H_2 starts to be inhibitory at 60 Pa (Angenent et al. 2004).

Several techniques have been employed to reduce the hydrogen partial pressure during fermentation, such as lowering the total pressure in the reactor (Ferraz et al. 2020, sparging with exterior gas or internal gas (Kraemer and Bagley 2008), modifying the biogas release (Esquivel-Elizondo et al. 2014), sequestration of H_2 (Bakonyi et al. 2017) and increase of the agitation speed (Montiel Corona and Razo-Flores 2018).

At the microbial community scale, the increase of H_2 and CO_2 concentrations tend to favour homoacetogenesis (Harper and Pohland 1986). Some authors reported a shift in the microbial community when applying such techniques. For instance, Buitrón et al. (2020) reported a microbial community shift from *C. carboxidivorans* and *C. ljungdahlii* (homoacetogens) to *Megasphaera elsdenii* (HPB) and improved the H_2 production, when applying biogas recirculation. Indeed, biogas recirculation increased the H_2 mass transfer at the liquid-gas interface and reduced the H_2 supersaturation, avoiding H_2 consumption by homoacetogenic bacteria. Nasr et al. (2015) also reported higher H_2 production rate, H_2 yield, bacterial richness and diversity and a diminution of non-H_2 producers, using a KOH trap for CO_2 sequestration. This improvement took place even if H_2 content in biogas increased from 57% to 100%, suggesting that bacterial selection is more important than enzymatic inhibition by H_2.

5. Metabolic and Microbial Community Engineering: A Tool to Improve Hydrogen Production

By applying different pre-treatments and modifying operational parameters, H_2 production can be enhanced. Another solution to improve H_2 production could be by engineering synthetic communities through the use of specific microorganisms, or by metabolic engineering of these microorganisms when operated in pure culture (Fig. 3).

5.1 Co-cultures

A co-culture is a reconstitution of a microbial consortium by mixing several isolated strains. The low diversity of the co-culture, regarding complex mixed cultures, helps to better control ecosystem functioning. According to Du et al. (2020), H_2 production based on co-cultures containing *Clostridium* sp. can be improved by commensalism or mutualism. A commensalism interaction implies the benefit for one of the partners without affecting the other. For example, co-culturing *Rhodobacter sphaeroides* with *Clostridium butyricum* on starch has been reported to improve H_2 production due to the transformation of acetate and butyrate (produced by *C. butyricum*) to H_2 by *R. sphaeroides* through photofermentation (Laurinavichene et al. 2018). Meanwhile, in a mutualistic interaction both partners benefit from the interaction. For instance, the cultivation of *Clostridium beijerinckii* and *Geobacter metallireducens* in the presence of an extracellular electron shuttle, the anthrahydroquinone-2, 6-disulfonate (AH2QDS), promoted the production of H_2, acetate and butyrate by *C. beijerinckii*, while *G. metallireducens* consumed the acetate produced by *C. beijerinckii* to regenerate AH2QDS (Zhang et al. 2013). In another context, Benomar et al. (2015) observed the formation of a physical interaction between *Clostridium acetobutylicum* and *Desulfovibrio vulgaris*. This interaction generated genetic expression shifts that changed the metabolic flux distribution, substantially increasing the H_2 production by *C. acetobutylicum* and supporting the growth of *D. vulgaris*. On the contrary, neutral and negative interactions in co-culture with *Clostridium* sp. or other species have also been reported in the literature and more investigations are required to better understand the impact of microbial interactions on fermentative pathways (Vatsala et al. 2008). Illustratively, Eder et al. (2020) co-cultivated *C. beijerinckii* NCIMB6444 and *C. butyricum* NCIMB9578, without improving the H_2 production from vinasses.

5.2 Bioaugmentation

Bioaugmentation consists in adding selected strain(s) into a microbial inoculum to improve its functionality, achieve better process performances and withstand process fluctuations (Yang and Wang 2018). It can be implemented by adding one specialized microorganism, several specialized strains, or a mixed enrichment culture. Adding one or several specialized well-known strains presents the advantage to address one specific problem regarding the functionality of the system and better predict the microbial interactions taking place (Kumar et al. 2016). Whilst, adding a complex specialized mixed-culture could be used to improve the degradability of a highly complex substrate or increase the robustness of the inoculum (Kumar et al. 2016).

The added strains can improve H_2 production through several mechanisms: Guo et al. (2010a) have observed a faster start-up and a higher H_2 yield on a continuous reactor fed with glucose and bioaugmented with *Ethanoligenens harbinense* B49 when compared with the non-bioaugmented control reactor. By adding *Ethanoligenens harbinense* B49, the lag phase was prevented and a supplementary H_2 producing pathway (ethanol-H_2 type fermentation) was added to the host inoculum leading to a higher H_2 yield. Bioaugmentation has also been reported to accelerate the recovery from perturbations, such as organic overloading (Goud et al. 2014), temperature fluctuations (Okonkwo et al. 2020) or oxygen concentration increase in the reactor (Kumar et al. 2015).

The imported strain(s) could improve the conversion of organic waste, undesired compounds, and other raw materials. As an example, the addition of a photosynthetic culture, in dark fermentation reactors, was observed to avoid HPB inhibition due to volatile fatty acids removal from the media, leading to 40% higher H_2 production and 10% increase in substrate consumption (Chandra and Mohan 2014), while the addition of *Clostridium* sp. *TXW1*, which is a cellulolytic bacteria, improved substrate degradability when using kitchen waste and napiergrass (Kuo et al. 2012). Indeed, simple sugars were more easily available to the HPB, promoting the increase in H_2 (Kuo et al. 2012).

The selection and concentration of the microorganisms to be added, as well as the mixed culture concentration are important factors affecting the bioaugmentation results. DF stability can be improved, even though the added strains remain at low concentration (Poirier et al. 2020), although in some cases, bioaugmentation can have a neutral or a negative effect on H_2 production (Yang and Wang 2018). For instance, Guo et al. (2014) bioaugmented a dark fermenter operated on cornstalk in batch at 36°C with an isolated high-efficiency hydrogen producing *Bacillus* sp. FS2011 at different doses ranging between 2 and 12%. Whilst the 12% dosage was observed not to improve the H_2 yield, the 10% dosage induced a 16% improvement.

5.3 Enhancing Hydrogen Production by Metabolic Engineering

During dark fermentation, only 20% of the energy contained in a substrate will be converted into H_2 and at most 4 moles of H_2 per mole of hexose will be achieved. This thermodynamic limitation cannot be outperformed by any of the previous exposed strategies. However, the use of genetic tools has been proposed to achieve or even surpass H_2 yield limitations. Several authors reviewed the use of metabolic engineering to improve hydrogen production (Arimi et al. 2015, Baeyens et al. 2020, Goyal et al. 2013, Hallenbeck and Ghosh 2012, Majidian et al. 2018, Mathews and Wang 2009, Wang and Yin 2019). Microorganisms such as *E. Coli, Clostridium* sp., *Enterobacter* sp., *Klebsiella* sp. are the most studied model microorganisms that have been genetically engineered to improve their H_2 producing capacities. In general, two strategies can be applied in order to genetically modify an organism: (i) modifying an existing pathway or (ii) adding a synthetic pathway.

The modification of an existing pathway consists in the optimization of the electrons flux to maximise the H_2 production in H_2 producing bacteria. For this purpose, researchers have focused either on suppressing genes coding for an H_2 competing pathway, or on increasing the expression of H_2 producing enzymes. In the first case, main targeted enzymes are the uptake hydrogenases, and the ones involved in competitive pathways that drain pyruvate or NADH from H_2 production. For instance, hydrogenase 1 and 2 inactivation in *E. Coli* was observed to induce a 3.2-fold increase in H_2 production when compared with the wild-type strain (Maeda et al. 2008). Succinate and lactate synthesis pathway suppression by fumarate reductase and lactate dehydrogenase inactivation resulted in a 1.4-fold increase in hydrogen production with regard to the wild-type strain (Ica et al. 2006), while for the second case, H_2 production improvement by hydrogenases and FHL overexpression, NADH production increase, or hydrogenases genetic modification have also been tested (Sekar et al. 2017, Wang and Yin 2019, Klein et al. 2010, Bisaillon et al. 2006).

Alternatively, the addition of a synthetic pathway consists in the introduction of genetic material to express a hydrogen producing pathway in a strain that naturally does not produce H_2. For instance, the transformation of *Enterobacter aerogenes* with a plasmid containing the gene coding for the hydrogenase from *Enterobacter cloacae* induced a 2-fold increase in H_2 production compared to the wild-type *Enterobacter aerogenes* (Song et al. 2010). Meanwhile, the addition of a synthetic pathway from pyruvate to H_2 in *Escherichia coli* BL21, a non-H_2-producing bacteria, was successful as the modified microorganism was able to produce H_2 (Akhtar and Jones 2009). *Clostridium thermocellum* was genetically modified to grow on xylose in addition to cellulose widening the ability to produce H_2 from other substrates, and increasing by 2-fold the H_2 production (Xiong et al. 2018).

To conclude, microbial genetic engineering allows improving H_2 yields and acquiring a better understanding of metabolic regulation. However, H_2 production beyond the natural limits still needs to be investigated.

6. Conclusion

Overall, DF represents a sustainable and feasible process to produce biohydrogen from organic biomass. Microbial communities and their stability over time appear as key factors to properly drive the DF process. Nevertheless, abiotic parameters such as temperature, HRT or pH should be carefully chosen to avoid metabolic or microbial community shifts that may lead to process inhibition and failure. Several strategies exist to increase the H_2 yield by pre-treating the biomass itself, by adding some bacteria or by selecting the best hydrogen producers to avoid non-H_2 producing pathways or H_2-consuming pathways. However, the diversity of bacteria involved in the DF process is huge and the knowledge about their role and the complex interactions between them still need to be further investigated to ensure future stable and efficient operations. Developing the coupling of DF with other processes such as anaerobic digestion, photofermentation or microbial electrolysis cells is also necessary to improve the overall conversion efficiency of the biomass and to make DF economically viable at industrial scale.

References

Abbad-Andaloussi, S., Manginot-Durr, C., Amine, J., Petitdemange, E. and Petitdemange, H. 1995. Isolation and characterization of Clostridium butyricum DSM 5431 mutants with increased resistance to 1,3-propanediol and altered production of acids. Appl. Environ. Microbiol. 61: 4413–4417. https://doi.org/10.1128/aem.61.12.4413-4417.1995.

Aguilar-Reynosa, A., Romaní, A., Ma. Rodríguez-Jasso, R., Aguilar, C.N., Garrote, G. and Ruiz, H.A. 2017. Microwave heating processing as alternative of pretreatment in second-generation biorefinery: An overview. Energy Convers. Manag. 136: 50–65. https://doi.org/10.1016/j.enconman.2017.01.004.

Akhlaghi, N. and Najafpour-Darzi, G. 2020. A comprehensive review on biological hydrogen production. Int. J. Hydrogen Energy 45: 22492–22512. https://doi.org/10.1016/j.ijhydene.2020.06.182.

Akhtar, M.K. and Jones, P.R. 2009. Construction of a synthetic YdbK-dependent pyruvate:H2 pathway in *Escherichia coli* BL21(DE3). Metab. Eng. 11: 139–147. https://doi.org/10.1016/j.ymben.2009.01.002.

Alvarado-Cuevas, Z.D., López-Hidalgo, A.M., Ordoñez, L.G., Oceguera-Contreras, E., Ornelas-Salas, J.T. and De León-Rodríguez, A. 2015. Biohydrogen production using psychrophilic bacteria isolated from Antarctica. Int. J. Hydrogen Energy 40: 7586–7592. https://doi.org/10.1016/j.ijhydene.2014.10.063.

Angenent, L.T., Karim, K., Al-Dahhan, M.H., Wrenn, B.A. and Domíguez-Espinosa, R. 2004. Production of bioenergy and biochemicals from industrial and agricultural wastewater. Trends Biotechnol. 22: 477–485. https://doi.org/10.1016/j.tibtech.2004.07.001.

Arimi, M.M., Knodel, J., Kiprop, A., Namango, S.S., Zhang, Y. and Geißen, S.U. 2015. Strategies for improvement of biohydrogen production from organic-rich wastewater: A review. Biomass and Bioenergy 75: 101–118. https://doi.org/10.1016/j.biombioe.2015.02.011.

Baeyens, J., Zhang, H., Nie, J., Appels, L., Dewil, R., Ansart, R. and Deng, Y. 2020. Reviewing the potential of bio-hydrogen production by fermentation. Renew. Sustain. Energy Rev. 131: 110023. https://doi.org/10.1016/j.rser.2020.110023.

Baghchehsaraee, B., Nakhla, G., Karamanev, D. and Margaritis, A. 2009. Effect of extrinsic lactic acid on fermentative hydrogen production. Int. J. Hydrogen Energy 34: 2573–2579. https://doi.org/10.1016/j.ijhydene.2009.01.010.

Bakonyi, P., Buitrón, G., Valdez-Vazquez, I., Nemestóthy, N. and Bélafi-Bakó, K. 2017. A novel gas separation integrated membrane bioreactor to evaluate the impact of self-generated biogas recycling on continuous hydrogen fermentation. Appl. Energy 190: 813–823. https://doi.org/10.1016/j.apenergy.2016.12.151.

Bastidas-Oyanedel, J.R., Mohd-Zaki, Z., Zeng, R.J., Bernet, N., Pratt, S., Steyer, J.P. and Batstone, D.J. 2012. Gas controlled hydrogen fermentation. Bioresour. Technol. 110: 503–509. https://doi.org/10.1016/j.biortech.2012.01.122.

Benomar, S., Ranava, D., Cárdenas, M.L., Trably, E., Rafrafi, Y., Ducret, A., Hamelin, J., Lojou, E., Steyer, J.P. and Giudici-Orticoni, M.T. 2015. Nutritional stress induces exchange of cell material and energetic coupling between bacterial species. Nat. Commun. 6. https://doi.org/10.1038/ncomms7283.

Bhat, J.V. and Barker, H.A. 1947. Clostridium lacto-acetophilum nov. spec. and the role of acetic acid in the butyric acid fermentation of lactate. J. Bacteriol. 54: 381–391.

Bisaillon, A., Turcot, J. and Hallenbeck, P.C. 2006. The effect of nutrient limitation on hydrogen production by batch cultures of Escherichia coli 31: 1504–1508. https://doi.org/10.1016/j.ijhydene.2006.06.016.

Blanco, V.M.C., Oliveira, G.H.D. and Zaiat, M. 2019. Dark fermentative biohydrogen production from synthetic cheese whey in an anaerobic structured-bed reactor: Performance evaluation and kinetic modeling. Renew. Energy 139: 1310–1319. https://doi.org/10.1016/j.renene.2019.03.029.

Bryant, M.P. and Burkey, L.A. 1956. The characteristics of lactate-fermenting sporeforming anaerobes from silage. J. Bacteriol. 71: 43–46.

Buitrón, G., Muñoz-Páez, K.M., Quijano, G., Carrillo-Reyes, J. and Albarrán-Contreras, B.A. 2020. Biohydrogen production from winery effluents: control of the homoacetogenesis through the headspace gas recirculation. J. Chem. Technol. Biotechnol. 95: 544–552. https://doi.org/10.1002/jctb.6263.

Bundhoo, M.A.Z., Mohee, R. and Hassan, M.A. 2015. Effects of pre-treatment technologies on dark fermentative biohydrogen production: A review. J. Environ. Manage. 157: 20–48. https://doi.org/10.1016/j.jenvman.2015.04.006.

Cabrol, L., Marone, A., Tapia-Venegas, E., Steyer, J.P., Ruiz-Filippi, G. and Trably, E. 2017. Microbial ecology of fermentative hydrogen producing bioprocesses: Useful insights for driving the ecosystem function. FEMS Microbiol. Rev. 41: 158–181. https://doi.org/10.1093/femsre/fuw043.

Cadoux, C. and Milton, R.D. 2020. Recent enzymatic electrochemistry for reductive reactions. ChemElectroChem 7: 1974–1986. https://doi.org/10.1002/celc.202000282.

Carosia, M.F., dos Reis, C.M., de Menezes, C.A., Sakamoto, I.K., Varesche, M.B.A. and Silva, E.L. 2021. Homoacetogenesis: New insights into controlling this unsolved challenge by selecting the optimal C/N ratio, C/P ratio and hydraulic retention time. Process Saf. Environ. Prot. 145: 273–284. https://doi.org/10.1016/j.psep.2020.08.009.

Carrillo-Reyes, J., Celis, L.B., Alatriste-Mondragón, F. and Razo-Flores, E. 2012. Different start-up strategies to enhance biohydrogen production from cheese whey in UASB reactors. Int. J. Hydrogen Energy 37: 5591–5601. https://doi.org/10.1016/j.ijhydene.2012.01.004.

Carrillo-Reyes, J., Celis, L.B., Alatriste-Mondragón, F., Montoya, L. and Razo-Flores, E. 2014. Strategies to cope with methanogens in hydrogen producing UASB reactors: Community dynamics. Int. J. Hydrogen Energy 39: 11423–11432.

Castelló, E., García y Santos, C., Iglesias, T., Paolino, G., Wenzel, J., Borzacconi, L. and Etchebehere, C. 2009. Feasibility of biohydrogen production from cheese whey using a UASB reactor: Links between microbial community and reactor performance. Int. J. Hydrogen Energy 34: 5674–5682. https://doi.org/10.1016/j.ijhydene.2009.05.060.

Castelló, E., Nunes Ferraz-Junior, A.D., Andreani, C., Anzola-Rojas, M. del P., Borzacconi, L., Buitrón, G., Carrillo-Reyes, J., Gomes, S.D., Maintinguer, S.I., Moreno-Andrade, I., Palomo-Briones, R., Razo-Flores, E., Schiappacasse-Dasati, M., Tapia-Venegas, E., Valdez-Vázquez, I., Vesga-Baron, A., Zaiat, M. and Etchebehere, C. 2020. Stability problems in the hydrogen production by dark fermentation: Possible causes and solutions. Renew. Sustain. Energy Rev. 119. https://doi.org/10.1016/j.rser.2019.109602.

Chandra, R. and Mohan, S.V. 2014. Enhanced bio-hydrogenesis by co-culturing photosynthetic bacteria with acidogenic process: Augmented dark-photo fermentative hybrid system to regulate volatile fatty acid inhibition. Int. J. Hydrogen Energy 39: 7604–7615. https://doi.org/10.1016/j.ijhydene.2014.01.196.

Chang, D.E., Jung, H.C., Rhee, J.S. and Pan, J.G. 1999. Homofermentative production of D- or L-lactate in metabolically engineered Escherichia coli RR1. Appl. Environ. Microbiol. 65: 1384–1389. https://doi.org/10.1128/aem.65.4.1384-1389.1999.

Chang, J.-J., Chou, C.-H., Ho, C.-Y., Chen, W.-E., Lay, J.-J. and Huang, C.-C. 2008a. Syntrophic co-culture of aerobic Bacillus and anaerobic Clostridium for bio-fuels and bio-hydrogen production. Int. J. Hydrogen Energy 33: 5137–5146. https://doi.org/10.1016/j.ijhydene.2008.05.021.

Chang, J.J., Wu, J.H., Wen, F.S., Hung, K.Y., Chen, Y.T., Hsiao, C.L., Lin, C.Y. and Huang, C.C. 2008b. Molecular monitoring of microbes in a continuous hydrogen-producing system with different hydraulic retention time. Int. J. Hydrogen Energy 33: 1579–1585. https://doi.org/10.1016/j.ijhydene.2007.09.045.

Chang, S., Li, J.Z. and Liu, F. 2011. Evaluation of different pretreatment methods for preparing hydrogen-producing seed inocula from waste activated sludge. Renew. Energy 36: 1517–1522. https://doi.org/10.1016/j.renene.2010.11.023.

Chatellard, L., Trably, E. and Carrère, H. 2016. The type of carbohydrates specifically selects microbial community structures and fermentation patterns. Bioresour. Technol. 221: 541–549. https://doi.org/10.1016/j.biortech.2016.09.084.

Chen, S. and Dong, X. 2004. Acetanaerobacterium elongatum gen. nov., sp. nov., from paper mill waste water. Int. J. Syst. Evol. Microbiol. 54: 2257–2262. https://doi.org/10.1099/ijs.0.63212-0.

Chen, Y., Yin, Y. and Wang, J. 2021. Influence of butyrate on fermentative hydrogen production and microbial community analysis. Int. J. Hydrogen Energy. https://doi.org/10.1016/j.ijhydene.2021.05.185.

Chou, C., Shockley, K.R., Conners, S.B., Lewis, D.L., Comfort, D.A., Adams, M.W.W. and Kelly, R.M. 2007. Impact of substrate glycoside linkage and elemental sulfur on bioenergetics of and hydrogen production by the hyperthermophilic Archaeon Pyrococcus furiosus[†]. Appl. Environ. Microbiol. 73: 6842–6853. https://doi.org/10.1128/AEM.00597-07.

Cieciura-Włoch, W., Borowski, S. and Otlewska, A. 2020. Biohydrogen production from fruit and vegetable waste, sugar beet pulp and corn silage via dark fermentation. Renew. Energy 153: 1226–1237. https://doi.org/10.1016/j.renene.2020.02.085.

Costa, K.C., Yoon, S.H., Pan, M., Burn, J.A., Baliga, N.S. and Leigh, J.A. 2013. Effects of H2 and formate on growth yield and regulation of methanogenesis in Methanococcus maripaludis. J. Bacteriol. 195: 1456–1462. https://doi.org/10.1128/JB.02141-12.

Dauptain, K., Schneider, A., Noguer, M., Fontanille, P., Escudie, R., Carrere, H. and Trably, E. 2021. Impact of microbial inoculum storage on dark fermentative H2 production. Bioresour. Technol. 319: 124234. https://doi.org/10.1016/j.biortech.2020.124234

Dauptain, K., Trably, E., Santa-Catalina, G., Bernet, N. and Carrere, H. 2020. Role of indigenous bacteria in dark fermentation of organic substrates. Bioresour. Technol. 313: 123665. https://doi.org/10.1016/j.biortech.2020.123665.

de Amorim, E.L.C., Barros, A.R., Rissato Zamariolli Damianovic, M.H. and Silva, E.L. 2009. Anaerobic fluidized bed reactor with expanded clay as support for hydrogen production through dark fermentation of glucose. Int. J. Hydrogen Energy 34: 783–790. https://doi.org/10.1016/j.ijhydene.2008.11.007.

De Gioannis, G., Friargiu, M., Massi, E., Muntoni, A., Polettini, A., Pomi, R. and Spiga, D. 2014. Biohydrogen production from dark fermentation of cheese whey: Influence of pH. Int. J. Hydrogen Energy 39: 20930–20941. https://doi.org/10.1016/j.ijhydene.2014.10.046.

Dessì, P., Porca, E., Frunzo, L., Lakaniemi, A.M., Collins, G., Esposito, G. and Lens, P.N.L. 2018a. Inoculum pretreatment differentially affects the active microbial community performing mesophilic and thermophilic dark fermentation of xylose. Int. J. Hydrogen Energy 43: 9233–9245. https://doi.org/10.1016/j.ijhydene.2018.03.117.

Dessì, P., Porca, E., Waters, N.R., Lakaniemi, A.M., Collins, G. and Lens, P.N.L. 2018b. Thermophilic versus mesophilic dark fermentation in xylose-fed fluidised bed reactors: Biohydrogen production and active microbial community. Int. J. Hydrogen Energy 43: 5473–5485. https://doi.org/10.1016/j.ijhydene.2018.01.158.

Detman, A., Laubitz, D., Chojnacka, A., Kiela, P.R., Salamon, A., Barberán, A., Chen, Y., Yang, F., Błaszczyk, M.K. and Sikora, A. 2021a. Dynamics of dark fermentation microbial communities in the light of lactate and butyrate production. Microbiome 9: 1–21. https://doi.org/10.1186/s40168-021-01105-x.

Detman, A., Laubitz, D., Chojnacka, A., Wiktorowska-Sowa, E., Piotrowski, J., Salamon, A., Kaźmierczak, W., Błaszczyk, M.K., Barberan, A., Chen, Y., Łupikasza, E., Yang, F. and Sikora, A. 2021b. Dynamics and complexity of dark fermentation microbial communities producing hydrogen from sugar beet molasses in continuously operating packed bed reactors. Front. Microbiol. 11: 1–19. https://doi.org/10.3389/fmicb.2020.612344.

Detman, A., Mielecki, D., Chojnacka, A., Salamon, A., Błaszczyk, M.K. and Sikora, A. 2019. Cell factories converting lactate and acetate to butyrate: Clostridium butyricum and microbial communities from dark fermentation bioreactors. Microb. Cell Fact. 18: 1–12. https://doi.org/10.1186/s12934-019-1085-1.

Dey, S., Das, P.K. and Dey, A. 2013. Mononuclear iron hydrogenase. Coord. Chem. Rev. 257: 42–63. https://doi.org/10.1016/j.ccr.2012.04.021.

Diez-Gonzalez, F., Russell, J.B. and Hunter, J.B. 1995. The role of an NAD-independent lactate dehydrogenase and acetate in the utilization of lactate by Clostridium acetobutylicum strain P262. Arch. Microbiol. 164: 36–42.

Dinamarca, C., Gañán, M., Liu, J. and Bakke, R. 2011. H2 consumption by anaerobic non-methanogenic mixed cultures. Water Sci. Technol. 63: 1582. https://doi.org/10.2166/wst.2011.214.

Dincer, I. and Acar, C. 2015. Science direct review and evaluation of hydrogen production methods for better sustainability. Int. J. Hydrogen Energy 40: 11094–11111. https://doi.org/10.1016/j.ijhydene.2014.12.035.

Drake, H.L. 1991. Acetogenesis, Acetogenic Bacteria, and the Acetyl-CoA "Wood/Ljungdahl" Pathway: Past and Current Perspectives.

Du, Y., Zou, W., Zhang, K., Ye, G. and Yang, J. 2020. Advances and applications of clostridium co-culture systems in biotechnology. Front. Microbiol. 11: 1–22. https://doi.org/10.3389/fmicb.2020.560223.

Eder, A.S., Magrini, F.E., Spengler, A., da Silva, J.T., Beal, L.L. and Paesi, S. 2020. Comparison of hydrogen and volatile fatty acid production by Bacillus cereus, Enterococcus faecalis and Enterobacter aerogenes singly, in co-cultures or in the bioaugmentation of microbial consortium from sugarcane vinasse. Environ. Technol. Innov. 18: 100638. https://doi.org/10.1016/j.eti.2020.100638.

Elbeshbishy, E., Dhar, B.R., Nakhla, G. and Lee, H.S. 2017. A critical review on inhibition of dark biohydrogen fermentation. Renew. Sustain. Energy Rev. 79: 656–668. https://doi.org/10.1016/j.rser.2017.05.075.

Esquivel-Elizondo, S., Chairez, I., Salgado, E., Aranda, J.S., Baquerizo, G. and Garcia-Peña, E.I. 2014. Controlled continuous bio-hydrogen production using different biogas release strategies. Appl. Biochem. Biotechnol. 173: 1737–1751. https://doi.org/10.1007/s12010-014-0961-8.

Etchebehere, C., Castelló, E., Wenzel, J., del Pilar Anzola-Rojas, M., Borzacconi, L., Buitrón, G., Cabrol, L., Carminato, V.M., Carrillo-Reyes, J., Cisneros-Pérez, C., Fuentes, L., Moreno-Andrade, I., Razo-Flores, E., Filippi, G.R., Tapia-Venegas, E., Toledo-Alarcón, J. and Zaiat, M. 2016. Microbial communities from 20 different hydrogen-producing reactors studied by 454 pyrosequencing. Appl. Microbiol. Biotechnol. 100: 3371–3384. https://doi.org/10.1007/s00253-016-7325-y.

Fang, H.H.P. and Liu, H. 2002. Effect of pH on hydrogen production from glucose by a mixed culture. Bioresour. Technol. 82: 87–93.

Favaro, L., Alibardi, L., Lavagnolo, M.C., Casella, S. and Basaglia, M. 2013. Effects of inoculum and indigenous microflora on hydrogen production from the organic fraction of municipal solid waste. Int. J. Hydrogen Energy 38: 11774–11779. https://doi.org/10.1016/j.ijhydene.2013.06.137.

Ferguson, T.J. and Mah, R.A. 1983. Effect of H2-CO2 on methanogenesis from acetate or methanol in methanosarcina spp. Appl. Environ. Microbiol. 46: 348–355. https://doi.org/10.1128/aem.46.2.348-355.1983.

Ferraz Júnior, A.D.N., Etchebehere, C. and Zaiat, M. 2015. High organic loading rate on thermophilic hydrogen production and metagenomic study at an anaerobic packed-bed reactor treating a residual liquid stream of a Brazilian biorefinery. Bioresour. Technol. 186: 81–88. https://doi.org/10.1016/j.biortech.2015.03.035.

Ferraz, N., Latrille, E., Bernet, N., Zaiat, M. and Trably, E. 2020. Biogas sequestration from the headspace of a fermentative system enhances hydrogen production rate and yield 5. https://doi.org/10.1016/j.ijhydene.2020.02.064.

Ferreira, T.B., Rego, G.C., Ramos, L.R., Soares, L.A., Sakamoto, I.K., de Oliveira, L.L., Varesche, M.B.A. and Silva, E.L. 2018. Selection of metabolic pathways for continuous hydrogen production under thermophilic and mesophilic temperature conditions in anaerobic fluidized bed reactors. Int. J. Hydrogen Energy 43: 18908–18917. https://doi.org/10.1016/j.ijhydene.2018.08.177.

Fiala, G. and Stetter, K.O. 1986. Pyrococcus furiosus sp. nov. represents a novel genus of marine heterotrophic archaebacteria growing optimally at 100°C. Arch. Microbiol. 145: 56–61.

Flores, M.G., Nakhla, G. and Hafez, H. 2017. Hydrogen production and microbial kinetics of Clostridium termitidis in mono-culture and co-culture with Clostridium beijerinckii on cellulose. AMB Express. https://doi.org/10.1186/s13568-016-0256-2.

Fontecilla-camps, J.C., Volbeda, A., Cavazza, C., Nicolet, Y. and Fourier, J. 2007. Structure/Function Relationships of [NiFe]- and [FeFe]-Hydrogenases. Chem. Rev. 107: 4273–4303.

François, E., Dumas, C., Gougeon, R.D., Alexandre, H., Vuilleumier, S. and Ernst, B. 2021. Unexpected high production of biohydrogen from the endogenous fermentation of grape must deposits. Bioresour. Technol. 320. https://doi.org/10.1016/j.biortech.2020.124334.

Fu, Q., Wang, D., Li, X., Yang, Q., Xu, Q., Ni, B.J., Wang, Q. and Liu, X. 2021. Towards hydrogen production from waste activated sludge: Principles, challenges and perspectives. Renew. Sustain. Energy Rev. 135: 110283. https://doi.org/10.1016/j.rser.2020.110283.

Gao, S., Fan, W., Liu, Y., Jiang, D. and Duan, Q. 2020. Artificial water-soluble systems inspired by [FeFe]-hydrogenases for electro- and photocatalytic hydrogen production. Int. J. Hydrogen Energy 45: 4305–4327. https://doi.org/10.1016/j.ijhydene.2019.11.206.

García-Depraect, O., Diaz-Cruces, V.F., Rene, E.R., Castro-Muñoz, R. and León-Becerril, E. 2020. Long-term preservation of hydrogenogenic biomass by refrigeration: Reactivation characteristics and microbial community structure. Bioresour. Technol. Reports 12: 1–5. https://doi.org/10.1016/j.biteb.2020.100587.

García-Depraect, O., Muñoz, R., Rodríguez, E., Rene, E.R. and León-Becerril, E. 2021. Microbial ecology of a lactate-driven dark fermentation process producing hydrogen under carbohydrate-limiting conditions. Int. J. Hydrogen Energy 46: 11284–11296. https://doi.org/10.1016/j.ijhydene.2020.08.209.

García-Depraect, O., Valdez-Vázquez, I., Rene, E.R., Gómez-Romero, J., López-López, A. and León-Becerril, E. 2019. Lactate- and acetate-based biohydrogen production through dark co-fermentation of tequila vinasse and nixtamalization wastewater: Metabolic and microbial community dynamics. Bioresour. Technol. 282: 236–244. https://doi.org/10.1016/j.biortech.2019.02.100.

Ghimire, A., Frunzo, L., Pirozzi, F., Trably, E., Escudie, R., Lens, P.N.L. and Esposito, G. 2015. A review on dark fermentative biohydrogen production from organic biomass: Process parameters and use of by-products. Appl. Energy 144: 73–95. https://doi.org/10.1016/j.apenergy.2015.01.045.

Goud, R.K., Sarkar, O., Chiranjeevi, P. and Venkata Mohan, S. 2014. Bioaugmentation of potent acidogenic isolates: A strategy for enhancing biohydrogen production at elevated organic load. Bioresour. Technol. 165: 223–232. https://doi.org/10.1016/j.biortech.2014.03.049.

Goyal, Y., Kumar, M. and Gayen, K. 2013. Metabolic engineering for enhanced hydrogen production: A review. Can. J. Microbiol. 59: 59–78. https://doi.org/10.1139/cjm-2012-0494.

Greses, S., Tomás-Pejó, E. and González-Fernández, C. 2021. Short-chain fatty acids and hydrogen production in one single anaerobic fermentation stage using carbohydrate-rich food waste. J. Clean. Prod. 284. https://doi.org/10.1016/j.jclepro.2020.124727.

Guangsheng, C., Plugge, C.M., Roelofsen, W., Houwen, F.P. and Stams, A.J.M. 1992. Selenomonas acidaminovorans sp. nov., a versatile thermophilic proton-reducing anaerobe able to grow by decarboxylation of succinate to propionate. Arch. Microbiol. 157: 169–175. https://doi.org/10.1007/BF00245286.

Guo, W., Ren, N., Wang, X. and Xiang, W. 2010a. Accelerated startup of biological hydrogen production process by addition of Ethanoligenens harbinense B49 in a biofilm-based column reactor. Int. J. Hydrogen Energy 35: 13407–13412. https://doi.org/10.1016/j.ijhydene.2009.11.115.

Guo, X.M., Trably, E., Latrille, E., Carrre, H., Steyer, J.-P.P., Carrère, H., Steyer, J.-P.P., Carrre, H. and Steyer, J.-P.P. 2010b. Hydrogen production from agricultural waste by dark fermentation: A review. Int. J. Hydrogen Energy 35: 10660–10673. https://doi.org/10.1016/j.ijhydene.2010.03.008.

Guo, Y.C., Dai, Y., Bai, Y.X., Li, Y.H., Fan, Y.T. and Hou, H.W. 2014. Co-producing hydrogen and methane from higher-concentration of corn stalk by combining hydrogen fermentation and anaerobic digestion. Int. J. Hydrogen Energy 39: 14204–14211. https://doi.org/10.1016/j.ijhydene.2014.02.089.

Hafez, H., Nakhla, G., El. Naggar, M.H., Elbeshbishy, E. and Baghchehsaraee, B. 2010. Effect of organic loading on a novel hydrogen bioreactor. Int. J. Hydrogen Energy 35: 81–92. https://doi.org/10.1016/j.ijhydene.2009.10.051.

Hallenbeck, P.C. 2009. Fermentative hydrogen production: Principles, progress, and prognosis. Int. J. Hydrogen Energy 34: 7379–7389. https://doi.org/10.1016/j.ijhydene.2008.12.080.

Hallenbeck, P.C., Abo-Hashesh, M. and Ghosh, D. 2012. Strategies for improving biological hydrogen production. Bioresour. Technol. 110: 1–9. https://doi.org/10.1016/j.biortech.2012.01.103.

Hallenbeck, P.C. and Benemann, J.R. 2002. Biological hydrogen production; Fundamentals and limiting processes. Int. J. Hydrogen Energy 27: 1185–1193. https://doi.org/10.1016/S0360-3199(02)00131-3.

Hallenbeck, P.C. and Ghosh, D. 2012. Improvements in fermentative biological hydrogen production through metabolic engineering. J. Environ. Manage. 95: S360–S364. https://doi.org/10.1016/j.jenvman.2010.07.021.

Harper, S.R. and Pohland, F.G. 1986. Management during anaerobic biological wastewater treatment. Biotechnol. Bioeng. 28: 585–602.

Hawkes, F.R., Hussy, I., Kyazze, G., Dinsdale, R. and Hawkes, D.L. 2007. Continuous dark fermentative hydrogen production by mesophilic microflora: Principles and progress. Int. J. Hydrogen Energy 32: 172–184. https://doi.org/10.1016/j.ijhydene.2006.08.014.

Heinekey, D.M. 2009. Hydrogenase enzymes: Recent structural studies and active site models. J. Organomet. Chem. 694: 2671–2680. https://doi.org/10.1016/j.jorganchem.2009.03.047.

Herrero, A.A., Gomez, R.F., Snedecor, B., Tolman, C.J. and Roberts, M.F. 1985. Growth inhibition of Clostridium thermocellum by carboxylic acids: A mechanism based on uncoupling by weak acids, 53–62.

Huang, Y., Zong, W., Yan, X., Wang, R., Hemme, C.L., Zhou, J. and Zhou, Z. 2010. Succession of the bacterial community and dynamics of hydrogen producers in a hydrogen-producing bioreactor[†]. Appl. Environ. Microbiol. 76: 3387–3390. https://doi.org/10.1128/AEM.02444-09.

Hung, C.H., Chang, Y.T. and Chang, Y.J. 2011. Roles of microorganisms other than Clostridium and Enterobacter in anaerobic fermentative biohydrogen production systems—A review. Bioresour. Technol. 102: 8437–8444. https://doi.org/10.1016/j.biortech.2011.02.084.

Ica, B., Ucts, P., Pro, N.D., Engine, C., Yoshida, A., Nishimura, T., Kawaguchi, H., Inui, M. and Yukawa, H. 2006. Enhanced hydrogen production from glucose using ldh- and frd-inactivated Escherichia coli strains, 67–72. https://doi.org/10.1007/s00253-006-0456-9.

Jo, J.H., Jeon, C.O., Lee, D.S. and Park, J.M. 2007. Process stability and microbial community structure in anaerobic hydrogen-producing microflora from food waste containing kimchi. J. Biotechnol. 131: 300–308. https://doi.org/10.1016/j.jbiotec.2007.07.492.

Kandler, O., Schillinger, U. and Weiss, N. 1983. Lactobacillus bifermentans sp. nov., nom. rev., an organism forming CO_2 and H_2 from lactic acid. Syst. Appl. Microbiol. 4: 408–412. https://doi.org/10.1016/S0723-2020(83)80025-3.

Kapdan, I.K. and Kargi, F. 2006. Bio-hydrogen production from waste materials. Enzyme Microb. Technol. 38: 569–582. https://doi.org/10.1016/j.enzmictec.2005.09.015.

Kawagoshi, Y., Hino, N., Fujimoto, A., Nakao, M., Fujita, Y., Sugimura, S. and Furukawa, K. 2005. Effect of inoculum conditioning on hydrogen fermentation and pH effect on bacterial community relevant to hydrogen production. J. Biosci. Bioeng. 100: 524–530. https://doi.org/10.1263/jbb.100.524.

Kim, D.H. and Kim, M.S. 2011. Hydrogenases for biological hydrogen production. Bioresour. Technol. 102: 8423–8431. https://doi.org/10.1016/j.biortech.2011.02.113.

Kim, D.H., Jang, S., Yun, Y.M., Lee, M.K., Moon, C., Kang, W.S., Kwak, S.S. and Kim, M.S. 2014. Effect of acid-pretreatment on hydrogen fermentation of food waste: Microbial community analysis by next generation sequencing. Int. J. Hydrogen Energy 39: 16302–16309. https://doi.org/10.1016/j.ijhydene.2014.08.004.

Kim, D.H., Kim, S.H., Ko, I.B., Lee, C.Y. and Shin, H.S. 2008. Start-up strategy for continuous fermentative hydrogen production: Early switchover from batch to continuous operation. Int. J. Hydrogen Energy 33: 1532–1541. https://doi.org/10.1016/j.ijhydene.2008.01.012.

Kim, D.H., Wu, J., Jeong, K.W., Kim, M.S. and Shin, H.S. 2011. Natural inducement of hydrogen from food waste by temperature control. Int. J. Hydrogen Energy 36: 10666–10673. https://doi.org/10.1016/j.ijhydene.2011.05.153.

Klein, M., Ansorge-schumacher, M.B., Fritsch, M. and Hartmeier, W. 2010. Enzyme and microbial technology influence of hydrogenase overexpression on hydrogen production of Clostridium acetobutylicum DSM 792. Enzyme Microb. Technol. 46: 384–390. https://doi.org/10.1016/j.enzmictec.2009.12.015.

Koo, J., Shiigi, S., Rohovie, M., Mehta, K. and Swartz, J.R. 2016. Characterization of [FeFe] hydrogenase O2 sensitivity using a new, physiological approach. J. Biol. Chem. 291: 21563–21570. https://doi.org/10.1074/jbc.M116.737122.

Kotay, S.M. and Das, D. 2007. Microbial hydrogen production with Bacillus coagulans IIT-BT S1 isolated from anaerobic sewage sludge. Bioresour. Technol. 98: 1183–1190. https://doi.org/10.1016/j.biortech.2006.05.009.

Kovanda, L. 2019. In-vitro-antimicrobial-activities-of-organic-acids-and-their-derivatives-on-several-species-of-Gramnegative-and-Grampositive-bacteria2019MoleculesOpen-Access.pdf.

Kraemer, J.T. and Bagley, D.M. 2008. Measurement of H2 consumption and its role in continuous fermentative hydrogen production. Water Sci. Technol. 57: 681–685. https://doi.org/10.2166/wst.2008.066.

Kumar, G., Bakonyi, P., Kobayashi, T., Xu, K.Q., Sivagurunathan, P., Kim, S.H., Buitrón, G., Nemestóthy, N. and Bélafi-Bakó, K. 2016. Enhancement of biofuel production via microbial augmentation: The case of dark fermentative hydrogen. Renew. Sustain. Energy Rev. 57: 879–891. https://doi.org/10.1016/j.rser.2015.12.107.

Kumar, G., Bakonyi, P., Sivagurunathan, P., Kim, S.H., Nemestóthy, N., Bélafi-Bakó, K. and Lin, C.Y. 2015. Enhanced biohydrogen production from beverage industrial wastewater using external nitrogen sources and bioaugmentation with facultative anaerobic strains. J. Biosci. Bioeng. 120: 155–160. https://doi.org/10.1016/j.jbiosc.2014.12.011.

Kuo, W., Chao, Y., Wang, Y. and Cheng, S. 2012. Bioaugmentation strategies to improve cellulolytic and hydrogen producing characteristics in CSTR intermittent fed with vegetable kitchen waste and napiergrass 29: 82–91. https://doi.org/10.1016/j.egypro.2012.09.011.

Kyazze, G., Martinez-Perez, N., Dinsdale, R., Premier, G.C., Hawkes, F.R., Guwy, A.J. and Hawkes, D.L. 2006. Influence of substrate concentration on the stability and yield of continuous biohydrogen production. Biotechnol. Bioeng. 93: 971–979. https://doi.org/10.1002/bit.20802.

Lanzillo, F., Ruggiero, G., Raganati, F., Russo, M.E. and Marzocchella, A. 2020. Batch syngas fermentation by clostridium carboxidivorans for production of acids and alcohols. Processes 8: 1–13. https://doi.org/10.3390/pr8091075.

Laurinavichene, T., Laurinavichius, K., Shastik, E. and Tsygankov, A. 2018. Long-term H2 photoproduction from starch by co-culture of Clostridium butyricum and Rhodobacter sphaeroides in a repeated batch process. Biotechnol. Lett. 40: 309–314. https://doi.org/10.1007/s10529-017-2486-z.

Lazaro, C.Z., Perna, V., Etchebehere, C. and Varesche, M.B.A. 2014. Sugarcane vinasse as substrate for fermentative hydrogen production: The effects of temperature and substrate concentration. Int. J. Hydrogen Energy 39: 6407–6418. https://doi.org/10.1016/j.ijhydene.2014.02.058.

Lee, Y.J., Miyahara, T. and Noike, T. 2002. Effect of pH on microbial hydrogen fermentation. J. Chem. Technol. Biotechnol. 77: 694–698. https://doi.org/10.1002/jctb.623.

Li, C. and Fang, H.H.P. 2007. Fermentative hydrogen production from wastewater and solid wastes by mixed cultures. Crit. Rev. Environ. Sci. Technol. 37: 1–39. https://doi.org/10.1080/10643380600729071.

Li, S.L., Lin, J.S., Wang, Y.H., Lee, Z.K., Kuo, S.C., Tseng, I.C. and Cheng, S.S. 2011. Strategy of controlling the volumetric loading rate to promote hydrogen-production performance in a mesophilic-kitchen-waste fermentor and the microbial ecology analyses. Bioresour. Technol. 102: 8682–8687. https://doi.org/10.1016/j.biortech.2011.02.067.

Li, Y.C., Nissilä, M.E., Wu, S.Y., Lin, C.Y. and Puhakka, J.A. 2012. Silage as source of bacteria and electrons for dark fermentative hydrogen production. Int. J. Hydrogen Energy 37: 15518–15524. https://doi.org/10.1016/j. ijhydene.2012.04.060.

Li, Z., Liu, B., Cui, H., Ding, J., Li, H., Xie, G., Ren, N. and Xing, D. 2019. The complete genome sequence of Ethanoligenens harbinense reveals the metabolic pathway of acetate-ethanol fermentation: A novel understanding of the principles of anaerobic biotechnology. Environ. Int. 131: 105053. https://doi.org/10.1016/j. envint.2019.105053.

Linger, J.G., Ford, L.R., Ramnath, K. and Guarnieri, M.T. 2020. Development of Clostridium tyrobutyricum as a microbial cell factory for the production of fuel and chemical intermediates from lignocellulosic feedstocks. Front. Energy Res. 8: 1–15. https://doi.org/10.3389/fenrg.2020.00183.

Liu, H. and Wang, G. 2012. Hydrogen production of a salt tolerant strain Bacillus sp. B2 from marine intertidal sludge 31–37. https://doi.org/10.1007/s11274-011-0789-0.

Lu, Y., Slater, F.R., Mohd-Zaki, Z., Pratt, S. and Batstone, D.J. 2011. Impact of operating history on mixed culture fermentation microbial ecology and product mixture. Water Sci. Technol. 64: 760–765. https://doi.org/10.2166/ wst.2011.699.

Łukajtis, R., Hołowacz, I., Kucharska, K., Glinka, M., Rybarczyk, P., Przyjazny, A. and Kamiński, M. 2018. Hydrogen production from biomass using dark fermentation. Renew. Sustain. Energy Rev. 91: 665–694. https:// doi.org/10.1016/j.rser.2018.04.043.

Luo, G., Karakashev, D., Xie, L., Zhou, Q. and Angelidaki, I. 2011. Long-term effect of inoculum pretreatment on fermentative hydrogen production by repeated batch cultivations: Homoacetogenesis and methanogenesis as competitors to hydrogen production. Biotechnol. Bioeng. 108: 1816–1827. https://doi.org/10.1002/bit.23122.

Luo, Y., Zhang, H., Salerno, M., Logan, B.E. and Bruns, M.A. 2008. Organic loading rates affect composition of soil-derived bacterial communities during continuous, fermentative biohydrogen production. Int. J. Hydrogen Energy 33: 6566–6576. https://doi.org/10.1016/j.ijhydene.2008.08.047.

Macías-muro, M., Arellano-garcía, L., Vel, J.B. and Marino-marmolejo, E.N. 2021. Continuous hydrogen production and microbial community profile in the dark fermentation of tequila vinasse: Response to increasing loading rates and immobilization of biomass 172. https://doi.org/10.1016/j.bej.2021.108049.

Madigan, M.T., Martinko, J.M., Bender, K.S., Buckley, D.H. and Stahl, D.A. 2015. Brock Biology of Microorganisms, 14th ed., Pearson Education. Pearson Education. https://doi.org/10.1017/cbo9780511549984.016.

Maeda, T., Sanchez-torres, V. and Wood, T.K. 2008. Metabolic engineering to enhance bacterial hydrogen production. Microbial Biotechnology 1: 30–39. https://doi.org/10.1111/j.1751-7915.2007.00003.x.

Majidian, P., Tabatabaei, M. and Zeinolabedini, M. 2018. Metabolic engineering of microorganisms for biofuel production. Renew. Sustain. Energy Rev. 82: 3863–3885. https://doi.org/10.1016/j.rser.2017.10.085.

Mallette, M.F., Reece, P. and Dawes, E.A. 1974. Culture of Clostridium pasteurianum in defined medium and growth as a function of sulfate concentration. Appl. Microbiol. 28: 999–1003. https://doi.org/10.1128/am.28.6.999-1003.1974.

Mariakakis, I., Bischoff, P., Krampe, J., Meyer, C. and Steinmetz, H. 2011. Effect of organic loading rate and solids retention time on microbial population during bio-hydrogen production by dark fermentation in large lab-scale. Int. J. Hydrogen Energy 36: 10690–10700. https://doi.org/10.1016/j.ijhydene.2011.06.008.

Mathews, J. and Wang, G. 2009. Metabolic pathway engineering for enhanced biohydrogen production. Int. J. Hydrogen Energy 34: 7404–7416. https://doi.org/10.1016/j.ijhydene.2009.05.078.

Matsumoto, M. and Nishimura, Y. 2007. Hydrogen production by fermentation using acetic acid and lactic acid. J. Biosci. Bioeng. 103: 236–241. https://doi.org/10.1263/jbb.103.236.

Monlau, F., Sambusiti, C., Barakat, A., Guo, X.M., Latrille, E. and Trably, E. 2012. Predictive models of biohydrogen and biomethane production based on the compositional and structural features of lignocellulosic materials. Enviromental Sci. Technol. 46(21): 12217–12225.

Monlau, F., Sambusiti, C., Barakat, A., Quéméneur, M., Trably, E., Steyer, J.-P. and Carrère, H. 2014. Do furanic and phenolic compounds of lignocellulosic and algae biomass hydrolyzate inhibit anaerobic mixed cultures? A comprehensive review. Biotechnol. Adv. 32: 934–51. https://doi.org/10.1016/j.biotechadv.2014.04.007.

Montiel Corona, V. and Razo-Flores, E. 2018. Continuous hydrogen and methane production from Agave tequilana bagasse hydrolysate by sequential process to maximize energy recovery efficiency. Bioresour. Technol. 249: 334–341. https://doi.org/10.1016/j.biortech.2017.10.032.

Moreno-Andrade, I., Carrillo-Reyes, J., Santiago, S.G. and Bujanos-Adame, M.C. 2015. Biohydrogen from food waste in a discontinuous process: Effect of HRT and microbial community analysis. Int. J. Hydrogen Energy 40: 17246–17252. https://doi.org/10.1016/j.ijhydene.2015.04.084.

Mota, V.T., Ferraz Júnior, A.D.N., Trably, E. and Zaiat, M. 2018. Biohydrogen production at pH below 3.0: Is it possible? Water Res. 128: 350–361. https://doi.org/10.1016/j.watres.2017.10.060.

Mugnai, G., Borruso, L., Mimmo, T., Cesco, S., Luongo, V., Frunzo, L., Fabbricino, M., Pirozzi, F., Cappitelli, F. and Villa, F. 2021. Dynamics of bacterial communities and substrate conversion during olive-mill waste dark fermentation: Prediction of the metabolic routes for hydrogen production. Bioresour. Technol. 319: 124157. https://doi.org/10.1016/j.biortech.2020.124157.

Nasr, N., Velayutham, P., Elbeshbishy, E., Nakhla, G., El Naggar, M.H., Khafipour, E., Derakhshani, H., Levin, D.B. and Hafez, H. 2015. Effect of headspace carbon dioxide sequestration on microbial biohydrogen communities. Int. J. Hydrogen Energy 40: 9966–9976. https://doi.org/10.1016/j.ijhydene.2015.06.077.

Nikolaidis, P. and Poullikkas, A. 2017. A comparative overview of hydrogen production processes. Renew. Sustain. Energy Rev. 67: 597–611. https://doi.org/10.1016/j.rser.2016.09.044.

Noblecourt, A., Christophe, G., Larroche, C. and Fontanille, P. 2018. Hydrogen production by dark fermentation from pre-fermented depackaging food wastes. Bioresour. Technol. 247: 864–870. https://doi.org/10.1016/j.biortech.2017.09.199.

Ohnishi, A., Hasegawa, Y., Fujimoto, N. and Suzuki, M. 2022. Biohydrogen production by mixed culture of Megasphaera elsdenii with lactic acid bacteria as Lactate-driven dark fermentation. Bioresour. Technol. 343: 126076. https://doi.org/10.1016/j.biortech.2021.126076.

Okonkwo, O., Escudie, R., Bernet, N., Mangayil, R., Lakaniemi, A.M. and Trably, E. 2020. Bioaugmentation enhances dark fermentative hydrogen production in cultures exposed to short-term temperature fluctuations. Appl. Microbiol. Biotechnol. 104: 439–449. https://doi.org/10.1007/s00253-019-10203-8.

Paillet, F., Barrau, C., Escudié, R. and Trably, E. 2020. Inhibition by the ionic strength of hydrogen production from the organic fraction of municipal solid waste. Int. J. Hydrogen Energy 45: 5854–5863. https://doi.org/10.1016/j.ijhydene.2019.08.019.

Palomo-Briones, R., Razo-Flores, E., Bernet, N. and Trably, E. 2017. Dark-fermentative biohydrogen pathways and microbial networks in continuous stirred tank reactors: Novel insights on their control. Appl. Energy 198: 77–87. https://doi.org/10.1016/j.apenergy.2017.04.051.

Parthiba Karthikeyan, O., Trably, E., Mehariya, S., Bernet, N., Wong, J.W.C. and Carrere, H. 2018. Pretreatment of food waste for methane and hydrogen recovery: A review. Bioresour. Technol. 249: 1025–1039. https://doi.org/10.1016/j.biortech.2017.09.105.

Patel, S.K.S., Kumar, P., Mehariya, S., Purohit, H.J., Lee, J. and Kalia, V.C. 2014. Enhancement in hydrogen production by co-cultures of Bacillus and Enterobacter. Int. J. Hydrogen Energy 39: 14663–14668. https://doi.org/10.1016/j.ijhydene.2014.07.084.

Pawar, S.S., Van Niel, E.W.J. and Niel, E.W.J. Van, 2013. Thermophilic biohydrogen production: How far are we? Appl. Microbiol. Biotechnol. 97: 7999–8009. https://doi.org/10.1007/s00253-013-5141-1.

Pecorini, I., Baldi, F. and Iannelli, R. 2019. Biochemical hydrogen potential tests using different inocula. Sustain. 11: 1–17. https://doi.org/10.3390/su11030622.

Perera, K.R.J., Ketheesan, B., Gadhamshetty, V. and Nirmalakhandan, N. 2010. Fermentative biohydrogen production: Evaluation of net energy gain. Int. J. Hydrogen Energy 35: 12224–12233. https://doi.org/10.1016/j.ijhydene.2010.08.037.

Poirier, S., Steyer, J.P., Bernet, N. and Trably, E. 2020. Mitigating the variability of hydrogen production in mixed culture through bioaugmentation with exogenous pure strains. Int. J. Hydrogen Energy 45: 2617–2626. https://doi.org/10.1016/j.ijhydene.2019.11.116.

Prabhu, R., Altman, E. and Eitemana, M.A. 2012. Lactate and acrylate metabolism by Megasphaera elsdenii under batch and steady-state conditions. Appl. Environ. Microbiol. 78: 8564–8570. https://doi.org/10.1128/AEM.02443-12.

Pu, Y., Tang, J., Wang, X.C., Hu, Y., Huang, J., Zeng, Y., Ngo, H.H. and Li, Y. 2019. Hydrogen production from acidogenic food waste fermentation using untreated inoculum: Effect of substrate concentrations. Int. J. Hydrogen Energy 44: 27272–27284. https://doi.org/10.1016/j.ijhydene.2019.08.230.

Rafieenia, R., Lavagnolo, M.C. and Pivato, A. 2018. Pre-treatment technologies for dark fermentative hydrogen production: Current advances and future directions. Waste Manag. 71: 734–748. https://doi.org/10.1016/j.wasman.2017.05.024.

Rafrafi, Y., Trably, E., Hamelin, J., Latrille, E., Meynial-Salles, I., Benomar, S., Giudici-Orticoni, M.T. and Steyer, J.P. 2013. Author's personal copy sub-dominant bacteria as keystone species in microbial communities producing bio-hydrogen. Int. J. Hydrogen Energy 38: 4975–4985. https://doi.org/10.1016/j.ijhydene.2013.02.008.

Ramírez-Morales, J.E., Tapia-Venegas, E., Toledo-Alarcón, J. and Ruiz-Filippi, G. 2015. Simultaneous production and separation of biohydrogen in mixed culture systems by continuous dark fermentation. Water Sci. Technol. 71: 1271–1285. https://doi.org/10.2166/wst.2015.104.

Ren, N., Xing, D., Rittmann, B.E., Zhao, L., Xie, T. and Zhao, X. 2007. Microbial community structure of ethanol type fermentation in bio-hydrogen production. Environ. Microbiol. 9: 1112–1125. https://doi.org/10.1111/j.1462-2920.2006.01234.x.

Rodríguez-Valderrama, S., Escamilla-Alvarado, C., Amezquita-Garcia, H.J., Cano-Gómez, J.J., Magnin, J.P. and Rivas-García, P. 2019. Evaluation of feeding strategies in upflow anaerobic sludge bed reactor for hydrogenogenesis at psychrophilic temperature. Int. J. Hydrogen Energy 4: 12346–12355. https://doi.org/10.1016/j.ijhydene.2018.09.215.

Rorke, D. and Gueguim Kana, E.B. 2016. Biohydrogen process development on waste sorghum (Sorghum bicolor) leaves: Optimization of saccharification, hydrogen production and preliminary scale up. Int. J. Hydrogen Energy 41: 12941–12952. https://doi.org/10.1016/j.ijhydene.2016.06.112.

Rosa, P.R.F., Santos, S.C., Sakamoto, I.K., Varesche, M.B.A. and Silva, E.L. 2014. Hydrogen production from cheese whey with ethanol-type fermentation: Effect of hydraulic retention time on the microbial community composition. Bioresour. Technol. 161: 10–19. https://doi.org/10.1016/j.biortech.2014.03.020.

Saady, N.M.C. 2013. Homoacetogenesis during hydrogen production by mixed cultures dark fermentation: Unresolved challenge. Int. J. Hydrogen Energy 38: 13172–13191. https://doi.org/10.1016/j.ijhydene.2013.07.122.

Salem, A.H., Brunstermann, R., Mietzel, T. and Widmann, R. 2018. Effect of pre-treatment and hydraulic retention time on biohydrogen production from organic wastes. Int. J. Hydrogen Energy 43: 4856–4865. https://doi.org/10.1016/j.ijhydene.2018.01.114.

Sambusiti, C., Bellucci, M., Zabaniotou, A., Beneduce, L. and Monlau, F. 2015. Algae as promising feedstocks for fermentative biohydrogen production according to a biorefinery approach: A comprehensive review. Renew. Sustain. Energy Rev. 44: 20–36. https://doi.org/10.1016/j.rser.2014.12.013.

Scheifinger, C.C., Linehan, B. and Wolin, M.J. 1975. H2 Production by Selenomonas ruminantium in the absence and presence of methanogenic bacteria. Appl. Microbiol. 29: 480–483. https://doi.org/10.1128/am.29.4.480-483.1975.

Schicho, R.N., Ma, K., Adams, M.W.W. and Kelly, R.M. 1993. Bioenergetics of sulfur reduction in the hyperthermophilic Archaeon Pyrococcus furiosus. J. Bacteriol. 175: 1823–1830.

Schuchmann, K. and Müller, V. 2016. Energetics and application of heterotrophy in acetogenic bacteria. Appl. Environ. Microbiol. 82: 4056–4069. https://doi.org/10.1128/AEM.00882-16.

Sedewitz, B., Schleifer, K.H. and Gotz, F. 1984. Physiological role of pyruvate oxidase in the aerobic metabolism of Lactobacillus plantarum. J. Bacteriol. 160: 462–465. https://doi.org/10.1128/jb.160.1.462-465.1984.

Sekar, B.S., Seol, E. and Park, S. 2017. Co-production of hydrogen and ethanol from glucose in Escherichia coli by activation of pentose-phosphate pathway through deletion of phosphoglucose isomerase (pgi) and overexpression of glucose-6-phosphate dehydrogenase (zwf) and 6-phosphog. Biotechnol. Biofuels, 1–12. https://doi.org/10.1186/s13068-017-0768-2.

Shafaat, H.S., Rüdiger, O., Ogata, H. and Lubitz, W. 2013. [NiFe] hydrogenases: A common active site for hydrogen metabolism under diverse conditions. Biochim. Biophys. Acta-Bioenerg. 1827: 986–1002. https://doi.org/10.1016/j.bbabio.2013.01.015.

Shen, G.J., Annous, B.A., Lovitt, R.W., Jain, M.K. and Zeikus, J.G. 1996. Biochemical route and control of butyrate synthesis in Butyribacterium methylotrophicum. Appl. Microbiol. Biotechnol. 45: 355–362. https://doi.org/10.1007/s002530050696.

Silva, P.J., Ban, E.C.D. Van Den, Wassink, H., Haaker, H., Castro, B. De, Robb, F.T. and Hagen, W.R. 2000. Enzymes of hydrogen metabolism in Pyrococcus furiosus. Eur. J. Biochem. 267: 6541–6551.

Sivagurunathan, P., Sen, B. and Lin, C.Y. 2014. Overcoming propionic acid inhibition of hydrogen fermentation by temperature shift strategy. Int. J. Hydrogen Energy 39: 19232–19241. https://doi.org/10.1016/j.ijhydene.2014.03.260.

Sivaramakrishnan, R., Shanmugam, S., Sekar, M., Mathimani, T., Incharoensakdi, A., Kim, S.H., Parthiban, A., Edwin Geo, V., Brindhadevi, K. and Pugazhendhi, A. 2021. Insights on biological hydrogen production routes and potential microorganisms for high hydrogen yield. Fuel 291: 120136. https://doi.org/10.1016/j.fuel.2021.120136.

Slonczewski, J.L., Fujisawa, M., Dopson, M. and Krulwich, T.A. 2009. Cytoplasmic pH measurement and homeostasis in bacteria and Archaea. Advances in Microbial Physiology. Elsevier. https://doi.org/10.1016/S0065-2911(09)05501-5.

Song, J.Z.Æ.W., Cheng, Æ.J. and Zhang, C. 2010. Heterologous expression of a hydrogenase gene in Enterobacter aerogenes to enhance hydrogen gas production 177–181. https://doi.org/10.1007/s11274-009-0139-7.

Song, Z.X., Li, W.W., Li, X.H., Dai, Y., Peng, X.X., Fan, Y.T. and Hou, H.W. 2013. Isolation and characterization of a new hydrogen-producing strain Bacillus sp. FS2011. Int. J. Hydrogen Energy 38: 3206–3212. https://doi.org/10.1016/j.ijhydene.2013.01.001.

Stams, A.J.M. 1994. Metabolic interactions between anaerobic bacteria in methanogenic environments. Antonie Van Leeuwenhoek 66: 271–294.

Tao, Y. 2016. Production of butyrate from lactate by a newly isolated Clostridium sp. BPY5. Appl. Biochem. Biotechnol. 361–374. https://doi.org/10.1007/s12010-016-1999-6.

Tapia-Venegas, E., Ramirez-Morales, J.E., Silva-Illanes, F., Toledo-Alarcón, J., Paillet, F., Escudie, R., Lay, C.H., Chu, C.Y., Leu, H.J., Marone, A., Lin, C.Y., Kim, D.H., Trably, E. and Ruiz-Filippi, G. 2015. Biohydrogen production by dark fermentation: scaling-up and technologies integration for a sustainable system. Rev. Environ. Sci. Biotechnol. 14: 761–785. https://doi.org/10.1007/s11157-015-9383-5.

Temudo, M.F., Muyzer, G., Kleerebezem, R. and Van Loosdrecht, M.C.M. 2008. Diversity of microbial communities in open mixed culture fermentations: Impact of the pH and carbon source. Appl. Microbiol. Biotechnol. 80: 1121–1130. https://doi.org/10.1007/s00253-008-1669-x.

Tholozan, J.L., Tozuel, J.P., Samain, E., Grivet, J.P., Prensier, G. and Albagnac, G. 1992. Clostridium neopropionicum sp. nov., a strict anaerobic bacterium fermenting ethanol to propionate through acrylate pathway. Arch. Microbiol. 157: 249–257.

Ueno, Y., Sasaki, D., Fukui, H., Haruta, S., Ishii, M. and Igarashi, Y. 2006. Changes in bacterial community during fermentative hydrogen and acid production from organic waste by thermophilic anaerobic microflora. J. Appl. Microbiol. 101: 331–343. https://doi.org/10.1111/j.1365-2672.2006.02939.x.

Valdez-Vazquez, I., Ríos-Leal, E., Esparza-García, F., Cecchi, F. and Poggi-Varaldo, H.M. 2005. Semi-continuous solid substrate anaerobic reactors for H2 production from organic waste: Mesophilic versus thermophilic regime. Int. J. Hydrogen Energy 30: 1383–1391. https://doi.org/10.1016/j.ijhydene.2004.09.016.

Van Ginkel, S. and Logan, B. 2005a. Increased biological hydrogen production with reduced organic loading. Water Res. 39: 3819–3826. https://doi.org/10.1016/j.watres.2005.07.021.

Van Ginkel, S. and Logan, B.E. 2005b. Inhibition of biohydrogen production by undissociated acetic and butyric acids. Environ. Sci. Technol. 39: 9351–9356. https://doi.org/10.1021/es0510515.

Van Niel, E.W.J., Claassen, P.A.M. and Stams, A.J.M. 2003. Substrate and product inhibition of hydrogen production by the extreme thermophile, Caldicellulosiruptor saccharolyticus. Biotechnol. Bioeng. 81: 255–262. https://doi.org/10.1002/bit.10463.

Vatsala, T.M., Raj, S.M. and Manimaran, A. 2008. A pilot-scale study of biohydrogen production from distillery effluent using defined bacterial co-culture. Int. J. Hydrogen Energy 33: 5404–5415. https://doi.org/10.1016/j.ijhydene.2008.07.015.

Veeravalli, S.S., Lalman, J.A., Chaganti, S.R. and Heath, D.D. 2017. Continuous hydrogen production using upflow anaerobic sludge blanket reactors: Effect of organic loading rate on microbial dynamics and H2 metabolism. J. Chem. Technol. Biotechnol. 92: 544–551. https://doi.org/10.1002/jctb.5032.

Verhaart, M.R.A.A., Bielen, A.A.M.M., Van Der Oost, J., Stams, A.J.M., Kengen, S.W.M.M., Oost, J. Van Der, Alfons, J.M., Kengen, S.W.M.M., Verhaart, M.R.A.A., Bielen, A.A.M.M., Oost, J. Van Der and Alfons, J.M. 2010. Hydrogen production by hyperthermophilic and extremely thermophilic bacteria and archaea: mechanisms for reductant disposal. Environ. Technol. 31: 993–1003. https://doi.org/10.1080/09593331003710244.

Vignais, P.M. 2008. Hydrogenases and H+-reduction in primary energy conservation. Results Probl. Cell Differ. 45: 223–252. https://doi.org/10.1007/400_2006_027.

Vignais, P.M. and Billoud, B. 2007. Occurrence, classification, and biological function of hydrogenases: An overview. Chem. Rev. 107: 4206–4272. https://doi.org/10.1021/cr050196r.

Wang, B., Wan, W. and Wang, J. 2008. Inhibitory effect of ethanol, acetic acid, propionic acid and butyric acid on fermentative hydrogen production. Int. J. Hydrogen Energy 33: 7013–7019. https://doi.org/10.1016/j.ijhydene.2008.09.027.

Wang, J. and Yin, Y. 2019. Progress in microbiology for fermentative hydrogen production from organic wastes. Crit. Rev. Environ. Sci. Technol. 49: 825–865. https://doi.org/10.1080/10643389.2018.1487226.

Wang, Q., Li, H., Feng, K. and Liu, J. 2020. Oriented fermentation of food waste towards high-value products: A review. Energies 13(21): 5638. https://doi.org/10.3390/en13215638.

Westerholm, M., Dolfing, J. and Schnürer, A. 2019. Growth characteristics and thermodynamics of syntrophic acetate oxidizers. Environ. Sci. Technol. 53: 5512–5520. https://doi.org/10.1021/acs.est.9b00288.

Whitham, J.M., Tirado-Acevedo, O., Chinn, M.S., Pawlak, J.J. and Grunden, A.M. 2015. Metabolic response of Clostridium ljungdahlii to oxygen exposure. Appl. Environ. Microbiol. 81: 8379–8391. https://doi.org/10.1128/AEM.02491-15.

Wong, Y.M., Wu, T.Y. and Juan, J.C. 2014. A review of sustainable hydrogen production using seed sludge via dark fermentation. Renew. Sustain. Energy Rev. 34: 471–482. https://doi.org/10.1016/j.rser.2014.03.008.

Wu, C.W., Whang, L.M., Cheng, H.H. and Chan, K.C. 2012. Fermentative biohydrogen production from lactate and acetate. Bioresour. Technol. 113: 30–36. https://doi.org/10.1016/j.biortech.2011.12.130.

Xiong, W., Reyes, L.H., Maness, W.E.M.P. and Chou, K.J. 2018. Engineering cellulolytic bacterium Clostridium thermocellum to co-ferment cellulose- and hemicellulose-derived sugars simultaneously, 1755–1763. https://doi.org/10.1002/bit.26590.

Xue, C. and Cheng, C. 2019. Butanol production by Clostridium, 1st ed., Advances in Bioenergy. Elsevier Inc. https://doi.org/10.1016/bs.aibe.2018.12.001.

Yang, G. and Wang, J. 2018. Various additives for improving dark fermentative hydrogen production: A review. Renew. Sustain. Energy Rev. 95: 130–146. https://doi.org/10.1016/j.rser.2018.07.029.

Zhang, T., Jiang, D., Zhang, H., Jing, Y., Tahir, N., Zhang, Y. and Zhang, Q. 2020. Comparative study on bio-hydrogen production from corn stover: Photo-fermentation, dark-fermentation and dark-photo co-fermentation. Int. J. Hydrogen Energy 45: 3807–3814. https://doi.org/10.1016/j.ijhydene.2019.04.170.

Zhang, X., Ye, X., Finneran, K.T., Zilles, J.L. and Morgenroth, E. 2013. Interactions between Clostridium beijerinckii and Geobacter metallireducens in co-culture fermentation with anthrahydroquinone-2, 6-disulfonate (AH2QDS) for enhanced biohydrogen production from xylose. Biotechnol. Bioeng. 110: 164–172. https://doi.org/10.1002/bit.24627.

Zhou, M., Yan, B., Wong, J.W.C. and Zhang, Y. 2018. Enhanced volatile fatty acids production from anaerobic fermentation of food waste: A mini-review focusing on acidogenic metabolic pathways. Bioresour. Technol. 248: 68–78. https://doi.org/10.1016/j.biortech.2017.06.121.

CHAPTER 5

Biohydrogen from Biomass
Fermentation Pathway and Economic Aspects

Md. Tanvir Ahad,[1] Munshi Md. Shafwat Yazdan,[2]
Thinesh Selvaratnam,[3,4] Zahed Siddique[1] and Ashiqur Rahman[4,]*

1. Introduction

In recent times, the decreasing rate of oil reserves with increased fuel demand and environmental concerns have invigorated the exploration of clean energy production. Cutting the carbon dioxide (CO_2) emission with dependable energy sources as an alternative clean and renewable energy (e.g., biofuels, bio-hydrogen) showed the pathways for a nation's sustainable economic growth (Sudhakar et al. 2012). Among the various sources of clean energy, bio-hydrogen is predicted to be a crucial renewable energy carrier in the near future. Previous data indicated that approximately 95% of the hydrogen production comes from fossil origin and from carbonaceous raw material (Binder et al. 2018). Nowadays, the hydrogen is being used in petrochemical, food and metallurgical industries, while only small fraction is used for energy. However, the growing demand of zero carbon emission fuels increases the demand of hydrogen (Milne et al. 2002). Bio-hydrogen production could be a propitious alternative source for future de-carbonized energy applications (Binder et al. 2018). Hydrogen market demand has continued growing expeditiously using biological method as a mainstream process through microorganism. Steady state performance and effective hydrogen production will need to keep the pace to mitigate the current demand.

Recently, United Nations Framework Convention on Climate Change (UNFCCC), for the first time, brought all nations together with goals to decrease the global temperature by 2°C (Agreement 2015). This agreement, widely known as the Paris Agreement, is a strong framework to develop more advanced alternative sustainable sources for the global industrial process. A strong possible substitution over fossil fuel could be hydrogen (Das et al. 2019, Park et al. 2021, Dunn 2002,

[1] School of Aerospace and Mechanical Engineering, University of Oklahoma, Norman, OK, USA.
[2] Department of Civil and Environmental Engineering, Idaho State University, Pocatello, ID 83209, USA.
[3] Department of Civil and Environmental Engineering, Lamar University, Beaumont, TX 77710, TX, USA.
[4] Center for Midstream Management and Science, Lamar University, Beaumont, TX 77705, TX, USA.
* Corresponding author: arahman2@lamar.edu

Hefner Iii 2002, Srivastava et al. 2019). While hydrogen is playing an important role in chemical industries and refineries, large-scale hydrogen production still depends on fossil fuel. To get rid of greenhouse gases, hydrogen production from renewable sources have been discussed among the research communities. Also, the scientific communities have considered this issue along with the world leaders aiming to decarbonize system to mitigate the future demand of energy (Binder et al. 2018, Kumar et al. 2019d, Usman et al. 2021).

Large scale hydrogen production from renewable sources such as biomass still has many challenges to overcome. Still there is an urge for an established technology rather than lab scale generation (Wang et al. 2019a). Hydrogen content is lower in biomass compared to methane (CH_4), also the energy content is low 40% due to high oxygen content in biomass. The cost for growing, transportation and harvesting is also high (Milne et al. 2002, Cao et al. 2020). Over the years, research efforts have been made to establish techno-economic analysis and life cycle assessment of hydrogen production from different biomass (Salkuyeh et al. 2018, Liu 2018, Wang et al. 2019b, Salam et al. 2018, Kumar et al. 2019b, Mehmeti et al. 2018, Xu et al. 2019). The development of biomass to hydrogen process needs to identify the optimum match of feedstock, end-use options and sustainable production technology. Dark fermentation technology showed promising result in the production of hydrogen without the need of light energy (Show et al. 2018). Along with the development of dark fermentation technology, algal photosynthetic capacity using molecular engineering approach has been explored-suggesting genetic engineering as a feasible solution. The installation of bioreactor systems to maximize hydrogen production rate paves the way for compact techno-economic sustainable hydrogen generators for large-scale applications. This chapter sheds light on the state-of-the-art bio-hydrogen production using microbial community and discusses the prospects and challenges in the biological production process.

2. Pathways of Bio-hydrogen Production

Biohydrogen production is a front-runner in the energy economy, considered an alternative to many conventional sources (Ahmed et al. 2021, Foong et al. 2021). Currently, biomass conversion technologies are divided into two categories: (i) direct production routes and (ii) conversion of storable intermediates (Milne et al. 2002). The hydrogen production processes can be classified into three categories: electro-chemical, biological and thermochemical methods. Each type of bio-hydrogen production process has their own advantages and flaws. Despite the fact, due to autonomy of non-renewable substrates, hydrogen production from microbial sources have been encouraged worldwide. Biofuel production through efficient conversion technologies employing thermochemical, electrochemical, or biochemical routes have been presented in Fig. 1. Based on the energy conversion efficiency (a ratio of output and input of energy conversion process), these pathways of hydrogen production are not fully commercial except gasification (thermochemical method) which has efficiency around 88.1%, as reported by Detchusananard et al. (Detchusananard et al. 2018). The common feedstock in all the processes is biomass.

Electro-chemical method is known as the electrolysis of water for power-to-hydrocarbon concept including hydrogen production via electrolysis. Marcelo and Dell'Era (Marcelo and Dell'Era 2008) reported two major types of electrolysis process: (i) the polymer electrolyte membrane (PEM) electrolyzer and (ii) alkaline electrolyzer. Depending on the size and type of the apparatus, the operational efficiencies of these electrolyzers range between 52–85% (Binder et al. 2018). Electricity into chemical energy conversion utilizing electrolysis illustrates a propitious technology. This concept answers why the development of power-to-gas facilities increment happens globally. Proton exchange membrane (PEM) electrolyzers, alkaline electrolyzers and electrolyzers (SOEC, MCEC) are widely known for electro-chemical method. Utilizing solar and wind energy

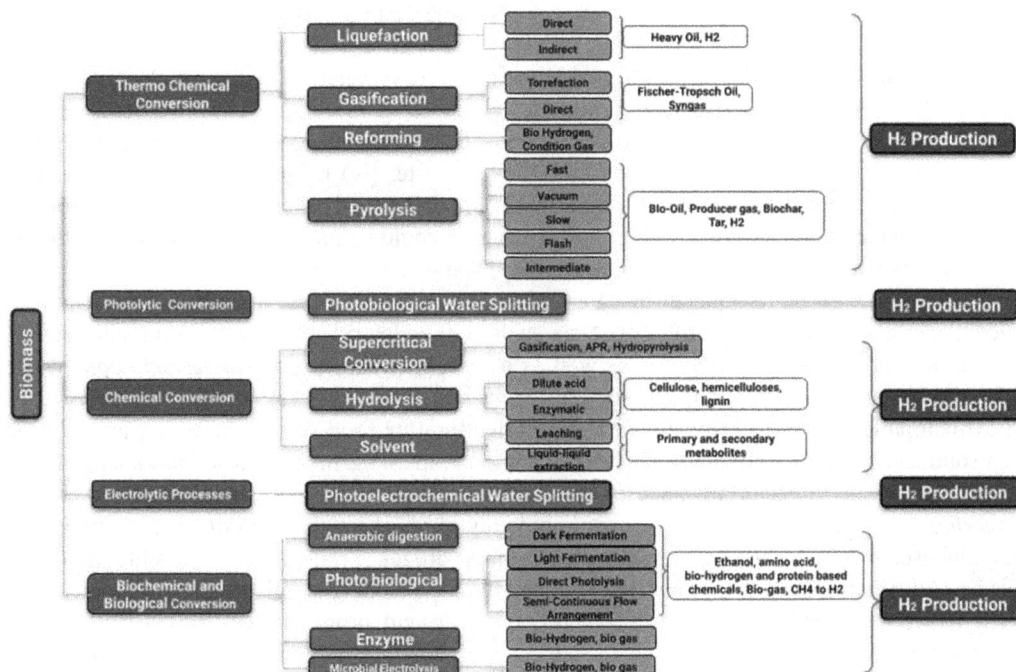

Fig. 1. Different conversion routes/pathways of biomass to bio-hydrogen production adopted from (Mohanty et al. 2015).

would be an auspicious best way for electrolysis and the technology is considered environment friendly. However, this method is not yet a good choice for commercial production of hydrogen due to its cost per unit calculation. More attention and investments are needed to make the technology economically viable.

Thermochemical approach, based on fossil fuels, is state of the art for industrial scale hydrogen production in terms of energy conversion efficiency and cost per unit. Thermal dissociation, thermal pretreatment (pyrolysis and gasification), and reforming are the technologies reported for hydrogen production through thermochemical process. All these technologies deliver a synthesis or syngas to produce pure hydrogen while focusing on carbon monoxide and hydrogen production. Thermal pretreatment method uses carbonaceous matter, thermal dissociation method uses direct splitting of water, and reforming refers to utilization of steam or oxygen to convert the biomass into gaseous products for hydrogen yield purpose. Reforming and pyrolysis process use the hydrocarbons as raw materials which are non-renewable, leading to a heavy discussion for not considering them as a source of green technology. Efficient technology development is needed for sustainable hydrogen production from biomass (Binder et al. 2018).

By using microorganisms, dark fermentation or photo-fermentation are considered state-of-the-art hydrogen production approaches. The operation at ambient temperature, pressure, and pH level along with the usage of renewable feedstock or solar energy puts biological methods on an advantageous position. Different biological routes for hydrogen production are (i) photo fermentation, dark fermentation, and hybrid systems (ii) bio-photolysis of water using green micro-algae, and (iii) biological water gas shift reaction. These biological processes are catalyzed by enzymes known as hydrogenase and nitrogenase. Previous studies reported microorganism's dark fermentation and photo fermentation as promising pathways of hydrogen production (Chaubey et al. 2013). However, current technology indicated that biological methods for hydrogen production will still take long term to be industrially scalable.

3. Microbial Sources of Bio-Hydrogen

Present bio-hydrogen technology uses a wide variety of raw materials. Till date, research efforts divided the materials into five categories: (i) animal feces, including cow, pig, and chicken manure, (ii) plant straw materials, such as rice straw, corn straw, deciduous tree, and energy grass, (iii) garbage, including landfill leachate and kitchen waste, (iv) residues produced by the food industry, such as apple pomace, lees, rapeseed cake, and cottonseed cakes, (v) biogas slurry, sewage sludge, sediment, and microalgae. Hydrogen production could be different because of the different nature of raw materials and chemical compositions (Wang et al. 2018).

Hydrogen can be produced by the metabolism of microbes including photosynthetic microalgae, cyanobacteria anaerobic bacteria such as the clostridia, nitrogen-fixing soil bacteria such as Rhizobium and *Azotobacter*, as well as enteric bacteria such as *Escherichia coli*. While used as an energy source, hydrogen oxidation is an important chemical reaction that takes place in extremophiles bacteria. Plant biomass and agricultural or food waste processing are selected as a source of biofuels technologies. Theoretically, maximum 12 molecules of hydrogen yield are possible from one molecule of glucose. However, in real practice, fermentative bacteria such as *Clostridium acetylbutylicum* or *E. coli* could possibly produce roughly two to three H_2 molecules per glucose. While a little improvement on bio-hydrogen production is possible by using *Caldicellulosiruptor saccharolyticus*, the process is less efficient (Łukajtis et al. 2018, Sargent and Kelly 2013). At the end of the fermentation process, acetic acid, propionic acid, butyric acid etc., are generated as the coproducts of hydrogen (Sudheer et al. 2020). Therefore, it is extremely necessary to pursue an in-depth research to find out efficient, environment friendly and economically prospective bio-hydrogen production sources from nature in terms of unit gas production and gas production rate.

4. Fermentation/Photo Fermentation and Dark Fermentation

A metabolic process of energy generation involving the chemical changes of organic waste materials through the action of microorganism is widely known as fermentation. Applied catalyst (isolated enzyme or microorganism producer), organic substrate (mostly carbohydrate or protein) and the process parameters control the fermentation process. Fermentation process could be either aerobic or anaerobic (Tomasik and Horton 2012). Fermentation can be divided into two ways based on the necessity of light for the microorganisms in the process: (a) dark fermentation and (b) photo fermentation. Dark fermentation is carried out in anaerobic condition to produce hydrogen along with organic acids and alcohols. However, following the similar process like dark fermentation, photo-fermentation additionally uses sunlight to produce carbon dioxide and hydrogen (Rizwan et al. 2019, Osman et al. 2020).

4.1 Dark Fermentation

Dark fermentation is a biological hydrogen production process which effectuates under dark anaerobic conditions and involves the acidogenic stage of anaerobic digestion process. Fermentation activity by anaerobic bacteria can produce hydrogen at temperature ranging 30°C to 80°C and pressures ranging from 25 to 102 bar without using any photo-energy. This puts dark fermentation process in an advantage for lower production cost approximately 340 times lower than the photosynthetic processes (Antonopoulou et al. 2011, Morimoto 2002, Osman et al. 2020). By maneuvering microbial cultures or a mixture of anaerobic microorganisms, this biological process can employ hydrogen-producing enzymes (hydrogenases) during operation. During the process, no oxygen is produced or consumed in the involved reactions. Depending on the reaction process and the substrate being used, the products are mostly hydrogen and carbon dioxide combined with other gases, such as CH_4 or hydrogen sulfide (H_2S). The dark fermentation process of hydrogen production entails several chemical reactions. Among the major chemical reactions' Equations 1 and 2, presented below are the key reactions. The proton reduction by generated electrons from C-source degradation results in hydrogen production, as explained by Equation 1. Maximum 4 mol H_2/mol is produced per mole glucose in the dark fermentation process when the end product is acetic acid. However, Equation 2 demonstrates a production of 12 mol H_2 per mole glucose (Sarangi and Nanda 2020). Fig. 2 illustrates the hydrogen production process through dark fermentation. However, in practice, hydrogen production is limited by the production of other by-products such as acetic acid, propionic acid and butyric acid. Equation 3 demonstrates the chemical reaction between glucose and water molecules to produce acetic acid. Similarly, Equation 4 shows the propionic acid pathway and Equation 5 presents the butyric acid pathway from glucose. In all of these pathways, carbon dioxide and hydrogen production could be achievable in different quantities.

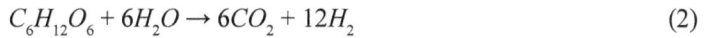

$$2H^+ + 2e^- \leftrightarrow H_2 \tag{1}$$

$$C_6H_{12}O_6 + 6H_2O \rightarrow 6CO_2 + 12H_2 \tag{2}$$

$$C_6H_{12}O_6 + 6H_2O \rightarrow 2CH_3COOH + 2CO_2 + 4H_2 [(Acetic\ acid\ pathway)] \tag{3}$$

$$C_6H_{12}O_6 \rightarrow CH_3COOH + CH_3CH_2COOH + CO_2 + H_2 [(Propionic\ acid\ pathway)] \tag{4}$$

$$C_6H_{12}O_6 + 6H_2O \rightarrow 2CH_3CH_2CH_2COOH + 2CO_2 + 2H_2 [(Butyric\ acid\ pathway)] \tag{5}$$

Fig. 2 illustrates the hydrogen production by dark fermentation with the technological challenges and future prospects. Hydrogen production by dark fermentation method is controlled by some important parameters such as pH value, hydraulic retention time (HRT) and gas partial pressure. The pH levels are maintained very carefully during the process to influence the activity of microorganisms to affect the substrate degradation (Azbar et al. 2009, Guo et al. 2010). Research efforts identified

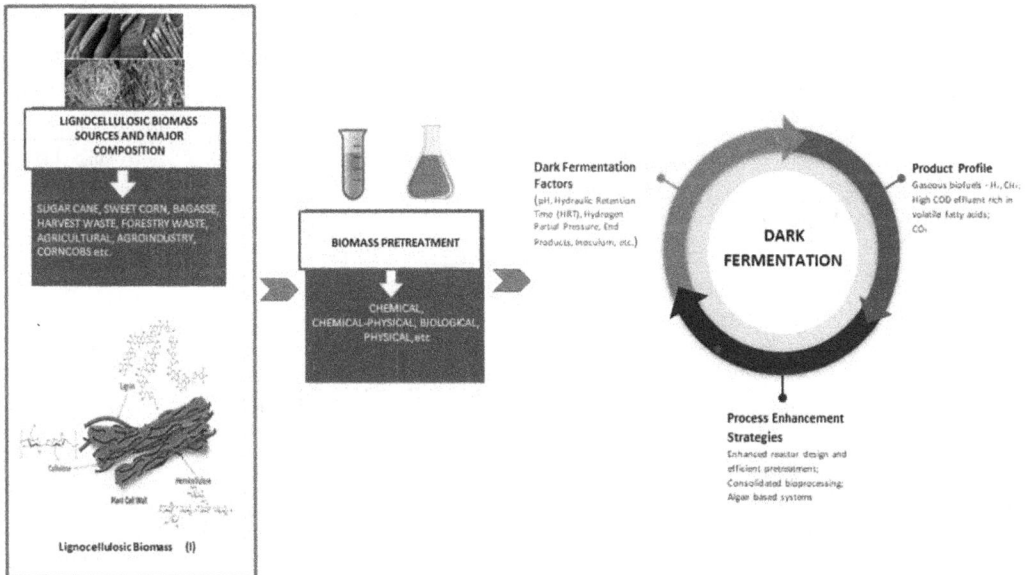

Fig. 2. Dark fermentation biohydrogen production. Lignocellulosic Biomass (I) (adopted from (Jensen et al. 2017)).

that the highest bio-hydrogen production occurs at pH 5.5 and pH 6.5, and also pointed out that pH level should be maintained between 5 and 6 for the optimal hydrogen production (Khanal et al. 2004, Li and Fang 2007, Kumar and Das 2000). Also, partial pressure plays an important role for the hydrogen production. Santiago et al. (Santiago et al. 2020) reported that HRT and solid retention time (SRT) have a great impact on the bio-hydrogen production through dark fermentation. Microorganisms capable of converting organic waste substrates in the dark fermentation are classified as thermophiles (45–65°C), mesophiles (25–45°C) and psychrophiles (0–25°C) based on their living temperatures (Osman et al. 2020). The most reported thermophilic microorganisms are *Thermoanaerobium* (*Thermoanaerobacterium thermosaccharolyticum*); however, the commonly used mesophilic cultures are *Clostridium* and *Enterobacter* (*Clostridium beijerinckii, Clostridium butyricum, Enterobacter aerogenes* and *Enterobacter asburiae*) for hydrogen production (Xing et al. 2010, Osman et al. 2020). Table 1 describes different studies of bio-hydrogen production using a dark fermentation pathway. A number of factors especially pre-treatment before fermentation can affect the production of hydrogen by dark fermentation. Till date, biohydrogen production through

Table 1. Bio-hydrogen production from wastes with the dark fermentation process.

Substrate	Microorganism	pH	Temperature (°C)	H_2 Productivity	H_2 Yield	Ref.
Kitchen waste	Mixed cultures	5.5	55	N.D.	72 cm³ H2/g VS	(Jayalakshmi et al. 2009)
Kitchen garbage	Anaerobic digester sludge	5	55	1.7 l H2/l/d	66 cm³ H2/g VS	(Chu et al. 2012)
Dairy manure	Mixed cultures	5.0	36	N. D	31.5 cm³ H2/g VS	(Xing et al. 2010)
Organic municipal solid waste	Mixed culture	5.5	50	5.7 l H2/l/d	N.D.	(Zahedi et al. 2013)
Organic municipal waste mixed with husbandry slaughterhouse waste	Mesophilic anaerobic sludge	6	34	N.D.	71.3 cm³ H_2/g VS	(Gomez et al. 2006)
Rice straw	*Clostridium pasteurianum*	7.5	37.0	N.D.	2.6 L/L hydrolysate	(Srivastava et al. 2017)
Sugarcane bagasse	*Enterobacter aerogenes*	5.5	30	N.D.	1000 mL/L hydrolysate	(Rai et al. 2014)
Synthetic food waste	Anaerobic sludge	6	37	0.9 l H_2/l/d	N.D.	(Nathao et al. 2013)
Potato steam peel 10 g glucose/l	Mixed culture	6.9	75	12.5 mM H_2/l/h	N.D.	(Mars et al. 2010)
Kitchen waste (café) 50 g COD/l	Anaerobic sludge	5.5	55	79 mM H_2/l medium/d	N.D.	(Yasin et al. 2011)
Brewery wastewater	*Klebsiella pneumoniae*	5.5	35	N.D.	1.7 mL/L hydrolysate	(Estevam et al. 2018)
Wheat straw	*Caldicellulosir ptor saccharolyticus*	6.5	70	N.D.	134 mmol H_2/L	(Soto et al. 2019)
Glucose	*Thermotoga neapolitana*	6.5	70	N.D.	1.7 mL/L hydrolysate	(Okonkwo et al. 2020)
Food waste hydrolysate	*A. awamori, A. oryzae*	4.0–4.6	37	N.D.	219.9 (39.1 mL/g food waste)	(Han et al. 2015)

dark fermentation is expensive. Therefore, extensive research efforts are needed towards the large scale production in order for the current small scale practices to become a competitive and viable technology.

4.2 Photo-fermentation

Photo fermentation occurs when organic substrates are oxidized through a microbial process in the presences of light, under anaerobic conditions to produce hydrogen and carbon dioxide (Antonopoulou et al. 2011). Prokaryotic microorganisms called purple non-sulfur bacteria (PNSB) are mainly used in the photo-fermentation hydrogen production process (Basak and Das 2007). Eukaryotic microorganisms have also been used for the photo fermentation process (Hemschemeier and Happe 2005). Proteobacteria such as PNSB, which develop under the anaerobic or microaerobic condition, do not generate oxygen in the system. As electron donor, PNSB typically uses a low concentration of sulfide, thus it is called "non-sulfur". PNSB are germ-negative with red colonies and comprise of carotenoids and bacteriochlorophyll pigments (Reungsang et al. 2018, Tao et al. 2008). With the help of scanning electron microscopy (SEM), detailed physical shapes and structures of PNSB have been obtained (Fig. 3). The aquatic environments (sediments and moist soils) and wastewater treatment sites are the primary sources for PNSB (Imhoff et al. 2005).

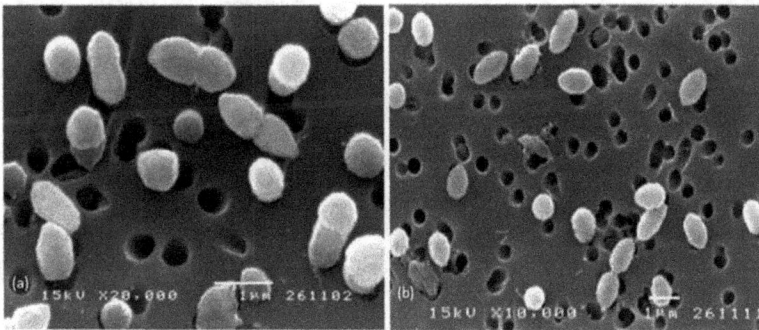

Fig. 3. Physical shape and structure of PNSB (a): *Rhodobacter* sp. KKU-PS1 (adapted from (Assawamongkholsiri and Reungsang 2015)); (b) *Rhodobacter sphaeroides* KKU-PS5 (adapted from (Laocharoen and Reungsang 2014)) under the scanning electron microscopy (SEM).

PNSB are able to degrade carbon substrates such as carbohydrate, organic matter, bio-wastes and organic acids for the production of hydrogen (Monroy and Buitrón 2020). Their hydrogen production pathway is supported by ATP-dependent nitrogenase where adenosine triphosphate (ATP) is instituted via photosynthesis. Bio-hydrogen production reaction via the photo fermentation process using glucose and acetic acid is presented by the Equations 6 and 7. The production of ATP uses light through photophosphorylation process which provides the necessary energy for the growth of microorganisms in the process.

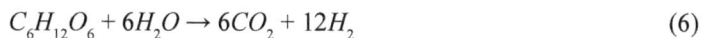

$$C_6H_{12}O_6 + 6H_2O \rightarrow 6CO_2 + 12H_2 \tag{6}$$

$$CH_3COOH + 2H_2O + h\nu \rightarrow 2CO_2 + 4H_2 [\Delta G^0 = +75.2 kJ/mol] \tag{7}$$

Unlike the hydrogenase enzymes used in the dark fermentation for hydrogen production, the photo-fermentative process of PNSB uses both nitrogenase and hydrogenase as the key enzyme (Table 2). In the active sites (metalloproteins), both enzymes carry metals. Based on metal cluster, nitrogenase enzyme can be categorized into three families. Three homologous nitrogenases are molybdenum (Mo), vanadium (V) and iron (Fe) nitrogenases (Eady 1996, Hu et al. 2012) and

Table 2. Properties of nitrogenase and hydrogenase (adopted from Koku et al. 2002).

Property	Hydrogenase	Nitrogenase
Number of proteins	One	Two (Fe and MoFe proteins)
Metal elements	Ni, Fe	Mo, Fe
Substrates	H_2	Electrons, protons, ATP (or N_2)
Products generated	Electrons, protons, ATP	H_2 (or NH_4^+)
Stimulators	H_2	Light
Inhibitors	Oxygen, CO, EDTA	Oxygen, ammonia
Optimum temperature	55°C (*Rhodospirillum rubrum*)	30°C (*Azotobacter vinelandii*)
Optimum pH	Ranged from 6.5 to 7.5	Ranged from 7.1 to 7.3

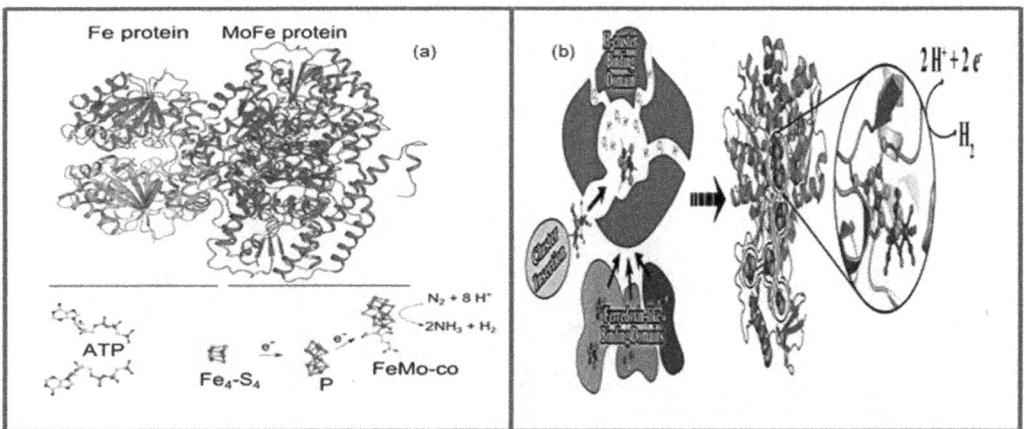

Fig. 4. The structure and mechanism of (a) nitrogenase adopted from (Seefeldt et al. 2013) and [FeFe]-hydrogenase adopted from (Mulder et al. 2011).

for photo-fermentative hydrogen production, Mo-nitrogenase is the main responsible cluster (Tan et al. 2009). Iron (Fe) protein and MoFe protein are the two component proteins consisting of Mo-nitrogenase (Burgess and Lowe 1996). Mo-Fe protein is electron acceptor while Fe protein (Fe_4–S_4 cluster) acts as electron donor in the enzyme function. Structure of nitrogenase enzyme is illustrated in Fig. 4. During the reduction of nitrogen to ammonia, hydrogen production occurs as a byproduct. This reduction reaction requires two ATP molecules for one electron transfer. Therefore, one mole hydrogen resulted from one mole ammonia actually requires 16 ATP molecules (Redwood et al. 2009). The presence of ammonia causes slower hydrogen production. Thus, the environment which provides saturating light intensity and electron donors leads to the most rapid photo-fermentative hydrogen production system (Redwood et al. 2009, Reungsang et al. 2018).

Based on the metals on the active sites, hydrogenases are categorized into three families: [NiFe]-hydrogenase, [FeFe]-hydrogenase, and [Fe]-hydrogenase (Meyer 2007). The most common hydrogenase is [NiFe]-hydrogenase because of its tolerance to CO and O_2. Fig. 4 illustrates the structure and mechanism of nitrogenase and hydrogenase. Photoautotrophs, i.e., cyanobacteria and microalgae, are the source of hydrogenase and nitrogenase. In cyanobacteria, nitrogenase consumes ATP and re-oxidizes the electron carriers for photo-fermentative hydrogen production, while hydrogenase reduces $2H^+$ to H_2 in microalgae without any ATP requirement. PNSB produces hydrogen under photoheterotrophic conditions via the nitrogenase-driven reaction.

Scientists have reported the most pure culture for photo fermentation process using *Rhodopseudomonas*, *Rhodobacter* and *Rhodospirillum*. However, mixed cultures (Fang et al. 2005) and other genera such as *Rubrivivax* (Li and Fang 2008) and *Rhodobium* (Kawaguchi et al. 2001) have also been reported. García-Sánchez et al. (García-Sánchez et al. 2018), in their efforts for hydrogen production, experimented *Rhodopseudomonas pseudopalustris* from tequila vinasses (VT) in photo fermentation process. They yielded double hydrogen with the help of VT. Mirza et al. (Mirza et al. 2019) used raw sugarcane bagasse in their experiment along with PNSB isolated from the paddy rice field to generate bio-hydrogen (148–513 mL H_2/L) by photo-fermentative process. Keskin and Hallenbeck (Keskin and Hallenbeck 2012) illustrated results for bio-hydrogen production from two major sugar mill waste—black strap and beet molasses in photo fermentation process. With the help of an artificial source of light or solar illumination, continuous photo fermentation process can be successfully operated as shown in Fig. 5.

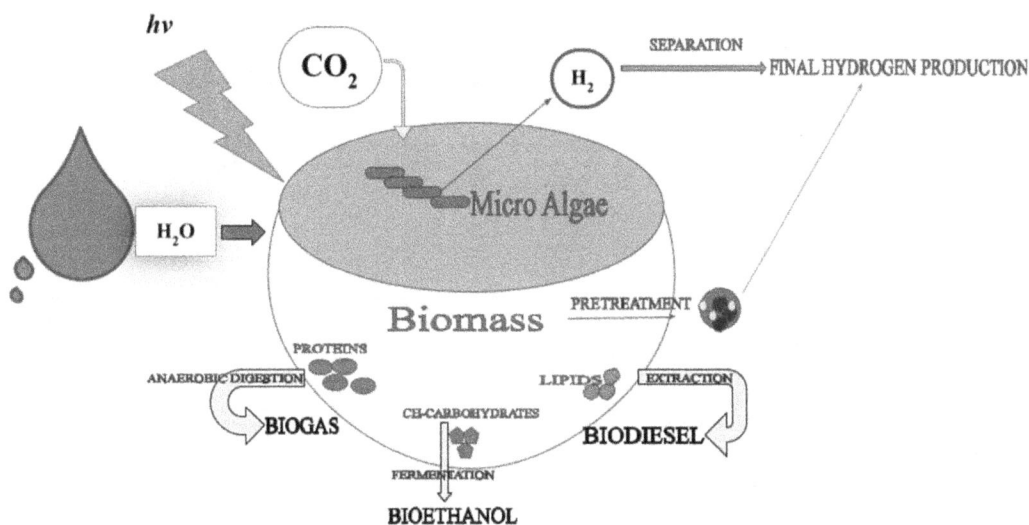

Fig. 5. Pathways of fermentative hydrogen production (Adopted from Wang and Yin 2018).

With the neutral pH value, maximum hydrogen production has been reported in many situations (Ghosh et al. 2017). Most of the experimental studies used an artificial light source which is a huge drawback for overall economic viability of a full-scale photo fermentation process. Sunlight, a free abundant alternative source of light needed for photo-fermentation, also have bottlenecks such as periodicity. For an economically viable photo-fermentative hydrogen production, many other factors such as nitrogen sources, carbon sources, pH level, light intensity, temperature and inoculum age and concentration should be taken into consideration. Table 3 presents different studies of biohydrogen production using a photo fermentation pathway. However, photo fermentation processes have to overcome three main limitations: (i) low solar energy conversion efficiency, (ii) demand for elaborate anaerobic photobioreactors covering large areas and (iii) the use of nitrogenase enzyme with high-energy demand to be considered as an economically viable large-scale bio-hydrogen production process.

5. Factors Affecting Bio-hydrogen Production

The hydrogen-production efficiency is firmly affiliated with the selected process factors, including inoculum, substrate, HRT, pH, hydrogen partial pressure, temperature, nitrogen, phosphate, the presence of inhibitors and hydrogen consuming enzymes or bacteria, nutrients concentration, the pretreatment conditions, raw materials, and the reactor configuration. Although a great number

Table 3. Bio-hydrogen production with photo fermentation.

Substrate	Microorganism	pH	Temperature (°C)	Light Intensity (W/m²)	H₂ Yield	Ref.
DF effluent of distillery wastewater	*R. capsulatus, R. sphaeroides*	7	30	30	3.2 mL/mL	(Laurinavichene et al. 2018)
Rotten apple batch	Mixed culture	7.1	30.5	24	112.0 mL/g TS	(Lu et al. 2016)
Palm oil mill effluent	*Rhodopseudomonas palustris*	5.5	30	55.3	2.3 mL H_2/mL POME	(Mishra et al. 2016)
DF effluent of sugarcane bagasse	*Rhodopseudomonas*	6.8	34	8.5	755.0 mL/L hydrolysate	(Rai et al. 2014)
DF effluent of corn stover	Mixed culture	7.8	30	23.7	4.7 m3/m3-d	(Zhang et al. 2018)
Sugar beet molasses	*R. sphaeroides, R. capsulatus, R. palustris*	7.5	30	114	9.4–19.0 mol/mom sucrose	(Sagir et al. 2017)
Cornstalk pith	Mixed culture	7	30	15.8	2.6 mol/mol sugar consumed	(Jiang et al. 2016)
Chlorella pyrenoidosa + cassava starch cellulose	*Clostridium butyricum*	7.1	30.1	47.4	388.0 ± 42.1 mL/g VS	(Xia et al. 2014)
	Cellulomonas fimi, R. palustris	N.D.	30	40	3.8 mol H_2/mol glucose	(Hitit et al. 2017)
Brewery wastewater	*R. sphaeroides*	7.4	30.2	126	408.3 mL H_2 L−1	(Al-Mohammedawi et al. 2019)
Cornstalk pith	*Rhodospirillum rubrum, Rhodopseudmonas capsulata, R. palustris, R. sphaeroides R. capsulatus*	7.8	30	15.8	211.9 mL/L-medium	(Jiang et al. 2020)
Agar embedded molasses	Heat-treated hot-spring sludge	7.4	37	39.5	226.2 mL H_2/g TS	(Mıynat et al. 2020)
Corn stover powder	*R. sphaeroides, Rhodospirillum rubrum, R. capsulatus R. palustris*	6.5	30	55.3	62.3 ± 0.8 mL/g VS	(Zhu et al. 2018)

of studies have reported the effects of these factors on fermentative hydrogen production, the optimum conditions of a given factor are not well established (Osman et al. 2020). For example, in an appropriate range, increasing HRT or temperature or pH could increase the ability of hydrogen-producing bacteria to produce hydrogen during fermentative hydrogen production process. According to Fang et al. (2002) and Calli et al. (2008), the pH range of 5–7.5 is mostly reported as the optimum level, while on the other hand, Lee et al. (Lee et al. 2002) proposed that pH of 4.5 (minimum) and 9.0 (maximum) apparently give the maximum yield for the hydrogen production. Even though the optimal temperature reported for fermentative hydrogen production was not always the same, it fell into the mesophilic range (around 37°C) and thermophilic range (around 55°C), respectively, although mesophilic hydrogen production is preferred to be more economic as it prevents the external requirement of heating.

Even for the same reactor, the optimal HRT for continuous fermentative hydrogen production has some expert disagreements. Zhang et al., reported that the optimal HRT for a CSTR was 0.5 h (Zhang et al. 2007). On the other hand, according to Arooj et al. (2008) it was 12 h. The possible factors for this disagreement were inoculum, substrate and HRT range as can be deciphered from these studies. It would be best to say that the real scale hydrogen production in a bioreactor at present can be predicted only either on the lab-scale or pilot-scale studies, while large-scale yield of bio-hydrogen would require stable and efficient engineering production processes.

6. Energy Analysis and Economic Feasibility of Bio-hydrogen

Implementation of a large-scale production unit requires an energy analysis to make the process financially viable. Energy calculation could be done by standard process (Association et al. 1912) by calculating input energy, output energy and net energy by using the following equations:

$$E_i = P * T * V * S \tag{8}$$

where, E_i refers to input energy (kWh); P denotes power utilized for the process (kW/kg); T, time used for disintegration (hours); V, reactor volume (m^3) and S, substrate (kg/m^3). Energy needed to disintegrate is expressed as the input energy by Equation 8. The output energy, in terms of energy gained as hydrogen, from pretreated sample could be achieved from Equation 9.

$$E_o = B * L * H * V * F \tag{9}$$

where, E_o, Output energy (kWh); B, Biodegradability of algal biomass (gCOD/gCOD); here, COD refers to chemical oxygen demand; L, COD load ($gCOD/m^3$); H, Hydrogen yield ($m^3/gCOD$); V, Reactor volume (m^3); F, Bio-hydrogen conversion factor (1 m^3 is equal to 3.5 kWh).

Net energy estimation could be done by the state-of-the-art calculation of difference between energy gained E_o versus energy spent E_i of the following Equation 10.

$$E_n = E_o - E_i \tag{10}$$

where, E_n, Net energy (kWh); E_o, Output energy (kWh); E_i, Input energy (kWh) (Kumar et al. 2019a).

A hydrogen economy, considered as the permanent goal of many countries, can possibly solve the energy security, along with economic and environmental prosperity. Nonetheless, the uncertainty of the evolution from a conventional petroleum-based energy system to a hydrogen production energy economy includes the development of highly efficient fuel technologies, problems associated with hydrogen production and stable infrastructure for distribution, and last but not the least the petroleum market's response. The advantages of bio-hydrogen processes are its clean and 'CO_2-neutral' features. Moreover, this is being fueled by carbohydrates derived from photosynthetic fixation of carbon dioxide. According to Macaskie et al. (2005), bio hydrogen is free of catalyst poisons (CO and H_2O) and does not even need any treatment for being used in fuel cells to produce electricity. This technique can be implemented on both large scale (e.g., industry) and the small or pilot scale level (e.g., vehicle). Due to the technological advances, the limitations of bio hydrogen processes are being reduced and the hydrogen economy could possibly be enhanced. There has not been a lot but a few existing literatures available about the hydrogen economy. As most of the studies have been performed at lab scale, harvesting or scaling up of the production is yet to be analyzed. Resnick et al. (2004) demonstrated a comparative economic study using a series of models to predict the future of capital and operating costs of the various approaches which were tested at a lab scale.

According to Resnick et al. (2004), the estimated capacity was 50 million SCFD (standard cubic feet per day). The size of each plant was assessed on the basis of specific hydrogen production rate mentioned in the existing literature for the numerous bio hydrogen processes. According to the experts, the market analysis of hydrogen production can be divided into three scales of application: small scale, medium scale and large scale (Sharma and Kaushik 2017, Binder et al. 2018). Hydrogen filling stations are considered as small-scale application where the capacity is 15 to 50 kg·h^{-1} (0.5 MW to 1.7 MW). Application on refineries is considered as medium scale where the capacity is 1 000 to 3 000 kg·h^{-1} (33 MW to 100 MW). In industrial areas, hydrogen is used on a larger scale where the capacity is 2,000 to 10,000 kg·h^{-1} (66 MW to 333 MW). According to the analysts, excluding the cost of CO and glucose would decrease the operating costs to 17.44, 5.60, 4.43 \$/GJ, respectively. Hydrogen production rate is very crucial before making any final analysis. Increase or decrease of any factors or values could drastically affect the economy of the production.

7. Challenges and Future Aspects of Bio-hydrogen Production

Till date, several research efforts have been made to establish the economic feasibility of hydrogen production from biomass. Having many advantages of this process over the conventional process, there are many challenges that need to be addressed for this process to be long-term feasible. Bio-hydrogen production process varies from process to process. Dark fermentation and photo fermentation using photosynthetic bacteria are the most popular methods for biological production of hydrogen. The efficiency of energy conversion is very low with 4.3% and 5.11% for dark- and photo fermentation processes, respectively (Zhang et al. 2017). Dark fermentation faces challenges such as high BOD (Biological oxygen demand) level in effluent. Additionally, after hydrogen production, separation of H$_2$ from CO$_2$+H$_2$ is needed and pre-treatment of lignocellulosic waste is necessary. The challenges for photo fermentation are noted as: low hydrogen production rate, low light conversion efficiency, not suitable for other wastes except VFA (volatile fatty acid) rich waste and specially required an external source of light (Osman et al. 2020). However, dark fermentation and photo fermentation have several advantages such as high COD removal rate and so on. Former hydrogen production process is mature which can only process specific feedstocks (food waste, sewage sludge and crops waste), while gasification can process the whole portion of the biomass and fermentation can process the non-edible cellulosic part of lignocellulosic biomass, but the technology is still not completely developed worldwide (Osman et al. 2020). Due to low bio-hydrogen production rate and the high cost of the raw feedstocks, hydrogen production is still dominated by fossil-based fuels. Using the organic waste materials with improved agricultural practices and breeding efforts may be useful to mitigate the challenges. However, large-scale production of this process still has to overcome lots of challenges to become economically viable.

In the vision of the zero-carbon economy for the future world, bio-hydrogen has showed promising alternative over fossil fuel. Though bio-hydrogen is considered as a promising alternative, future research and development (R&D) is needed on biomass to hydrogen conversion technology and the markets for hydrogen in areas where the feedstocks are available. Future research should concentrate on feedstock preparation and feeding process, modular systems development, gasification gas conditioning, valuable co-product integration, integration of more than one production process along with different biomass waste streams to identify the most promising large-scale technology. The cost of hydrogen regardless of the production technology must be below \$4/gallon gasoline for being competitive in the market. Another future aspect could be the development of hydrogen storage facility as hydrogen is difficult to store for long time in conventional cylinder. Thus, bio-hydrogen can alleviate the challenges associated with fossil-based fuel and promote environmental benefits towards the zero-carbon economy.

8. Conclusion

Bio-hydrogen is recognized as one of the most promising energy carriers for fossil fuel replacement due to its zero-carbon emission. Hydrogen production from biomass has been investigated since last decades. Biomass is potentially a reliable energy resource for hydrogen production. Also, biomass is renewable, abundant, and easy to use. Different methods for bio-hydrogen production offers distinctive advantages. Fermentation methods are greener and environment friendly. However, the process has low efficiency till date and the production cost is high. However, pretreatment of the raw materials adding bacteria can improve the hydrogen and in addition the process of anaerobic fermentation, the sludge/wastewater is economical on inoculum, pH level and controlled temperature conditions. Despite having diverse benefits, fermentation production technique can be affected by factors such as substrate, HRT, pH level, hydrogen partial pressure, temperature, hydrogen consuming enzymes or bacteria and the reactor configuration etc. Also, owing to the cost associated to microalgae cultivation system, reactor design, and the use of metabolic route, the production cost remains high and noncompetitive to other fuels. To compete with gasoline, bio-hydrogen price should be reduced and competitive enough. Thus, to make the sustainable and renewable bio-hydrogen production from biomasses, more research work should be focused on the standardization of the various operational parameters with in-depth understanding of system biology of bio-hydrogen production, metabolic engineering, and genetic manipulation. Nevertheless, most economic assessments are optimistic but dimmed for not considering cost factors like storage, handling, and transportation of hydrogen. With further development of technologies, biomass will play an important role in the development of sustainable hydrogen economy and consequently working toward the zero-carbon economy.

References

Agreement, P. 2015. Paris agreement. In Report of the Conference of the Parties to the United Nations Framework Convention on Climate Change (21st Session, 2015: Paris). Retrived December, volume 4, page 2017. HeinOnline.

Al-Mohammedawi, H.H., Znad, H. and Eroglu, E. 2019. Improvement of photofermentative biohydrogen production using pre-treated brewery wastewater with banana peels waste. International Journal of Hydrogen Energy 44(5): 2560–2568.

Antonopoulou, G., Ntaikou, I., Stamatelatou, K. and Lyberatos, G. 2011. Biological and fermentative production of hydrogen. *In*: Handbook of Biofuels Production, pp. 305–346. Elsevier.

Aqsha, A., Tijani, M.M., Moghtaderi, B. and Mahinpey, N. 2017. Catalytic pyrolysis of straw biomasses (wheat, flax, oat and barley) and the comparison of their product yields. Journal of Analytical and Applied Pyrolysis 125: 201–208.

Arooj, M.F., Han, S.-K., Kim, S.-H., Kim, D.-H. and Shin, H.-S. 2008. Continuous biohydrogen production in a cstr using starch as a substrate. International Journal of Hydrogen Energy 33(13): 3289–3294.

Association, A.P.H., Association, A.W.W., Federation, W.P.C. and Federation, W.E. 1912. Standard Methods for the Examination of Water and Wastewater, volume 2. American Public Health Association.

Azbar, N., Dokgöz, F., Keskin, T., Eltem, R., Korkmaz, K., Gezgin, Y., Akbal, Z., Öncel, S., Dalay, M., Gönen, Ç. et al. 2009. Comparative evaluation of bio-hydrogen production from cheese whey wastewater under thermophilic and mesophilic anaerobic conditions. International Journal of Green Energy 6(2): 192–200.

Banu, J.R., Eswari, A.P., Kavitha, S., Kannah, R.Y., Kumar, G., Jamal, M.T., Saratale, G.D., Nguyen, D.D., Lee, D.-G. and Chang, S.W. 2019. Energetically efficient microwave disintegration of waste activated sludge for biofuel production by zeolite: Quantification of energy and biodegradability modelling. International Journal of Hydrogen Energy 44(4): 2274–2288.

Basak, N. and Das, D. 2007. The prospect of purple non-sulfur (pns) photosynthetic bacteria for hydrogen production: The present state of the art. World Journal of Microbiology and Biotechnology 23(1): 31–42.

Binder, M., Kraussler, M., Kuba, M. and Luisser, M. 2018. Hydrogen from biomass gasification. IEA Bioenergy. https://www.ieabioenergy.com/wp-content/uploads/2019/01/Wasserstoffstudie_IEA-final.pdf; accessed June 28, 2022.

Bundhoo, Z.M. 2019. Potential of bio-hydrogen production from dark fermentation of crop residues: A review. International Journal of Hydrogen Energy 44(32): 17346–17362.

Calli, B., Schoenmaekers, K., Vanbroekhoven, K. and Diels, L. 2008. Dark fermentative h2 production from xylose and lactose—Effects of on-line ph control. International Journal of Hydrogen Energy 33(2): 522–530.

Cao, L., Iris, K., Xiong, X., Tsang, D.C., Zhang, S., Clark, J.H., Hu, C., Ng, Y.H., Shang, J. and Ok, Y.S. 2020. Biorenewable hydrogen production through biomass gasification: A review and future prospects. Environmental Research, 186: 109547.

Chaubey, R., Sahu, S., James, O.O. and Maity, S. 2013. A review on development of industrial processes and emerging techniques for production of hydrogen from renewable and sustainable sources. Renewable and Sustainable Energy Reviews 23: 443–462.

Chu, C.-F., Xu, K.-Q., Li, Y.-Y. and Inamori, Y. 2012. Hydrogen and methane potential based on the nature of food waste materials in a two-stage thermophilic fermentation process. International Journal of Hydrogen Energy 37(14): 10611–10618.

Das, D., Khanna, N. and Dasgupta, C.N. 2019. Biohydrogen Production: Fundamentals and Technology Advances. CRC Press.

Detchusananard, T., Im-orb, K., Ponpesh, P. and Arpornwichanop, A. 2018. Biomass gasification integrated with CO2 capture processes for high-purity hydrogen production: Process performance and energy analysis. Energy Conversion and Management 171: 1560–1572.

Dunn, S. 2002. Hydrogen futures: Toward a sustainable energy system. International Journal of Hydrogen Energy 27(3): 235–264.

Estevam, A., Arantes, M.K., Andrigheto, C., Fiorini, A., da Silva, E.A. and Alves, H.J. 2018. Production of biohydrogen from brewery wastewater using klebsiella pneumoniae isolated from the environment. International Journal of Hydrogen Energy 43(9): 4276–4283.

Fang, H.H. and Liu, H. 2002. Effect of ph on hydrogen production from glucose by a mixed culture. Bioresource Technology 82(1): 87–93.

Fang, H.H., Liu, H. and Zhang, T. 2005. Phototrophic hydrogen production from acetate and butyrate in wastewater. International Journal of Hydrogen Energy 30(7): 785–793.

Franco, C., Pinto, F., Gulyurtlu, I. and Cabrita, I. 2003. The study of reactions influencing the biomass steam gasification process. Fuel 82(7): 835–842.

Gañan, J., Abdulla, A.A.-K., Miranda, A., Turegano, J., Correia, S. and Cuerda, E. 2005. Energy production by means of gasification process of residuals sourced in extremadura (spain). Renewable Energy 30(11): 1759–1769.

García-Sánchez, R., Ramos-Ibarra, R., Guatemala Morales, G., Arriola-Guevara, E., Toriz-Gonzalez, G. and Corona-González, R.I. 2018. Photofermentation of tequila vinasses by *Rhodopseudomonas pseudopalustris* to produce hydrogen. International Journal of Hydrogen Energy 43(33): 15857–15869.

Ghosh, S., Dairkee, U.K., Chowdhury, R. and Bhattacharya, P. 2017. Hydrogen from food processing wastes via photofermentation using purple non-sulfur bacteria (pnsb)—A review. Energy Conversion and Management 141: 299–314.

Gomez, X., Moran, A., Cuetos, M. and Sanchez, M. 2006. The production of hydrogen by dark fermentation of municipal solid wastes and slaughterhouse waste: A two-phase process. Journal of Power Sources 157(2): 727–732.

Guo, X.M., Trably, E., Latrille, E., Carrere, H. and Steyer, J.-P. 2010. Hydrogen production from agricultural waste by dark fermentation: A review. International Journal of Hydrogen Energy 35(19): 10660–10673.

Han, W., Ye, M., Zhu, A.J., Zhao, H.T. and Li, Y.F. 2015. Batch dark fermentation from enzymatic hydrolyzed food waste for hydrogen production. Bioresource Technology 191: 24–29.

Hefner Iii, R.A. 2002. The age of energy gases. International Journal of Hydrogen Energy 27(1): 1–9.

Hein, D. and Karl, J. 2006. Conversion of biomass to heat and electricity. Energy Technologies 3: 374–413.

Hemschemeier, A. and Happe, T. 2005. The exceptional photofermentative hydrogen metabolism of the green alga chlamydomonas reinhardtii. Biochemical Society Transaction 33(1): 39–41.

Hitit, Z.Y., Lazaro, C.Z. and Hallenbeck, P.C. 2017. Single stage hydrogen production from cellulose through Photo fermentation by a co-culture of cellulomonas fimi and rhodopseudomonas palustris. International Journal of Hydrogen Energy 42(10): 6556–6566.

Hu, B., Li, Y., Zhu, S., Zhang, H., Jing, Y., Jiang, D., He, C. and Zhang, Z. 2020. Evaluation of biohydrogen yield potential and electron balance in the photofermentation process with different initial pH from starch agricultural leftover. Bioresource Technology 305: 122900.

Jayalakshmi, S., Joseph, K. and Sukumaran, V. 2009. Bio hydrogen generation from kitchen waste in an inclined plug flow reactor. International Journal of Hydrogen Energy 34(21): 8854–8858.

Jiang, D., Ge, X., Lin, L., Zhang, T., Liu, H., Hu, J. and Zhang, Q. 2020. Continuous photo-fermentative hydrogen production in a tubular photobioreactor using corn stalk pith hydrolysate with a consortium. International Journal of Hydrogen Energy 45(6): 3776–3784.

Jiang, D., Ge, X., Zhang, T., Liu, H. and Zhang, Q. 2016. Photofermentative hydrogen production from enzymatic hydrolysate of corn stalk pith with a photosynthetic consortium. International Journal of Hydrogen Energy 41(38): 16778–16785.

Kannah, R.Y., Kavitha, S., Sivashanmugham, P., Kumar, G., Nguyen, D.D., Chang, S.W. and Banu, J.R. 2019. Biohydrogen production from rice straw: Effect of combinative pretreatment, modelling assessment and energy balance consideration. International Journal of Hydrogen Energy 44(4): 2203–2215.

Kawaguchi, H., Hashimoto, K., Hirata, K. and Miyamoto, K. 2001. H2 production from algal biomass by a mixed culture of rhodobium marinum a-501 and lactobacillus amylovorus. Journal of Bioscience and Bioengineering 91(3): 277–282.

Keskin, T. and Hallenbeck, P.C. 2012. Hydrogen production from sugar industry wastes using single-stage photofermentation. Bioresource Technology 112: 131–136.

Khanal, S.K., Chen, W.-H., Li, L. and Sung, S. 2004. Biological hydrogen production: Effects of pH and intermediate products. International Journal of Hydrogen Energy 29(11): 1123–1131.

Kumar, D., Eswari, A.P., Park, J.-H., Adishkumar, S. and Banu, J.R. 2019a. Biohydrogen generation from macroalgal biomass, chaetomorpha antennina through surfactant aided microwave disintegration. Frontiers in Energy Research 7: 78.

Kumar, M., Oyedun, A.O. and Kumar, A. 2019b. A comparative analysis of hydrogen production from the thermochemical conversion of algal biomass. International Journal of Hydrogen Energy 44(21): 10384–10397.

Kumar, M.D., Kaliappan, S., Gopikumar, S., Zhen, G. and Banu, J.R. 2019c. Synergetic pretreatment of algal biomass through H2O2 induced microwave in acidic condition for biohydrogen production. Fuel 253: 833–839.

Kumar, M.D., Tamilarasan, K., Kaliappan, S., Banu, J.R., Rajkumar, M. and Kim, S.H. 2018. Surfactant assisted disperser pretreatment on the liquefaction of ulva reticulata and evaluation of biodegradability for energy efficient biofuel production through nonlinear regression modelling. Bioresource Technology 255: 116–122.

Kumar, N. and Das, D. 2000. Enhancement of hydrogen production by enterobacter cloacae iit-bt 08. Process Biochemistry 35(6): 589–593.

Kumar, S., Sharma, S., Thakur, S., Mishra, T., Negi, P., Mishra, S., Hesham, A.E.-L., Rastegari, A.A., Yadav, N. and Yadav, A.N. 2019d. Bioprospecting of microbes for biohydrogen production: Current status and future challenges. Bioprocessing for Biomolecules Production, pp. 443–471.

Laurinavichene, T., Tekucheva, D., Laurinavichius, K. and Tsygankov, A. 2018. Utilization of distillery wastewater for hydrogen production in one-stage and two-stage processes involving photofermentation. Enzyme and Microbial Technology 110: 1–7.

Lee, Y.J., Miyahara, T. and Noike, T. 2002. Effect of pH on microbial hydrogen fermentation. Journal of Chemical Technology & Biotechnology 77(6): 694–698.

Li, C. and Fang, H.H. 2007. Fermentative hydrogen production from wastewater and solid wastes by mixed cultures. Critical Reviews in Environmental Science and Technology 37(1): 1–39.

Li, R.Y. and Fang, H.H. 2008. Hydrogen production characteristics of photoheterotrophic rubrivivax gelatinosus l31. International Journal of Hydrogen Energy 33(3): 974–980.

Liu, Z. 2018. Economic analysis of energy production from coal/biomass upgrading; part 1: Hydrogen production. Energy Sources, Part B: Economics, Planning, and Policy 13(2): 132–136.

Lu, C., Zhang, Z., Ge, X., Wang, Y., Zhou, X., You, X., Liu, H. and Zhang, Q. 2016. Bio-hydrogen production from apple waste by photosynthetic bacteria hau-m1. International Journal of Hydrogen Energy 41(31): 13399–13407.

Łukajtis, R., Hołowacz, I., Kucharska, K., Glinka, M., Rybarczyk, P., Przyjazny, A. and Kaminski, M. 2018. Hydrogen production from biomass using dark fermentation. Renewable and Sustainable Energy Reviews 91: 665–694.

Macaskie, L., Baxter-Plant, V., Creamer, N., Humphries, A., Mikheenko, I., Mikheenko, P., Penfold, D. and Yong, P. 2005. Applications of bacterial hydrogenases in waste decontamination, manufacture of novel bionanocatalysts and in sustainable energy. Biochemical Society Transactions 33(1): 76–79.

Marcelo, D. and Dell'Era, A. 2008. Economical electrolyser solution. International Journal of Hydrogen Energy 33(12): 3041–3044.

Mars, A.E., Veuskens, T., Budde, M.A., Van Doeveren, P.F., Lips, S.J., Bakker, R.R., De Vrije, T. and Claassen, P.A. 2010. Biohydrogen production from untreated and hydrolyzed potato steam peels by the extreme thermophiles caldicellulosiruptor saccharolyticus and thermotoga neapolitana. International Journal of Hydrogen Energy 35(15): 7730–7737.

Mehmeti, A., Angelis-Dimakis, A., Arampatzis, G., McPhail, S.J. and Ulgiati, S. 2018. Life cycle assessment and water footprint of hydrogen production methods: from conventional to emerging technologies. Environments 5(2): 24.

Milne, T.A., Elam, C.C. and Evans, R.J. 2002. Hydrogen from biomass: State of the art and research challenges. United States: N. p. 2002.

Miranda, M., Arranz, J., Román, S., Rojas, S., Montero, I., López, M. and Cruz, J. 2011. Characterization of grape pomace and pyrenean oak pellets. Fuel Processing Technology 92(2): 278–283.

Mirza, S.S., Qazi, J.I., Liang, Y. and Chen, S. 2019. Growth characteristics and photofermentative biohydrogen production potential of purple non sulfur bacteria from sugar cane bagasse. Fuel 255: 115805.

Mishra, P., Thakur, S., Singh, L., Ab Wahid, Z. and Sakinah, M. 2016. Enhanced hydrogen production from palm oil mill effluent using two stage sequential dark and photo fermentation. International Journal of Hydrogen Energy 41(41): 18431–18440.

Mıynat, M.E., Ören, İ., Özkan, E. and Argun, H. 2020. Sequential dark and photo-fermentative hydrogen gas production from agar embedded molasses. International Journal of Hydrogen Energy 45(60): 34730–34738.

Mohanty, P., Pant, K.K. and Mittal, R. 2015. Hydrogen generation from biomass materials: Challenges and opportunities. Wiley Interdisciplinary Reviews: Energy and Environment 4(2): 139–155.

Molino, A., Larocca, V., Chianese, S. and Musmarra, D. 2018. Biofuels production by biomass gasification: A review. Energies 11(4): 811.

Monroy, I. and Buitrón, G. 2020. Production of polyhydroxybutyrate by pure and mixed cultures of purple non-sulfur bacteria: A review. Journal of Biotechnology 317: 39–47.

Morimoto, M. 2002. Why is the anaerobic fermentation in the production of the biohydrogen attractive. The proceedings of conversion of biomass into bioenergy. Organized by New energy and Industrial Technology Development Organization (NEPO), Japan and Malaysian Palm oil Board (MPOP).

Nathao, C., Sirisukpoka, U. and Pisutpaisal, N. 2013. Production of hydrogen and methane by one and two stage fermentation of food waste. International Journal of Hydrogen Energy 38(35): 15764–15769.

Okonkwo, O., Papirio, S., Trably, E., Escudie, R., Lakaniemi, A.-M. and Esposito, G. 2020. Enhancing thermophilic dark fermentative hydrogen production at high glucose concentrations via bioaugmentation with thermotoga neapolitana. International Journal of Hydrogen Energy 45(35): 17241–17249.

Osman, A.I., Deka, T.J., Baruah, D.C. and Rooney, D.W. 2020. Critical challenges in biohydrogen production processes from the organic feedstocks. Biomass Conversion and Biorefinery, pp. 1–19.

Park, J.-H., Chandrasekhar, K., Jeon, B.-H., Jang, M., Liu, Y. and Kim, S.-H. 2021. State-of-the-art technologies for continuous high-rate biohydrogen production. Bioresource Technology 320: 124304.

Rai, P. K., Singh, S., Asthana, R. and Singh, S. 2014. Biohydrogen production from sugarcane bagasse by integrating dark-and photo fermentation. Bioresource Technology 152: 140–146.

Resnick, R.J. 2004. The economics of biological methods of hydrogen production. PhD thesis, Massachusetts Institute of Technology.

Rizwan, M., Shah, S.H., Mujtaba, G., Mahmood, Q., Rashid, N. and Shah, F.A. 2019. Ecofuel feedstocks and their prospect. *In*: Advanced Biofuels, pp. 3–16. Elsevier.

Roy, M.M. and Corscadden, K.W. 2012. An experimental study of combustion and emissions of biomass briquettes in a domestic wood stove. Applied Energy 99: 206–212.

Sagir, E., Ozgur, E., Gunduz, U., Eroglu, I. and Yucel, M. 2017. Single-stage photofermentative biohydrogen production from sugar beet molasses by different purple non-sulfur bacteria. Bioprocess and Biosystems Engineering 40(11): 1589–1601.

Salam, M.A., Ahmed, K., Akter, N., Hossain, T. and Abdullah, B. 2018. A review of hydrogen production via biomass gasification and its prospect in Bangladesh. International Journal of Hydrogen Energy 43(32): 14944–14973.

Salkuyeh, Y.K., Saville, B.A. and MacLean, H.L. 2018. Techno-economic analysis and life cycle assessment of hydrogen production from different biomass gasification processes. International Journal of Hydrogen Energy 43(20): 9514–9528.

Santiago, S.G., Morgan-Sagastume, J.M., Monroy, O. and Moreno-Andrade, I. 2020. Biohydrogen production from organic solid waste in a sequencing batch reactor: An optimization of the hydraulic and solids retention time. International Journal of Hydrogen Energy 45(47): 25681–25688.

Sarangi, P.K. and Nanda, S. 2020. Biohydrogen production through dark fermentation. Chemical Engineering & Technology 43(4): 601–612.

Sargent, F. and Kelly, C.L. 2013. Biohydrogen-a clean fuel but a dirty business. Biofuels. https://microbiologysociety.org/publication/past-issues/biofuels-past-issue/article/biohydrogen-a-clean-fuel-but-a-dirty-business.html (Accessed June 28, 2022).

Serrano, C., Monedero, E., Lapuerta, M. and Portero, H. 2011. Effect of moisture content, particle size and pine addition on quality parameters of barley straw pellets. Fuel Processing Technology 92(3): 699–706.

Sharara, M.A., Holeman, N., Sadaka, S.S. and Costello, T.A. 2014. Pyrolysis kinetics of algal consortia grown using swine manure wastewater. Bioresource Technology 169: 658–666.

Sharma, M. and Kaushik, A. 2017. Biohydrogen economy: Challenges and prospects for commercialization. *In*: Biohydrogen Production Sustainability of Current Technology and Future Perspective. Springer, pp. 253–267.

Show, K.-Y., Yan, Y., Ling, M., Ye, G., Li, T. and Lee, D.-J. 2018. Hydrogen production from algal biomass–advances, challenges and prospects. Bioresource Technology 257: 290–300.

Soto, L.R., Byrne, E., van Niel, E.W., Sayed, M., Villanueva, C.C. and Hatti-Kaul, R. 2019. Hydrogen and polyhydroxybutyrate production from wheat straw hydrolysate using caldicellulosiruptor species and ralstonia eutropha in a coupled process. Bioresource Technology 272: 259–266.

Srivastava, N., Srivastava, M., Kushwaha, D., Gupta, V.K., Manikanta, A., Ramteke, P. and Mishra, P. 2017. Efficient dark fermentative hydrogen production from enzyme hydrolyzed rice straw by clostridium pasteurianum (mtcc116). Bioresource Technology 238: 552–558.

Srivastava, N., Srivastava, M., Malhotra, B.D., Gupta, V.K., Ramteke, P., Silva, R.N., Shukla, P., Dubey, K.K. and Mishra, P. 2019. Nanoengineered cellulosic biohydrogen production via dark fermentation: A novel approach. Biotechnology Advances 37(6): 107384.

Sudhakar, K., Premalatha, M. et al. 2012. Techno economic analysis of micro algal carbon sequestration and oil production. Int. J. Chem. Tech. Res. 4: 974–4290.

Sudheer, P.D., Chauhan, S. and Velramar, B. 2020. Biohydrogen: Technology developments in microbial fuel cells and their future prospects. *In*: Biotechnology for Biofuels: A Sustainable Green Energy Solution, pp. 61–94. Springer.

Tomasik, P. and Horton, D. 2012. Enzymatic conversions of starch. Advances in Carbohydrate Chemistry and Biochemistry 68: 59–436.

Usman, M., Kavitha, S., Kannah, Y., Yogalakshmi, K., Sivashanmugam, P., Bhatnagar, A., Kumar, G. et al. 2021. A critical review on limitations and enhancement strategies associated with biohydrogen production. International Journal of Hydrogen Energy 46(31): 16565–16590.

Wang, H., Xu, J., Sheng, L., Liu, X., Lu, Y. and Li, W. 2018. A review on bio-hydrogen production technology. International Journal of Energy Research 42(11): 3442–3453.

Wang, J. and Yin, Y. 2018. Fermentative hydrogen production using various biomass-based materials as feedstock. Renewable and Sustainable Energy Reviews 92: 284–306.

Wang, M., Wang, G., Sun, Z., Zhang, Y. and Xu, D. 2019a. Review of renewable energy-based hydrogen production processes for sustainable energy innovation. Global Energy Interconnection 2(5): 436–443.

Wang, Y., Li, G., Liu, Z., Cui, P., Zhu, Z. and Yang, S. 2019b. Techno-economic analysis of biomass-to-hydrogen process in comparison with coalto-hydrogen process. Energy 185: 1063–1075.

Wang, Y., Wang, D., Chen, F., Yang, Q., Li, Y., Li, X. and Zeng, G. 2019c. Effect of triclocarban on hydrogen production from dark fermentation of waste activated sludge. Bioresource Technology 279: 307–316.

Xia, A., Cheng, J., Ding, L., Lin, R., Song, W., Zhou, J. and Cen, K. 2014. Enhancement of energy production efficiency from mixed biomass of chlorella pyrenoidosa and cassava starch through combined hydrogen fermentation and methanogenesis. Applied Energy 120: 23–30.

Xing, Y., Li, Z., Fan, Y. and Hou, H. 2010. Biohydrogen production from dairy manures with acidification pretreatment by anaerobic fermentation. Environmental Science and Pollution Research 17(2): 392–399.

Xu, L., Wang, Y., Shah, S.A.A., Zameer, H., Solangi, Y.A., Walasai, G.D. and Siyal, Z.A. 2019. Economic viability and environmental efficiency analysis of hydrogen production processes for the decarbonization of energy systems. Processes 7(8): 494.

Yang, X., Wang, H., Strong, P.J., Xu, S., Liu, S., Lu, K., Sheng, K., Guo, J., Che, L., He, L. et al. 2017. Thermal properties of biochars derived from waste biomass generated by agricultural and forestry sectors. Energies 10(4): 469.

Yasin, N.H.M., Man, H.C., Yusoff, M.Z.M., Hassan, M.A. et al. 2011. Microbial characterization of hydrogen-producing bacteria in fermented food waste at different ph values. International Journal of Hydrogen Energy 36(16): 9571–9580.

Zahedi, S., Sales, D., Romero, L. and Solera, R. 2013. Hydrogen production from the organic fraction of municipal solid waste in anaerobic thermophilic acidogenesis: Influence of organic loading rate and microbial content of the solid waste. Bioresource Technology 129: 85–91.

Zhang, Q., Zhang, Z., Wang, Y., Lee, D.-J., Li, G., Zhou, X., Jiang, D., Xu, B., Lu, C., Li, Y. et al. 2018. Sequential dark and photo fermentation hydrogen production from hydrolyzed corn stover: A pilot test using 11 m3 reactor. Bioresource Technology 253: 382–386.

Zhang, T., Jiang, D., Zhang, H., Jing, Y., Tahir, N., Zhang, Y. and Zhang, Q. 2020. Comparative study on bio-hydrogen production from corn stover: photo fermentation, dark-fermentation and dark-photo co-fermentation. International Journal of Hydrogen Energy 45(6): 3807–3814.

Zhang, Z., Li, Y., Zhang, H., He, C. and Zhang, Q. 2017. Potential use and the energy conversion efficiency analysis of fermentation effluents from photo and dark fermentative bio-hydrogen production. Bioresource Technology 245: 884–889.

Zhang, Z.-P., Show, K.-Y., Tay, J.-H., Liang, D.T., Lee, D.-J. and Jiang, W.-J. 2007. Rapid formation of hydrogen-producing granules in an anaerobic continuous stirred tank reactor induced by acid incubation. Biotechnology and Bioengineering 96(6): 1040–1050.

Zhu, S., Zhang, Z., Li, Y., Tahir, N., Liu, H. and Zhang, Q. 2018. Analysis of shaking effect on photo-fermentative hydrogen production under different concentrations of corn stover powder. International Journal of Hydrogen Energy 43(45): 20465–20473.

CHAPTER 6

Microbial Roles in Second Generation Bioethanol

Abu Yousuf,[1] *Md. Shahadat Hossain*[1,2,]* and *Md. Anisur Rahman*[1,3]

1. Introduction

1.1 Fossil Fuels and Drawbacks

Fossil fuels are mainly used in energy generation and transportation sector. Comparatively a smaller percentage of fossil fuels are used in non-energy usage and various chemicals' production (Brockway et al. 2019). Overall use of the fossil fuels is presented in Fig. 1, but these fossil fuels reserves are declining continuously. Studies have estimated that fossil fuels reserves could get exhausted between 2070 and 2090 based on different economic growth (Fig. 2) (Stephens et al. 2010).

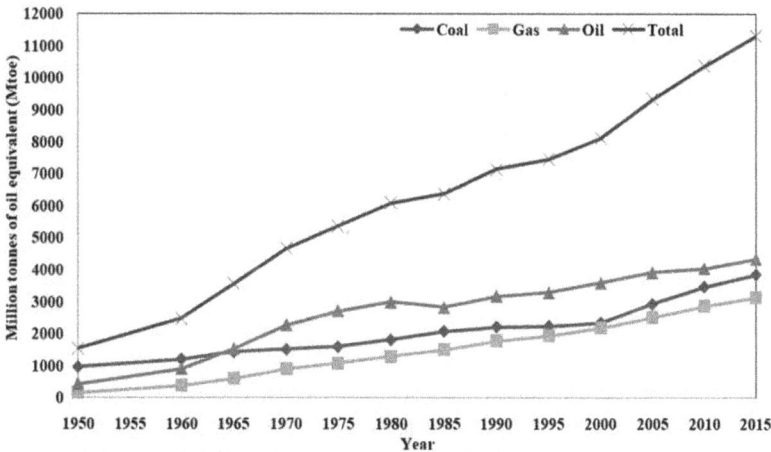

Fig. 1. Fossil fuels consumption between 1950 and 2015 (Pirani 2018).

[1] Department of Chemical Engineering and Polymer Science, Shahjalal University of Science and Technology, Sylhet - 3114, Bangladesh.
[2] Department of Chemical Engineering, State University of New York College of Environmental Science and Forestry, Syracuse, NY 13210, USA.
[3] Department of Chemical Engineering, University of Massachusetts Amherst, Amherst, Massachusetts, USA.
* Corresponding author: mhossain2@esf.edu

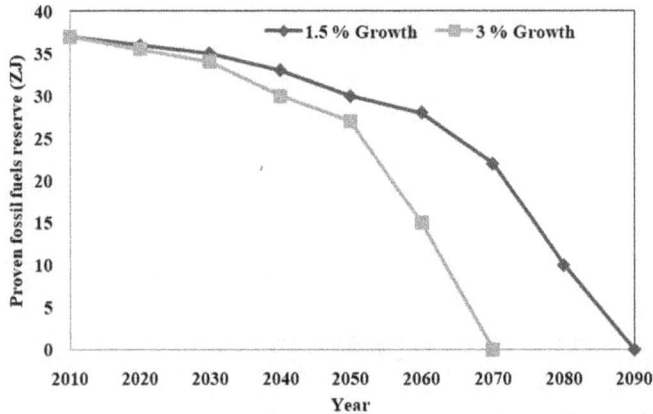

Fig. 2. Fossil fuels depletion trend (on the basis of 6.7 billion people in 2008 and 9.2 billion in 2050) (Stephens et al. 2010).

In addition, political instability in the middle eastern nations and the Organization of the Petroleum Exporting Countries (OPEC) frequently hampers both the crude oil production and exportation. This instability badly impacts the fossil fuels' price and use worldwide (Khan et al. 2021). Moreover, the increasing rate of greenhouse gases (GHG) emissions (Fig. 3) from the production processes is another critical concern. To reduce those adverse environmental changes and limit the temperature increase below 2°C, it was agreed to cut the GHG emissions 25–40 percent in 2020 and 80–90 percent in 2050 at the Copenhagen Climate Change Summit (2009) (Altman and Jordan 2018, Stephens et al. 2010).

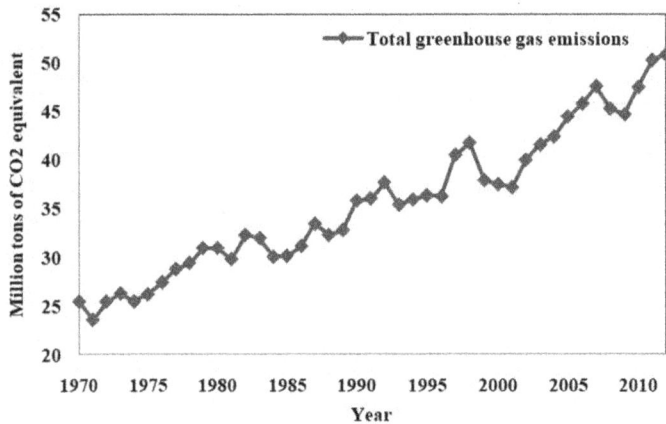

Fig. 3. Global GHG emissions between 1970–2012 (Group 2021).

Several countries across the globe have already planned to reduce fossil fuel burning and concentrate efforts on renewable energy harnessing. So far, most of the effort centers around electricity production from the solar, wind, photovoltaic, and geothermal sources but this electricity represents only one-third of the total world energy consumption (Brandon and Kurban 2017, Clark II and Rifkin 2006). Liquid fuels use in various sectors represents the other two-thirds of the energy consumption (Li et al. 2018). Biofuels production from renewable feedstocks can be a suitable alternative for fossil fuel-based liquid fuels.

1.2 Biofuels (Bioethanol) Production from the Renewable Feedstocks

Over the last few decades, significant efforts have been made to develop and investigate several biofuels such as biodiesel, bioethanol, biobutanol, and biomethane. All of these biofuels provide environmental benefits and are high potential alternatives to fossil fuels. However, bioethanol only has gained commercial success because of its excellent blending capability with the existing liquid fuels at various ratios. Bioethanol production research started in the early 80s with the processing of different annual food crops such as sugarcane and corn. Later, commercial exploration of bioethanol reached to about 29,000 million gallons in 2019 compared to only around 12,000 million gallons in 2009 (Fig. 4a). In this time period, both the United States and Brazil remained larger producer of bioethanol. For instance, United States and Brazil produced 54 percent and 30 percent bioethanol correspondingly in 2019 while European Union, China, Canada, and India produced smaller proportion of bioethanol at the same time (Fig. 4b). Although the bioethanol production from food crops (first-generation ethanol) was attractive from the yields and economics point of view, several challenges limit the long-term sustainability of these processes. These limitations compelled researchers to shift the research focus on biofuel production from lignocellulosic biomass, also known as second-generation biofuels. Lignocellulosic feedstocks, such as agricultural residues, forest residues, municipal waste, dedicated energy crops, are abundant and low-cost feedstock for biofuel production.

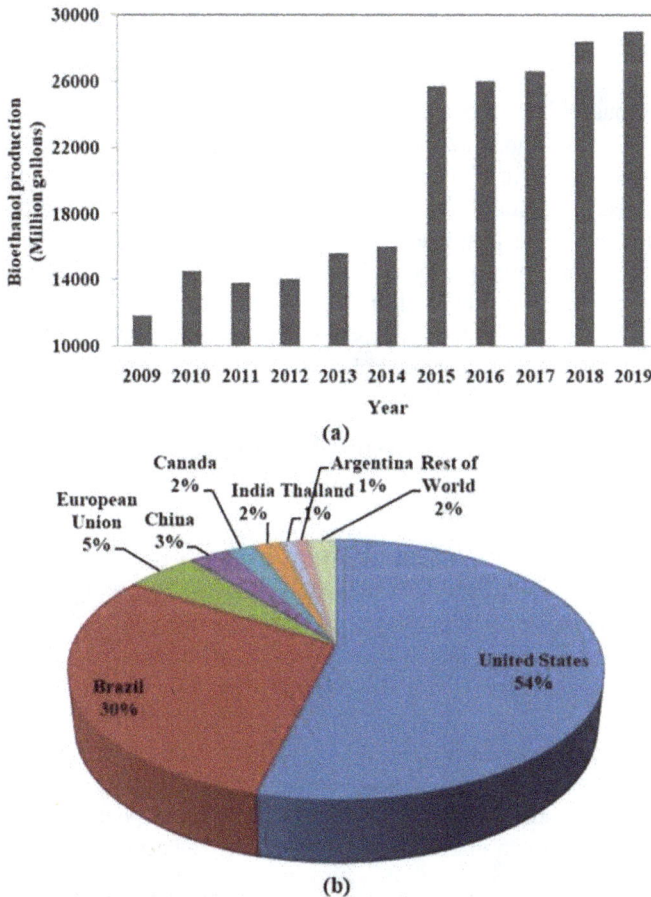

Fig. 4. (a) Annual bioethanol production from 2009–2019 and (b) regional share in bioethanol production in 2019 (Association; Zabed et al. 2017).

1.3 First Generation Biorefinery

As discussed previously, the commercial scale bioethanol is produced mainly from the food crops (Mohanty and Swain 2019). Although the economic and sustainable production route has already been defined for first-generation bioethanol, it creates skepticism in energy sector research due to several reasons. The first and foremost one is its competition with food and fiber production for fertilizer and water use (Wenger and Stern 2019). With some uncertainties and controversy, many institutions and literature have reported rapid price increment of food products for both humans and animals, which is due to first generation biorefineries (Javed et al. 2019). In addition, many governments are providing subsidies to make large production and processing costs of first-generation biorefineries competitive with petroleum refineries. This also hinders the development of other essential sectors such as health, education, and social security (Pino et al. 2018). Additionally, varying assessment of net greenhouse gas emissions and disruption in the nearby ecosystems by distillation residues of first-generation biorefineries drives the biorefinery research to utilize alternative renewable feedstock.

1.4 Second Generation Biorefinery

Second-generation biorefineries utilize non-food feedstocks, mainly lignocellulosic biomass, to produce liquid transportation fuels and bioproducts. The use of these lignocellulosic biomass in second-generation biorefineries offers several advantages (Hoang et al. 2021). Second-generation biofuels production does not compete with food for fertile land while previously barren and infertile land can be used for dedicated energy crops plantation; they reduce the greenhouse gas emissions and improve the soil quality for future cultivation (Redondo-Gómez et al. 2020). As discussed in the previous section, first-generation biorefineries use only the grain or seed part of the annual food crops while second-generation biorefineries can produce biofuels from the agricultural wastes from food crops. Previously, those agricultural residues that remained as leftover in the field, or burnt in some cases, produced GHG emissions (Sur et al. 2021). The utilization of agricultural residues in second-generation biorefineries for biofuels, chemicals, and electricity generation without using additional agricultural land offers both societal and environmental benefits. The utilization of microalgal biomass for biofuels production is classified as third-generation biorefinery but this remains out of scope for this chapter. This chapter discusses about only second generation biorefinery.

2. Feedstock for Second Generation Biorefinery

2.1 Sources of Biomass

Biomass for the second generation biorefinery is broadly categorized as woody and non-woody biomass in Table 1 (Kumar and Verma 2021). Both types of biomass contain mainly lignocellulosic material which is a complex matrix of cellulose, hemicellulose, and lignin. Potential sources of both woody and non-woody biomass are forest, agricultural land, and different waste materials (Vallejos et al. 2017). Agricultural land and wastes provide both woody and non-woody biomass whereas forest provides only woody biomass. Forests supply whole tree chips, bole and log chips, stalks, and stems as woody biomass. In contrast, agricultural land is the main and attractive source of lignocellulosic biomass since it provides a wide variety of biomass, for example, fast growing trees and energy crops (Nanda et al. 2020). Different fast-growing hardwood trees and perennial energy crops (Table 1) are grown in agricultural lands which consequently serve as lignocellulosic biomass in both the short and long-term (Pablo et al. 2020). Moreover, agricultural residues leftover in agricultural lands after collecting grain parts of annual crops can be used as feedstocks for the second generation biorefineries that provide an abundant supply of carbon at low cost. In addition,

Table 1. Source of lignocellulosic feedstock for second generation biorefinery.

Source of Biomass	Type of Biomass	
	Woody Biomass	**Non-woody Biomass**
Forests	Logs	–
	Bole chips	–
	Bark chips	–
	Tree chips	–
Agricultural lands	**Fast-growing hard wood tree** e.g., Poplars (*Populus* spp.) Willows (*Salix* spp.) Sycamore (*Platanus occidentalis* L.) Sweetgum (*Liquidambar styraciflua* L.) Yellow poplar (*Liriodendron tulipifera* L.) Eucalyptus (*Eucalyptus* spp.)	**Perennial lignocellulosic energy crops** e.g., Miscanthus (*Miscanthus giganteus*) Giant reedgrass (*Arundo donax*) Reed canary grass (*Phalaris arundinaces*) Elephant grass (*Pennisetum purpureum*) Switchgrass (*Panicum virgatum*)
		Lignocellulosic agricultural residue e.g., rice straw, corn stover, wheat straw, sorghum stalks
Waste materials	Sugarcane bagasse	Municipal waste Kitchen waste
	Rice husk	
	Saw dust	
	Pulp and Paper waste	

different industrial waste streams, for instance, sugarcane bagasse, pulp and paper waste, rice husk, and saw dust as well as municipal solid waste materials can be used feedstock for the second generation biorefineries (Liu et al. 2021).

2.2 *Composition of Lignocellulosic Biomass*

Lignocellulosic biomass mainly consists of carbohydrate polymer cellulose and hemicellulose and phenolic polymer lignin (Senatore et al. 2021). The proportion of cellulose, hemicellulose, and lignin in biomass varies and this variation depends on a specific plant and land type, fertilizer used, and surrounding ecosystem (Xin-Qing Zhao et al. 2011).

a. Cellulose is a polysaccharide and consists of linear polymer chains of cellobiose (glucose-glucose dimer) which are linked together by $\beta - 1, 4$ – glycosidic bonds. In addition, hydrogen bond and Van der Waals forces also exist between polymer chains (Wang et al. 2017) which creates a rigid and crystalline structure of cellulose inside the lignocellulosic biomass.

b. Hemicellulose is an amorphous polysaccharide and composed of various hexose and pentose sugars and a small amount of acetyl group. Hexose sugars present in the hemicellulose are mannose, galactose, and glucose whereas pentose sugars are xylose and arabinose (Avanthi et al. 2017). Sugar molecules are linked together by $\beta - 1, 4$ linkages and form a highly branched short amorphous polymeric chain.

c. Lignin is a long non-sugar polymeric chain and it along with the hemicellulose creates a protective cover for cellulose. Lignin forms the cell wall and reinforces cell together. Although it is a residue for the biofuel production process, it contains a significant portion of the total lignocellulosic biomass. Hence, it is economical as well as environmentally friendly in biorefinery operation to use lignin in the combined heat and power production (CHP) process (Amiri et al. 2019).

In addition to the above constituents, lignocellulosic biomass contains moisture, ash, acids, minerals, and extractives (Demirbas 2017).

3. Conversion Process for Second-Generation Biorefinery

A specific conversion process is required to convert specific lignocellulosic biomass into desired biofuels and value-added chemicals in second-generation biorefinery operation. Overall biomass conversion processes in second-generation biorefinery can be divided into four categories: physical, thermochemical, biochemical, and hybrid conversion processes (Yamakawa et al. 2018). Since physical conversion processes—briquetting, pelletizing, and fiber extraction–produce only solid biofuels, those processes are excluded from the discussion in this chapter.

3.1 Thermochemical Conversion Process

Controlled biomass conversion can be carried out in thermochemical conversion process, pyrolysis, and torrefaction, to produce biofuels (Lewandowski et al. 2020). However, those thermochemical conversion processes cannot utilize biomass energy potential properly; hence, those processes are often combined with other conversion processes for effective biomass conversion. Different gasification technologies are also included as thermochemical conversion processes. The gasification process converts the lignocellulosic biomass mainly into a mixture of gas and synthesis gas. It is carried out at a wide range of elevated temperature and pressure, 500–1400°C and 1–33 bar, in presence of an oxidizer (Inayat et al. 2019). Air, steam, pure oxygen, and mixture these gases can be used as an oxidizer for synthesis gas production.

3.2 Biochemical Conversion Process

Biochemical conversion processes use microorganisms to exploit biomass energy into both liquid and gaseous biofuels. Two biochemical conversion processes—anaerobic digestion and fermentation - are most widely used for biofuel production (Hossain 2019). Anaerobic digestion is a combination of several microbial steps such as hydrolysis, acidogenesis, acetogenesis, and methanogenesis for biofuels and chemicals co-production in absence of oxygen from the biomass feedstock (Maneein et al. 2018). Another biochemical conversion process, fermentation, has been used for liquid fuel (alcohol) production. Fermentation can be defined as a biochemical conversion process where both natural and genetically modified microorganisms act on various hydrocarbons for fuels and chemicals production in absence of oxygen (Inui et al. 2017). These microorganisms can be bacteria, yeast, and fungi (Geertje van Hooijdonk 2005) which can ferment both the starchy and lignocellulosic sugars to gain the cellular energy, but present day fermentation process mainly depends on the effective utilization of lignocellulosic sugars (Francois et al. 2020). Several potential microbial species–yeast (*Saccharomyces cerevisiae*) (Liu and Hu 2010), bacteria (*Escherichia coli*), and fungi (*E. oxy*) - have already been identified for lignocellulosic sugars fermentation process for bioethanol production (Lin and Tanaka 2006, Xiao-Jun Ji et al. 2012).

3.3 Hybrid Conversion Process

Although some conversion processes can be used standalone, the integrated processing approach (using multiple processes in a defined sequence) is found more effective in achieving high fuel yields. For instance, physical conversion processes can be used for size reduction, and then desired sized biomass can be processed in a thermochemical conversion process for biofuel production. Even physical, thermochemical, and biochemical conversion processes could be employed altogether for single biomass conversion into biofuel. For example, physical and thermochemical conversion processes are used for size and moisture reduction correspondingly. Then processed biomass is converted into synthesis gas by thermochemical conversion process (gasification), which is subsequently converted into bioethanol through fermentation process. Such types of other hybrid conversion processes are summarized in Table 2.

Table 2. Conversion processes for second generation biorefinery (Aguilar-Reynosa et al. 2017, Yamakawa et al. 2018).

Conversion Type	Conversion Process	Conversion Product
Physical conversion process	Briquetting	Highly dense solid fuel block
	Pelletizing	Solid fuel
	Fiber extraction	Fiber
Thermal conversion process	Direct burning	Heat and electricity
	Pyrolysis	Tar and bio-char
	Torrefaction	Bio-char
Thermochemical conversion process	Indirect circulating fluidized bed gasification (iCFBG)	Synthesis gas
	Entrained-flow gasification (EFG)	Synthesis gas
	Moving bed gasification	Synthesis gas
Biochemical conversion process	Anaerobic digestion	Biomethane, biohydrogen, biofertilizer
	Fermentation	Bioethanol, and chemicals (furfural, acetic acid etc.)
Hybrid conversion process	Gasification and fermentation	Bioethanol and chemicals

4. Biochemical Conversion – Yeast Based Bioethanol Production

The biochemical conversion of lignocellulosic biomass contains two major steps: (i) hydrolysis of structural carbohydrates (cellulose and hemicellulose) to sugar monomers using enzymes and (ii) fermentation of sugar monomers to ethanol. However, the natural recalcitrance in the lignocellulosic biomass structure does not allow the direct conversion of cellulose and hemicellulose to sugar monomers. A pretreatment step is used to reduce recalcitrance by breaking the complex matrix of cellulose, hemicellulose, and lignin. Microbial roles for the pretreatment of such type of lignocellulosic biomass are discussed first in details in this section. Then separate enzymatic hydrolysis and fermentation of sugars (using genetically modified *S. cerevisiae*) are discussed.

4.1 Biomass Pretreatment

Considering a critical step in the biochemical conversion process, a large number of pretreatment processes have been developed and investigated on lignocellulosic biomass. The pretreatment processes can be broadly classified as: physical pretreatment, chemical pretreatment, physicochemical pretreatment, and biological pretreatment.

Due to stringent operating conditions and increased environmental concern, chemical pretreatment is generally avoided for lignocellulosic biomass pretreatment (Kumar et al. 2020). Similarly, higher capital cost is associated with physical pretreatment processes because of sophisticated equipment for creating mechanical forces for biomass depolymerization (Kumar et al. 2020). Besides, this chapter discusses various microbial roles only for lignocellulosic ethanol production; therefore, biological pretreatment, utilizing different microbial species, is discussed in this section.

In biological pretreatment, a wide variety of bacteria, fungi, and some other microorganisms are used to depolymerize the complex structure of lignocellulosic biomass. Those microbial communities secrete the extracellular cellulolytic, hemicellulolytic, and ligninolytic enzymes (on the basis of specific microorganism used) for biomass depolymerization. A subsequent enzymatic hydrolysis and fermentation process then converts the lignocellulosic sugars into the bioethanol. Several bacterial species—*Cellulomonas* spp., *Thermomonospora* spp., *Clostridium* spp., *Bacillus* spp. etc.—and fungi species, for example, *Trichoderma* spp., *Schizophyllum* spp., *Orpinomyces* spp. etc., are widely known to be used in the biological pretreatment of lignocellulosic biomass (Sharma et al. 2019).

Among the bacterial species, *Cellulomonas fimi* and *Thermomonospora fusca* are widely used in cellulase enzyme secretion for biological pretreatment. Both of the species produce enzyme in higher quantity and concentration (Sharma et al. 2019). Another cellulolytic bacterium, *Paenibacillus campinasensis*, is also extensively used because of sustainability of this species in severe pretreatment conditions. Rumen bacteria—*Fibrobacter succinogenes*, *Ruminococcus flavefaciens*, and *Ruminococcus albus* etc.—have notable mechanism to attach with the cellulosic fraction of biomass and subsequent effective hydrolysis into simple sugars (Liang et al. 2020). *Clostridium thermocellum* and *Bacteroides cellulosolvens* are anaerobic bacteria which produce higher quantity of cellulase enzyme but the concentration of such enzyme is not high enough for the effective depolymerization of biomass. However, *Zymomonas mobilis* is an exceptional anaerobic species which has been extensively studied in biomass pretreatment due to producing cellulase enzyme both in higher amount and concentration (Yang et al. 2018). Several gram positive bacteria, *Firmicutes* spp., and gram negative bacterial, *Pseudomonas*, *Rahnella*, and *Buttiauxella* strains have also been identified as potential cellulolytic microorganism for biomass pretreatment (Sharma et al. 2019).

Fungal species are well known for lignocellulosic decrystallization since those species have extracellular enzymes for cellulose, hemicellulose and lignin (in some cases) fraction of biomass hydrolysis. Various ascomycetes, basidiomycetes, and some anaerobic fungal species have been reported for their lignocellulolytic activity (Zabed et al. 2019). One of the ascomycetes fungi species, *Trichoderma reesei,* secretes xylanases and β-glucosidase in sufficient amount with comparatively higher cellulase activity for efficient biomass pretreatment (Ummalyma et al. 2019). Another promising ascomycetes fungi is *Trichoderma longibrachiatum*. This soil fungi species secretes all three types of cellulase enzymes (endoglucanases, exoglucanases, and β-glucosidases) which work synergistically for releasing simple sugars from the lignocellulosic biomass (Sharma et al. 2019). *Schizophyllum* spp., *Phanerochaete chrysosporium*, and *Fomitopsis palustris* are the notable basidiomycetes fungi species for the lignocellulosic biomass biological pretreatment (Andlar et al. 2018). *Orpinomyces* spp. is an example of anaerobic fungi species for a similar type of biomass pretreatment. Since lignin creates a protective cover for the cellulosic and hemicellulosic sugars, delignification is a crucial step of the lignocellulosic bioethanol production. White rot fungi species are reported in literatures for their natural capabilities of lignin degradation. This microbial species secretes variety of ligninolytic enzymes such as lignin peroxidases and manganese peroxidases for natural depolymerization of lignin. Among them, *Phellinus pini-2, Pholiota mutabilis,* and *Phlebia brevispora-1* white-rot fungi are reported for higher lignocellulosic biomass hydrolysis rate (Sahay 2022).

Microbial activities depend on the various physical, chemical, and biological parameters. Those parameters largely affect the microbial growth and their extracellular enzyme activity which finally influence the depolymerization of lignocellulosic biomass. Temperature is one of the important physical parameters and bacterial and fungal species maintain their activity over a wide array of temperature. Mostly those species are at their optimal activity state in the mesophilic temperature range (20 to 45°C) (Baruah et al. 2018). Very few bacteria can maintain their activity in the psychrophilic temperature range (−15 to 10°C). Moisture content is also a crucial physical parameter for the microbial growth. Very low moisture content hinders the growth while high moisture content is responsible for creating anaerobiosis. 40–80 percent moisture content is considered as optimal for most of the bacteria and fungi species (Baruah et al. 2018). Retention time for the biological pretreatment varies depending on the biomass type and microorganism involved. Longer retention time increases the delignification, at the same time it also reduces the amount of cellulosic and hemicellulosic sugar depolymerization. Hence, optimization of retention time is carried out for maximum monomeric sugar and bioethanol yield. pH of the culture medium is a critical chemical parameter as microbial metabolic activity as well as enzymatic activity depends on it. It is seen that pH value decreases after startup of biological pretreatment which increases the efficiency of ligninolytic enzymes (Bhutto et al. 2017). Generally, ligninolytic enzymes work better at lower pH value, for example, most of the white rot fungi prefer acidic conditions. However, much lower and higher pH value decreases the efficiency of the enzyme. Efficiency of the cellulase enzyme reduces at lower pH while this enzyme dissolves at higher pH. Apart from others those, structural complexity, loss of polysaccharides during pretreatment, and utilization of microbial co-culture are considered as other important parameters for microorganisms during biological pretreatment.

4.2 Separate Hydrolysis and Fermentation (SHF) Process

This biomass processing area consists of two different steps: enzymatic hydrolysis and fermentation. In the enzymatic hydrolysis area, mainly cellulose fraction of biomass is hydrolyzed into monomeric glucose sugar by cellulase enzyme and smaller hemicellulose fraction is also hydrolyzed into xylose, mannose, and galactose etc., sugar. In contrast, in the fermentation area, mainly glucose, but other sugars (in smaller extent) such as xylose, mannose, galactose etc., produced in the preceding steps, are fermented into bioethanol in presence of a suitable microorganism. Since co-fermentation of both pentose and hexose sugars are advantageous in many aspects, genetically modified *Saccharomyces cerevisiae*'s role is described in this section for that co-fermentation purpose. Some commonly used genetically modified *S. cerevisiae* species for bioethanol production are listed in Table 3. Genetically engineered *S. cerevisiae* can metabolize xylose sugar in pentose phosphate pathway (Cunha et al. 2019). Xylose is converted into xylulose at first in this pathway and subsequent phosphorylation of xylulose produces xylose-5-phosphate, which are further metabolized in the bioethanol production process. This xylose-assimilating pathway is totally absent in the native *S. cerevisiae* species but these types of pathways are available in the fungal and bacterial species (Kwak and Jin 2017). Oxidoreductase xylose-assimilating pathway is found in the fungi species while isomerase xylose-assimilating pathway is available in the bacterial species. Oxidoreductase pathway is a two-step enzymatic process where xylose is firstly reduced by the xylose reductase (XR) enzyme into xylitol in presence of NADPH cofactor. In the second step, xylitol is further oxidized by xylitol dehydrogenase (XDH) enzyme to xylulose, in the presence of NADH cofactor. There is always a possibility of cofactor imbalance in this pathway, resulting in xylitol accumulation and bioethanol production inhibition (Cunha et al. 2019). In contrast, isomerase xylose-assimilating pathway is a one-step enzymatic process for xylose to xylulose metabolism by xylose isomerase enzyme without any cofactor use. The xylose-assimilating pathways enable the xylose metabolism of native *S. cerevisiae* species that result in bioethanol production. This metabolism rate in some cases is still lower than the glucose metabolism rate which requires further rational and inverse metabolic engineering of the *S. cerevisiae* species (Zhu et al. 2021). Through the rational metabolism endogenous hexose and heterologous xylose transporters, cofactors, one or more type of specific genes (*XKS1* and *TAL1*), and certain type of metabolic pathways (acetate biosynthesis and ammonia assimilation) are overexpressed in the native *S. cerevisiae* for inhibitor tolerance and improved bioethanol yield. By the inverse metabolic engineering, specific types of genes, such as *GRE3*, *PHO13*, and *YLR042C*, are deleted from the native *S. cerevisiae* for inhibitor

Table 3. *S. cerevisiae* used in lignocellulosic sugar fermentation for bioethanol production.

Lignocellulosic Biomass	Genetically Modified *S. cerevisiae* Use		Bioethanol Production	Reference
	S. cerevisiae Species	*S. cerevisiae* Loading		
Olive tree trimmings	*S. cerevisiae* CECT1170	1.0 g/L	39.00 g/L	(Requejo et al. 2011)
Corn stover	*S. cerevisiae* strain Y73[a]	1.0 OD$_{600}$	28.00 g/L	(Liu et al. 2014)
	S. cerevisiae strain Y128[b]	1.0 OD$_{600}$	30.00 g/L	
Japanese red pine	*S. cerevisiae* ATCC 26603	2% (v/v)	16.7 g/L	(Kalyani et al. 2013)
	S. cerevisiae ATCC 26603 and *P. stipitis* KCCM 12009	2% (v/v) at 1:1 ratio	21.6 g/L	
Almond, walnut, and pine wood mixture	*S. cerevisiae* NS 22273	–	27.7 g/L	(Barcelos et al. 2021)
Spruce wood	*S. cerevisiae* TMB3400	–	32.9 g/L	(Olofsson et al. 2010)
Yellow poplar	*S. cerevisiae* ATTC 26603	1 g/L	42.80 g/L	(Kim et al. 2015)
Silver fir wood	*S. cerevisiae* YSC2	–	52.0 g/L	(Senila et al. 2020)

[a] use of xyitol dehydrogenase and xylose reductase pathway, [b] use of xylose isomerase pathway.

accumulation reduction and tolerance improvements as well as specific set of genes are expressed in native species for bioethanol yield enhancement.

In biorefinery operation, enzymatic hydrolysis of biomass slurry (after pre-treatment) is mostly carried out in two reactors—firstly in continuous high solid reactors and then in batch bioreactors. Cellulase and xylanase enzymes are added in the continuous high solids reactors for partial hydrolysis of cellulosic and hemicellulosic sugars. Then further saccharification or hydrolysis is carried out in the batch bioreactors. Varying amounts of enzymes are used (Table 4) based on the biomass processed for the bioethanol production and other crucial operating parameters, such as temperature, residence time, pH, solid loading etc. Enzymatic hydrolysis mostly produces monomeric glucose and xylose sugar. Although monomeric sugar yield increases with the increased amount of enzyme loading, extra enzyme usage incurs additional operating cost (Humbird et al. 2010, Maslova et al. 2019). Optimum enzyme loading along with other operating conditions of enzymatic hydrolysis from several studies are listed in Table 4.

Table 4. Operating conditions of enzymatic hydrolysis in SHF process.

Parameters	Enzymatic Hydrolysis Conditions					
	(Chu et al. 2018)	(Requejo et al. 2011)	(Liu et al. 2014)	(Kalyani et al. 2013)	(Barcelos et al. 2021)	(Olofsson et al. 2010)
Cellulase loading	20 FPU/g	70.1 FPU/mL	11.55 mg/g	22.5 FPU/g	40 FPU/g	42 FPU/g
β-glucosidase loading	10 U/g	630 IU/mL	–	–	–	20 IU/g
Xylanase loading	–	–	3.45 mg/g	–	95 U/g	340 IU/g
Temperature	50°C	35°C	50°C	30°C	50°C	34°C
Residence time	48 hr	–	72 h	–	-	–
pH	4.8	5.0	–	5.2	72 h	–
Stirring	180 rpm	120 rpm	45 rpm	150 rpm	45 rpm	–
Solid loading	10% (w/v)	10% (w/v)	–	15% (w/v)	–	8% (w/v)
Biomass feedstock	Eucalyptus wood chips	European olive trimmings	Corn stover	Japanese red pine wood	Almond, walnut, and pine wood	Spruce wood chips

Saccharified sugars from the batch bioreactors after enzymatic hydrolysis are split into two streams. A small percentage (10 percent) of sugar is sent to *S. cerevisiae* yeast culture and the optimum conditions and sugar metabolism involved with the culture medium are summarized in Table 5 and Table 6 accordingly.

Table 5. *S. cerevisiae* yeast culture conditions (Geberekidan et al. 2019).

Saccharified sugar level	10 vol % after enzymatic hydrolysis
Batch time	24 h
Fermentor turnaround time	12 h
Number of trains	2
Number of fermentor stages	5
Maximum fermentor volume	757 m^3
Nutrient loading	0.50 (w/w) %
DAP loading	0.67 g/L fermentation broth

Table 6. Reaction and conversion of yeast culture and fermenter (Bouaziz et al. 2020, Cunha et al. 2020, Humbird et al. 2011).

Reaction	Reactant	Reactant Conversion
Glucose $\rightarrow$ 2 Ethanol + 2 CO_2	Glucose	90.0%
Glucose + 0.047 Nutrients + 0.018 DAP $\rightarrow$ 6 *S. cerevisiae* + 2.4 H_2O	Glucose	4.0%
Glucose + 2 H_2O $\rightarrow$ 2 Glycerol + O_2	Glucose	0.4%
Glucose + 2 CO_2 $\rightarrow$ 2 Succinic Acid + O_2	Glucose	0.6%
3 Xylose $\rightarrow$ 5 Ethanol + 5 CO_2	Xylose	80.0%
Xylose + 0.039 Nutrients + 0.015 DAP $\rightarrow$ 5 *S. cerevisiae* + 2 H_2O	Xylose	4.0%
3 Xylose + 5 H_2O $\rightarrow$ 5 Glycerol + 2.5 O_2	Xylose	0.3%
Xylose + H_2O $\rightarrow$ Xylitol + 9.5 O_2	Xylose	4.6%
3 Xylose + 5 CO_2 $\rightarrow$ 5 Succinic Acid + 2.5 O_2	Xylose	0.9%

Rest of the sugars (90 percent of the total saccharified sugars), along with nutrients, diammonium phosphate (DAP), and co-fermenting *S. cerevisiae*, are fed into a batch fermenter. Reactions involved with the fermentation process and operating conditions of the fermenter are represented in Table 6 and 7 respectively. *S. cerevisiae* co-ferment most of the soluble hexose and small amount of pentose sugars mainly into bioethanol. Among the hexose sugars, *S. cerevisiae* mainly converts glucose sugars through fermentation process. Up to 90 percent of the available glucose sugar present in the fermentation broth can be converted into bioethanol. Some authors also reported 4.0 percent of the available sugar usage in the yeast growth while very low percentages of the glucose are converted into glycerol (0.4 percent) and succinic acid (0.6 percent). Rest of the glucose remains unconverted during the fermentation process which is recycled for sugar recovery. Although naturally available *S. cerevisiae* cannot ferment the pentose sugar, for example, xylose, genetically modified *S. cerevisiae* able to ferment xylose sugar into bioethanol, maximum 80 percent conversion of that sugar is reported in various studies. Significant amounts of xylose (4.6 percent) are converted into xylitol and 4 percent of the xylose sugar are used up in yeast growth. Smaller amounts of glycerol and succinic acid are also produced from the xylose fermentation. In addition, huge amounts of dissolved CO_2 are also produced during this xylose fermentation process which are been separated later in a flash tank. Produced CO_2 is released into the atmosphere after treatment (scrubbing) while raw bioethanol is sent to the product recovery area.

Table 7. Operating conditions of fermentation reactor (Jansen et al. 2017, Sato et al. 2016).

Parameter	Fermenter
Organism	*S. cerevisiae* GLBRCY87
Genetic modifications	XR/XDH pathway addition from *Scheffersomyces stipitis* fungi species
Key improvements	Glucose-Xylose sugar co-fermentation and hydrolysate inhibition resistant
Temperature	32°C
Initial fermentation solids level	19.8% total solids
Residence time	1.5 days
Inoculum level	10 vol %
Nutrients level	0.25 (w/w) %
DAP level	0.33 g/L fermentation broth

4.3 Fuel Grade Bioethanol Recovery Process

Bioethanol product recovery area separates fermenter product into bioethanol, solid lignin, and CO_2. This processing area contains two distillation columns, one molecular sieve adsorption column, and water scrubber (Bhatia et al. 2019). Fermenter product is fed into the first distillation column (beer column) which separates all carbon dioxide and water as top product and lignin as bottom product. Bioethanol is recovered as vapor side-stream from this distillation column. Then bioethanol vapor stream enters into the second distillation column—rectification column; this increases the concentration of bioethanol up to azeotropic level (92.5 percent) (Haigh et al. 2018). Finally, a molecular sieve adsorption column is used to produce fuel grade bioethanol (99.5 percent) from the azeotropic concentration (Bezerra et al. 2020). Without this, CO_2 streams from beer column and SHF area are scrubbed into a water scrubber to recover bioethanol; it is recycled back to beer column. Design specifications of both beer and rectification columns for bioethanol production are summarized in Table 8.

Table 8. Design specification of beer and rectification distillation column (Bezerra et al. 2020, Hossain et al. 2019).

Parameter	Beer Column	Rectification Column
Number of stages	32	45
Efficiency of stage (%)	48	76
Feed stage location	4 (from the top)	33 (from top)
Reflux ratio	3	3.5
Operating pressure (atm)	2	1.6

5. *Clostridium* Acetogenic Bacteria Based Hybrid Conversion Route

Woody biomass and wood-based residues are the largest source of lignocellulosic biomass and there is a microbial pathway named Wood-Ljungdahl pathway (WLP) for bioethanol production from this biomass. This pathway requires gaseous feed—cleaned and conditioned synthesis gas— for bioethanol production. Physical pretreatment processes, gasification process, and subsequent synthesis gas conditioning are required for such synthesis gas production which is outlined in Fig. 5. Pre-treatment processes are used for woody biomass size reduction and drying. Then biomass gasification is carried out in various types of gasifiers for synthesis gas production. Since this gas contains several types of contaminants, extensive cleaning is required for synthesis gas conditioning before using in fermentation process. However, as this chapter solely describes microbial roles for bioethanol production, several biomass pretreatment, gasification, and synthesis gas conditioning

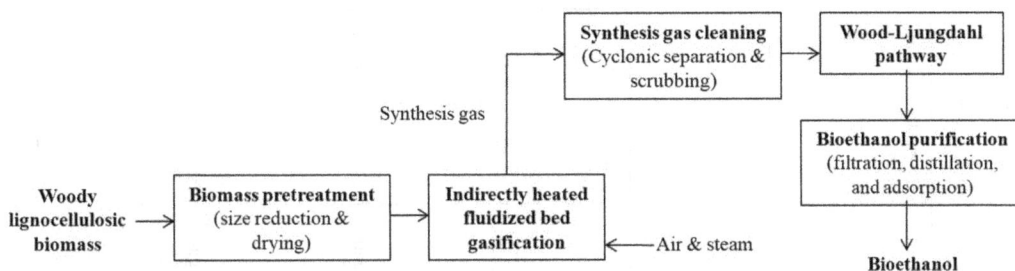

Fig. 5. Gasification coupled with Wood-Ljungdahl pathway for bioethanol production.

processes (based on physico-chemical mechanisms) for synthesis gas production are summarized in Table 9, instead of detailed discussion. Later, in synthesis gas fermentation process, acetogenic bacteria can be used for bioethanol production. This type of acetogens uses CO or CO_2 and H_2 as their primary source of carbon and produce acetyl-CoA for further bioethanol production. A wide variety of *Clostridium* species (tabulated in Table 10) have been studied and reported for bioethanol production, following the WLP. Table 10 shows that bioethanol concentration during synthesis gas fermentation increases at relatively lower pH. This lower pH has detrimental effect for most of the microbial species except acetogenic bacteria because of lower substrate and electron flow toward the microbial cell (Yasin et al. 2019). In case of acetogenic bacteria, acidogenesis shifts towards the solventogenesis at lower pH which results in higher ethanol concentration in the synthesis gas fermentation (Fernández-Naveira et al. 2017). Synthesis gas fermentation frequently encounters low mass transfer, and can result from the following processing steps: synthesis gas transfer to the gas-liquid interface, synthesis gas diffusion through the fermentation medium, and synthesis gas diffusion to the microbial cell. Mass transfer limitations in those processing steps can be improved by either increasing the solubility of the synthesis gas or reducing the surface tension of the fermentation medium (Shen et al. 2017). To lower the surface tension of the medium, most of the studies in Table 10 carried out synthesis gas fermentation at relatively higher stirring through CSTR reactors. On the contrary, synthesis gas solubility can be increased by lowering the fermentation medium temperature (Gaide 2017, Mohammadi et al. 2011). Therefore, Table 10 shows that studies reported higher bioethanol concentration because of improved synthesis gas solubility at lower fermentation temperature. Studies reported that lower bioethanol concentration in Table 10 could have higher concentration of CO_2 and/or CO in the synthesis gas used for the fermentation. Those components of the synthesis gas generate organic acids, mostly accompanied with the hydrogen gas production. Produced hydrogen can accumulate in the fermentation medium headspace and reverse the carbon flow in the *Clostridium* acetogen's metabolic cycle which could impede the ethanol production (Mohammadi et al. 2011).

5.1 Synthesis Gas Fermentation and Bioethanol Recovery

Clostridium species are mostly reported acetogenic bacteria for bioethanol production (Table 10). Cleaned and conditioned synthesis gas is fermented into bioethanol in presence of a *Clostridium* species. The microbial pathway *Clostridium* species follow for bioethanol production is WLP which is also known as acetyl-CoA pathway (Fig. 6). Synthesis gas is used as both carbon and energy source for this pathway (Monir et al. 2020b). Nutrients such as amino acids, vitamins, and metal ions are also present in the fermentation medium. Fermentation is carried out mainly in submerged anaerobic conditions. Operating conditions for synthesis gas fermentation processes typically range as following: temperature (35–42°C), pressure (0–5 bar), pH (4–6) (Bengelsdorf et al. 2018).

Before starting the WLP, 2 moles of reducing equivalent [H] are produced from the H_2 via a hydrogenease (HYA) enzyme. Same amount of reducing equivalents are produced from the CO and

Table 9. Woody and non-woody biomass source, pretreatment and gasification conditions for cleaned and conditioned synthesis gas production.

	Biomass	Biomass Pretreatment	Biomass Gasification	Synthesis Gas Conditioning	Reference
Woody biomass	Norway spruce, grey alder, and scots pine	Debarking, milling, and sieving; Pretreatment conditions Biomass particle size < 1 mm	Batch drop-tube gasifier, updraft gasifier, and fixed bed gasifier; Gasification conditions 750–850°C, isothermal operation, N_2 and O_2 flow rate - 0.3 L/min, equivalence ratio (ER) - 0.38	Nitrogen purging (5 min), cooling and condensation (4-6°C)	(Lyons Cerón et al. 2021)
	Japanese cedar and cypress trees timber waste	Pulverizing and drying; Pretreatment conditions Particle size - 2.0 to 2.8 mm, Drying - 105°C, 24 h	Fixed bed downdraft gasifier; Gasification conditions N_2 flow rate - 200 mL/min, steam (200°C) flow rate - 0.01 to 1.0 mL/min, steam to carbon (S/C) ratio - 1	Water and acetone - isopropanol scrubbing	(Koido et al. 2021)
	Wood chips and sewage sludge	Drying; Pretreatment conditions Feedstock moisture - 10 wt.%	Downdraft gasifier; Gasification conditions 642°C, air flow - 250 Nm³/h, ER - 0.351	–	(Gabbrielli et al. 2021)
	Wood	Steam drying; Pretreatment conditions Biomass humidity < 20%	Directly heated pressurized gasifier; Gasification conditions 800–900°C, O_2 flow rate - 1.3 kg/s, steam flow rate - 2.0 kg/s	Hot synthesis gas cleaning	(Codina Gironès et al. 2018)
Non-woody biomass	Palm oil mill waste and forest residues	Drying, milling, and sieving; Pretreatment conditions Biomass particle size < 250 μm, Drying - 100°C, 24 h	Entrained flow gasifier; Gasification conditions 700–900°C, ER - 0.2 to 0.4, biomass flow rate - 1.02 g/s, gasifying agent - air, carrier gas - N_2, occurrent flow	Cyclonic separation and condensation	(Ismail et al. 2019)
	Corn stover	Biomass drying; Pretreatment conditions Moisture content < 5 wt.%	Indirectly heated fluidized bed gasifier; Gasification conditions 800°C, circulating silicon dioxide, gasifying agent - 20% excess air, carrier gas - steam	Reforming (890°C), cooling (60°C), acid gas scrubbing	(Hossain et al. 2019)

Table 10. *Clostridium* species for synthesis gas (from lignocellulosic biomass) fermentation for bioethanol production.

Clostridium Species	Synthesis Gas Fermentation Condition	Bioethanol Concentration (mmol/L)	Reference
Clostridium ljungdahlii	37°C, 200 rpm, pH 6.8, 24 h, and use of nanoparticles	6.65	(Kim et al. 2014)
Clostridium butyricum	37°C, 200 rpm, pH 4.0–6.0, and 16 days	29.94	(Monir et al. 2020a)
Clostridium thermocellum	35°C, 120 rpm, pH 6.0, and 3 days	146.51	(Gupta et al. 2014)
Co-culture of *Clostridium thermocellum* and *Thermoanaerobacter pseudethanolicus*	–	60.00	(He et al. 2011)
Clostridium strain DBT-IOC-DC21	70°C, pH 7.0, and 96 h	19.48	(Singh et al. 2018)
Clostridium ljungdahlii (ATCC 55383)	37°C, 200 rpm, 60 h, and use of magnetic nanoparticles	10.61	(Kim and Lee 2016)

H_2O via carbon monoxide dehydrogenase (CODH) through biochemical water gas shift reaction (Gencic and Grahame 2020). Those reducing equivalents are utilized by the *Clostridium* to fix carbon molecules inside their cell biomass from the synthesis gas components during WLP. This pathway consists of two branches; one is the Eastern or Methyl Branch and another one is the Western and Carbonyl Branch.

Eastern or Methyl branch provides the methyl fraction of the acetyl-CoA and six electrons are used for CO_2 reduction in this branch (Song et al. 2018). In the first step, CO_2 is reduced to formate in presence of formate dehydrogenase (FDH) enzyme while second step is the one mole ATP utilizing step. Second step converts formate by tetrahydrofolate (THF) enzyme into formyl-THF. Once formyl-THF is formed, it undergoes subsequent reducing steps for methyl-THF formation. Then methyl fraction is transferred to the Cobalt (Co) center of the corrinoid-iron sulfur-protein (Co-FeS-P) to form the organometalic intermediate, [CH₃]-Co-FeS-P, by methyltransferase (MTR) while the THF is recycled back to another mol of formate reduction (Song et al. 2020).

The western branch provides the carbonyl fraction of acetyl-CoA. Here either 1 mole of CO_2 is reduced to carbonyl group or CO can be directly converted in carbonyl group in presence of CODH enzyme (Zhao et al. 2019). Bifunctional CODH then combines the carbonyl group with acetyl-CoA synthase (ACS) for CODH/ACS complex formation. After that CH₃ is separated from the CH₃-Co-FeS-P (CO-FeS-P is leaving to be used as carrier again) and combined with CO group of the CODH/ACS complex. As a result, acetyl metal is formed which is then combined with the coenzyme (CoA) to form acetyl-CoA while coenzyme SH-CoA is used. Acetyl-CoA can be converted into acetate, ethanol and other value-added products.

Acetate is formed mainly during the initial rapid cell growth stage of the microorganism when one mole of ATP is also formed, but this ATP is used up in the formate reduction step in the methyl branch (Abubackar et al. 2019). When cell growth enters into the stationary phase due to lower amount of nutrients in the fermentation broth and lower pH, ethanol is produced in two steps. In the first step, acetyl-CoA is reduced into acetaldehyde with the presence of acetaldehyde dehydrogenase (AAD). Then in the second step, acetaldehyde is reduced further to ethanol by alcohol dehydrogenase (ADH). However, acetate can also be converted into ethanol; acetate is firstly reduced by aldehyde ferredoxin oxidoreductase (AOR) to acetaldehyde and then into ethanol by ADH. This step for ethanol production is comparatively more preferable due to its lower concentration of undissociated acetic acid in the fermentation broth (Istiqomah et al. 2021). As mentioned earlier, this later step of ethanol production is preferred at lower microbial growth and nutrient conditions which is evident by the presence of at least 18 proteins (responsible for amino acid, sulfur, thiamine etc., transfer and metabolism) and 27 intracellular metabolites at higher concentrations, greater than 2000 and 1000 µmol/L

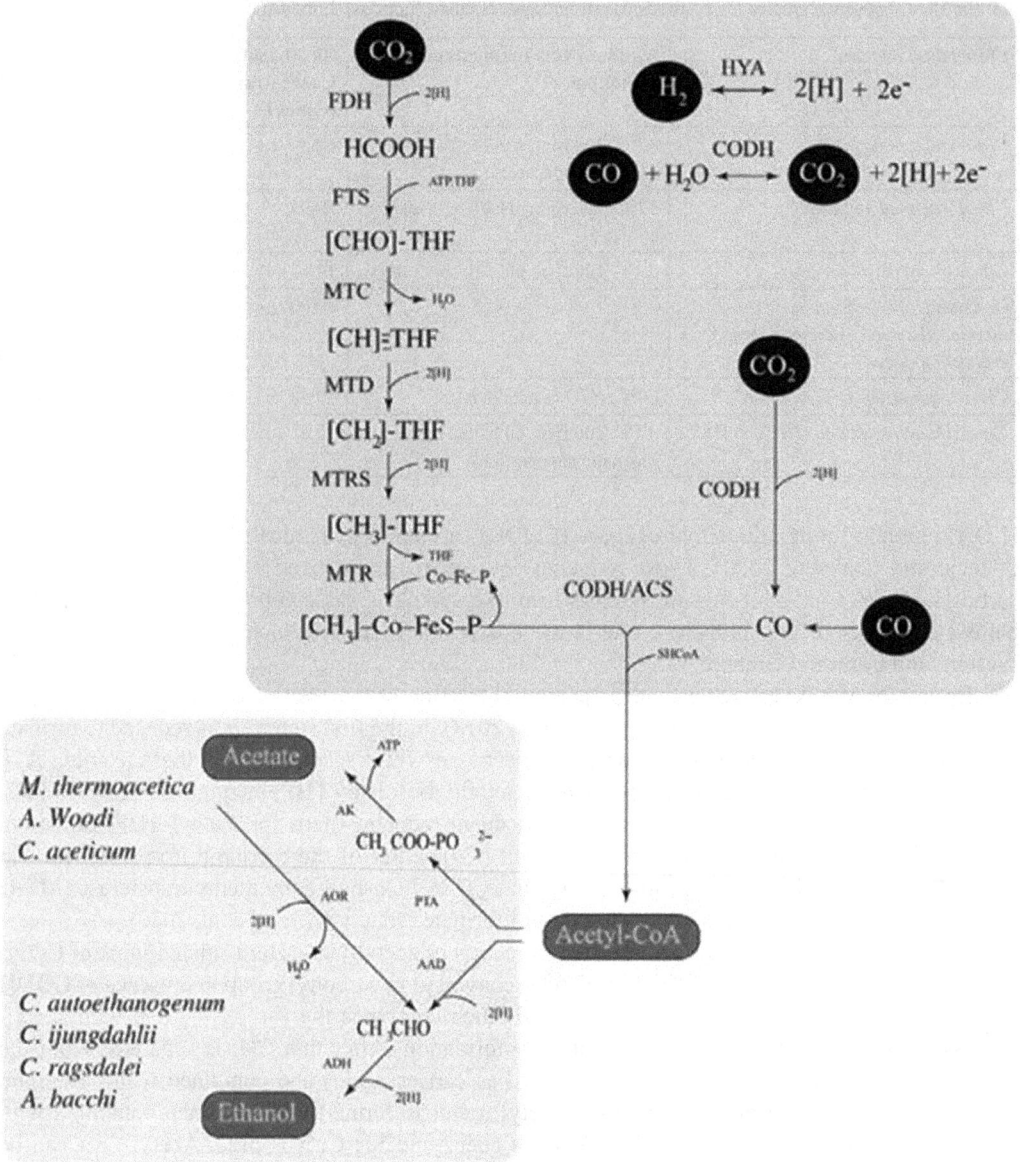

Fig. 6. Wood-Ljungdahl pathway or acetyl-CoA pathway for bioethanol production by *Clostridium* species (Monir et al. 2020b, Ragsdale and Pierce 2008). (Reprinted with the permission from Monir et al. (2020b)).

accordingly (Richter et al. 2016). At the same time, oxidation of synthesis gas continues, resulting in continuous production of reducing agents as well as WLP assimilates carbon for Acetyl-CoA formation while large amounts of acetate are also formed from carbon. This large acetate becomes intracellular by diffusion and undissociated acetic acid at lower pH. While acetic acid and reducing agents reach a critical concentration, both are transferred into ethanol production (because microbial growth is already at the stationary phase). In addition, abundance of ADH and AOR enzymes is coupled with reducing agent and acetic acid's critical concentration for ethanol production, resulting in conversion of corresponding acid during the acetate production. Thus, lower pH, nutrients level, and stationary microbial growth conditions prefer bioethanol production than the acetate generation.

After fermentation, total fermentation broth is filtered to separate bacteria and nutrients (which are recycled back to fermentation process again) from fermentation liquid. Notable nutrients that are recycled can be categorized as amino acids, vitamins (vitamin B12), mineral salts (ammonium ion ($NH4^+$), phosphate ($PO4^{3+}$), sulfide (S^{2-}), and magnesium (Mg^{2+})), and trace metals (nickel (Ni), tungstate (W), iron (Fe), cobalt (Co), molybdate (Mo), zinc (Zn), selenite (Se)) (Sun et al. 2019). Then the fermented liquid is distilled and further concentrated by molecular sieve for fuel grade bioethanol production. Although bioethanol production thorough synthesis gas fermentation route has been successfully carried out both in laboratory and small pilot scales, only three companies (INEOS Bio, Coskata, and LanzaTech) so far have carried out large scale bioethanol production in this route (Abanades et al. 2021). Various challenges for this route have been identified for commercial scale bioethanol production. Slow gas-liquid reaction rate due to lower mass transfer between gaseous feed and liquid fermentation medium is one of them (Asimakopoulos et al. 2018). Stringent synthesis gas composition requirements (to avoid fermentation medium contamination) result from extensive and additional synthesis gas clean-up and conditioning processes. Very few microbial species have been identified up to now for the synthesis gas fermentation; genetical modifications can broaden the microbial species for commercial scale synthesis of gas-to-bioethanol production (Yasin et al. 2019). At the same time, limited number of products (ethanol, butanol, acetate, and butyrate) can be produced by the synthesis gas fermentation, and limits the commercial biorefinery concept for a wide range of biofuels and value-added chemicals coproduction.

6. Conclusion

Second-generation bioethanol production via microbial communities has become an attractive and alternative biorefinery operation for many reasons. It removes the difficulties of the uncontrolled biomass conversion of thermochemical processes for biofuel production. In addition, it lowers the operating costs and increases the flexibility in biorefinery operation compared to the thermochemical conversion processes. For instance, genetically modified *S. cerevisiae* yeast has been employed for bioethanol production from the woody lignocellulosic biomass. This microbe can co-ferment all the hexose and small amount of pentose sugars from the cellulosic and hemicellulosic fractions of biomass. Similarly, synthesis gas fermentation by *Clostridium* bacteria species removes the requirement of the use of high temperature and pressurized Fischer-Tropsch process for bioethanol production as well as lowers the strict requirement of synthesis gas composition for Fischer-Tropsch catalysts. However, microbial community-based bioethanol production has some limitations as well. Specific and specialized microorganism's requirement for certain type of biomass sugar conversion is notable. Inefficient gas liquid mass transfer and subsequent lower microbial cell growth and biofuel yield can also hamper the microbes based biorefinery operation. Fortunately, genetically modified microorganisms, designing of certain type of bioreactors, and integrated biorefinery operation can lower the difficulties of the microbes-based bioethanol production. Thus, sustainable second-generation bioethanol production process will minimize the future energy scarcity.

References

Abanades, S., Abbaspour, H., Ahmadi, A., Das, B., Ehyaei, M., Esmaeilion, F., Assad, M.E.H., Hajilounezhad, T., Jamali, D. and Hmida, A. 2021. A critical review of biogas production and usage with legislations framework across the globe. International Journal of Environmental Science and Technology, 1–24.

Abubackar, H.N., Veiga, M.C. and Kennes, C. 2019. Syngas fermentation for bioethanol and bioproducts. *In*: Sustainable Resource Recovery and Zero Waste Approaches, Elsevier, pp. 207–221.

Aguilar-Reynosa, A., Romani, A., Rodriguez-Jasso, R.M., Aguilar, C.N., Garrote, G. and Ruiz, H.A. 2017. Microwave heating processing as alternative of pretreatment in second-generation biorefinery: An overview. Energy Conversion and Management 136: 50–65.

Altman, J. and Jordan, K. 2018. Impact of climate change on indigenous Australians: Submission to the Garnaut climate change review.

Amiri, M.T., Dick, G.R., Questell-Santiago, Y.M. and Luterbacher, J.S. 2019. Fractionation of lignocellulosic biomass to produce uncondensed aldehyde-stabilized lignin. Nature Protocols 14(3): 921–954.

Andlar, M., Rezić, T., Marđetko, N., Kracher, D., Ludwig, R. and Šantek, B. 2018. Lignocellulose degradation: An overview of fungi and fungal enzymes involved in lignocellulose degradation. Engineering in Life Sciences 18(11): 768–778.

Asimakopoulos, K., Gavala, H.N. and Skiadas, I.V. 2018. Reactor systems for syngas fermentation processes: A review. Chemical Engineering Journal 348: 732–744.

Association, R.R.F. 2021. Annual Fuel Ethanol Production, Vol. 2021. Washington D.C., USA.

Avanthi, A., Kumar, S., Sherpa, K.C. and Banerjee, R. 2017. Bioconversion of hemicelluloses of lignocellulosic biomass to ethanol: An attempt to utilize pentose sugars. Biofuels 8(4): 431–444.

Barcelos, C.A., Oka, A.M., Yan, J., Das, L., Achinivu, E.C., Magurudeniya, H., Dong, J., Akdemir, S., Baral, N.R., Yan, C., Scown, C.D., Tanjore, D., Sun, N., Simmons, B.A., Gladden, J. and Sundstrom, E. 2021. High-efficiency conversion of ionic liquid-pretreated woody biomass to ethanol at the pilot scale. ACS Sustainable Chemistry & Engineering 9(11): 4042–4053.

Baruah, J., Nath, B.K., Sharma, R., Kumar, S., Deka, R.C., Baruah, D.C. and Kalita, E. 2018. Recent trends in the pretreatment of lignocellulosic biomass for value-added products. Frontiers in Energy Research 6: 141.

Bengelsdorf, F.R., Beck, M.H., Erz, C., Hoffmeister, S., Karl, M.M., Riegler, P., Wirth, S., Poehlein, A., Weuster-Botz, D. and Duerre, P. 2018. Bacterial anaerobic synthesis gas (syngas) and CO2+ H2 fermentation. Advances in Applied Microbiology 103: 143–221.

Bezerra, P.X.O., Silva, C.E.D.F., Soletti, J.I. and de Carvalho, S.H.V. 2020. Cellulosic ethanol from sugarcane straw: A discussion based on industrial experience in the northeast of Brazil. BioEnergy Research, 1–13.

Bhatia, L., Garlapati, V.K. and Chandel, A.K. 2019. Scalable technologies for lignocellulosic biomass processing into cellulosic ethanol. *In*: Horizons in Bioprocess Engineering, Springer, pp. 73–90.

Bhutto, A.W., Qureshi, K., Harijan, K., Abro, R., Abbas, T., Bazmi, A.A., Karim, S. and Yu, G. 2017. Insight into progress in pre-treatment of lignocellulosic biomass. Energy 122: 724–745.

Bouaziz, F., Abdeddayem, A.B., Koubaa, M., Barba, F.J., Jeddou, K.B., Kacem, I., Ghorbel, R.E. and Chaabouni, S.E. 2020. Bioethanol production from date seed cellulosic fraction using *Saccharomyces cerevisiae*. Separations 7(4): 67.

Brandon, N. and Kurban, Z. 2017. Clean energy and the hydrogen economy. Philosophical Transactions of the Royal Society A: Mathematical, Physical and Engineering Sciences 375(2098): 20160400.

Brockway, P.E., Owen, A., Brand-Correa, L.I. and Hardt, L. 2019. Estimation of global final-stage energy-return-on-investment for fossil fuels with comparison to renewable energy sources. Nature Energy 4(7): 612–621.

Chu, Q., Song, K., Bu, Q., Hu, J., Li, F., Wang, J., Chen, X. and Shi, A. 2018. Two-stage pretreatment with alkaline sulphonation and steam treatment of Eucalyptus woody biomass to enhance its enzymatic digestibility for bioethanol production. Energy Conversion and Management 175: 236–245.

Clark II, W.W. and Rifkin, J. 2006. A green hydrogen economy. Energy Policy 34(17): 2630–2639.

Codina Gironès, V., Peduzzi, E., Vuille, F. and Maréchal, F. 2018. On the assessment of the CO2 mitigation potential of woody biomass. Frontiers in Energy Research 5.

Cunha, J.T., Soares, P.O., Romaní, A., Thevelein, J.M. and Domingues, L. 2019. Xylose fermentation efficiency of industrial Saccharomyces cerevisiae yeast with separate or combined xylose reductase/xylitol dehydrogenase and xylose isomerase pathways. Biotechnology for Biofuels 12(1): 1–14.

Cunha, J.T., Soares, P.O., Baptista, S.L., Costa, C.E. and Domingues, L. 2020. Engineered Saccharomyces cerevisiae for lignocellulosic valorization: A review and perspectives on bioethanol production. Bioengineered 11(1): 883–903.

Demirbas, A. 2017. Higher heating values of lignin types from wood and non-wood lignocellulosic biomasses. Energy Sources, Part A: Recovery, Utilization, and Environmental Effects 39(6): 592–598.

Fernández-Naveira, Á., Veiga, M.C. and Kennes, C. 2017. Effect of pH control on the anaerobic H-B-E fermentation of syngas in bioreactors. Journal of Chemical Technology & Biotechnology 92(6): 1178–1185.

Francois, J.M., Alkim, C. and Morin, N. 2020. Engineering microbial pathways for production of bio-based chemicals from lignocellulosic sugars: Current status and perspectives. Biotechnology for Biofuels 13(1): 1–23.

Gabbrielli, R., Frigo, S. and Bressan, L. 2021. Oxy-steam co-gasification of sewage sludge and woody biomass for bio-methane production: An experimental and numerical approach, Vol. 238, EDP Sciences. Les Ulis.

Gaide, T. 2017. Sustainable Process Development for Olefin Carbonylation Reactions, Technische Universität Dortmund.

Geberekidan, M., Zhang, J., Liu, Z.L. and Bao, J. 2019. Improved cellulosic ethanol production from corn stover with a low cellulase input using a β-glucosidase-producing yeast following a dry biorefining process. Bioprocess and Biosystems Engineering 42(2): 297–304.

Geertje van Hooijdonk, A.P.F. and Carlo N. Hamelinck. 2005. Ethanol from lignocellulosic biomass: Techno-economic evaluation. Biomass and Bioenergy, 384–410.

Gencic, S. and Grahame, D.A. 2020. Diverse energy-conserving pathways in Clostridium difficile: Growth in the absence of amino acid Stickland acceptors and the role of the Wood-Ljungdahl pathway. Journal of Bacteriology 202(20): e00233–20.

Group, T.W.B. 2021. Total Greenhouse Gas Emissions, Vol. 2021.

Gupta, A., Das, S.P., Ghosh, A., Choudhary, R., Das, D. and Goyal, A. 2014. Bioethanol production from hemicellulose rich Populus nigra involving recombinant hemicellulases from Clostridium thermocellum. Bioresource Technology 165: 205–213.

Haigh, K.F., Petersen, A.M., Gottumukkala, L., Mandegari, M., Naleli, K. and Görgens, J.F. 2018. Simulation and comparison of processes for biobutanol production from lignocellulose via ABE fermentation. Biofuels, Bioproducts and Biorefining 12(6): 1023–1036.

He, Q., Hemme, C.L., Jiang, H., He, Z. and Zhou, J. 2011. Mechanisms of enhanced cellulosic bioethanol fermentation by co-cultivation of Clostridium and Thermoanaerobacter spp. Bioresource Technology 102(20): 9586–9592.

Hoang, A.T., Nizetic, S., Ong, H.C., Chong, C.T., Atabani, A.E. and Pham, V.V. 2021. Acid-based lignocellulosic biomass biorefinery for bioenergy production: Advantages, application constraints, and perspectives. Journal of Environmental Management 296: 113194.

Hossain, M.S., Theodoropoulos, C. and Yousuf, A. 2019. Techno-economic evaluation of heat integrated second generation bioethanol and furfural coproduction. Biochemical Engineering Journal 144: 89–103.

Hossain, S.Z. 2019. Biochemical conversion of microalgae biomass into biofuel. Chemical Engineering & Technology 42(12): 2594–2607.

Humbird, D., Davis, R., Tao, L., Kinchin, C., Hsu, D., Aden, A., Schoen, P., Lukas, J., Olthof, B., Worley, M.J.B.B. and Sexton, D. 2011. Process design and economics for biochemical conversion of lignocellulosic biomass to ethanol: dilute-acid pretreatment and enzymatic hydrolysis of corn stover (No. NREL/TP-5100-47764). National Renewable Energy Lab. (NREL), Golden, CO (United States).

Humbird, D., Mohagheghi, A., Dowe, N. and Schell, D. 2010. Economic impact of total solids loading on enzymatic hydrolysis of dilute-acid pretreated corn stover. Biotechnology Progress, 1245–1251.

Inayat, M., Sulaiman, S., Kurnia, J. and Naz, M. 2019. Catalytic and noncatalytic gasification of wood–coconut shell blend under different operating conditions. Environmental Progress & Sustainable Energy 38(2): 688–698.

Inui, H., Ishikawa, T. and Tamoi, M. 2017. Wax ester fermentation and its application for biofuel production. Euglena: Biochemistry, Cell and Molecular Biology, 269–283.

Ismail, W.M.S.W., Mohd Thaim, T. and Abdul Rasid, R. 2019. Biomass gasification of oil palm fronds (OPF) and Koompassia malaccensis (Kempas) in an entrained flow gasifier: A performance study. Biomass and Bioenergy 124: 83–87.

Istiqomah, N., Kresnowati, M. and Setiadi, T. 2021. Syngas fermentation for production of ethanol. IOP Conference Series: Materials Science and Engineering. IOP Publishing, pp. 012014.

James Daniell, M.K.a.S.D.S. 2012. Commercial biomass syngas fermentation. Energies, 5372–5417.

Jansen, M.L., Bracher, J.M., Papapetridis, I., Verhoeven, M.D., de Bruijn, H., de Waal, P.P., van Maris, A.J., Klaassen, P. and Pronk, J.T. 2017. Saccharomyces cerevisiae strains for second-generation ethanol production: From academic exploration to industrial implementation. FEMS Yeast Research 17(5).

Javed, F., Aslam, M., Rashid, N., Shamair, Z., Khan, A.L., Yasin, M., Fazal, T., Hafeez, A., Rehman, F., Rehman, M.S.U., Khan, Z., Iqbal, J. and Bazmi, A.A. 2019. Microalgae-based biofuels, resource recovery and wastewater treatment: A pathway towards sustainable biorefinery. Fuel 255: 115826.

Kalyani, D., Lee, K.-M., Kim, T.-S., Li, J., Dhiman, S.S., Kang, Y.C. and Lee, J.-K. 2013. Microbial consortia for saccharification of woody biomass and ethanol fermentation. Fuel 107: 815–822.

Khan, K., Su, C.-W., Umar, M. and Yue, X.-G. 2021. Do crude oil price bubbles occur? Resources Policy 71: 101936.

Kim, H.Y., Hong, C.Y., Kim, S.H., Yeo, H. and Choi, I.G. 2015. Optimization of the organosolv pretreatment of yellow poplar for bioethanol production by response surface methodology. Journal of the Korean Wood Science and Technology 43(5): 600–612.

Kim, Y.-K. and Lee, H. 2016. Use of magnetic nanoparticles to enhance bioethanol production in syngas fermentation. Bioresource Technology 204: 139–144.

Kim, Y.-K., Park, S.E., Lee, H. and Yun, J.Y. 2014. Enhancement of bioethanol production in syngas fermentation with Clostridium ljungdahlii using nanoparticles. Bioresource Technology 159: 446–450.

Koido, K., Iwasaki, T., Kurosawa, K., Takaku, R., Ohashi, H. and Sato, M. 2021. Cesium-catalyzed hydrogen production by the gasification of woody biomass for forest decontamination. ACS Omega 6(8): 5233–5243.

Kumar, B., Bhardwaj, N., Agrawal, K., Chaturvedi, V. and Verma, P. 2020. Current perspective on pretreatment technologies using lignocellulosic biomass: An emerging biorefinery concept. Fuel Processing Technology 199: 106244.

Kumar, B. and Verma, P. 2021. Biomass-based biorefineries: An important architype towards a circular economy. Fuel 288: 119622.

Kwak, S. and Jin, Y.-S. 2017. Production of fuels and chemicals from xylose by engineered Saccharomyces cerevisiae: A review and perspective. Microbial Cell Factories 16(1): 1–15.

Lewandowski, W.M., Ryms, M. and Kosakowski, W. 2020. Thermal biomass conversion: A review. Processes 8(5): 516.

Li, J., He, Y., Tan, L., Zhang, P., Peng, X., Oruganti, A., Yang, G., Abe, H., Wang, Y. and Tsubaki, N. 2018. Integrated tuneable synthesis of liquid fuels via Fischer-Tropsch technology. Nature Catalysis 1(10): 787–793.

Liang, J., Nabi, M., Zhang, P., Zhang, G., Cai, Y., Wang, Q., Zhou, Z. and Ding, Y. 2020. Promising biological conversion of lignocellulosic biomass to renewable energy with rumen microorganisms: A comprehensive review. Renewable and Sustainable Energy Reviews 134: 110335.

Lin, Y. and Tanaka, S. 2006. Ethanol fermentation from biomass resources: Current state and prospects. Applied Microbiol Biotechnology, 627–642.

Liu, E. and Hu, Y. 2010. Construction of a xylose–fermenting Saccharomyces cerevisiae strain by combined approaches of genetic engineering, chemical mutagenesis and evolutionary adaption. Biochemical Engineering Journal, 204–210.

Liu, T., Williams, D.L., Pattathil, S., Li, M., Hahn, M.G. and Hodge, D.B. 2014. Coupling alkaline pre-extraction with alkaline-oxidative post-treatment of corn stover to enhance enzymatic hydrolysis and fermentability. Biotechnology for Biofuels 7(1): 48.

Liu, Y., Lyu, Y., Tian, J., Zhao, J., Ye, N., Zhang, Y. and Chen, L. 2021. Review of waste biorefinery development towards a circular economy: From the perspective of a life cycle assessment. Renewable and Sustainable Energy Reviews 139: 110716.

Lyons Cerón, A., Konist, A., Lees, H. and Järvik, O. 2021. Effect of woody biomass gasification process conditions on the composition of the producer gas. Sustainability 13(21): 11763.

Maneein, S., Milledge, J.J., Nielsen, B.V. and Harvey, P.J. 2018. A review of seaweed pre-treatment methods for enhanced biofuel production by anaerobic digestion or fermentation. Fermentation 4(4): 100.

Maslova, O., Stepanov, N., Senko, O. and Efremenko, E. 2019. Production of various organic acids from different renewable sources by immobilized cells in the regimes of separate hydrolysis and fermentation (SHF) and simultaneous saccharification and fermentation (SFF). Bioresource Technology 272: 1–9.

Mohammadi, M., Najafpour, G.D., Younesi, H., Lahijani, P., Uzir, M.H. and Mohamed, A.R. 2011. Bioconversion of synthesis gas to second generation biofuels: A review. Renewable and Sustainable Energy Reviews 15(9): 4255–4273.

Mohanty, S.K. and Swain, M.R. 2019. Chapter 3—Bioethanol production from corn and wheat: food, fuel, and future. pp. 45–59. *In*: Ray, R.C. and Ramachandran, S. (eds.). Bioethanol Production from Food Crops. Academic Press.

Monir, M.U., Aziz, A.A., Khatun, F. and Yousuf, A. 2020a. Bioethanol production through syngas fermentation in a tar free bioreactor using Clostridium butyricum. Renewable Energy 157: 1116–1123.

Monir, M.U., Yousuf, A. and Aziz, A.A. 2020b. Chapter 6—Syngas fermentation to bioethanol. pp. 195–216. *In*: Yousuf, A., Pirozzi, D. and Sannino, F. (eds.). Lignocellulosic Biomass to Liquid Biofuels. Academic Press.

Nanda, S., Vo, D.-V.N. and Sarangi, P.K. 2020. Biorefinery of Alternative Resources: Targeting Green Fuels and Platform Chemicals. Springer Nature.

Olofsson, K., Wiman, M. and Lidén, G. 2010. Controlled feeding of cellulases improves conversion of xylose in simultaneous saccharification and co-fermentation for bioethanol production. Journal of Biotechnology 145(2): 168–175.

Pablo, G., Domínguez, V.D., Domínguez, E., Gullón, P., Gullón, B., Garrote, G. and Romaní, A. 2020. Comparative study of biorefinery processes for the valorization of fast-growing Paulownia wood. Bioresource Technology 314: 123722.

Pardo-Planas, O., Atiyeh, H.K., Phillips, J.R., Aichele, C.P. and Mohammad, S. 2017. Process simulation of ethanol production from biomass gasification and syngas fermentation. Bioresource Technology 245: 925–932.

Pino, M.S., Rodríguez-Jasso, R.M., Michelin, M., Flores-Gallegos, A.C., Morales-Rodriguez, R., Teixeira, J.A. and Ruiz, H.A. 2018. Bioreactor design for enzymatic hydrolysis of biomass under the biorefinery concept. Chemical Engineering Journal 347: 119–136.

Pirani, S. 2018. Burning Up: A Global History of Fossil Fuel Consumption. Pluto Press.

Ragsdale, S.W. and Pierce, E. 2008. Acetogenesis and the Wood-Ljungdahl pathway of CO2 fixation. Biochim. Biophys., 1873–1898.

Redondo-Gómez, C., Rodríguez Quesada, M., Vallejo Astúa, S., Murillo Zamora, J.P., Lopretti, M. and Vega-Baudrit, J.R. 2020. Biorefinery of biomass of agro-industrial banana waste to obtain high-value biopolymers. Molecules 25(17): 3829.

Requejo, A., Peleteiro, S., Rodríguez, A., Garrote, G. and Parajó, J.C. 2011. Second-generation bioethanol from residual woody biomass. Energy & Fuels 25(10): 4803–4810.

Richter, H., Molitor, B., Wei, H., Chen, W., Aristilde, L. and Angenent, L.T. 2016. Ethanol production in syngas-fermenting Clostridium ljungdahlii is controlled by thermodynamics rather than by enzyme expression. Energy & Environmental Science 9(7): 2392–2399.

Sahay, S. 2022. Deconstruction of lignocelluloses: Potential biological approaches. *In*: Handbook of Biofuels, Elsevier, pp. 207–232.

Sato, T.K., Tremaine, M., Parreiras, L.S., Hebert, A.S., Myers, K.S., Higbee, A.J., Sardi, M., McIlwain, S.J., Ong, I.M. and Breuer, R.J. 2016. Directed evolution reveals unexpected epistatic interactions that alter metabolic regulation and enable anaerobic xylose use by Saccharomyces cerevisiae. PLoS Genetics 12(10): e1006372.

Senatore, A., Giorgianni, G., Dalena, F. and Giglio, E. 2021. Lignocellulosic biomass conversion into bioenergy: Feedstock overview. Journal of Phase Change Materials 1(1).

Senila, L., Costiug, S., Becze, A., Kovacs, D., Kovacs, E., Scurtu, D.A., Todor-Boer, O. and Senila, M. 2020. Bioethanol production from Abies Alba wood using adaptive neural fuzzy interference system mathematical modeling. Cellulose Chemistry and Technology 54(1-2): 53–64.

Sharma, H.K., Xu, C. and Qin, W. 2019. Biological pretreatment of lignocellulosic biomass for biofuels and bioproducts: An overview. Waste and Biomass Valorization 10(2): 235–251.

Shen, Y., Brown, R.C. and Wen, Z. 2017. Syngas fermentation by Clostridium carboxidivorans P7 in a horizontal rotating packed bed biofilm reactor with enhanced ethanol production. Applied Energy 187: 585–594.

Singh, N., Puri, M., Tuli, D.K., Gupta, R.P., Barrow, C.J. and Mathur, A.S. 2018. Bioethanol production by a xylan fermenting thermophilic isolate Clostridium strain DBT-IOC-DC21. Anaerobe 51: 89–98.

Song, Y., Lee, J.S., Shin, J., Lee, G.M., Jin, S., Kang, S., Lee, J.-K., Kim, D.R., Lee, E.Y. and Kim, S.C. 2020. Functional cooperation of the glycine synthase-reductase and Wood–Ljungdahl pathways for autotrophic growth of Clostridium drakei. Proceedings of the National Academy of Sciences 117(13): 7516–7523.

Song, Y., Shin, J., Jin, S., Lee, J.-K., Kim, D.R., Kim, S.C., Cho, S. and Cho, B.-K. 2018. Genome-scale analysis of syngas fermenting acetogenic bacteria reveals the translational regulation for its autotrophic growth. BMC Genomics 19(1): 1–15.

Stephens, E., Ross, I., Mussgnug, J., Wagner, L., Borowitzka, M., Posten, C., Kruse, O. and Hankamer, B. 2010. Future prospects of microalgal biofuel production systems. Trends in Plant Science 15: 554–64.

Sun, X., Atiyeh, H.K., Huhnke, R.L. and Tanner, R.S. 2019. Syngas fermentation process development for production of biofuels and chemicals: A review. Bioresource Technology Reports 7: 100279.

Sur, S., Dave, V., Prakesh, A. and Sharma, P. 2021. Expansion and scale up of technology for ethanol production based on the concept of biorefinery. Journal of Food Process Engineering 44(2): e13582.

Ummalyma, S.B., Supriya, R.D., Sindhu, R., Binod, P., Nair, R.B., Pandey, A. and Gnansounou, E. 2019. Biological pretreatment of lignocellulosic biomass—Current trends and future perspectives. *In*: Second and Third Generation of Feedstocks, Elsevier, pp. 197–212.

Vallejos, M.E., Kruyeniski, J. and Area, M.C. 2017. Second-generation bioethanol from industrial wood waste of South American species.

Wang, F.-L., Li, S., Sun, Y.-X., Han, H.-Y., Zhang, B.-X., Hu, B.-Z., Gao, Y.-F. and Hu, X.-M. 2017. Ionic liquids as efficient pretreatment solvents for lignocellulosic biomass. RSC Advances 7(76): 47990–47998.

Wenger, J. and Stern, T. 2019. Reflection on the research on and implementation of biorefinery systems—A systematic literature review with a focus on feedstock. Biofuels, Bioproducts and Biorefining 13(5): 1347–1364.

Xiao-Jun Ji, H.H., Zhi-Kui Nie, Liang Qu, Qing Xu and George T. Tsao. 2012. Fuels and Chemicals from hemicellulose sugars. Advanced Biochemical Engineering/Biotechnology, 199–224.

Xin-Qing Zhao, L.-H.Z., Feng-Wu Bai, Hai-Long Lin, Xiao-Ming Hao, Guo-Jun Yue and Nancy, W.Y. Ho. 2011. Bioethanol from lignocellulosic biomass. Advanced Biochemical Engineering/Biotechnology, 25–51.

Yamakawa, C.K., Qin, F. and Mussatto, S.I. 2018. Advances and opportunities in biomass conversion technologies and biorefineries for the development of a bio-based economy. Biomass and Bioenergy 119: 54–60.

Yang, Y., Hu, M., Tang, Y., Geng, B., Qiu, M., He, Q., Chen, S., Wang, X. and Yang, S. 2018. Progress and perspective on lignocellulosic hydrolysate inhibitor tolerance improvement in Zymomonas mobilis. Bioresources and Bioprocessing 5(1): 1–12.

Yasin, M., Cha, M., Chang, I.S., Atiyeh, H.K., Munasinghe, P. and Khanal, S.K. 2019. Syngas fermentation into biofuels and biochemicals. Biofuels: Alternative Feedstocks and Conversion Processes for the Production of Liquid and Gaseous Biofuels, 301–327.

Zabed, H., Sahu, J., Suely, A., Boyce, A. and Faruq, G. 2017. Bioethanol production from renewable sources: Current perspectives and technological progress. Renewable and Sustainable Energy Reviews 71: 475–501.

Zabed, H.M., Akter, S., Yun, J., Zhang, G., Awad, F.N., Qi, X. and Sahu, J. 2019. Recent advances in biological pretreatment of microalgae and lignocellulosic biomass for biofuel production. Renewable and Sustainable Energy Reviews 105: 105–128.

Zhao, R., Liu, Y., Zhang, H., Chai, C., Wang, J., Jiang, W. and Gu, Y. 2019. CRISPR-Cas12a-mediated gene deletion and regulation in Clostridium ljungdahlii and its application in carbon flux redirection in synthesis gas fermentation. ACS Synthetic Biology 8(10): 2270–2279.

Zhu, L., Xu, S., Li, Y. and Shi, G. 2021. Improvement of 2-phenylethanol production in Saccharomyces cerevisiae by evolutionary and rational metabolic engineering. Plos One 16(10): e0258180.

CHAPTER 7

Production of Bioethanol from Lignocellulosic Materials by Non-conventional Microorganisms

*Juan Carlos López-Linares, Juan Miguel Romero-García,
María del Mar Contreras, Inmaculada Romero* and *Eulogio Castro**

1. Introduction

Since the start of the Industrial Revolution, the energy supply across the world has generally become dominated by fossil fuels. Per capita usage has increased more than 800 percent, which is a reflex of the population growth and an increased energy use (Heinberg 2017). However, energy has a key role for the global climate since it leads to 60% of the total greenhouse gases emission (United Nations 2021).

To resolve this energy and environmental challenge, renewable energy sources have emerged as environmentally affable alternatives to achieve a low-carbon society. Moreover, renewable energy is closely related to the circular bioeconomy concept, which has attracted increasing scientific and policy attention (Chen et al. 2019).

Among them, the production of liquid (bioethanol, biobutanol, biodiesel, etc.) and gas biofuels (e.g., hydrogen and methane) from different biomasses is currently being investigated and implemented (Robak and Balcerek 2020). Biofuels, including bioethanol, are classified into four generations taking into account the biomass type. First-generation biofuels are made from food-based crops rich in starch (e.g., corn, wheat, sugarcane, etc.), second-generation are based on lignocellulosic forestry and agri-food biomasses, while third- and fourth-generation biofuels are produced using algal biomasses. In the latter case, genetically modified algae, which generate high lipid content, are applied (Robak and Balcerek 2020). Fig. 1 shows the biofuel energy production across the world in 2019, highlighting that United States of America and Brazil are the major producers.

Department of Chemical, Environmental and Materials Engineering, Center for Advanced Studies in Earth Sciences, Energy and the Environment, University of Jaén, 23071 Jaén, Spain.
* Corresponding author: ecastro@ujaen.es

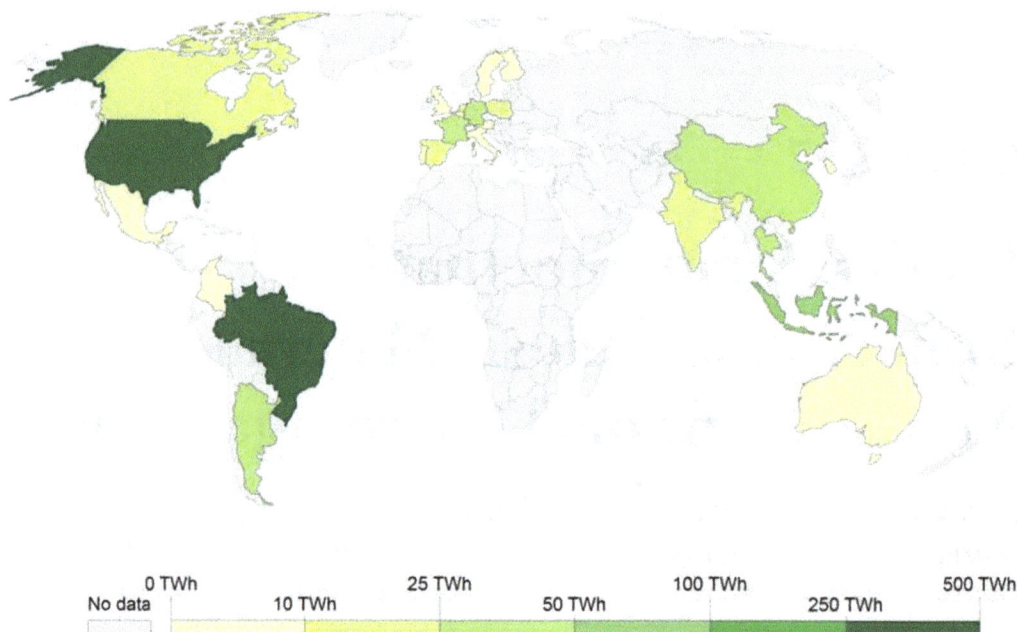

Fig. 1. Global biofuel energy production (bioethanol and biodiesel) in 2019, obtained from Ritchie and Roser 2020.

The major part of renewable energy consumed in transport (around 92% in 2016) was policy driven and came in the form of biofuels, mainly first-generation bioethanol and biodiesel generally blended with fossil fuels. In 2016, their consumption in transport was closer to 2% and 1%, respectively (IEA et al. 2019). Bioethanol presents 68% lower energy content compared to petrol, but its use makes the combustion cleaner with lower emission of toxic substances due to its high oxygen content (Aditiya et al. 2016).

The estimated global bioethanol production was about 100 billion liters in 2011, 110 billion liters in 2018 and it is expected to increase around 3% in 2022 (Robak and Balcerek 2020). In line with Fig. 1, the main producers are United States of America (54%), Brazil (30%), and European Union (5%) using corn, sugarcane, and wheat and sugar beet, respectively (Susmozas et al. 2020). Thus, bioethanol is almost exclusively produced from food crops, which competes over the utilization of arable land to produce food and feed (Robak and Balcerek 2020, Susmozas et al. 2020).

Alternatively, second generation bioethanol is produced from lignocellulosic materials (LCM) including low-cost forestry and agricultural residues, as commented before. Its exploitation is in line with moving towards using energy-smart agri-food systems to take better advantage of the relationship between energy and food (FAOSTAT 2021) and to develop biorefinery models as a driving force towards a circular bioeconomy.

However, second-generation bioethanol has technological limitations, which is the main barrier to its industrial production. This explains that few companies are currently producing second-generation bioethanol, such as GranBio and Raizen in Brazil, Beta Renewables (now Versalis) in Italy, and DuPont (now Verbio) in United States, which generally use sugar cane and sugar beet residues (Eni 2021, Verbio 2021).

In order to maximize the extraction of sugars, the general sequential operations to fractionate LCM and to produce second-generation bioethanol are the following: (1) conditioning, (2) pretreatment to open up the lignocellulosic polysaccharides, (3) hydrolysis to produce fermentable sugars, (4) fermentation to produce ethanol and (5) ethanol recovery (distillation and dehydration) (Fig. 2) (Kumar et al. 2016). The production yield depends on the biomass type and the production

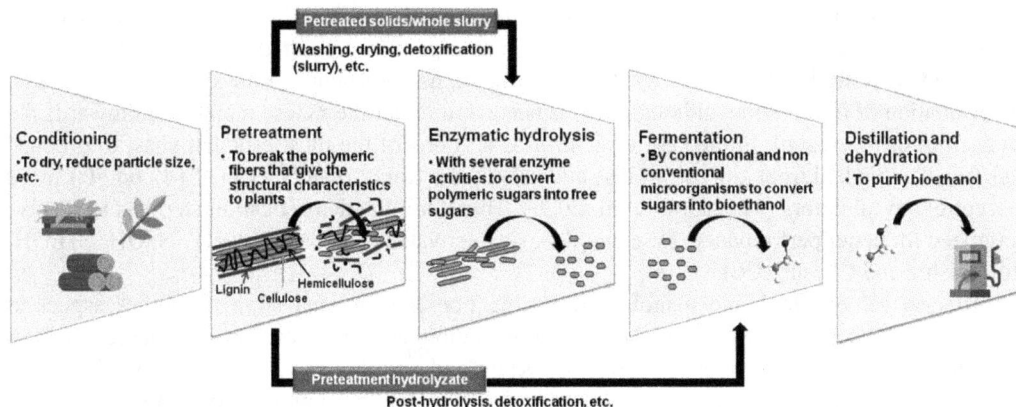

Fig. 2. Processing steps applied to fractionate lignocelullosic materials and to produce bioethanol. Some icons have been designed using resources from Flaticon.com.

route and thereby the selected method will result in different overall production costs (Contreras et al. 2020). The main critical processing steps are the steps 2, 3, and 4, particularly, the fermentation of pentoses, while process integration is also desirable to reduce the number of these steps (Mishra and Ghosh 2020). Concerning the two latter steps, different configurations of enzymatic hydrolysis and fermentation can be applied, including hydrolysis and sequential or simultaneous fermentation, and co-fermentation. In fermentation, glucose is the only sugar fermented, generally, using the yeast *Saccharomyces cerevisiae*, which is not natively capable of fermenting xylose, the second major sugar in LCM. In the co-fermentation, the fermentation of cellulosic glucose and hemicellulosic sugars is performed by a non-conventional microorganism able to ferment both types of sugars or using two compatible microorganisms (mixed fermentation). In simultaneous configurations, sugars are produced and fermented in a single step, and thus thermotolerance of the microorganisms is looked for to provide high yields. In addition, the consolidated bioprocessing combines enzymes production, enzymatic saccharification, and fermentation in a single step (Contreras et al. 2020).

According to several authors (Kumar et al. 2016, Mishra and Ghosh 2020), operational efficiency improvements to ferment xylose have been addressed to:

(1) Develop genetically modified *S. cerevisiae* strains able to ferment glucose and xylose.

(2) Apply non-conventional microorganisms able to produce ethanol from different type of sugars, including pentoses. Examples of yeasts are *Scheffersomyces* (or *Pichia*) *stipitis*, *Candida* (or *Scheffersomyces*) *shehatae*, *Kluyveromyces marxianus*, and *Hansenula polymorpha*. Among bacteria, *Escherichia coli*, *Zymomonas mobilis*, and *Klebsiella oxytoca* have been applied.

(3) Apply non-conventional microorganisms as co-production systems to obtain bioethanol and other co-products to diversify the production chain (e.g., *Candida tropicalis*).

(4) Intensifying hexose and pentose fermentation process by using microorganism co-cultures (e.g., using *Z. mobilis* and *C. shehatae* or *S. cerevisiae* and *S. stipitis*).

In the first case, xylose assimilation is due to the conversion of xylose into xylulose (*via* oxidoreductase or isomerase pathway), then the phosphorylation to xylulose-5-phosphate occurs and finally it is metabolized through the pentose phosphate pathway. To make these routes possible, the efficient expression of fungal and bacterial enzymatic complexes in *S. cerevisiae* is required, which is not always successful (Cunha et al. 2019).

In all cases, microbial strains should present tolerance to inhibitor compounds and osmotic pressure to provide high production yields (Yamakawa et al. 2020. For example, the latter authors have explored the capacity of *H. polymorpha* to utilize xylose in synthetic media and in a hemicellulosic hydrolysate of wheat straw. This microorganism presented high resistance to toxic compounds and ability to produce xylitol and ethanol, as co-product, from xylose. Nonetheless, the evaluation of the aeration, although low, is needed to orient the xylose metabolism towards the production of bioethanol. In the case of *S. stipitis*, it is one of the most efficient yeast to produce high bioethanol yield from glucose, xylose and cellobiose (Karagöz and Özkan 2014), but it is more susceptible to inhibitors (Yamakawa et al. 2020). Therefore, the detoxification step has to be also optimized for better performance, for example, using activated charcoal, $Ca(OH)_2$, NaOH, NH_4OH, etc. (Mishra and Ghosh 2020).

To also achieve high bioethanol yields using non-conventional strains, another aspect to be considered is the enzymatic hydrolysis-fermentation configuration; as commented before, there are several available options. Fernandes-Klajn and co-workers (2018) have found that the detoxified slurry from extracted and pretreated olive tree pruning achieved good bioethanol yield (13.9 g ethanol/100 g raw biomass) when simultaneous saccharification and cofermentation with *E. coli* were applied at 37°C (without aeration) to ferment glucose and xylose. Also, if two microorganisms are used in co-fermentation configurations, e.g., *S. cerevisiae* and a non-conventional microorganism, providing optimal environmental conditions to grow the two strains is a prerequisite (e.g., inoculation ratio, aeration level, etc.) (Ashoor et al. 2015). In addition, the absence of inhibitory effects between the microorganisms is desirable since it can lead to low ethanol yields (Farias and Maugeri Filho 2019).

Therefore, to tackle these challenges and provide more profitable processes by fermenting sugars other than glucose, this chapter firstly gives an overview of LCM applied to produce sugars, including glucose and xylose. Secondly, the main characteristics and fermenting properties of these so-called non conventional microorganisms for producing bioethanol are reviewed. Finally, to explore their applicability, this chapter provides examples based on olive-derived LCM as starting feedstock.

2. Lignocellulosic Materials as a Source of Sugars

The term LCM encompasses a wide range of non-food crops and forest and agricultural residues mainly composed of cellulose, hemicelluloses and lignin. Due to their residual character, LCM constitute the most interesting type of biomass for conversion into biofuels and other chemicals that can be used as platform molecules for different synthesis routes to high-added value renewable products (Robak and Balcerek 2020).

The term lignocellulosic structure is related to the part of the plant that forms the cell wall, composed of fibrous structures, basically made up of polysaccharides (Sun and Cheng 2002). In this case, it is worth highlighting the greater structural complexity of these materials, which makes them more difficult to use for obtaining monomeric sugars in comparison to sugary or starchy raw materials. In LCM, these sugars are found in the form of more complex polysaccharides (cellulose and hemicelluloses) which are difficult to hydrolyze, due to their structural and defensive function (Kumar et al. 2020).

Considering its composition, two types of components can be found in lignocellulosic biomass: structural and non-structural components. Among structural components, cellulose, hemicelluloses and lignin account generally for more than 75% by weight. Cellulose and hemicelluloses are high molecular weight polysaccharides, which represent between 60–80% of the lignocellulosic materials' total weight. However, lignin is a three-dimensional non-polysaccharide polymer of phenylpropane units, representing between 20–35% of the total. Nevertheless, it is worth mentioning that the content of these main components varies depending on the raw material (Robak and Balcerek 2020).

On the other hand, regarding the non-structural components, extractive compounds and ash are found, which are minor fractions within the lignocellulosic material. In addition, they have no structural function (Kumar et al. 2020). The extractives are a very heterogeneous group with about 10–15% of average values with respect to biomass dry weight, although it could reach values of up to 32% (Cara et al. 2008); among the different types of compounds, there are fats, terpenes, alkaloids, proteins, phenols, pectins, gums, or resins, acting as metabolic intermediaries, energy reserve or as part of the defense mechanisms against microbial attacks. Likewise, they are responsible for the color, smell and resistance to wilting of the plant (Kumar et al. 2020). With regards to the ashes in the biomass, they mainly contain Ca, K, Mg and Si, which are inorganic compounds that remain after the incineration of the material, its general proportion being less than 1% of biomass dry weight.

Table 1 shows the typical composition of different LCM, including agricultural and forestry (hardwoods and softwoods) residues. In general, the proportion of the main components of LCM (cellulose, hemicellulose and lignin) varies depending on the type of material, woody biomass being the material with the highest cellulosic content, while agricultural residues usually present lower lignin content (Satari et al. 2019).

Agricultural residues, such as rice straw, wheat straw, corn stover or rapeseed straw, are those generated in the agricultural activity, which are left in the fields or burned to avoid the spread of pest. Then, this type of residue lacks utility, representing around 60% of the total crop by weight. Comparing with forestry residues, agricultural residues can be pretreated much more easily (under lower temperatures and times), the fermentation process also being more efficient and economically profitable. Moreover, this category also includes the agro-industries wastes (for instance, those

Table 1. Chemical composition of different lignocellulosic materials (% dry basis).

Biomass Type	Name of the Biomass	Composition of Lignocellulosic Biomass			References
		Cellulose	Hemicellulose	Lignin	
Forestry residues	Poplar	43.2	14.7	25.6	(Tian et al. 2020)
	Willow	42.4	20.6	16.9	(Mussatto and Dragone 2016)
	Spruce	37.6	17.6	32.6	(Matsakas et al. 2019)
	Aspen	49.0	18.2	25.6	(Goshadrou et al. 2013)
	Eucalyptus	44.9	28.9	26.2	(Muranaka et al. 2017)
Agricultural residues	Sugarcane bagasse	34.8	25.0	24.6	(Fan et al. 2020)
	Rice husk	28.7	12.0	15.4	(Mussatto and Dragone 2016)
	Vine shoots	28.8	17.3	25.9	(Senila et al. 2020)
	Corn cobs	35.9	32.7	18.8	(Xie et al. 2014)
	Corn stover	30.6	19.1	16.7	(Huang et al. 2016)
	Rice straw	29.2	23.0	17.0	(Mussatto and Dragone 2016)
	Wheat straw	35.1	23.4	21.1	(Chen et al. 2018)
	Barley straw	36.0	24.0	6.3	(Mussatto and Dragone 2016)
	Rapeseed straw	31.6	17.4	17.8	(López-Linares et al. 2015)
	Sunflower stalks	29.6	20.7	13.3	(Díaz et al. 2011)
	Spent coffee grounds	16.3	27.7	39.2	(López-Linares et al. 2021)
	Olive tree pruning	21.6	14.5	17.7	(Martínez-Patiño et al. 2017b)
	Extracted olive pomace	10.1	11.3	21.9	(Manzanares et al. 2017)
	Olive leaves	9.3	9.5	17.7	(Manzanares et al. 2017)
	Olive stones	19.2	28.6	37.2	(Romero-García et al. 2016)
	Brewer's spent grains	17.9	28.7	25.8	(López-Linares et al. 2019)

generated in the olive oil, nuts, and wine industries) and horticultural residues, such as tree trunks, branches and trimmings originated from the pruning of crops such as olive trees, fruit trees and vineyards (Duque et al. 2021).

Regarding forestry residues, such as eucalyptus, poplar, pinewood or spruce, they are generated in forest pruning and cleaning tasks as well as in processing activities of woody products, which involve up to 30% of this category of residues (5–8% sawdust and 10–15% bark) (Duque et al. 2021).

On the other hand, regarding the most important agricultural and forestry residues generated, Table 2 shows the total world production by year for each of them, which has been calculated considering the harvested area in the world (Agencia Extremeña de la Energía 2020, FAOSTAT 2021) or the production (Antar et al. 2021, Tye et al. 2016) for each of these agricultural crops and forest harvest in the year 2019, and the average yield of residual biomass generated (Antar et al. 2021, Gómez-García 2021, Junta de Andalucía 2008, Tye et al. 2016).

Finally, it is worth highlighting the need to take advantage of both cellulose and hemicellulose sugars, due to the high content of hemicellulose (mainly pentoses, such as xylose) contained in some lignocellulosic residues such as eucalyptus, almond shell, corn cobs, olive stones, or brewer's spent grains (about 30–35 g/100 g raw material, Table 1). Therefore, the use of microorganisms able to consume both cellulosic and hemicellulosic sugars is greatly interesting in order to produce fuels such as bioethanol. As an example, considering only corn stover, more than 850 million tons of ethanol per year could be produced only from the hemicelluloses contained in this lignocellulosic residue, based on an ethanol theoretical yield of 0.51 g ethanol/g sugar.

Table 2. Harvested area (millions of ha) in the world of different agricultural crops and forest harvest in 2019, average yield of residual biomass generated (t/ha·year), and production of lignocellulosic residue by year (millions of t/year).

Lignocellulosic Residue	Harvested Area (millions ha)	Average Yield of Residual Biomass (t/ha·year)	Residue Production (millions t/year)
Wheat straw	215.90	2.74	591.57
Corn stover	197.20	22.62	4460.66
Barley straw	51.15	2.03	103.83
Sunflower stalks	27.37	1.52	41.60
Rapeseed straw	70.51[1]	1.6[2]	112.82
Rice straw	473.1[1]	1.39[2]	657.5
Olive tree pruning	10.58	1.49	15.76
Vine shoots	6.93	3.50	24.26
Empty fruit bunches	277.7[1]	0.23[2]	63.9
Sugarcane bagasse	26.78	16.57	443.74
Sorghum straw	61.9[1]	0.19[2]	12.0
Rice husk	162.06	12.41	2011.16
Oat straw	22.9[1]	0.45[2]	10.4
Almond shell	2.13	1.30	2.77
Miscanthus	–	–	256.4
Switchgrass	–	–	282.9
Pinewook	7.27	7	50.89
Eucalyptus	10.9	11	119.9

[1] Expressed as million ton of seed (for example: rapeseed, oat, rice, or sorghum, among others).
[2] Expressed as t residual biomass/t seed.

3. Production of Bioethanol from Pentoses

A great variety of microorganisms (yeasts, fungi and bacteria) are capable of fermenting pentoses (mostly xylose) to ethanol naturally, Table 3. The use of wild-type bacteria and fungi has been practically ruled out due to the low productivity and/or low yield of these types of microorganisms, together with the fact that yeasts with better performance have been identified. Within wild yeasts, the three species identified that could best carry out xylose fermentation are *Scheffersomyces (Pichia) stipitis*, *Pachysolen tannophilus* and *Candida shehatae* (Robak and Balcerek 2020). These wild yeasts have limitations such as relatively low fermentation yields, sensitivity to inhibitors and high concentrations of ethanol (> 5%) and the need for microaerophilic conditions (Wirawan et al. 2020). Given these limitations, microorganisms capable of co-fermenting both glucose and pentoses from hemicellulose are being developed through genetic engineering. To this end, microorganisms that have the quality of fermenting both pentoses and hexoses can be genetically improved (*E. coli* or *K. oxytoca*) to produce ethanol, or genes involved in the metabolism of pentoses can be introduced in hexose-fermenting microorganisms, as was the case in *S. cerevisiae* or *Z. mobilis* (Sun and Jin 2021).

Table 3. Natural-type and recombinant genera capable of fermentation of pentoses to ethanol (adapted from Olsson and Hahn-Hägerdal 1996).

Type	Genera	
	Natural	**Recombinant**
Yeast	*Brettanomyces* *Candida* *Clavispora* *Kluyveromyces* *Pachysolen* *Pichia* *Schizosaccharomyces*	*Saccharomyces* *Schizosaccharomyces*
Fungi	*Aeurobasiclium* *Fusarium* *Monilia* *Mucor* *Neurospora* *Paecilomyces* *Polyporus* *Rhizopus*	
Bacteria	*Aerobacter* *Aeromonas* *Bacillus* *Bacteroides* *Clostridium* *Erwinia* *Klebsiella* *Thermoanaerobacter*	*Erwinia* *Escherichia* *Klebsiella* *Zymomonas*

3.1 Natural Microorganisms

In natural bacteria, yeasts, and fungi, the pathways for fermentation of xylose to ethanol are similar with differences in transport, regulation, cofactor requirements, and products of pyruvate fermentation. Ethanol production from xylose is believed to mostly follow the pentose phosphate (PP) and Embden-Meyerhof-Parnas (EMP) pathways, once transported into the cell and converted to xylulose-5-phosphate (Fig. 3) with pyruvate finally converted to ethanol. The method by which

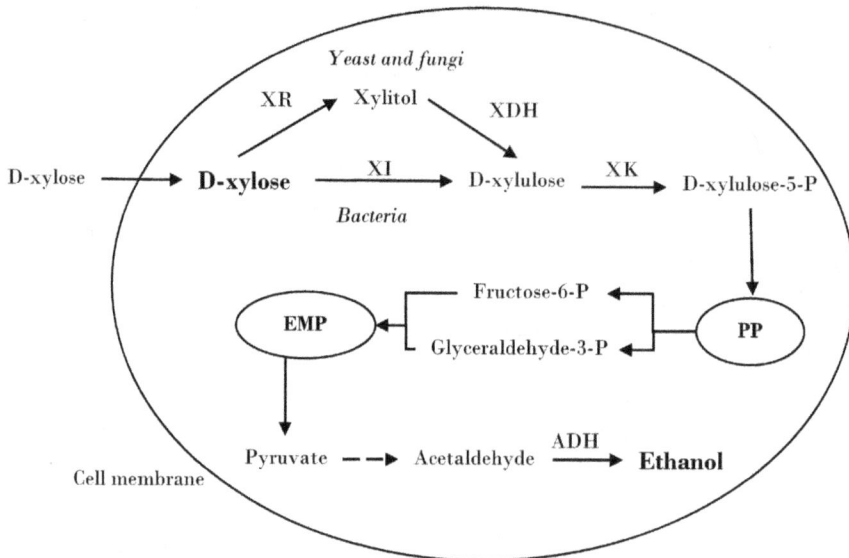

Fig. 3. D-Xylose to ethanol general metabolic pathway. PP, pentose phosphate pathway; EMP, Embden-Meyerhof-Parnas pathways; XR, xylose reductase; XDH, xylitol dehydrogenase; XI, xylose isomerase; XK, xylulokinase; ADH, alcohol dehydrogenase.

pyruvate is converted to acetaldehyde is different in most bacteria and fungi than in yeast, although all microorganisms use alcohol dehydrogenase to convert acetaldehyde into ethanol (Jagtap and Rao 2018).

In the general fermentation reaction, 3 moles of xylose are required to produce 5 moles of ethanol as shown in Equation (1) (NAD(P)H equilibrium is neglected). Theoretical ethanol based on this stoichiometry is 0.51 g ethanol/g xylose.

$$3\text{Xylose} + 5\text{ADP} + 5\text{Pi} \rightarrow 5\text{Ethanol} + 5\text{CO}_2 + 5\text{ATP} + 5\text{H}_2\text{O} \qquad (1)$$

Table 4 lists examples in which the yeasts *S. stipitis*, *P. tannophilus*, *H. polymorpha*, *K. marxianus* and *C. shehatae* have been used for the production of ethanol. Depending on the operational conditions, a wide range of ethanol yield, concentrations and productivities are obtained. In most cases, the ethanol concentration is not higher than 25 g/l and the productivities are low (< 1 g/lh).

3.2 Recombinant Microorganisms

The three most commonly used genetically modified microorganisms for the production of ethanol from xylose are *E. coli*, *S. cerevisiae* and *Z. mobilis*. Some recent reports on ethanol production using hydrolysates from lignocellulosic materials are summarized in Table 5.

Wild *E. coli* strains under anaerobic conditions are capable of metabolizing a wide variety of hexoses (including glucose, mannose, fructose, etc.) and pentoses (including xylose, arabinose, ribose, etc.). By means of metabolic engineering techniques with *E. coli*, several ethanologenic strains have been constructed in which their metabolism has been redirected to the production of ethanol (Ohta et al. 1991). Among them, one of the pioneering examples that has been used at industrial scale is the *E. coli*-KO11. The strain has continued to be improved to avoid the use of supplements in the media, increase its stability and its tolerance to ethanol concentration and resistance to inhibitors (furfural, hydroxymethylfurfural, etc.) (Orencio-Trejo et al. 2010).

Table 4. Ethanol concentration, productivity and yield reported from lignocellulosic hydrolysates by selected natural microorganisms.

Strain	Hydrolysate (g/l)	Ethanol Conc. (g/l)	Ethanol Productivity (g/l·h)	Ethanol Yield (g/g)	Ref.
C. shehatae NCIM 3501	Sugarcane bagasse: 22 g/l glucose, and 16 g/l xylose	15.5	0.16	0.41	Prajapati et al. 2020
C. shehatae ATCC 22984	Napier Grass: 90 g/l glucose, and 20 g/l xylose	40.8	0.56	0.37	Kongkeitkajorn et al. 2020
H. polymorpha ATCC 34438	Sunflower stalks: 1.5 g/l glucose and 12 g/l xylose	1.9	0.03	0.14	Martínez-Cartas et al. 2019
K. marxianus CCT 7735	Sweet sorghum: 42 g/l glucose	19.1	0.79	0.45	Tinôco et al. 2021
K. marxianus MM III-41	Coconut-tree leaf stalk: 45.4 g/l glucose and 7.9 g/l xylose	12.9	1.62	0.25	Gomes et al. 2021
K. marxianus ATCC 12424	*Quercus aegilops*: 5.7 g/l glucose and 13 g/l xylose	6.9	0.24	0.37	Tahir et al. 2020
P. stipitis CBS 6054	Brewer's spent grain: 9 g/l glucose, and 24 g/l xylose	11.4	0.38	0.27	Rojas-Chamorro et al. 2020
P. stipitis KCTC 17574	Kariba weed: 22.5 g/l glucose, and 12.5 g/l xylose	15.9	0.33	0.45	Kityo et al. 2021
P. stipitis NCIM 3499	Rice straw: 40 g/l glucose, and 17.7 g/l xylose	25.3	0.63	0.44	Prasad et al. 2020
P. tannophilus ATCC 32691	Palm trees: 38 g/l glucose, and 0.6 g/l xylose	12.4	–	0.38	Antit et al. 2021
P. tannophilus (IMTECH)	SSF Rice straw: 73 g/l glucose, and 16 g/l xylose	23.1	0.32	0.26	Goel and Wati 2016
P. tannophilus P-01	Corn stover: 106 g/l glucose, and 19 g/l xylose	24.3	0.68	0.19	Xie et al. 2015

S. cerevisiae and *Z. mobilis* can produce ethanol with high yields and productivities, but not from pentoses (Zhang et al. 2019). *S. cerevisiae* can convert xylulose to xylulose 5-phosphate, an intermediate of the pentose phosphate (PP) pathway. The modification of *S. cerevisiae* is aimed at the heterologous expression of genes that encode key enzymes in microorganisms that use pentoses and the overexpression of xylulokinase (Fig. 4). *S. stipitis* genes have been used to construct a functional XR-XDH (xylose reductase-xylitol dehydrogenase) pathway in *S. cerevisiae* for the utilization of xylose. The other alternative is to express genes that directly transform xylose into xylulose (xylose isomerase, XI) mainly from bacteria, which improves ethanol yield but worsens xylose assimilation (Kwak and Jin 2017).

In the case of *Z. mobilis*, the modification is directed to the heterologous expression of genes that encode key enzymes in microorganisms that use pentoses and genes that encode enzymes of the pentose phosphate pathway (Fig. 4). *Z. mobilis* CP4 (pZB5) was the first recombinant strain capable of producing ethanol from xylose, developed at the National Renewable Energy Laboratory (NREL) in 1995 (Xia et al. 2019). In this strain, *E. coli* genes were expressed for the expression of the enzymes responsible for the assimilation of xylose and the pentose phosphate pathway enzymes such as xylose isomerase (XI), xylulokinase (XK), transaldolase (TAL), and transketolase (TKT) achieving an ethanol yield of 86% with respect to the theoretical one (Zhang et al. 1995). Genes from *E. coli* have also been used to metabolize another pentose such as arabinose in *Z. mobilis* (Zhang et al. 2019).

Table 5. Ethanol concentration, productivity and yield for different modified *E. coli, S. cerevisiae* and *Z. mobilis* from lignocellulosic hydrolysates.

Strain	LCM-hydrolysate	Ethanol Concentration (g/l)	Ethanol Productivity (g/l·h)	Ethanol Yield (g/g)	Ref.
E coli SL100	Brewer's spent grain: 57 g/l glucose, and 32 g/l xylose	38.6	0.39	0.41	Rojas-Chamorro et al. 2019
E coli MM160	Olive tree pruning: 32 g/l glucose, and 22 g/l xylose	26.2	0.55	0.45	Martínez-Patiño et al. 2018
E. coli FBR5	Corn stover: 33 g/l glucose, and 22 g/l xylose	28.9	0.72	0.48	Saha et al. 2015
Z. mobilis 8b	Corn stover: 26 g/l glucose, and 66 g/l xylose	38	1.1	0.41	Jennings and Schell 2011
Z. mobilis 2032	Switchgrass: 90 g/l glucose, and 60 g/l xylose	53	0.96	0.35	Zhang et al. 2020
Z. mobilis ZMT2	Manure: 20 g/l glucose, and 9 g/l xylose	10.6	0.22	0.37	You et al. 2017
Z. mobilis [sucZE2::manA, pZA22-xt]	Japanese cedar: 59 g/l glucose, 19 g/l mannose and 11 g/l xylose	35.1	0.49	0.48	Yanase et al. 2012
S. cerevisiae	Barley straw: 60 g/l glucose, and 25 g/l xylose	38	0.53	0.45	Duque et al. 2020
S. cerevisiae TISTR 5339	Napier grass: 91 g/l glucose, and 19 g/l xylose	44.7	1.9	0.41	Kongkeitkajorn et al. 2020
S. cerevisiae INVSc1	Oil palm empty fruit bunch: 50 g/l glucose, and 32 g/l xylose	31.3	1.0	0.38	Liu et al. 2020
S. cerevisiae XUSE	*Miscanthus*: 40 g/l glucose, and 23 g/l xylose	30.1	0.84	0.48	Hoang Nguyen Tran et al. 2020

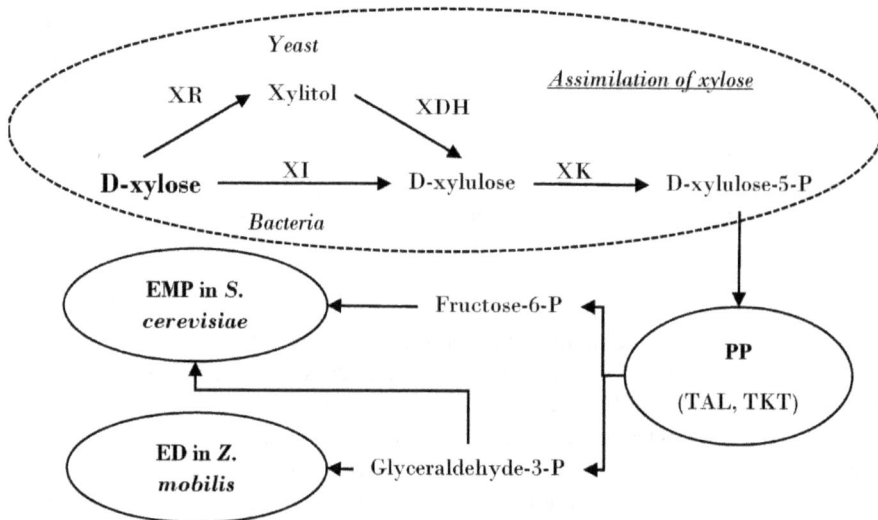

Fig. 4. Strategies for engineering *Saccharomyces cerevisiae* and *Zymomonas mobilis* with xylose metabolism (Xia et al. 2019). XR, xylose reductase; XI, xylose isomerase; XDH, xylitol dehydrogenase; XK, xylulokinase; PP, pentose phosphate pathway; TAL, transaldolase; TKL, transketolase; ED, Entner–Doudoroff pathway; EMP, Embden–Meyerhof–Parnas pathway.

4. Production of Bioethanol by Non-conventional Microorganisms from Lignocellulosic Residues. Case study: Olive Derived Biomass

4.1 Biomass Produced from Olive Tree Crop and Olive Oil Production Processes

The olive tree crop and its associated industries represent an important economic activity in the Mediterranean countries (Cardoza et al. 2021). According to FAOSTAT, more than 10.6 million hectares was dedicated to the olive tree crop in the world in 2019, 48% in the European Union countries. This sector produces different lignocellulosic wastes both in the crop fields and in the industries where the olive oil and the olive pomace oil are produced. In fact, according to estimations, one hectare of olive tree crop generates annually 1.5 tons of olive tree pruning biomass (OTPB), 0.4 tons of olive stones (OS), 0.50 tons of defatted olive pomace (DOP) and 0.16 tons of olive mill leaves (Romero-García et al. 2014, Cardoza et al. 2021).

The pruning of the olive trees generates large amounts of OTPB, a residual biomass composed mainly of thin branches, thick branches and leaves (Negro et al. 2017). Currently, this agricultural waste is burnt in the crop fields or it is chopped and spread on the field as organic input for the soil with risk of spreading pests. Both options mean environmental problems and both imply an economic cost for the farmers (Ruiz et al. 2017).

The olive oil production process generates wastes such as olive leaves and by-products such as olive pomace (OP) and olive stones (OS) (Romero-García et al. 2014). OP is composed of stones, skin and residual oil and it is traditionally used as raw material in the extracting industries for obtaining olive pomace oil. This process generates DOP as main waste, which is used as biofuel in the same industries where it is generated. In addition, OS represents about 10% by weight of the olive fruit, and it is generated from the olive pomace pitting. In general, there is no industrial application for these LCMs (Manzanares et al. 2017, Negro et al. 2017). Therefore, their use as feedstock for sugar production that can be bioconverted into ethanol is an attractive option to be explored.

4.2 Ethanol Production from Olive Derived Biomass

Ethanol is a renewable fuel able to replace gasoline and it can be obtained from LCM containing sugars. Their biochemical conversion into ethanol means a challenge and an alternative to ethanol production processes based on starchy and sugary feedstock. The carbohydrates are forming a complex matrix, which requires a pretreatment step to disorganize it and subsequently to recover them (Nandal et al. 2020). After pretreatment, enzymatic hydrolysis of the pretreated biomass and the fermentation of the resulting hydrolysate are steps required for ethanol production (Fig. 5). These steps mean the bioconversion of the cellulose fraction into bioethanol using mainly the industrial yeast *S. cerevisiae* characterized by a high specific ethanol productivity and for being very tolerant to high ethanol concentrations (Robak and Balcerek 2020). Different strategies can be used for these latter process steps. Hydrolysis and fermentation can be carried out simultaneously or separately.

In fact, the lignocellulosic composition of the olive wastes converts them in interesting feedstock for production of bioethanol and renewable chemicals in a biorefinery strategy. In this sense, OTPB is especially attractive to obtain bioethanol due to its carbohydrate content, around 40% (Martínez-Patiño et al. 2015) (Table 1). The bioconversion of its cellulosic glucose into ethanol using a simultaneous configuration process with *S. cerevisiae* has been reported after steam explosion (Cara et al. 2008b), liquid hot water (Manzanares et al. 2011), sulfuric acid pretreatment (Martínez-

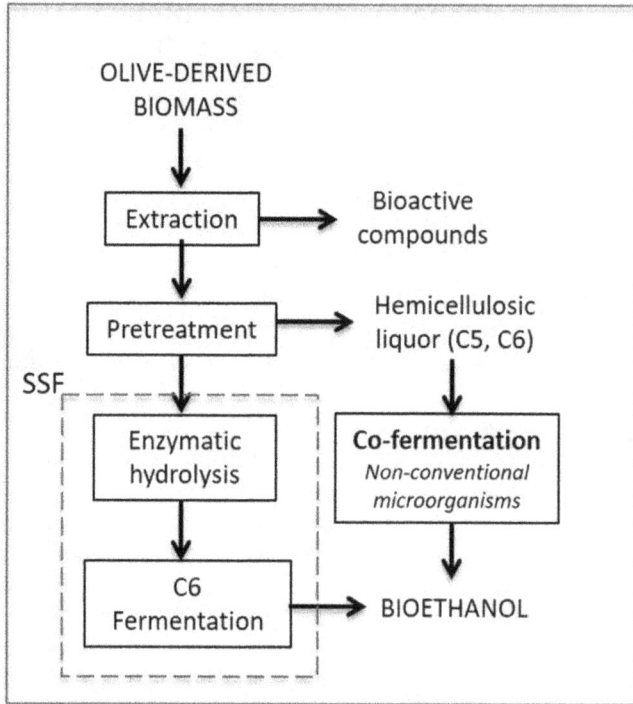

Fig. 5. Bioconversion process of sugars into ethanol for olive-derived biomass. SSF, simultaneous saccharification fermentation.

Patiño et al. 2018) and acid/oxidative-alkaline pretreatment (Martínez-Patiño et al. 2017a). A maximum ethanol concentration of 5.5% (v/v) was obtained after the sequential acid/alkaline-peroxide pretreatment and the subsequent simultaneous process at 20% solids (Martínez-Patiño et al. 2017a). Manzanares et al. (2011) reached an ethanol concentration of 3.7% (v/v) after a liquid hot water pretreatment of OTPB followed by a prehydrolysis and simultaneous saccharification and fermentation at 23% solids.

For olive LCM rich in soluble components such as extractives, a previous extraction step could be included in the bioethanol production process in order to achieve a more efficient fractionation (Fig. 5). The effectiveness of this step has been tested for OTPB, which resulted in a better performance of the enzymatic hydrolysis after a partial removal of its extractive fraction (Ballesteros et al. 2011).

In addition to the cellulose to ethanol conversion, the bioconversion of the hemicellulosic sugars solubilized during the pretreatment can be crucial, especially for feedstock with high content of xylans, like OTPB and OS. Nevertheless, the traditional yeast *S. cerevisiae* is unable to utilize pentose sugars during the fermentation and consequently, non-conventional microorganisms, capable of assimilating both pentose and hexose sugars present in the hemicelluloses, are needed (Robak and Balcerek 2020).

4.3 Non-Conventional Microorganisms for Bioethanol Production from Olive Biomass

The utilization of sugars from hemicellulose in biomass using non-conventional microorganisms would contribute to the economic feasibility of the bioethanol process integrated in a biorefinery system based on LCM feedstock (Yamakawa et al. 2020). This implies the use of pentose-fermenters microorganisms. As commented before, *P. tannophilus, S. stipitis or C. shehatae* have been reported as efficient xylose fermenters. Nevertheless, their ethanol productivity is lower than traditional *S. cerevisiae* (Robak and Balcerek 2020). Some of these microorganisms have been tested to produce bioethanol from olive LCM because the presence of hemicellulosic is noticeable, especially in OS and OTPB.

Therefore, the valorisation of these sugars, pentoses and hexoses, is key for an integral utilization of these olive biomasses. Table 6 shows the non-conventional microorganisms used for bioethanol production from different olive-derived biomass. The hemicellulose fraction of OS accounts for around 26%–29% of its composition compared to 19%–20% of cellulose (Padilla-Rascón et al. 2020, Romero-García et al. 2016). In the case of OTPB, with 20% of hemicellulose, around 40% of its sugar content corresponds to hemicellulosic sugars (Martínez-Patiño et al. 2017b) (Table 1). In addition, the presence of a high glucose content in the extractive fraction of OTPB, higher than 5%, has been reported (Martínez-Patiño et al. 2017b). This non-structural glucose is solubilised easily during the pretreatment if the biomass has not been previously extracted and consequently, the pretreatment liquor will have a higher presence of glucose together with the hemicellulosic sugars. The bioconversion of all these sugars from OTPB into ethanol is an interesting option that has been studied using different non-conventional microorganisms. Thus, ethanologenic yeasts such as *P. tannophilus (*Díaz-Villanueva et al. 2012, Romero et al. 2007), *S. stipitis* (Díaz et al. 2009) and *C. tropicalis* (García-Martín et al. 2010) have been used to ferment the liquors resulting from the acid pretreatment of OTPB, without previous extraction. The hemicellulosic liquors from steam explosion pretreatment of phosphoric acid impregnated OTPB after an aqueous extraction were fermented using two non-conventional ethanologenic yeasts, *S. stipitis* (Negro et al. 2014) and a genetically modified strain *S. cerevisiae* F12 (Oliva et al. 2020); higher ethanol yield was obtained using the former microorganism (Table 6).

Different ethanologenic strains of *E. coli* have also been tested to ferment pretreatment liquors of OTPB and yields close to the theoretical maximum yield have been reported (Martínez-Patiño et al. 2015, 2017a, Fernandes-Klajn et al. 2018). Martínez-Patiño et al. (2018) studied a novel strategy of hydrolysis and fermentation for OTPB that included the enzymatic hydrolysis of the pretreatment slurry followed by the co-fermentation of the resulting hydrolysate with *E. coli* MM160 after detoxification with overliming. This process alternative strategy allowed the fermentation of all sugars content in biomass in a single step. The bioconversion of cellulosic glucose from OTPB after phosphoric acid pretreatment has been also reported using the modified microbial strain, *E. coli* MS04, and it resulted in the complete conversion of glucose into ethanol (Martínez-Patiño et al. 2015).

In addition to OTPB, other olive-derived biomasses have been used as raw materials for bioethanol production. In this sense, Romero-García et al. (2016) reported the fermentation of hemicellulosic liquors from OS with *E. coli* MM160. OP and DOP were also used as raw materials for bioethanol production using *E. coli* SL100 (López-Linares et al. 2020) and *E. coli* FBR5 (El Asli and Qatibi 2009), respectively, obtaining similar ethanol yields (Table 6).

Overall, the integration of all these olive LCM in a biorefinery strategy to produce a liquid fuel such as bioethanol and other renewable chemicals would contribute to the development of a model based on the bioeconomy. This idea is especially interesting in the rural areas where the olive tree cultivation, the olive mills and the extracting industries are located (Cardoza et al. 2021).

Table 6. Co-fermentation of mixed sugar hydrolysates from different olive-derived biomass.

Biomass	Extraction	Pretreatment	Configuration Process	Microorganism	Ethanol Concentration (g/L)	Ethanol Yield (g/g)	Ref.
OTPB	–	Acid pretreatment (1% H_2SO_4, 180°C, 10 min)	Hemicellulosic liquor fermentation	*P. tannophilus* CECT 12920	12.3	0.44	Díaz-Villanueva et al. 2012
OTPB	–	Acid pretreatment (1% H_2SO_4, 190°C, 10 min)	Hemicellulosic liquor fermentation	*S. stipitis* CECT 1922	12	0.42	Díaz et al. 2009
OTPB	Water extraction (121°C, 60 min)	Steam explosion (175°C, 1% H_3PO_4 impregnation)	Hemicellulosic liquor fermentation	*S. stipitis* CBS 6054	8.9	0.37	Negro et al. 2014
OTPB	Water extraction (120°C, 60 min)	Acid pretreatment (0.5% H_3PO_4, 170°C)	Hemicellulosic liquor fermentation	*E. coli* MS04	23	0.46	Martínez-Patiño et al. 2015
OTPB	Water extraction (120°C, 60 min)	Combined: Acid (2.4% H_2SO_4, 130°C, 84 min) + Alkaline-peroxide (7% H_2O_2, 80°C, 90 min)	Hemicellulosic liquor fermentation	*E. coli* MM160	19	0.43	Martínez-Patiño et al. 2017a
OTPB	Water extraction (120°C, 60 min)	Acid pretreatment 160°C, 1.7% H_2SO_4	Hemicellulosic liquor fermentation	*E. coli* MM160	14.5	0.45	Martínez-Patiño et al. 2017b
OTPB	Water extraction (120°C, 60 min)	Acid pretreatment (0.9% H_2SO_4, 164°C, 10 min)	Slurry saccharification + fermentation	*E. coli* MM160	26.2	0.45 g/g	Martínez-Patiño et al. 2018
OTPB	Alkaline extraction (1% NaOH, 110°C, 30 min)	Acid pretreatment (0.9% H_2SO_4, 164°C, 10 min)	Presaccharification + hydrolysate fermentation	*E. coli* SL100	–	0.48	Fernandes-Klajn et al. 2018
OTPB	Water extraction (121°C, 60 min)	Steam explosion (175°C, 1% H_3PO_4 impregnation)	Hemicellulosic liquor fermentation	*S. cerevisiae* F12	55	0.33	Oliva et al. 2020
OTPB	–	Liquid hot water (200°C)	Hemicellulosic liquor fermentation	*C. tropicalis* NBRC 0618	–	0.44	García-Martín et al. 2010
OTPB	–	Acid pretreatment (0.5 N H_3PO_4, 90°C, 240 min)	Hemicellulosic liquor fermentation	*P. tannophilus* ATCC 32691	5	0.38	Romero et al. 2007
OS	–	Acid pretreatment (2% H_2SO_4, 130°C, 60 min)	Hemicellulosic liquor fermentation	*E. coli* MM160	25	0.35	Romero-García et al. 2016
OP	–	Acid pretreatment (1.75% H_2SO_4, 160°C, 10 min)	Hemicellulosic liquor fermentation	*E. coli* FBR5	8.1	0.45	El Asli and Qatibi 2009
DOP	Water extraction (100°C, 30 min)	Acid pretreatment (2% H_2SO_4, 170°C)	Hemicellulosic liquor fermentation	*E. coli* SL100	14.5	0.46	López-Linares et al. 2020

OTPB: olive tree pruning biomass; OS: olive stones; OP: olive pomace; DOP: defatted olive pomace.

5. Conclusions

The economic feasibility of the production of bioethanol from lignocellulosic feedstocks requires the conversion of the hemicellulosic sugars, in addition to the glucose coming out from the cellulose. To achieve this, non-conventional microorganisms are to be used. Depending on the raw material, a number of factors will be relevant for a good performance of the microorganism. As outlined here, genetic modifications play an essential role but also process conditions and configurations need to be carefully assessed for better results. In general terms, ethanol yields produced from hemicellulosic sugars are lower than those from glucose, although their contribution to the global economy of the conversion can be essential.

Acknowledgements

Financial support from Agencia Estatal de Investigación (MICINN, Spain) and European Regional Development Fund, reference project ENE2017-85819-C2-1-R and FEDER UJA project 1260905 is gratefully acknowledged. MdM Contreras thanks the Ministry of Science and Innovation of Spain for the Ramón y Cajal grant (RYC2020-030546-I/AEI/10.13039/501100011033) and the European Social Fund. J.M. Romero-García expresses his gratitude to the Junta de Andalucía for financial support (Postdoctoral researcher R-29/12/2020).

References

Aditiya, H.B., Mahlia, T.M.I., Chong, W.T., Nur, H. and Sebayang, A.H. 2016. Second generation bioethanol production: A critical review. Renew. Sust. Energ. Rev. 66: 631–653.

Agencia Extremeña de la Energía. 2020. La biomasa forestal. Available online: https://www.agenex.net/images/stories/deptos/la-biomas-forestal.pdf (accessed on 1 June 2021).

Antar, M., Lyu, D., Nazari, M., Shah, A., Zhou, X. and Smith, D.L. 2021. Biomass for a sustainable bioeconomy: An overview of world biomass production and utilization. Renew. Sustain. Energy 139: 110691.

Antit, Y., Olivares, I., Hamdi, M. and Sánchez, S. 2021. Biochemical conversion of lignocellulosic biomass from date palm of *Phoenix dactylifera* L. into ethanol production. Energies 14: 1887.

Ashoor, S., Comitini, F. and Ciani, M. 2015. Cell-recycle batch process of *Scheffersomyces stipitis* and *Saccharomyces cerevisiae* co-culture for second generation bioethanol production. Biotechnol. Lett. 37(11): 2213–2218.

Ballesteros, I., Ballesteros, M., Cara, C., Sáez, F., Castro, E., Manzanares, P. et al. 2011. Effect of water extraction on sugars recovery from steam exploded olive tree pruning. Bioresour. Technol. 102: 6611–6616.

Cara, C., Ruiz, E., Oliva, J.M., Sáez, F. and Castro, E. 2008. Conversion of olive tree biomass into fermentable sugars by dilute acid pretreatment and enzymatic saccharification. Bioresour. Technol. 99: 1869–1876.

Cardoza, D.I., Romero, I., Martínez, T., Ruiz, E., Gallego, F.J., López-Linares, J.C. et al. 2021. Location of biorefineries based on olive-derived biomass in Andalusia, Spain. Energies 14: 3052.

Chen, J., Wu, Y., Xu, C., Song, M. and Liu, X. 2019. Global non-fossil fuel consumption: driving factors, disparities, and trends. Management Decision 57(4): 791–810.

Chen, Y.A., Zhou, Y., Liu, D., Zhao, X. and Qin, Y. 2018. Evaluation of the action of Tween 20 non-ionic surfactant during enzymatic hydrolysis of lignocellulose: Pretreatment, hydrolysis conditions and lignin structure. Bioresour. Technol. 269: 329–338.

Contreras, M.d.M., Romero, I., Moya, M. and Castro, E. 2020. Olive-derived biomass as a renewable source of value-added products. Process Biochem. 97: 43–56.

Cunha, J.T., Soares, P.O., Romaní, A., Thevelein, J.M. and Domingues, L. 2019. Xylose fermentation efficiency of industrial *Saccharomyces cerevisiae* yeast with separate or combined xylose reductase/xylitol dehydrogenase and xylose isomerase pathways. Biotechnol. Biofuels 12(1): 1–14.

Díaz, M.J., Ruiz, E., Romero, I., Cara, C., Moya, M. and Castro, E. 2009. Inhibition of *Pichia stipitis* fermentation of hydrolysates from olive tree cuttings. World J. Microbiol. Biotechnol. 25: 891–899.

Díaz, M.J., Cara, C., Ruiz, E., Pérez-Bonilla, M. and Castro, E. 2011. Hydrothermal pre-treatment and enzymatic hydrolysis of sunflower stalks. Fuel 90: 3225–3229.

Díaz-Villanueva, M.J., Cara, C., Ruiz, E., Romero, I. and Castro, E. 2012. Olive tree pruning as an agricultural residue for ethanol production. Fermentation of hydrolysates from dilute acid pretreatment. Span. J. Agric. Res. 10: 643–648.

Duque, A., Álvarez, C., Doménech, P., Manzanares, P. and Moreno, A.D. 2021. Advanced bioethanol production: From novel raw materials to integrated biorefineries. Processes 9(2): 206.

Duque, A., Doménech, P., Álvarez, C., Ballesteros, M. and Manzanares, P. 2020. Study of the bioprocess conditions to produce bioethanol from barley straw pretreated by combined soda and enzyme-catalyzed extrusion. Renew. Energy 158: 263–270.

El Asli, A. and Qatibi, A.I. 2009. Ethanol production from olive cake biomass substrate. Biotechnol. Bioproc. Eng. 14: 118–122.

Eni, V. 2021. Versalis (Eni) starts the production of Invix, bioethanol-based disinfectant at Crescentino. Available online: https://www.eni.com/en-IT/media/press-release/2020/07/pr-versalis-invix-21-july-2020.html (accessed on 25 May 2021).

Fan, Z., Lin, J., Wu, J., Zhang, L., Lyu, X., Xiao, W. et al. 2020. Vacuum-assisted black liquor-recycling enhances the sugar yield of sugarcane bagasse and decreases water and alkali consumption. Bioresour. Technol. 309: 123349.

FAOSTAT. 2021. Food and Agriculture Organization of the United Nations. Available online: http://faostat3.fao.org/ (accessed on 2 June 2021).

Farias, D. and Maugeri Filho, F. 2019. Co-culture strategy for improved 2G bioethanol production using a mixture of sugarcane molasses and bagasse hydrolysate as substrate. Biochem. Eng. J. 147: 29–38.

Fernandes-Klajn, F., Romero-García, J.M., Díaz, M.J. and Castro, E. 2018. Comparison of fermentation strategies for ethanol production from olive tree pruning biomass. Ind. Crop. Prod. 122: 98–106.

Goel, A. and Wati, L. 2016. Ethanol production from rice (*Oryza sativa*) straw by simultaneous saccharification and cofermentation. Indian J. Exp. Biol. 54(8): 525–529.

Gomes, M.A., dos Santos Rocha, M.S.R., Barbosa, K.L., Abreu, I.B.S.d., Pimentel, W.R.d.O., Silva, C.E.D.F., et al. 2021. Agricultural coconut cultivation wastes as feedstock for lignocellulosic ethanol production by *Kluyveromyces marxianus*. Waste Biomass Valor. 12: 4943–4951.

Gómez-García, E. 2021. Estimation of primary forest harvest residues and potential bioenergy production from fast-growing tree species in NW Spain. Biomass Bioenerg. 148: 106055.

González, R., Tao, H., Shanmugam, K.T., York, S.W. and Ingram, L.O. 2002. Global gene expression differences associated with changes in glycolytic flux and growth rate in *Escherichia coli* during the fermentation of glucose and xylose. Biotechnol. Prog. 18(1): 6–20.

Goshadrou, A., Karimi, K. and Lefsrud, M. 2013. Characterization of ionic liquid pretreated aspen wood using semi-quantitative methods for ethanol production. Carbohydr. Polym. 96: 440–449.

Heinberg, R. 2017. The energy crisis: from fossil fuel abundance to renewable energy constraints. pp. 65–78. *In*: Lerch, D. (ed.). The Community Resilience Reader. Washington, DC: Island Press.

Hoang Nguyen Tran, P., Ko, J.K., Gong, G., Um, Y. and Lee, S.-M. 2020. Improved simultaneous co-fermentation of glucose and xylose by *Saccharomyces cerevisiae* for efficient lignocellulosic biorefinery. Biotechnol. Biofuels 13(1): 12.

Huang, Y.F., Chiueh, P.T. and Lo, S.L. 2016. A review on microwave pyrolysis of lignocellulosic biomass. Sustain. Environ. Res. 26: 103–109.

IEA, IRENA, UNSD, WB and WHO. 2019. The Energy Progress Report SDG7: The Energy Progress Report. Washington, DC. Available online: www.worldbank.org (accessed on 10 June 2021).

Jagtap, S.S. and Rao, C.V. 2018. Microbial conversion of xylose into useful bioproducts. Appl. Microbiol. Biotechnol. 102: 9015–9036.

Jennings, E.W. and Schell, D.J. 2011. Conditioning of dilute-acid pretreated corn stover hydrolysate liquors by treatment with lime or ammonium hydroxide to improve conversion of sugars to ethanol. Bioresour. Technol. 102: 1240–1245.

Junta de Andalucía. 2008. Potencial energético de la biomasa residual agrícola y ganadera en Andalucía. Available online: https://www.juntadeandalucia.es/export/drupaljda/Potencial%20energetico%20de%20la%20biomasa.pdf (accessed on 1 June 2021).

Karagöz, P. and Özkan, M. 2014. Ethanol production from wheat straw by *Saccharomyces cerevisiae* and *Scheffersomyces stipitis* co-culture in batch and continuous system. Bioresour. Technol. 158: 286–293.

Kityo, M.K., Sunwoo, I., Kim, S.H., Park, Y.R., Jeong, G.-T. and Kim, S.-K. 2021. Enhanced bioethanol fermentation by sonication using three yeasts species and Kariba Weed (*Salvinia molesta*) as biomass collected from Lake Victoria, Uganda. Appl. Biochem. Biotechnol. 192: 180–195.

Kongkeitkajorn, M.B., Sae-Kuay, C. and Reungsang, A. 2020. Evaluation of Napier grass for bioethanol production through a fermentation process. Processes 8(5): 567.

Kumar, B., Bhardwaj, N., Agrawal, K., Chaturvedi, V. and Verma, P. 2020. Current perspective on pretreatment technologies using lignocellulosic biomass: An emerging biorefinery concept. Fuel Process. Technol. 199: 106244.

Kumar, R., Tabatabaei, M., Karimi, K. and Horváth, I.S. 2016. Recent updates on lignocellulosic biomass derived ethanol—A review. Biofuel Res. J. 3(1): 347–356.

Kwak, S. and Jin, Y.-S. 2017. Production of fuels and chemicals from xylose by engineered *Saccharomyces cerevisiae*: A review and perspective. Microb. Cell Factories 16: 82.

Liu, T., Peng, B., Huang, S. and Geng, A. 2020. Recombinant xylose-fermenting yeast construction for the co-production of ethanol and cis,cis-muconic acid from lignocellulosic biomass. Bioresour. Technol. Rep. 9: 100395.

López-Linares, J.C., Ballesteros, I., Tourán, J., Cara, C., Castro, E., Ballesteros, M. et al. 2015. Optimization of uncatalyzed steam explosion pretreatment of rapeseed straw for biofuel production. Bioresour. Technol. 190: 97–105.

López-Linares, J.C., García-Cubero, M.T., Lucas, S., González-Benito, G. and Coca, M. 2019. Microwave assisted hydrothermal as greener pretreatment of brewer's spent grains for biobutanol production. Chem. Eng. J. 368: 1045–1055.

López-Linares, J.C., García-Cubero, M.T., Coca, M. and Lucas, S. 2021. Efficient biobutanol production by acetone-butanol-ethanol fermentation from spent coffee grounds with microwave assisted dilute sulfuric acid pretreatment. Bioresour. Technol. 320: 124348.

López-Linares, J.C., Gómez-Cruz, I., Ruiz, E., Romero, I. and Castro, E. 2020. Production of ethanol from hemicellulosic sugars of exhausted olive pomace by *Escherichia coli*. Processes 8: 533.

Manzanares, P., Negro, M.J., Oliva, J.M., Sáez, F., Ballesteros, I., Ballesteros, M. et al. 2011. Different process configurations for bioethanol production from pretreated olive pruning biomass. Chem. Technol. Biotechnol. 86: 881–887.

Manzanares, P., Ruiz, E., Ballesteros, M., Negro, M.J., Gallego, F.J., López-Linares, J.C. et al. 2017. Residual biomass potential in olive tree cultivation and olive oil industry in Spain: Valorization proposal in a biorefinery context. Spanish J. Agric. Res. 15: e0206.

Martínez-Cartas, M.L., Olivares, M.I. and Sánchez, S. 2019. Production of bioalcohols and antioxidant compounds by acid hydrolysis of lignocellulosic wastes and fermentation of hydrolysates with *Hansenula polymorpha*. Eng. Life Sci. 19: 522–536.

Martínez-Patiño, J.C., Romero, I., Ruiz, E., Cara, C., Romero-García, J.M. and Castro, E. 2017b. Design and optimization of sulfuric acid pretreatment of extracted olive tree biomass using response surface methodology. BioResources 12: 1779–1797.

Martínez-Patiño, J.C., Ruiz, E., Romero, I., Cara, C., López-Linares, J.C. and Castro, E. 2017a. Combined acid/alkaline-peroxide pretreatment of olive tree biomass for bioethanol production. Bioresour. Technol. 239: 326–335.

Martínez-Patiño, J.C., Ruiz, E., Cara, C., Romero, I. and Castro, E. 2018. Advanced bioethanol production from olive tree biomass using different bioconversion schemes. Biochem. Eng. J. 137: 172–181.

Martínez-Patiño, J.C., Romero-García, J.M., Ruiz, E., Oliva, J.M., Álvarez, C., Romero, I. et al. 2015. High solids loading pretreatment of olive tree pruning with dilute phosphoric acid for bioethanol production by *Escherichia coli*. Energy Fuels 29: 1735–1740.

Matsakas, L., Raghavendran, V., Yakimenko, O., Persson, G., Olsson, E., Rova, U. et al. 2019. Lignin-first biomass fractionation using a hybrid organosolv—Steam explosion pretreatment technology improves the saccharification and fermentability of spruce biomass. Bioresour. Technol. 273: 521–528.

Mishra, A. and Ghosh, S. 2020. Saccharification of Kans grass biomass by a novel fractional hydrolysis method followed by co-culture fermentation for bioethanol production. Renew. Energy 146: 750–759.

Moya, A.J., Peinado, S., Mateo, S., Fonseca, B.G. and Sánchez, S. 2016. Improving bioethanol production from olive pruning biomass by deacetylation step prior acid hydrolysis and fermentation processes. Bioresour. Technol. 220: 239–245.

Muranaka, Y., Nakagawa, H., Hasegawa, I., Maki, T., Hosokawa, J., Ikuta, J. et al. 2017. Lignin-based resin production from lignocellulosic biomass combining acidic saccharification and acetone-water treatment. Chem. Eng. J. 308: 754–759.

Mussatto, S.I. and Dragone, G.M. 2016. Biomass pretreatment, biorefineries, and potential products for a bioeconomy development. pp. 1–22. *In*: Mussatto, S.I. (ed.). Biomass Fractionation Technologies for a Lignocellulosic Feedstock Based Biorefinery. Elsevier Inc. All.

Nandal, P., Sharma, S. and Arora, A. 2020. Bioprospecting non-conventional yeasts for ethanol production from rice straw hydrolysate and their inhibitor tolerance. Renew. Energy 147: 1694–1703.

Negro, M.J., Álvarez, C., Ballesteros, I., Romero Pulido, I., Ballesteros, M., Castro, E. et al. 2014. Ethanol production from glucose and xylose obtained from steam exploded water-extracted olive tree pruning using phosphoric acid as catalyst. Bioresour. Technol. 153: 101–107.

Negro, M.J., Manzanares, P., Ruiz, E., Castro, E. and Ballesteros, M. 2017. The biorefinery concept for the industrial valorization of residues from olive oil industry. pp. 57–78. *In*: Galanakis, C.M. (ed.). Olive Mill Waste. Oxford: Academic Press.

Ohta, K., Beall, D.S., Mejia, J.P., Shanmugam, K.T. and Ingram, L.O. 1991. Genetic improvement of *Escherichia coli* for ethanol production: Chromosomal integration of *Zymomonas mobilis* genes encoding pyruvate decarboxylase and alcohol dehydrogenase II. Appl. Environ. Microbiol. 57: 893–900.

Oliva, J., Negro, M.J., Álvarez, C., Manzanares, P. and Moreno, A.D. 2020. Fermentation strategies for the efficient use of olive tree pruning biomass from a flexible biorefinery approach. Fuel 277: 118171.

Olsson, L. and Hahn-Hägerdal, B. 1996. Fermentation of lignocellulosic hydrolysates for ethanol production. Enzyme Microb. Technol. 18(5): 312–331.

Orencio-Trejo, M., Utrilla, J., Fernández-Sandoval, M.T., Huerta-Beristain, G., Gosset, G. and Martinez, A. 2010. Engineering the *Escherichia coli* fermentative metabolism. Adv. Biochem. Eng. Biotechnol. 121: 71–107.

Padilla-Rascón, C., Ruiz, E., Romero, I., Castro, E., Oliva, J.M., Ballesteros, I. et al. 2020. Valorisation of olive stone by-product for sugar production using a sequential acid/steam explosion pretreatment. Ind. Crop. Prod. 148: 112279.

Prajapati, B.P., Jana, U.K., Suryawanshi, R.K. and Kango, N. 2020. Sugarcane bagasse saccharification using *Aspergillus tubingensis* enzymatic cocktail for 2G bio-ethanol production. Renew. Energy 152: 653–663.

Prasad, S., Kumar, S., Yadav, K.K., Choudhry, J., Kamyab, H., Bach, Q.V. et al. 2020. Screening and evaluation of cellulytic fungal strains for saccharification and bioethanol production from rice residue. Energy 190: 116422.

Ritchie, H. and Roser, M. 2020. Renewable Energy. Available online: OurWorldInData.org. https://ourworldindata. org/renewable-energy (accessed on 15 June 2021).

Robak, K. and Balcerek, M. 2020. Current state-of-the-art in ethanol production from lignocellulosic feedstocks. Microbiol. Res. 240: 126534.

Rojas-Chamorro, J.A., Romero, I., López-Linares, J.C. and Castro, E. 2020. Brewer's spent grain as a source of renewable fuel through optimized dilute acid pretreatment. Renew. Energy 148: 81–90.

Rojas-Chamorro, J.A., Romero-García, J.M., Cara, C., Romero, I. and Castro, E. 2019. Improved ethanol production from the slurry of pretreated brewers' spent grain through different co-fermentation strategies. Bioresour. Technol. 296: 122367.

Romero, I., Moya, M., Sánchez, S., Ruiz, E., Castro, E. and Bravo, V. 2007. Ethanolic fermentation of phosphoric acid hydrolysates from olive tree pruning. Ind. Crop. Prod. 25(2): 160–168.

Romero-García, J.M., Martínez-Patiño, C., Ruiz, E., Romero, I. and Castro, E. 2016. Ethanol production from olive stone hydrolysates by xylose fermenting microorganisms. Bioethanol 2: 51–64.

Romero-García, J.M., Niño, L., Martínez-Patiño, C., Álvarez, C., Castro, E. and Negro, M.J. 2014. Biorefinery based on olive biomass. State of the art and future trends. Bioresour. Technol. 159: 421–432.

Ruiz, E., Romero-García, J.M., Romero, I., Manzanares, P., Negro, M.J. and Castro, E. 2017. Olive-derived biomass as a source of energy and chemicals. Biofuel. Bioprod. Biorefin. 11: 1077–1094.

Saha, B.C., Qureshi, N., Kennedy, G.J. and Cotta, M.A. 2015. Enhancement of xylose utilization from corn stover by a recombinant *Escherichia coli* strain for ethanol production. Bioresour. Technol. 190: 182–188.

Satari, B., Karimi, K. and Kumar, R. 2019. Cellulose solvent-based pretreatment for enhanced second-generation biofuel production: A review. Sustain. Energy Fuels 3: 11–62.

Senila, L., Kovacs, E., Scurtu, D.A., Cadar, O., Becze, A., Senila, M. et al. 2020. Bioethanol production from vineyard waste by autohydrolysis pretreatment and chlorite delignification via simultaneous saccharification and fermentation. Molecules 25: 2606.

Sun, L. and Jin, Y.-S. 2021. Xylose assimilation for the efficient production of biofuels and chemicals by engineered *Saccharomyces cerevisiae*. Biotechnol. J. 16: 2000142.

Sun, Y. and Cheng, J. 2002. Hydrolysis of lignocellulosic materials for ethanol production: A review. Bioresour. Technol. 83: 1–11.

Susmozas, A., Martín-Sampedro, R., Ibarra, D., Eugenio, M.E., Iglesias, R., Manzanares, P. et al. 2020. Process strategies for the transition of 1G to advanced bioethanol production. Processes 8(10): 1–45.

Tahir, B. and Mezori, H.A. 2020. Bioethanol production from *Quercus aegilops* using *Pichia stipitis* and *Kluyveromyces marxianus*. Biomass Conv. Bioref. (in press).

Tian, D., Guo, Y., Hu, J., Yang, G., Zhang, J., Luo, L. et al. 2020. Acidic deep eutectic solvents pretreatment for selective lignocellulosic biomass fractionation with enhanced cellulose reactivity. Int. J. Biol. Macromol. 142: 288–297.

Tinôco, D., Genier, H.L.A. and Silveira, W.B.d. 2021. Technology valuation of cellulosic ethanol production by *Kluyveromyces marxianus* CCT 7735 from sweet sorghum bagasse at elevated temperatures. Renew. Energy 173: 188–196.

Tye, Y.Y., Lee, K.T., Wan Abdullah, W.N. and Leh, C.P. 2016. The world availability of non-wood lignocellulosic biomass for the production of cellulosic ethanol and potential pretreatments for the enhancement of enzymatic saccharification. Renew. Sustain. Energy Rev. 60: 155–172.

United Nations. 2021. Sustainable Development Goals. Available online: www.un.org/sustainabledevelopment/energy/ (accessed on 17 June 2021).

VERBIO. 2021. Bioethanol. Available online: www.verbio.us/technology/bioethanol/ (accessed on 17 June 2021).

Wirawan, F., Cheng, C.-L., Lo, Y.-C., Chen, C.-Y., Chang, J.-S., Leu, S.-Y. et al. 2020. Continuous cellulosic bioethanol co-fermentation by immobilized *Zymomonas mobilis* and suspended *Pichia stipitis* in a two-stage process. Appl. Energy 266: 114871.

Xie, H., Wang, F., Yin, S., Ren, T. and Song, A. 2015. The preparation and ethanol fermentation of high-concentration sugars from steam-explosion corn stover. Appl. Biochem. Biotechnol. 176: 613–624.

Xie, N., Jiang, N., Zhang, M., Qi, W., Su, R. and He, Z. 2014. Effect of different pretreatment methods of corncob on bioethanol production and enzyme recovery. Cellul. Chem. Technol. 48: 313–319.

Yamakawa, C.K., Kastella, L., Mahler, M.R., Martinez, J.L. and Mussatto, S.I. 2020. Exploiting new biorefinery models using non-conventional yeasts and their implications for sustainability. Bioresour. Technol. 309: 12337.

Yanase, H., Miyawaki, H., Sakurai, M., Kawakami, A., Matsumoto, M., Haga, K. et al. 2012. Ethanol production from wood hydrolysate using genetically engineered *Zymomonas mobilis*. Appl. Microbiol. Biotechnol. 94(6): 1667–1678.

You, Y., Liu, S., Wu, B., Wang, Y.-W., Zhu, Q.-L., Qin, H. et al. 2017. Bio-ethanol production by *Zymomonas mobilis* using pretreated dairy manure as a carbon and nitrogen source. RSC Adv. 7(7): 3768–3779.

Zhang, K., Lu, X., Li, Y., Jiang, X., Liu, L. and Wang, H. 2019. New technologies provide more metabolic engineering strategies for bioethanol production in *Zymomonas mobilis*. Appl. Microbiol. Biotechnol. 103: 2087–2099.

Zhang, M., Eddykristine, C., Finkelsteinand, D. and Picataggio, S. 1995. Metabolic engineering of a pentose metabolism pathway in ethanologenic *Zymomonas mobilis*. Science 267: 240–243. https://doi.org/10.1126/science.267.5195.240.

Zhang, Y., Serate, J., Xie, D., Gajbhiye, S., Kulzer, P., Sanford, G. et al. 2020. Production of hydrolysates from unmilled AFEX-pretreated switchgrass and comparative fermentation with *Zymomonas mobilis*. Bioresour. Technol. Rep. 11: 100517.

CHAPTER 8

Strategies to Enhance Biobutanol Production from Lignocellulosic Biomass

*Baranitharan Ethiraj,[1] Topu Raihan,[2] Sumaya Sarmin,[3]
Ahasanul Karim,[4] Md. Maksudur Rahman Khan,[3]
Kirupa Sankar Muthuvelu,[5] Shanmugaprakash Muthusamy,[6]
Abu Yousuf[7] and M. Amirul Islam[8],**

1. Introduction

The acute energy crisis and rapid depletion of fossil fuels accelerated the search for renewable eco-friendly fossil fuel substitutes (Escobar et al. 2009). Reports on the exhaustion of available fossil fuels in the next 40–50 years further highlighted the necessity of alternative resources (Vohra et al. 2014). Bio-based fuels, i.e., biofuels, have numerous advantages compared to fossil fuels. Biofuels provide economic benefits (i.e., sustainability, higher jobs in the manufacturing sector, escalation in income taxes, expansion of contribution towards the sector of plants and equipment, agricultural development, and international competitiveness), environmental benefits (i.e., downscale in greenhouse gas (GHG) emissions, reduction of air pollution, biodegradability,

[1] Department of Biotechnology, Saveetha School of Engineering, Saveetha Institute of Medical and Technical Sciences (SIMATS), Chennai, India.
[2] Department of Genetic Engineering and Biotechnology, Shahjalal University of Science and Technology, Sylhet 3114, Bangladesh.
[3] Department of Chemical Engineering, College of Engineering, Universiti Malaysia Pahang, Lebuhraya Tun Razak, 26300 Kuantan, Pahang, Malaysia.
[4] Department of Soil Sciences and Agri-Food Engineering, Université Laval, Quebec, QC G1V 0A6, Canada.
[5] Department of Biotechnology, Bannari Amman Institute of Technology, Sathyamangalam.
[6] Department of Biotechnology, Kumaraguru College of Technology, Coimbatore.
[7] Department of Chemical Engineering & Polymer Science, Shahjalal University of Science and Technology, Sylhet, Bangladesh.
[8] Interdisciplinary Institute for Technological Innovation (3IT), CNRS UMI-3463, Laboratory for Quantum Semiconductors and Photon-based Bionanotechnology, Department of Electrical and Computer Engineering, Université de Sherbrooke, 3000, boul. de Université, Sherbrooke, Québec J1K 0A5, Canada.
* Corresponding author: amirul.geb@gmail.com

higher combustion efficiency, improved usage of land and water resources and carbon sequestration) and secure investment in the field of energy supply (i.e., targets in the domestic level, dependability and reliable nature of supply of products, reduction in the usage of fossil, non-renewable fuels, higher availability, domestic distribution and the renewable nature of products) (Balat 2011).

Lignocellulosic biomass (LCB) is gaining momentum as an excellent alternative to meet the high biofuel requirement. The LCB, composed of carbohydrate and aromatic polymers, is an abundant raw material for biofuel production. Furthermore, LCBs are inexpensive, renewable, and result in biofuels with a minor CO_2 footprint compared to petroleum fuels (Chintagunta et al. 2021). Biofuels, obtained by biological fixation of carbon (process involved in the conversion of inorganic carbon to organic molecules), are found to be a better alternative to the fossil fuels. Various biofuels are being produced as solids, liquids, or gases (Sánchez et al. 2019). Solid fuels fall under the primary category as they use solid materials for fuel production. At the same time, liquid or gaseous fuels come under secondary fuels, categorized into different generations, namely, first generation, second generation, and third generation fuels (Balan 2014).

The direct processing of food crops leads to first generation biofuel production. They are ethanol and biodiesel, produced from various substrates including sugars, vegetable oil, seeds, and grains; however, this leads to a rise in the cost of crops and foods (Pandey et al. 2015). The structure of biofuel production does not change within the generations. However, the source serves as the main difference between them. Corn, wheat, and sugar cane are considered widely utilized first generation feed stock. In a world with one billion people to feed, the production of fuel from food leads to various ethical issues.

Biofuels belonging to the category of second generations are made from non-food crops or sustainable feedstocks such as lignocellulosic biomass and agricultural wastes. Lignocellulosic feedstock requires several processing steps, such as thermochemical conversion and biochemical conversion, before being fermented into ethanol (Robak and Balcerek 2018). A second-generation biofuel can either be combined with petroleum-based fuels or combusted in internal combustion engines and used in slightly adapted vehicles (Naik et al. 2010). Biofuels of the third generation are based on algae or specifically engineered crops as the source of energy. These are of minimal cost, immense energy, and serve as an undividedly renewable feedstock (Fatma et al. 2018). It is estimated that algae can provide excess energy in an acre compared to conventional crops. Moreover, algae-based biofuels can be used as diesel, petrol, and jet fuel (Khan et al. 2017).

Countries like India and China are rich in LCB sources. However, the biomass residues are burnt directly, leading to low energy efficacy and environmental problems (Mosier et al. 2005). Recent technological advances are quite promising in converting LCB to high quality bioenergy with the least pollution emitted into the ecosystem. However, challenges exist in converting biomass to high energy products such as bioethanol and biobutanol from LCB due to its chemical composition and sequential steps, from recalcitrance to degradation (García et al. 2011).

Bioethanol is sustainable, efficient, cost effective, and non-toxic alternative to traditional fuel (Zabed et al. 2014). However, knowledge of efficient pretreatment, conversion, and fermentation strategies is a major requirement for the effective production of bioethanol and biobutanol. This chapter provides a deep insight into the basic requirements and strategies for better production and utilization of biofuels, viz., bioethanol, and biobutanol from LCB.

2. Structure and Composition of Lignocellulosic Biomass

LCB majorly encompasses cellulose, hemicellulose, and lignin (90% of dry matter) (Fig. 1) and materials such as extractives and ash in the remaining volume (Isikgor and Becer 2015). Cellulose is an insoluble glucose polymer available as crystalline microfibrils. Hemicellulose is composed of pentose and hexose sugars bound to cellulose microfibrils. Further, lignin, a phenyl propane

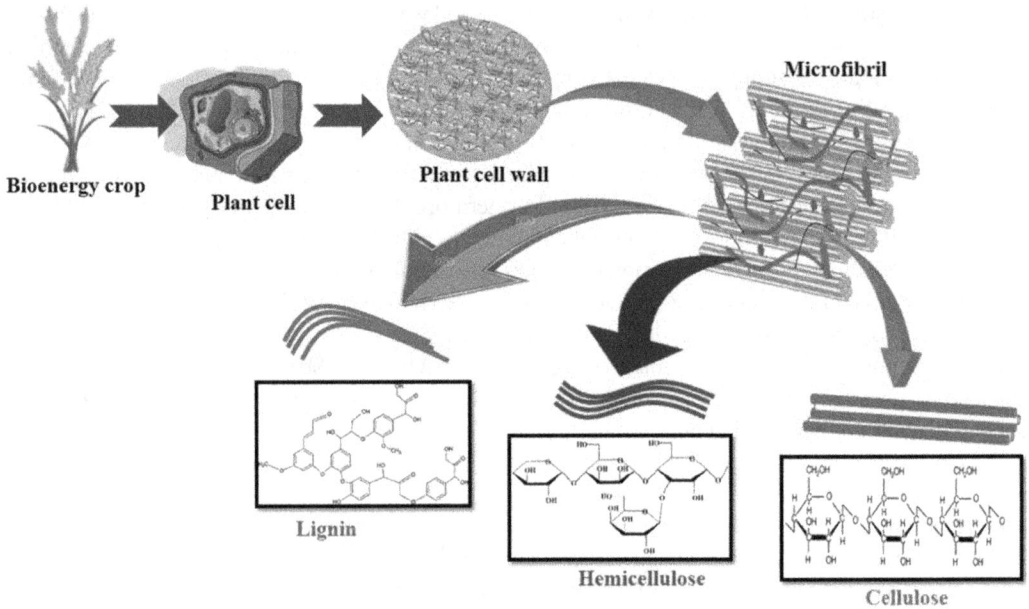

Fig. 1. Structure of lignocellulosic biomass.

polymer, makes a complicated mesh and cross-links with cellulose and hemicellulose (Tarasov et al. 2018). Hence, the breakdown of the lignocellulosic framework and bioconversion to produce higher fuel requires pretreatment.

LCB is classified as forest residues, municipal solid waste, waste paper residue and crop residue based on the available resources. Other resources are waste residues such as corn stover (Zabed et al. 2014), switchgrass (Keshwani and Cheng 2009), and palm bagasse. LCB is composed of 8% C, 6% H, and 45% O by weight with a lower composition of inorganic matter (Karimi et al. 2006). Nevertheless, the arrangement and composition of the chemical components of LCB raw material greatly vary based on environmental and genetic factors. For instance, straw from sources such as rice and wheat exhibit variation in their composition containing 65.47% and 75.27% volatile compounds, 15.86% and 17.71% fixed carbon, and 18.67%, 7.02% ash content (Karimi et al. 2006), respectively. The composition of woody biomass differs in their basic polymers with different species and distinctly between hard and soft woods (Balat 2009). A scheme of the structure and composition of various LCB is shown in Fig. 1.

LCB is considered a valuable feedstock due to its carbohydrate content favoring biofuel production. Lignocellulosic feedstocks are cheap and circumvent the food-energy competition compared to molasses and starch. However, the challenges faced in the production process include effective utilization of LCB, including the removal of lignin, slow conversion of cellulose to sugars, as well as the low concentration of sugar in some types of LCB (Damayanti et al. 2021). The composition of various LCB materials is depicted in Table 1.

3. Bioethanol from LCB

Bioethanol is a well-known biofuel used worldwide for transportation and is a perfect alternative to meet the fuel requirement sustainably. Bioethanol has been in use since 1894 in Germany and France (Balat 2011), as it renders significant benefits such as reduced GHG emission and contains high octane number (Celik 2008). Bioethanol is an ethyl alcohol chemically noted as C_2H_5OH. Due to its physico-chemical nature, it increases the compression ratio for gasoline engines and

Table 1. Composition of LCB.

Lignocellulosic Material	Cellulose (%)	Lignin (%)	Hemicellulose (%)	Reference
Bagasse	39	24.4	24.8	(Kim and Day 2011, Templeton et al. 2010)
Barley hull	33.6	37.2	19.3	(Kim et al. 2008)
Corn fiber	14.3	16.8	8.4	(Mosier et al. 2005)
Corn pericarp	22.5	23.7	4.7	(Kim et al. 2017)
Corn stover	37.0	22.7	18.6	(Kim et al. 2016)
Wheat straw	30.2	21.0	17	(Ballesteros et al. 2006)
Red maple	41.0	15.0	29.1	(Ximenes et al. 2013)
Rice straw	31.1	22.3	13.3	(Chen et al. 2011)
Rye straw	30.9	21.5	22.1	(García-Cubero et al. 2009)
Switch grass	39.5	20.3	17.8	(Li et al. 2010a)
Sugarcane bagasse	43.1	31.1	11.4	(Martín et al. 2007)
Sweet sorghum bagasse	27.3	13.1	14.3	(Li et al. 2010b)
Olive tree pruning	25.0	11.1	16.2	(Cara et al. 2008, Kumar et al. 2009)
Poplar	43.8	14.8	29.1	(Kumar et al. 2009)
Pinewood	40.0	28.5	27.7	(Du et al. 2010)
Spruce	43.8	6.3	28.3	(Shafiei et al. 2010)
Hardwood stem	20–25	45–50	20–25	(Lestander et al. 2012, McKendry 2002)
Softwood stem	27–30	35–40	25–35	(Lestander et al. 2012, McKendry 2002)
Napier grass	20–31.5	24.5–42	3–4.2	(Xia et al. 2013)

produces lower emission (Celik 2008). Bioethanol, an oxygenate fuel, renders reasonable antiknock value (Balat 2011).

Bioethanol of about 442 billion liters per year can be produced with LCBs as raw material (Hossain et al. 2017) (Fig. 2). Leading producers of bioethanol with rice straw as LCB are Africa, Asia, Europe, and America (Binod et al. 2010). Rice straw as a single raw LCB can end up producing about 205 billion liters of bioethanol per year (Karimi et al. 2006). Bioconversion of LCB to bioethanol involves a sequence of steps—(i) pretreatment, (ii) hydrolysis, (iii) fermentation, (iv) product separation/distillation.

The basic reason behind the involvement of these steps is the difficulties in (1) breakdown of biomass (2) breakdown of hemicellulose and cellulose that results in the release of various sugars and the effective fermentation of these sugars either with available microbes or by genetically engineering microbes and, (3) cost involved in the collection and preprocessing of low density LCBs. However, the major drawback with implementing bioethanol as biofuel is the low energy density obtained when compared to gasoline, adding to which, corrosiveness, decreased vapor pressure, low flame luminosity, and consumption of two-third of gasoline energy. The tendency to absorb water and toxicity in ecosystems makes it even more disadvantageous (Celik 2008).

4. Advantages of Biobutanol Over Bioethanol

Biobutanol has numerous astounding potentials in comparison to bioethanol. Biobutanol can replace gasoline in gasoline internal combustion engines, termed as biogasoline. The superior properties of biobutanol, including high-energy content, excellent blending ability, and moisture affinity, increase its preference among the biofuels (Dürre 2007).

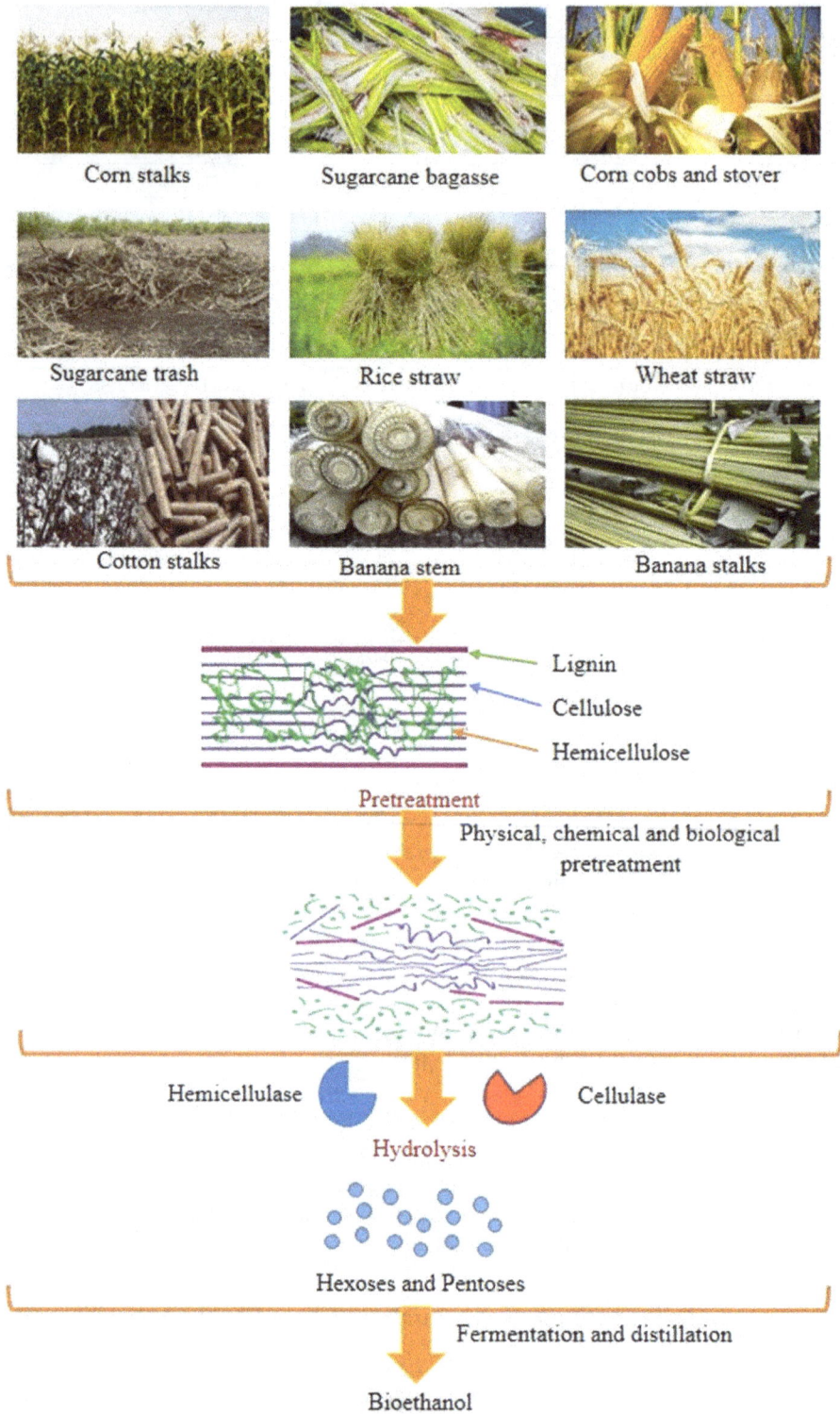

Fig. 2. Schematic representation of lignocellulosic biomass bio-conversion into ethanol.

Biobutanol has numerous industrial applications such as being a solvent in rubber production, biomolecule extractant in pharmaceuticals, domestic and industrial cleansing additive, additive in de-icing fluid in textile industries, mobile phase in chromatographic techniques, and as a precursor in the production of butyl acetate and glycol ethers (Mahapatra and Kumar 2017). Compared to bioethanol, it has a greater energy density and heat of vaporization (Mahapatra and Kumar 2017). Butanol combined with fuel is utilized in internal combustion, yielding CO_2 exclusively, thus making the fuel an eco-sustainable fuel. This fuel has been proved to be an excellent biofuel due to its direct use in gasoline driven engines without modifications showcasing superiority in energy density and heat of vaporization.

Biobutanol produced with LCB as a raw material was reported to significantly reduce GHG emissions of about 32–48% compared to commercial gasoline (Wu et al. 2008). The leading producer of biobutanol, viz., Cobalt technologies, stated that they have reduced GHG by about 70–90% in comparison to gasoline (Bankar et al. 2013). However, biobutanol production from LCB is still not implemented at commercial scale.

5. Comparison of Biobutanol and Bioethanol

The major techno-commercial limitations of existing biofuels have catalyzed the development of advanced biofuels such as cellulosic ethanol, biobutanol, and mixed alcohols. Biobutanol is generating a good deal of interest as a potential green alternative to petroleum fuels. It is increasingly being considered as a superior automobile fuel in comparison to bioethanol as its energy content is higher. The problem of demixing that is encountered with ethanol-petrol blends is considerably less severe with biobutanol-petrol blends. Besides, it reduces harmful emissions substantially. It is less corrosive and can be blended in any concentration with petrol (gasoline). Several research studies suggest that butanol can be blended into either petrol or diesel to as much as 45% without engine modifications or severe performance degradation (Fernández-Rodríguez et al. 2021).

Oxygenated fuel contains an increased oxygen content. Biobutanol contains 21.59% oxygen in comparison to bioethanol with 34.73% oxygen, promoting accelerated complete combustion and lower exhaust emissions (Patakova et al. 2011, Szulczyk 2010). Bioethanol has higher octane number compared to biobutanol (Szulczyk 2010). This property of the fuel prevents premature ignition, which may lead to engine damage due to knocking. The higher octane rating than biobutanol gives bioethanol advantages in improving thermal efficiency. However, it emits 2–4 times higher levels of acetaldehyde than biobutanol and hence is highly corrosive (Rasskazchikova et al. 2004). Both bioethanol and biobutanol have lower Reid Vapour Pressure (RVP) than gasoline, leading to disadvantages in the initiation of cold engines in cold weather (Bajpai 2020, Szulczyk et al. 2010). However, evaporation of bioethanol is easier when compared to biobutanol resulting in the emission of volatile organic compounds into the atmosphere as pollution, especially during hot summer days. These volatile organic compounds, along with NOx gases, are converted by ultraviolet radiation into ground ozone pollution (Szulczyk 2010). Thus, the lower vapor pressure of bioethanol and biobutanol brings both benefits and disadvantages to the performance. Carbon and hydrogen are known to raise the heating value due to the decline of oxygen during combustion. Therefore, both bioethanol and biobutanol contain a higher heating value than gasoline. The density of bioethanol and biobutanol is 794 kg m³ and 809 kg m³, respectively, which results in enhancing the volumetric fuel economy. The length of the carbon chain and the boiling point of alcohol are directly proportional. The boiling point of biobutanol and bioethanol are 117.7°C and 78.3°C, respectively. The specific values of each fuel directly influence the evaporative characteristics. The heat of vaporization (HoV) of bioethanol and biobutanol is slightly high, leading to reduced air–fuel mixture temperature during the intake stroke. This aids in improving the knock resistance, achieving better volumetric efficiency of the engine. However, high HoV of bioethanol and biobutanol may be disadvantageous due to the cooling effect of the air–fuel mixture at ambient temperature during engine initiation in cold weather (Patakova et al. 2011). Besides, high latent HoV promotes higher emissions of organic gases (Chiba et al. 2010). The viscosity of bio-butanol and bioethanol is higher than gasoline. These properties may adversely affect the fuel injection system due to higher flow resistance at a lower temperature (Patakova et al. 2011). The comparison of various properties of bioethanol and bio-butanol are depicted in Table 2.

Table 2. Comparison of bioethanol and biobutanol.

Parameter	Bioethanol	Biobutanol	References
Molecular weight	46.07	74.11	(Anderson et al. 2010, Bankar et al. 2013, He et al. 2019, Jin et al. 2011, Patakova et al. 2011, Pugazhendhi et al. 2019)
Boiling Point (°C)	78.3	117.7	
Density (kg.m^{-3})	794	809	
Flash point [°C]	14	35	
Water solubility at 25°C [g/L]	∞	73	
Flammability [% (V/V)]	3.3–19	1.4–11.2	
Energy density [MJ/L]	25	29.2	
Viscosity [mm^2/s] at 25°C	1.07	2.63	
Kinematic Viscosity (mm^2.s^{-1})	1.5	3.6	
Lower Heating Value (MJ.kg^{-1})	28.9	33.1	
Heat of vaporisation (MJ.kg^{-1})	0.92	0.71	
Research Octane Number RON	106–130	94	
Motor Octane Number MON	89–103	80	
Reid Vapor Pressure (kPa)	17	2.3	
Stoichiometric air/fuel ratio	9	11.1	
Oxygen Content (% w/w)	34.7	21.6	

6. Microbial Strains Involved in Biobutanol Production

6.1 *Clostridium* Species

High production of bio-butanol is obtained with the genus *Clostridium*. The only known species to produce butanol during fermentation was initially *Clostridium acetobutylicum*. However, in later years, three more species, viz., *Clostridium beijerinckii, Clostridium saccaroperbutylacetonicum,* and *Clostridium saccharobutylicum* present in a mixed culture were observed to produce a high yield of butanol (Mahapatra and Kumar 2017). Table 3 depicts the major strains involved in bio-butanol production.

DNA fingerprinting and 16S rRNA gene sequencing were later implemented to reclassify this four butanol producing strains. Later, *Clostridia* species were identified to produce butanol.

6.2 *Metabolism*

The Embden-Meyerhof-Parnas (EMP) pathway is involved in metabolism in acidogenesis and solventogenesis, in the conversion of sugars to pyruvates. Activities of pyruvate kinase (EC 2.7.1.40), glucose-6-phosphate isomerase (EC 5.3.1.9), and hexokinase (EC 2.7.1.2) were identified in various strain types of saccharolytic *Clostridium* spp. Two molecules of pyruvate, with the net formation of 2 molecules each of adenosine triphosphate (ATP) and nicotinamide adenine dinucleotide (NADH), were formed by the utilization of one hexose molecule. The phosphogluconate pathway parallel to glycolysis also serves as a usage for solvent-producing clostridia. Altogether, 5 molecules each of

Table 3. Production of butanol from different type of substrates.

Microorganism	Substrate	Yield/Production	Technology	Reference
C. acetobutylicum (immobilized)	Cheese whey (lactose)	Yield: 15% to 0.54 h^{-1} of dilution and 28% to 0.97 h^{-1} of dilution	Reactor (PBR) with immobilized clostridium	(Napoli et al. 2010)
C. beijerinckii ATCC 55025	Hydrolysate of wheat bran	Yield: 32%/Production: 8.8 g/L of biobutanol	Acid hydrolysis	(Liu et al. 2010)
C. beijerinckii	Cassava flour	Production: 23.98 g/L of butanol	Enzymatic treatment with a yield of 9.12% to reduce sugar	(Lépiz-Aguilar et al. 2011)
C. beijerinckii P260	Wheat straw	Yield: 42%	Acid pretreatment and enzymatic hydrolysis	(Lépiz-Aguilar et al. 2011)
	Barley straw	Yield: 43%/Production: 26,64 g/L of total solvents	Dilute sulfuric acid hydrolysis/overliming	(Qureshi et al. 2010)
	Corn stover	Yield: 43%/Production: 18.04 g/L of total solvents	Acid and enzymatic steps of hydrolysis/ overliming	
	Switchgrass	Yield: 37%/Production: 8.91 g/L of total solvents		
	Glucose	Production: 17.54 g/L of butanol	Intermittent vacuum application	(Mariano et al. 2011)
C. saccharobutylicum DSM 13864	Sago starch	Yield: 29%	Free microorganism fermentation	(Kumar and Gayen 2011)
C. acetobutylicum	Cassava bagasse	Yield: 32%/Production: 76.4 g/L of butanol	Hydrolyze by enzymes fibrous bed bioreactor/Gas stripping	(Lu et al. 2012)
	Palm empty fruit bunches	Production: 1.262 g/L of butanol	Acid pretreatment/enzymatic hydrolysis	(Noomtim and Cheirsilp 2011)
C. beijerinckii BA101	Liquefied corn starch	Butanol production: 81, 3 g/L (with gas stripping)/18.6 g/L (without gas stripping)	Bath reactor/gas stripping/enzymatic hydrolysis	(Ezeji et al. 2007a)
C. acetobutylicum ATCC 824 and *Bacillus subtilis* DSM 4451	Spoilage date palm fruits	Yield: 42%/Production: 1.56 g/L of Solvents	Bacterial consortium (anaerobic conditions)	(Abd-Alla and El-Enany 2012)
C. beijerinckii NCIMB 8052	Tropical maize stalk juice	Production: 0.27 g-butanol/g-sugar	Optimization of pH, agitation, sugar concentration	(Wang and Blaschek 2011)
C. acetobutylicum ATCC824	Sugar maple hemicellulosic material	Production: 7 g/L of butanol	Alkali pretreatment/acid hydrolysis/ overliming	(Sun and Liu 2012)

Table 3 contd. ...

...Table 3 contd.

Microorganism	Substrate	Yield/Production	Technology	Reference
C. saccharoperbutylacetonicum N1-4	Rice bran	Yield: 57% to sugar generated	Acid hydrolysis	(Al-Shorgani et al. 2012)
	De-oiled rice bran	Yield: 44% to sugar generated	Acid pretreatment/enzymatic hydrolysis	
C. acetobutylicum XY16	Glucose	Production: 20.3 g/L of butanol	pH steps in the fermentation	(Guo et al. 2012)
C. sporogenes BE01	Rice straw	Production of 3.49 g/L and 5.32 g/L of butanol and total solvents, respectively	Acid pretreatment/enzymatic hydrolysis/ overliming	(Gottumukkala et al. 2013)
C. saccharoperbutylacetonicum N1-4	Rice straw	Maximum butanol production of 6.6 g/L and butanol yield 0.2 g/g of total sugar	Absence of pretreatment/enzymatic hydrolysis/Non-sterile conditions	(Chen et al. 2013)
C. pasteurianum	Glycerol	Maximum butanol production of 8.8 g/L and butanol yield 0.35 g/g of glycerol at an initial substrate concentration of 25 g/L	Immobilized cells/Bath fermentation	(Khanna et al. 2013)
C. acetobutylicum NCIM 2337	Rice straw	Butanol production of 13.5 g/L and butanol yield 0.34 g/g of total sugar generated	Acid treatment with shear stress	(Ranjan et al. 2013)
C. acetobutylicum MTCC 481	Rice straw	Butanol production of 1.72 g/L	Steam explosion	(Ranjan et al. 2013)
		Butanol production of 1.6 g/L	Acid treatment	
		Butanol production of 2.1 g/L	Acid pre-treatment/enzymatic hydrolysis	
C. beijerinckii NCIMB 8052	Corncob	Butanol production of 8.2 g/L	Alkali pretreatment/enzymatic hydrolysis/ overliming	(Zhang et al. 2012)
C. acetobutylicum MTCC 481	Rice straw	Butanol production of 1.72 g/L Butanol production of 1.6 g/L Butanol production of 2.1 g/L	Steam explosion Acid treatment Acid pre-treatment/enzymatic hydrolysis	(Ranjan and Moholkar 2013)
C. acetobutylicum JB200	Glucose	Yield: 21%/Production: 172 g/L of solvents	Gas stripping	(Xue et al. 2012)
C. beijerinckii ATCC 10132	Glucose	Production: 20 g/L of butanol	Bath reactor	(Isar and Rangaswamy 2012)
C. acetobutylicum CICC 8008	Corn straw	Production: 6.20 g/L of butanol	Enzymatic hydrolysis/bath reactor	(Lin et al. 2011)
C. acetobutylicum P262	Whey permeate medium	Yield: 44%/Production: 98.97 g/L of solvents	Perstraction/bath reactor	(Qureshi and Maddox 2005)

ATP and NADH are produced by the use of three pentose molecules to make pyruvate. Therefore, it can be inferred that pyruvate serves as a major compound in the *Clostridium* spp. metabolism. The activity of three oxidoreductases—NADPH-ferredoxin reductase, NADH-ferredoxin, and [FeFe] hydrogenase, reductase—determines the fermentation's nature. Ferredoxin, with the ability to reduce electrons, is critical to the fermentation process. It works by the donation of electrons to hydrogen or pyridine nucleotides using hydrogenase or ferredoxin. The acetyl Co-A generated by Pyruvate-ferredoxin oxidoreductase may be transferred to various products such as CO_2, acetate, or acetone (products that can be oxidized, or to ethanol, butanol, or butyrate) limited products (Patakova et al. 2019).

7. Hydrolysis and Fermentation

7.1 Hydrolysis

The hydrolysis of LCB for effective biobutanol production involves various methods. These include the production of ideal conditions for either acid or alkali treatment of LCB or conditions to start the process of enzymatic hydrolysis. Both concentrated and dilute acid hydrolysis are carried out during the hydrolysis process; however, dilute hydrolysis is preferred in specific conditions for the prevention of degradation of monosaccharides and formation of inhibitors (Taherzadeh and Karimi 2007). Glycoside hydrolases, natural catalysts obtained from fungal species, are the major options for enzymatic hydrolysis of cellulose and hemicellulose. These enzymes depolymerise hemicellulose along with cellulose (brown and soft rot fungi) and lignin (white rot fungi) (Martínez et al. 2005). Since enzyme hydrolysis is a significant factor to be considered during the fermentation process, substrate concentration and its quality, enzyme loading, pH, and temperature are considered to be the key factors in butanol production.

The pretreated biomass is exposed to enzymatic hydrolysis using cellulase enzyme to initiate the conversion of cellulose to fermentable sugars. This reaction is heterogeneous in nature. The first step is enzyme-substrate binding by the process of adsorption. Cellulose is converted to cellobiose by the binding fraction of endoglucanase and exoglucanase. However, the vice versa conversion occurs by the unbound part of β-glucosidase. This suggests that the bound fractions of endoglucanase and exoglucanase are vital to the development of cellobiose. In contrast, the free fraction of β-glucosidase is essential for the creation of glucose (Du et al. 2014). The chemical approach necessitates an increasing working temperature and high values of acid concentration, rendering the procedure unprofitable; moreover, the neutralization and extraction of acid are also costly (Kucharska et al. 2018).

7.2 Fermentation Processes

Earlier manufacturing of butanol involves acetone-butanol-ethanol (ABE) fermentation process (3:6:1) using *Clostridium acetobutylicum*. ABE fermentation (Fig. 3) was considered the 2nd industrial fermentation procedure for butanol production; ethanol was the first fermentation process with yeast as its suitable microorganism. The butanol yield with ABE fermentation process is about 3% apart from its primary products such as acetone and ethanol (Jones and Woods 1986, Pfromm et al. 2010, Qureshi et al. 2006). At present, butanol is made via a fermentation process dependent on petrochemical products. The methods for the fermentation of butanol include batch, fed-batch and continuous processes including, simultaneous saccharification and co-fermentation (SSCF), separate hydrolysis and fermentation (SHF) and consolidated bioprocessing (CBP), simultaneous saccharification and fermentation (SSF).

However, the aim is to reduce the butanol toxicity to butanol-producing cells. The butyric acid and other end products inhibit the cellular growth during the fermentation process. A simple

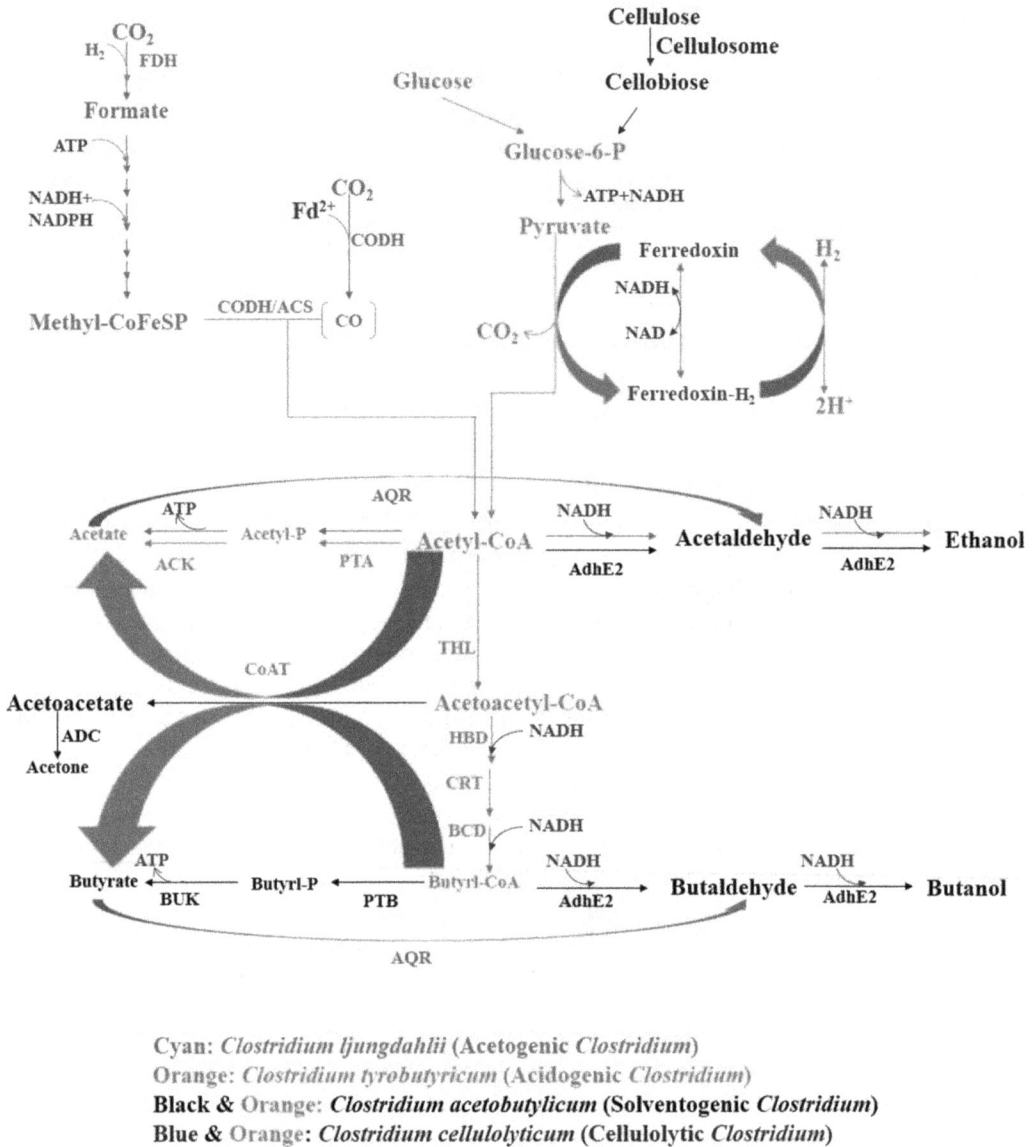

Fig. 3. Butanol production pathway for ABE (Acetone-Butanol-Ethanol) fermentation in several species of *Clostridium*. ACK, acetate kinase; ACS, acetyl-CoA synthase; ADC, acetoacetate decarboxylase; AdhE2, aldehyde/alcohol dehydrogenase; AOR, aldehyde: ferredoxin oxidoreductase; BCD, butyrylCoA dehydrogenase; BUK, butyrate kinase; CoAT, CoA transferase; CODH, carbon monoxide dehydrogenase; CRT, crotonase; FDH, formate dehydrogenase; HB D, 3-hydroxybutyryl-CoA dehydrogenase; PTA, phosphotransacetylase; PTB, phosphotransbutyrylase; TH L, thiolase (Xue and Cheng 2019).

biobutanol production involves a batch reactor (Dürre 2011, García et al. 2011), whereas alternate reactors are required to minimize the residence time of butanol which could ultimately increase the yield or butanol production.

SHF is a predominantly used technique to produce butanol from LCB, where cellulose is hydrolyzed to glucose molecules. The resulting glucose obtained after hydrolysis at 50°C is transferred to a fermentation reactor to produce butanol at temperatures around 35°C. Cheng et al. 2012 assessed the output of butanol from sugarcane bagasse and rice straw with the help of a mixed culture using SHF. The major reason to use SHF is the maintenance of different optimum conditions for fermentation and hydrolysis, respectively. Aggregation in glucose which drastically inhibits the cellulose activity, the cost involved in the maintenance of two separate equipment are the major hike in using SHF for butanol production (Cao et al. 2016).

The fermentation of glucose simultaneously with hydrolysis is known as SSF, whereas the same process with the co-fermentation of C5 and C6 sugars is known as SSCF (Ranjan and Moholkar 2013). In comparison, the SSF process is advantageous in low equipment cost and the reduction of end-product inhibition caused by sugar molecules. The optimum temperature for cellulose hydrolysis is 50°C whereas the optimum temperature for butanol-producing strains is around 35°C. It is, therefore impractical to optimize the two stages at the same time. The SSF method for butanol production is typically run at cooler temperatures to permit microbial growth and butanol production. As a result, the effectiveness of the enzymatic hydrolysis is necessarily impaired, and indeed the hydrolysis takes far longer to accomplish.

Though SHF and SSF are advantageous, higher utilization of cellulase enzymes which leads to a rise in the cost of butanol manufacturing, is the major barrier in its usage (Qureshi et al. 2008). Therefore, the factor for efficient cellulosic manufacturing of butanol is the establishment of cost-effective and productive saccharification and fermentation techniques. The substitutive technology combining cellulase synthesis, fermentation and cellulose hydrolysis aiming to eliminate this critical cost increasing factor is known as consolidated bioprocessing (CBP). CBP or direct microbial conversion is a single step butanol production process. Moreover, it involves up to 50% cost reduction becoming the most attractive methodology in butanol production (Cao et al. 2014, Talluri et al. 2013). The perfect CBP microorganism for the efficient degradation of LCB together with the production of butanol at desirable yield has not been discovered. Co-cultures were extensively researched to tackle the disadvantages for usage of the substrate by individual strains. The co-culture of *Bacillus* sp. SGP1 and *Clostridium tyrobutyricum* ATCC 25755T, for instance, have been described in the manufacturing of butyric acid from sucrose. Apparently, in order to benefit certain metabolic abilities, the co-culture of different microorganisms provides a dependable method of substrate conversion optimization and increase in product yield (Nakayama et al. 2011). For example, Nakayama et al., researched the efficacy of the making of butanol using crystal-based cellulose via CBP by co-cultivation of cellulolytic *Clostridium thermocellum* JN4 and the butanol producing strain, *Clostridium saccharoperbutylacetonicum* N1-4 (Nakayama et al. 2011). On 4% Avicel cellulose, a butanol yield of 7.9 g/L was produced after 9 days of fermentation. Therefore, the development of CBP microorganisms is very much required for the success of the CBP process.

8. Strategies for Increased Butanol Production Through Microbial Strains

8.1 Coculture for Biobutanol Production

In the near future, compared to butanol production with the anaerobic organism, co-culturing of aerobic microbes with anaerobes will result in a higher biobutanol yield. *Bacillus subtilis* TISTR 1032 (aerobic) together with *C. butylicum* WD 161 (anaerobic) brought about 5.4- and 6.5-folds upturn production of butanol compared to traditional methods (monocultures). It was observed that the saccharification process, specifically in the presence of *B. subtilis*, led to increased amylase activity, thereby leading to an increase in product yield (Tran et al. 2010). Currently, a platform has been introduced for the production of butanol based on the *Escherichia coli* co-culturing system (Saini et al. 2016). This system consists of two strains: strain BuT-8L-ato, which enables the conversion of butyryl-CoA and acetate into butyrate, and strain BuT-3E, leading to the formation of n-butanol associated with acetate from butyrate. The complete CoA-dependent synthetic pathway consisting of an internal cycle of acetate and butyrate was established by the combination of the pathways distributed in these two strains.

However, this led to the combined pathway of butanol in the redox-balanced state. The co-culturing system generated 4.9 g/L of butanol after 30 h, with an increase of 75% in comparison with a single strain BuT-8 encoding formate dehydrogenase (FDH) of *S. cerevisiae* (Sc-fdh) (Saini et al. 2016). Microbial co-culturing systems of *C. acetobutylicum* and *S. cerevisiae* were set up for the improvement of butanol production. The final concentrations of ABE and butanol obtained were 24.8 and 16.3 g/L, respectively, with increments of 37.8 and 46.8% when compared to those using only *C. acetobutylicum* (Luo et al. 2017). The mechanisms found to aid in ABE production are: (1) *S. cerevisiae* works by the secretion of amino acids under stressful environments, which may be favorable for the survival of *C. acetobutylicum* and butanol synthesis; (2) *C. acetobutylicum* has to compete with yeast cells for survival. During growth periods, the consumption of higher volumes of substrate leads to intracellular NADH production, required for butanol synthesis (Luo et al. 2015). This process indicates the alleviation of burden in multi bacterial synthetic biology of bacterial metabolic process to achieve improvement of product titer. The major advantage of co culture is to bring down the use of high cost reducing agents used in the maintenance of anaerobic conditions during the fermentation process (Mahapatra and Kumar 2017). In addition, co-culturing systems endure more changeable environments, thus improving the stability and robustness of co-culturing systems than a single culture. Nevertheless, bacteriophage infections during subculturing or transfer of microbes led to a major pitfall in co-culturing techniques (Mahapatra and Kumar 2017), which was later overcome by implementing sterile conditions and other control measures during disinfection and immunization of resistant strains (Mahapatra and Kumar 2017). However, current co-culturing systems are majorly involved in the exchange of intermediate metabolites, such as sugars and intermediates. The mechanism of cell–cell communication and the exchange of energy or signal remains unelaborated, thus restricting the design and construction of microorganism consortia. For example, negative interactions happen when two microorganisms compete for the same resource, such as space or limiting nutrients. In addition, due to different requirement of growth, the optimization of fermentation conditions become more complex (including the inoculation timing, the inoculation ratio, etc.). The majorly faced disadvantage is balancing the inoculation ratio and maintaining the co-existence of constituent strains in co-culturing systems. The compatibility of strain pairs is a significant element in setting up promising microbial co-culturing systems (Du et al. 2020).

8.2 Strain Enhancement

Various factors are involved in the selection of a particular strain in butanol production, viz., specific nutrient requirement, types of feedstocks, butanol tolerance, bacteriophage resistance, and targeted productivity. Hence, novel microbes have to be identified, and further strain enhancement has to be done with available technologies (Table 4). *C. beijerinckii* BA101, a mutant relative of the *C. beijerinckii* NCIMB 8052 strain, is the most studied hyper-butanol producer and is reported to produce 19 g/L butanol in the synthetic medium in the absence of an integrated product recovery system (Formanek et al. 1997). Transcriptomic analysis of BA101 has revealed the elevated expression of primary metabolic and motility genes compared to NCIMB 8052. Moreover, the maximum induction of sporulation genes in BA 101 was two to eight times lower than NCIMB 8052 (Shi and Blaschek 2008).

   C. acetobutylicum JB 200 is a hyper-butanol producing mutant strain of *C. acetobutylicum* ATCC 55025. *C. acetobutylicum* ATCC 55025 is an asporogenic (unable to produce spore) mutant strain derived from *C. acetobutylicum* ATCC 4259 (Fig. 4). *C. acetobutylicum* JB 200 is capable of producing 19.1 g/L butanol from glucose in the absence of product removal during fermentation

Table 4. Metabolic engineering in microorganisms for biobutanol production.

Feedstock	Microorganisms	Butanol Production (g/l)/Productivity (g/l/h)	Remarks	References
Crystalline cellulose	*C. cellulovorans* DSM 743B	1.42/0.0056	Overexpression of *adhE2* gene, directly utilize cellulose from crystalline cellulose	(Yang et al. 2015)
	C. cellulolyticum ATCC 35319	0.12/0.00025	Introduction of CoA-dependent pathway from *C. acetobutylicum*	(Gaida et al. 2016)
Alkali extracted corn cobs	*C. cellulovorans* DSM 743B and *C. beijerinckii* NCIMB 8052	11.5	Overexpression of *buk*, *xylR*, *xylT* and *ctfAB*, mesophilic co-culture and cellulosic butanol production	(Wen et al. 2017)
Glucose	*C. acetobutylicum* BEKW_ E1AB-atoB	55.7/2.64	Overexpression of *atoB*, *adhE1*, *ctfAB* genes	(Lee et al. 2016)
	C. beijerinckii CC101-SV6	12	Overexpression of *ctfAB*, *ald* and *adhE2* genes, improved capacity for acid assimilation and robustness to resist inhibitors	(Lu et al. 2017)
Sugarcane bagasse	*C. beijerinckii* CC101-SV6	7.6	Overexpression of *ctfAB*, *ald* and *adhE2* genes, robustness to resist inhibitors	(Lu et al. 2017)
Glucose and xylose	*C. tyrobutyricum* Ct(Δ*ack*)-pTBA	12/0.17	Co-overexpression of *xylT*, *xylA* and *xylB* genes with *adhE2* from *C. acetobutylicum* ATCC 824 to *C. tyrobutyricum*	(Yu et al. 2015)
Soybean hull	*C. tyrobutyricum* Ct(Δ*ack*)-pTBA	15.7/0.29	–	(Yu et al. 2015)
	C. pasteurianum	–	Deletion of genes hydA, rex, dhaBCE resulted in lower acid formation and increased n-butanol production	(Schwarz et al. 2017)
Sucrose	*C. tyrobutyricum* Ct(Δ*ack*)-pscrBAK	16/0.33	Overexpression of genes *scrA*, *scrB*, *scrK* and *adhE2*	(Zhang et al. 2017)

adhE1 and adhE2, aldehyde-alcohol dehydrogenase; ald, Co-A acylating aldehyde dehydrogenase; atoB, thiolase; buk, butyrate kinase; ctfAB, CoA transferase; dhaBCE, glycerol dehydratase; hydA, hydrogenase; rex, redox response regulator; sucrose catabolism (scrA, phosphoenolpyruvate phosphotransferase, scrB, sucrose 6-phosphate hydrolase, scrK, fructokinase); xylA, xylose isomerase; xylB, xylulokinase; xylR, D-xylose repressor protein; xylT, xylose proton-symport.

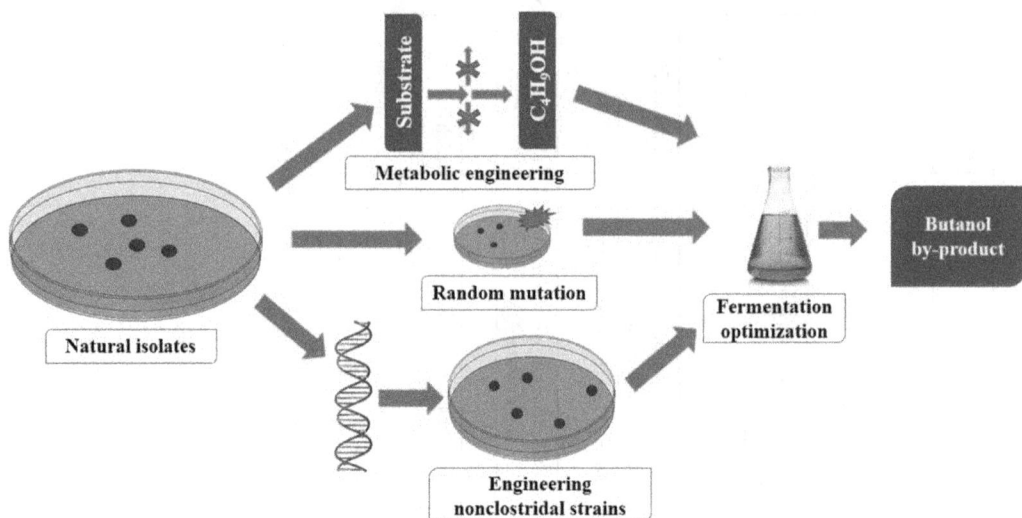

Fig. 4. Scheme of hyper butanol producer strain development.

(Xue et al. 2012). Comparative genomic analysis of JB 200 and ATCC 50525 has shown the presence of 170 gene variations, among which, 29 variations were related to sporulation, solventogenesis, and butanol tolerance. In this study, it was identified that a single-base deletion in *cac3319*, a histidine kinase gene, resulted in an increased butanol tolerance and production. *cac3319* gene was disrupted in ATCC 55025 using the ClosTron group II intron-based gene inactivation system, which resulted in an increase in butanol production by 44%, confirming the role of *cac3319* in solvent regulation (Xu et al. 2015).

　　C. acetobutylicum BKM19 was developed by random mutagenesis of *C. acetobutylicum* PJC4BK. The mutant strain produced 32.5 g/L of total solvents (ABE), which was 30.5% higher than the parent strain. The total solvent produced consists of 17 g/L butanol, 10.5 g/L ethanol, and 4.4 g/L acetone (Jang et al. 2013). The genomic analysis of parent and mutant strain depicted the presence of 13 single-nucleotide variants (SNVs), one deletion, and one back mutation SNV, among which, a mutation in the thiolase gene was identified, which could be responsible for increased butanol production (Cho et al. 2017).

9. Downstream Processing

As the butanol concentration increases in the solventogenic phase, there exists a need to rapidly separate the end product to avoid cell death due to end product toxicity. This remains a challenge since the periodical separation of butanol is difficult due to the lower concentration of butanol (~ 20 g/l) and its higher boiling point (117°C) (Yakovlev et al. 2013). Common methods such as distillation, adsorption, along with liquid-liquid extraction are used to overcome this problem. Technological advances led to the development of methods like adsorption, pervaporation, gas stripping, membrane solvent extraction, etc. to separate the end products (Table 5).

　　The carbon-neutral quality of bioethanol is negated by the distillation process that needs a substantial quantity of heat, and negates its carbon-neutral value (Kiuchi et al. 2015). Traditional distillation, liquid-liquid extraction (LLX), pervaporation, adsorption, reverse osmosis (RO), and other procedures could be used to separate the diluted ABE combination (Dimian et al. 2019). Along with its fuel efficiency and lack of adverse effects on organisms, pervaporation (PV) is regarded as the most promising separation process. In comparison to distillation, pervaporation is a more developed and cost-effective method for separating the butanol/water azeotrope (Liu et al. 2014).

Table 5. Recovery techniques used in the production of Biobutanol.

Recovery Method	Energy Requirement (Kcal/Kg butanol)	Technology Status	Advantage	Limitation	Reference
Adsorption	1948	Research	Energy-efficient recovery	Low material cost, absorbent regeneration, low selectivity	(Qureshi et al. 2005)
Pervaporation	3295	Research	Selective removal of butanol	Fouling, high cost	(Ezeji et al. 2004)
Gas stripping	5220	Research	Simple, no fouling, easy to operate, no harm to the culture	Low selectivity, low efficiency	(Ezeji et al. 2007b)
Freeze crystallization	NA	Research	Thermodynamically favorable due to lower enthalpy of fusion than the enthalpy of vaporization of butanol and water	Additional cost of solids handling	(Morone and Pandey 2014)
Solvent extraction	2119	Research	Easy extraction and easy recovery process, high selectivity	Extractant toxicity, layer formation	(Ezeji et al. 2004)
Flash fermentation	NA	Research	Fermentation at atmospheric pressure and no harm to culture	Cost associated with a separate flash tank. Low selectivity	(Mariano et al. 2010)

10. Conclusion

Biobutanol is considered to be a better substitute for gasoline. It emerged as an effective enhancer, blending up to 95% with gasoline, demonstrating its superiority above other fuels. LCB plays a vital part in the manufacturing process of butanol. Biotechnology offers a feasible and cost-effective approach to generate bioconversion of LCB to butanol from basic materials. The persisting challenge that must be addressed is the removal of undesirable products that directly influence the growth of cells and butanol production. Numerous research approaches and *in situ* techniques have come up till date, but the need for producing butanol with maximum yield is still a challenging task. In this chapter, we have proposed a few strategies, including the process of adsorption over the surface of materials, extraction on the basis of liquid-liquid interaction, and stripping of gas facilitates in purification and maximum extraction of fuel. In addition, the techniques for increased production of bio-butanol through novel microorganisms, co-cultures, and genetic engineering-based strain improvement techniques were well discussed in this chapter.

References

Abd-Alla, M.H. and El-Enany, A.-W.E. 2012. Production of acetone-butanol-ethanol from spoilage date palm (*Phoenix dactylifera* l.) fruits by mixed culture of *Clostridium acetobutylicum* and *Bacillus subtilis*. Biomass and Bioenergy 42: 172–178.

Al-Shorgani, N.K.N., Kalil, M.S. and Yusoff, W.M.W. 2012. Biobutanol production from rice bran and de-oiled rice bran by *Clostridium saccharoperbutylacetonicum* n1-4. Bioprocess and Biosystems Engineering 35: 817–826.

Bajpai, P. 2020. Developments in Bioethanol: Springer Nature.

Balan, V. 2014. Current challenges in commercially producing biofuels from lignocellulosic biomass. International Scholarly Research Notices, 2014: 1–31.

Balat, M. 2009. Gasification of biomass to produce gaseous products. Energy Sources, Part A 31: 516–526.

Balat, M. 2011. Production of bioethanol from lignocellulosic materials via the biochemical pathway: A review. Energy Conversion and Management 52: 858–875.

Ballesteros, I., Negro, M.J., Oliva, J.M., Cabañas, A., Manzanares, P. and Ballesteros, M. 2006. Ethanol production from steam-explosion pretreated wheat straw. Paper Presented at the Twenty-seventh Symposium on Biotechnology for Fuels and Chemicals.

Bankar, S.B., Survase, S.A., Ojamo, H. and Granström, T. 2013. Biobutanol: The outlook of an academic and industrialist. Rsc Advances 3: 24734–24757.

Binod, P., Sindhu, R., Singhania, R.R., Vikram, S., Devi, L., Nagalakshmi, S., Kurien, N., Sukumaran, R.K. and Pandey, A. 2010. Bioethanol production from rice straw: An overview. Bioresource Technology 101: 4767–4774.

Cao, G.-L., Zhao, L., Wang, A.-J., Wang, Z.-Y. and Ren, N.-Q. 2014. Single-step bioconversion of lignocellulose to hydrogen using novel moderately thermophilic bacteria. Biotechnology for Biofuels 7: 1–13.

Cao, G., Sheng, Y., Zhang, L., Song, J., Cong, H. and Zhang, J. 2016. Biobutanol production from lignocellulosic biomass: Prospective and challenges. Journal of Bioremediation and Biodegradation 7: 363.

Celik, M.B. 2008. Experimental determination of suitable ethanol–gasoline blend rate at high compression ratio for gasoline engine. Applied Thermal Engineering 28: 396–404.

Chen, W.-H., Chen, Y.-C. and Lin, J.-G. 2013. Evaluation of biobutanol production from non-pretreated rice straw hydrolysate under non-sterile environmental conditions. Bioresource Technology 135: 262–268.

Chen, W.-H., Pen, B.-L., Yu, C.-T. and Hwang, W.-S. 2011. Pretreatment efficiency and structural characterization of rice straw by an integrated process of dilute-acid and steam explosion for bioethanol production. Bioresource Technology 102: 2916–2924.

Cheng, C.L., Che, P.Y., Chen, B.Y., Lee, W.J., Lin, C.Y. and Chang, J.S. 2012 Biobutanol production from agricultural waste by an acclimated mixed bacterial microflora. Applied Energy 100: 3–9.

Chiba, F., Ichinose, H., Morita, K., Yoshioka, M., Noguchi, Y. and Tsukagoshi, T. 2010. High concentration ethanol effect on si engine emission. SAE International Journal of Engines 3: 1033–1041.

Chintagunta, A.D., Zuccaro, G., Kumar, M., Kumar, S.J., Garlapati, V.K., Postemsky, P.D., Kumar, N.S., Chandel, A.K. and Simal-Gandara, J. 2021. Biodiesel production from lignocellulosic biomass using oleaginous microbes: Prospects for integrated biofuel production. Frontiers in Microbiology 12: 658284.

Cho, C., Choe, D., Jang, Y.S., Kim, K.J., Kim, W.J., Cho, B.K., Papoutsakis, E.T., Bennett, G.N., Seung, D.Y. and Lee, S.Y. 2017. Genome analysis of a hyper acetone-butanol-ethanol (abe) producing *Clostridium acetobutylicum* bkm19. Biotechnology Journal 12: 1600457.

Damayanti, D., Supriyadi, D., Amelia, D., Saputri, D.R., Devi, Y.L.L., Auriyani, W.A. and Wu, H.S. 2021. Conversion of lignocellulose for bioethanol production, applied in bio-polyethylene terephthalate. Polymers 13: 2886.

Dimian, A.C., Bildea, C.S. and Kiss, A.A. 2019. Applications in Design and Simulation of Sustainable Chemical Processes, 285–327. Elsevier.

Du, B., Sharma, L.N., Becker, C., Chen, S.F., Mowery, R.A., van Walsum, G.P. and Chambliss, C.K. 2010. Effect of varying feedstock–pretreatment chemistry combinations on the formation and accumulation of potentially inhibitory degradation products in biomass hydrolysates. Biotechnology and Bioengineering 107: 430–440.

Du, J., Li, Y., Zhang, H., Zheng, H. and Huang, H. 2014. Factors to decrease the cellulose conversion of enzymatic hydrolysis of lignocellulose at high solid concentrations. Cellulose 21: 2409–2417.

Du, Y., Zou, W., Zhang, K., Ye, G. and Yang, J.-G. 2020. Advances and applications of *Clostridium* co-culture systems in biotechnology. Frontiers in Microbiology 11: 2842.

Dürre, P. 2007. Biobutanol: An attractive biofuel. Biotechnology Journal: Healthcare Nutrition Technology 2: 1525–1534.

Dürre, P. 2011. Fermentative production of butanol-the academic perspective. Current Opinion in Biotechnology 22: 331–336.

Escobar, J.C., Lora, E.S., Venturini, O.J., Yáñez, E.E., Castillo, E.F. and Almazan, O. 2009. Biofuels: Environment, technology and food security. Renewable and Sustainable Energy Reviews 13: 1275–1287.

Ezeji, T.C., Qureshi, N. and Blaschek, H.P. 2004. Butanol fermentation research: Upstream and downstream manipulations. Chem. Rec. 4: 305–314.

Ezeji, T.C., Qureshi, N. and Blaschek, H.P. 2007a. Production of acetone butanol (ab) from liquefied corn starch, a commercial substrate, using *Clostridium beijerinckii* coupled with product recovery by gas stripping. Journal of Industrial Microbiology and Biotechnology 34: 771–777.

Ezeji, T.C., Qureshi, N. and Blaschek, H.P. 2007b. Bioproduction of butanol from biomass: From genes to bioreactors. Current Opinion in Biotechnology 18: 220–227.

Fatma, S., Hameed, A., Noman, M., Ahmed, T., Shahid, M., Tariq, M., Sohail, I. and Tabassum, R. 2018. Lignocellulosic biomass: A sustainable bioenergy source for the future. Protein and Peptide Letters 25: 148–163.

Fernández-Rodríguez, D., Lapuerta, M. and German, L. 2021. Progress in the use of biobutanol blends in diesel engines. Energies 14: 3215.

Formanek, J., Mackie, R. and Blaschek, H.P. 1997. Enhanced butanol production by *Clostridium beijerinckii* ba101 grown in semidefined p2 medium containing 6 percent maltodextrin or glucose. Appl. Environ. Microbiol. 63: 2306–2310.

Gaida, S.M., Liedtke, A., Jentges, A.H.W., Engels, B. and Jennewein, S. 2016. Metabolic engineering of *Clostridium cellulolyticum* for the production of n-butanol from crystalline cellulose. Microbial Cell Factories 15: 1–11.

García-Cubero, M.T., González-Benito, G., Indacoechea, I., Coca, M. and Bolado, S. 2009. Effect of ozonolysis pretreatment on enzymatic digestibility of wheat and rye straw. Bioresource Technology 100: 1608–1613.

García, V., Päkkilä, J., Ojamo, H., Muurinen, E. and Keiski, R.L. 2011. Challenges in biobutanol production: How to improve the efficiency? Renewable and Sustainable Energy Reviews 15: 964–980.

Gottumukkala, L.D., Parameswaran, B., Valappil, S.K., Mathiyazhakan, K., Pandey, A. and Sukumaran, R.K. 2013. Biobutanol production from rice straw by a non acetone producing *Clostridium sporogenes* be01. Bioresource Technology 145: 182–187.

Guo, T., Sun, B., Jiang, M., Wu, H., Du, T., Tang, Y., Wei, P. and Ouyang, P. 2012. Enhancement of butanol production and reducing power using a two-stage controlled-ph strategy in batch culture of *Clostridium acetobutylicum* xy16. World Journal of Microbiology and Biotechnology 28: 2551–2558.

He, Z., Liu, G., Li, Z., Jiang, C., Qian, Y. and Lu, X. 2019. Comparison of four butanol isomers blended with diesel on particulate matter emissions in a common rail diesel engine. Journal of Aerosol Science 137: 105434.

Hossain, Z., Sahu, J. and Suely, A. 2017. Bioethanol production from lignocellulosic biomass: An overview of pretreatment, hydrolysis, and fermentation. Sustainable Utilization of Natural Resources, 145–186.

Isar, J. and Rangaswamy, V. 2012. Improved n-butanol production by solvent tolerant *Clostridium beijerinckii*. Biomass and Bioenergy 37: 9–15.

Isikgor, F.H. and Becer, C.R. 2015. Lignocellulosic biomass: A sustainable platform for the production of bio-based chemicals and polymers. Polymer Chemistry 6: 4497–4559.

Jang, Y.S., Malaviya, A. and Lee, S.Y. 2013. Acetone-butanol-ethanol production with high productivity using *Clostridium acetobutylicum* bkm19. Biotechnol. Bioeng. 110: 1646–1653.

Jin, C., Yao, M., Liu, H., Chia-fon, F.L. and Ji, J. 2011. Progress in the production and application of n-butanol as a biofuel. Renewable and Sustainable Energy Reviews 15: 4080–4106.

Jones, D.T. and Woods, D.R. 1986. Acetone-butanol fermentation revisited. Microbiological Reviews 50: 484–524.

Karimi, K., Emtiazi, G. and Taherzadeh, M.J. 2006. Ethanol production from dilute-acid pretreated rice straw by simultaneous saccharification and fermentation with *Mucor indicus, Rhizopus oryzae*, and *Saccharomyces cerevisiae*. Enzyme and Microbial Technology 40: 138–144.

Keshwani, D.R. and Cheng, J.J. 2009. Switchgrass for bioethanol and other value-added applications: A review. Bioresource Technology 100: 1515–1523.

Khan, S., Siddique, R., Sajjad, W., Nabi, G., Hayat, K.M., Duan, P. and Yao, L. 2017. Biodiesel production from algae to overcome the energy crisis. HAYATI Journal of Biosciences 24: 163–167.

Khanna, S., Goyal, A. and Moholkar, V.S. 2013. Production of n-butanol from biodiesel derived crude glycerol using *Clostridium pasteurianum* immobilized on amberlite. Fuel 112: 557–561.

Kim, D., Orrego, D., Ximenes, E.A. and Ladisch, M.R. 2017. Cellulose conversion of corn pericarp without pretreatment. Bioresource Technology 245: 511–517.

Kim, D., Ximenes, E.A., Nichols, N.N., Cao, G., Frazer, S.E. and Ladisch, M.R. 2016. Maleic acid treatment of biologically detoxified corn stover liquor. Bioresource Technology 216: 437–445.

Kim, M. and Day, D.F. 2011. Composition of sugar cane, energy cane, and sweet sorghum suitable for ethanol production at louisiana sugar mills. Journal of Industrial Microbiology and Biotechnology 38: 803–807.

Kim, T.H., Taylor, F. and Hicks, K.B. 2008. Bioethanol production from barley hull using SAA (soaking in aqueous ammonia) pretreatment. Bioresource Technology 99: 5694–5702.

Kiuchi, T., Yoshida, M. and Kato, Y. 2015. Energy saving bioethanol distillation process with self-heat recuperation technology. Journal of the Japan Petroleum Institute 58: 135–140.

Kucharska, K., Rybarczyk, P. and Hołowacz, I. 2018. Pretreatment of lignocellulosic materials as substrates for fermentation processes 23: 2937.

Kumar, M. and Gayen, K. 2011. Developments in biobutanol production: New insights. Applied Energy 88: 1999–2012.

Kumar, R., Mago, G., Balan, V. and Wyman, C.E. 2009. Physical and chemical characterizations of corn stover and poplar solids resulting from leading pretreatment technologies. Bioresource Technology 100: 3948–3962.

Lee, S.-H., Kim, S., Kim, J.Y., Cheong, N.Y. and Kim, K.H. 2016. Enhanced butanol fermentation using metabolically engineered *Clostridium acetobutylicum* with *ex situ* recovery of butanol. Bioresource Technology 218: 909–917.

Lépiz-Aguilar, L., Rodríguez-Rodríguez, C.E., Arias, M.L., Lutz, G. and Ulate, W. 2011. Butanol production by *Clostridium beijerinckii* ba101 using cassava flour as fermentation substrate: Enzymatic versus chemical pretreatments. World Journal of Microbiology and Biotechnology 27: 1933–1939.

Lestander, T.A., Finell, M., Samuelsson, R., Arshadi, M. and Thyrel, M. 2012. Industrial scale biofuel pellet production from blends of unbarked softwood and hardwood stems-the effects of raw material composition and moisture content on pellet quality. Fuel Processing Technology 95: 73–77.

Li, B.-Z., Balan, V., Yuan, Y.-J. and Dale, B.E. 2010b. Process optimization to convert forage and sweet sorghum bagasse to ethanol based on ammonia fiber expansion (afex) pretreatment. Bioresource Technology 101: 1285–1292.

Li, C., Knierim, B., Manisseri, C., Arora, R., Scheller, H.V., Auer, M., Vogel, K.P., Simmons, B.A. and Singh, S. 2010a. Comparison of dilute acid and ionic liquid pretreatment of switchgrass: Biomass recalcitrance, delignification and enzymatic saccharification. Bioresource Technology 101: 4900–4906.

Lin, Y., Wang, J., Wang, X. and Sun, X. 2011. Optimization of butanol production from corn straw hydrolysate by *Clostridium acetobutylicum* using response surface method. Chinese Science Bulletin 56: 1422–1428.

Liu, G., Wei, W. and Jin, W. 2014. Pervaporation membranes for biobutanol production. ACS Sustainable Chemistry & Engineering 2: 546–560.

Liu, Z., Ying, Y., Li, F., Ma, C. and Xu, P. 2010. Butanol production by *Clostridium beijerinckii* atcc 55025 from wheat bran. Journal of Industrial Microbiology and Biotechnology 37: 495–501.

Lu, C., Yu, L., Varghese, S., Yu, M. and Yang, S.-T. 2017. Enhanced robustness in acetone-butanol-ethanol fermentation with engineered *Clostridium beijerinckii* overexpressing adhe2 and ctfab. Bioresource Technology 243: 1000–1008.

Lu, C., Zhao, J., Yang, S.-T. and Wei, D. 2012. Fed-batch fermentation for n-butanol production from cassava bagasse hydrolysate in a fibrous bed bioreactor with continuous gas stripping. Bioresource Technology 104: 380–387.

Luo, H., Ge, L., Zhang, J., Zhao, Y., Ding, J., Li, Z., He, Z., Chen, R. and Shi, Z. 2015. Enhancing butanol production under the stress environments of co-culturing *Clostridium acetobutylicum/Saccharomyces cerevisiae* integrated with exogenous butyrate addition. PLoS One 10: e0141160.

Luo, H., Zeng, Q., Han, S., Wang, Z., Dong, Q., Bi, Y. and Zhao, Y. 2017. High-efficient n-butanol production by co-culturing *Clostridium acetobutylicum* and *Saccharomyces cerevisiae* integrated with butyrate fermentative supernatant addition. World J. Microbiol. Biotechnol. 33: 76.

Mahapatra, M.K. and Kumar, A. 2017. A short review on biobutanol, a second generation biofuel production from lignocellulosic biomass. J. Clean Energy Technol. 5: 27–30.

Mariano, A.P., Costa, C.B., Maciel, M.R., Maugeri Filho, F., Atala, D.I., de Angelis Dde, F. and Maciel Filho, R. 2010. Dynamics and control strategies for a butanol fermentation process. Applied Biochem. Biotechnol. 160: 2424–2448.

Mariano, A.P., Qureshi, N., Filho, R.M. and Ezeji, T.C. 2011. Bioproduction of butanol in bioreactors: New insights from simultaneous *in situ* butanol recovery to eliminate product toxicity. Biotechnology and Bioengineering 108: 1757–1765.

Martín, C., Klinke, H.B. and Thomsen, A.B. 2007. Wet oxidation as a pretreatment method for enhancing the enzymatic convertibility of sugarcane bagasse. Enzyme and Microbial Technology 40: 426–432.

Martínez, Á.T., Speranza, M., Ruiz-Dueñas, F.J., Ferreira, P., Camarero, S., Guillén, F., Martínez, M.J., Gutiérrez Suárez, A. and Río Andrade, J.C.d. 2005. Biodegradation of lignocellulosics: Microbial, chemical, and enzymatic aspects of the fungal attack of lignin.

Morone, A. and Pandey, R.A. 2014. Lignocellulosic biobutanol production: Gridlocks and potential remedies. Renewable and Sustainable Energy Reviews 37: 21–35.

Mosier, N., Wyman, C., Dale, B., Elander, R., Lee, Y., Holtzapple, M. and Ladisch, M. 2005. Features of promising technologies for pretreatment of lignocellulosic biomass. Bioresource Technology 96: 673–686.

Naik, S.N., Goud, V.V., Rout, P.K. and Dalai, A.K. 2010. Production of first and second generation biofuels: A comprehensive review. Renewable and Sustainable Energy Reviews 14: 578–597.

Nakayama, S., Kiyoshi, K., Kadokura, T. and Nakazato, A. 2011. Butanol production from crystalline cellulose by cocultured *Clostridium thermocellum* and *Clostridium saccharoperbutylacetonicum* n1-4. Applied Environ. Microbiol. 77: 6470–6475.

Napoli, F., Olivieri, G., Russo, M.E., Marzocchella, A. and Salatino, P. 2010. Butanol production by *Clostridium acetobutylicum* in a continuous packed bed reactor. Journal of Industrial Microbiology and Biotechnology 37: 603–608.

Noomtim, P. and Cheirsilp, B. 2011. Production of butanol from palm empty fruit bunches hydrolyzate by *Clostridium acetobutylicum*. Energy Procedia 9: 140–146.

Pandey, A., Bhaskar, T., Stöcker, M. and Sukumaran, R. (eds.). 2015. Recent advances in thermochemical conversion of biomass. Elsevier, Amsterdam, Netherlands.

Patakova, P., Branska, B., Sedlar, K., Vasylkivska, M., Jureckova, K., Kolek, J., Koscova, P. and Provaznik, I. 2019. Acidogenesis, solventogenesis, metabolic stress response and life cycle changes in *Clostridium beijerinckii* nrrl b-598 at the transcriptomic level. Scientific Reports 9: 1–21.

Patakova, P., Maxa, D., Rychtera, M., Linhova, M., Fribert, P., Muzikova, Z., Lipovsky, J., Paulova, L., Pospisil, M. and Sebor, G. 2011. Perspectives of biobutanol production and use. Biofuel's Engineering Process Technology 11: 243–261.

Pfromm, P.H., Amanor-Boadu, V., Nelson, R., Vadlani, P. and Madl, R. 2010. Bio-butanol vs. Bio-ethanol: A technical and economic assessment for corn and switchgrass fermented by yeast or clostridium acetobutylicum. Biomass and Bioenergy 34: 515–524.

Pugazhendhi, A., Mathimani, T., Varjani, S., Rene, E.R., Kumar, G., Kim, S.-H., Ponnusamy, V.K. and Yoon, J.-J. 2019. Biobutanol as a promising liquid fuel for the future-recent updates and perspectives. Fuel 253: 637–646.

Qureshi, N., Hughes, S., Maddox, I.S. and Cotta, M.A. 2005. Energy-efficient recovery of butanol from model solutions and fermentation broth by adsorption. Bioprocess and Biosystems Engineering 27: 215–222.

Qureshi, N., Li, X.L., Hughes, S., Saha, B.C. and Cotta, M.A. 2006. Butanol production from corn fiber xylan using clostridium acetobutylicum. Biotechnology Progress 22: 673–680.

Qureshi, N. and Maddox, I. 2005. Reduction in butanol inhibition by perstraction: Utilization of concentrated lactose/whey permeate by clostridium acetobutylicum to enhance butanol fermentation economics. Food and Bioproducts Processing 83: 43–52.

Qureshi, N., Saha, B.C., Dien, B., Hector, R.E. and Cotta, M.A. 2010. Production of butanol (a biofuel) from agricultural residues: Part I–use of barley straw hydrolysate. Biomass and Bioenergy 34: 559–565.

Qureshi, N., Saha, B.C., Hector, R.E., Hughes, S.R. and Cotta, M.A. 2008. Butanol production from wheat straw by simultaneous saccharification and fermentation using clostridium beijerinckii: Part I-batch fermentation. Biomass and Bioenergy 32: 168–175.

Ranjan, A. and Moholkar, V.S. 2013. Comparative study of various pretreatment techniques for rice straw saccharification for the production of alcoholic biofuels. Fuel 112: 567–571.

Ranjan, A., Khanna, S. and Moholkar, V. 2013. Feasibility of rice straw as alternate substrate for biobutanol production. Applied Energy 103: 32–38.

Rasskazchikova, T., Kapustin, V. and Karpov, S. 2004. Ethanol as high–octane additive to automotive gasolines. Production and use in russia and abroad. Chemistry and Technology of Fuels and Oils 40: 203–210.

Robak, K. and Balcerek, M. 2018. Review of second generation bioethanol production from residual biomass. Food Technology and Biotechnology 56: 174–187.

Saini, M., Chiang, C.-J., Li, S.-Y. and Chao, Y.-P. 2016. Production of biobutanol from cellulose hydrolysate by the *Escherichia coli* co-culture system. FEMS Microbiology Letters 363: fnw008.

Sánchez, J., Curt, M.D., Robert, N. and Fernández, J. 2019. Biomass Resources. The Role of Bioenergy in the Bioeconomy, pp. 25–111. Elsevier.

Schwarz, K.M., Grosse-Honebrink, A., Derecka, K., Rotta, C., Zhang, Y. and Minton, N.P. 2017. Towards improved butanol production through targeted genetic modification of *Clostridium pasteurianum*. Metabolic Engineering 40: 124–137.

Shafiei, M., Karimi, K. and Taherzadeh, M.J. 2010. Pretreatment of spruce and oak by n-methylmorpholine-n-oxide (nmmo) for efficient conversion of their cellulose to ethanol. Bioresource Technology 101: 4914–4918.

Shi, Z. and Blaschek, H.P. 2008. Transcriptional analysis of *Clostridium beijerinckii* ncimb 8052 and the hyper-butanol-producing mutant ba101 during the shift from acidogenesis to solventogenesis. Applied and Environmental Microbiology 74: 7709–7714.

Sun, Z. and Liu, S. 2012. Production of n-butanol from concentrated sugar maple hemicellulosic hydrolysate by *Clostridia acetobutylicum* atcc824. Biomass and Bioenergy 39: 39–47.

Szulczyk, K.R. 2010. Which is a better transportation fuel-butanol or ethanol? International Journal of Energy & Environment 1: 501–512.

Szulczyk, K.R., McCarl, B.A. and Cornforth, G. 2010. Market penetration of ethanol. Renewable and Sustainable Energy Reviews 14: 394–403.

Taherzadeh, M.J. and Karimi, K. 2007. Acid-based hydrolysis processes for ethanol from lignocellulosic materials: A review. BioResources 2: 472–499.

Talluri, S., Raj, S.M. and Christopher, L.P. 2013. Consolidated bioprocessing of untreated switchgrass to hydrogen by the extreme thermophile *Caldicellulosiruptor saccharolyticus* dsm 8903. Bioresource Technology 139: 272–279.

Tarasov, D., Leitch, M. and Fatehi, P. 2018. Lignin–carbohydrate complexes: Properties, applications, analyses, and methods of extraction: A review. Biotechnology for Biofuels 11: 1–28.

Tran, H.T.M., Cheirsilp, B., Hodgson, B. and Umsakul, K. 2010. Potential use of *Bacillus subtilis* in a co-culture with *Clostridium butylicum* for acetone–butanol–ethanol production from cassava starch. Biochemical Engineering Journal 48: 260–267.

Vohra, M., Manwar, J., Manmode, R., Padgilwar, S. and Patil, S. 2014. Bioethanol production: Feedstock and current technologies. Journal of Environmental Chemical Engineering 2: 573–584.

Wang, Y. and Blaschek, H.P. 2011. Optimization of butanol production from tropical maize stalk juice by fermentation with *Clostridium beijerinckii* ncimb 8052. Bioresource Technology 102: 9985–9990.

Wen, Z., Minton, N.P., Zhang, Y., Li, Q., Liu, J., Jiang, Y. and Yang, S. 2017. Enhanced solvent production by metabolic engineering of a twin-clostridial consortium. Metabolic Engineering 39: 38–48.

Wu, M., Wang, M., Liu, J. and Huo, H. 2008. Assessment of potential life-cycle energy and greenhouse gas emission effects from using corn-based butanol as a transportation fuel. Biotechnology Progress 24: 1204–1214.

Xia, Y., Fang, H.H. and Zhang, T. 2013. Recent studies on thermophilic anaerobic bioconversion of lignocellulosic biomass. RSC Advances 3: 15528–15542.

Ximenes, E., Kim, Y. and Ladisch, M.R. 2013. Biological conversion of plants to fuels and chemicals and the effects of inhibitors. Aqueous Pretreatment of Plant Biomass for Biological and Chemical Conversion to Fuels and Chemicals, 39–60.

Xu, M., Zhao, J., Yu, L., Tang, I.C., Xue, C. and Yang, S.T. 2015. Engineering *Clostridium acetobutylicum* with a histidine kinase knockout for enhanced n-butanol tolerance and production. Applied Microbiol. Biotechnol. 99: 1011–1022.

Xue, C. and Cheng, C. 2019. Chapter two-butanol production by clostridium. pp. 35–77. *In*: Li, Y. and Ge, X. (eds.). Advances in Bioenergy (Vol. 4). Elsevier.

Xue, C., Zhao, J., Lu, C., Yang, S.T., Bai, F. and Tang, I.C. 2012. High-titer n-butanol production by *Clostridium acetobutylicum* jb200 in fed-batch fermentation with intermittent gas stripping. Biotechnol. Bioeng. 109: 2746–2756.

Yakovlev, A.V., Shalygin, M.G., Matson, S.M., Khotimskiy, V.S. and Teplyakov, V.V. 2013. Separation of diluted butanol–water solutions via vapor phase by organophilic membranes based on high permeable polyacetylenes. Journal of Membrane Science 434: 99–105.

Yang, X., Xu, M. and Yang, S.-T. 2015. Metabolic and process engineering of *Clostridium cellulovorans* for biofuel production from cellulose. Metabolic Engineering 32: 39–48.

Yu, L., Xu, M., Tang, I.C. and Yang, S.T. 2015. Metabolic engineering of *Clostridium tyrobutyricum* for n-butanol production through co-utilization of glucose and xylose. Biotechnology and Bioengineering 112: 2134–2141.

Zabed, H., Faruq, G., Sahu, J.N., Azirun, M.S., Hashim, R. and Nasrulhaq Boyce, A. 2014. Bioethanol production from fermentable sugar juice. The Scientific World Journal 2014: 1–11.

Zhang, J., Yu, L., Lin, M., Yan, Q. and Yang, S.-T. 2017. N-butanol production from sucrose and sugarcane juice by engineered *Clostridium tyrobutyricum* overexpressing sucrose catabolism genes and adhe2. Bioresource Technology 233: 51–57.

Zhang, W., Liu, Z., Liu, Z. and Li, F. 2012. Butanol production from corncob residue using *Clostridium beijerinckii* ncimb 8052. Letters in Applied Microbiology 55: 240–246.

CHAPTER 9

Bioconversion of Agro-Industrial Residues into Biobutanol

Mónica Coca,[1,2,]* *M. Teresa García-Cubero,*[1,2] *Susana Lucas*[1,2]
and *Valeria Reginatto*[3]

1. Introduction

1.1 Properties and Uses of Butanol

There is a growing interest in the production of biobutanol from renewable resources to solve problems associated with the fluctuations in oil prices, as well as environmental concerns (Abo et al. 2019). Biobutanol is commonly used as a bulk chemical and as a solvent. The global butanol market will reach US$ 7.7 billion by 2024 with an annual growth rate of 3.5% for the 2019–2024 period (Birgen et al. 2021). Biobutanol is also considered a drop-in fuel. The advantages of butanol with respect to ethanol are described hereafter:

- Butanol can be used in engines in pure form or blended in any proportion with gasoline, while ethanol can be blended up to a maximum of 85%.
- Unlike ethanol, butanol can be used in a petrol engine without modification.
- The energy content of butanol (29.2 MJ/L) is similar to that of gasoline (32 MJ/L) and higher than that of ethanol (19.6 MJ/L).
- Butanol has a lower vapor pressure (0.43 MJ/kg) than ethanol (0.92 MJ/kg) and is therefore safer to handle and provides an easier engine start.
- Butanol is less hygroscopic, a characteristic that could prevent groundwater contamination in case of spills.
- It is less corrosive than ethanol, which means that existing fossil fuel infrastructures can be used without prior modification.

[1] Institute of Sustainable Processes. University of Valladolid, Spain.
[2] Department of Chemical Engineering and Environmental Technology, School of Industrial Engineering, University of Valladolid, Dr. Mergelina, s/n, Valladolid, Spain.
[3] Department of Chemistry, University of São Paulo, Av. Bandeirantes, 3900, CEP 14040-901 Ribeirão Preto, Brazil.
* Corresponding author: monica@iq.uva.es

Other 4-carbon branched-chain alcohols, including iso-butanol, 2-methyl-1-butanol and 3-methyl-1-butanol, all of which are n-butanol derivatives, have higher octane ratings and can be used as fuel additives (Veza et al. 2021). In view of the properties of butanol and its isomers, the growing interest and market potential of these bio-based butanol products is evident, either as eco-friendly biofuels/fuel additives or as feedstock/solvents in synthesizing resins and specialty chemicals (Lamani et al. 2017, Kattela et al. 2019).

1.2 Industrial Production of Biobutanol

Butalco, Butamax Advanced Biofuels, Eastman Chemical Company, GEVO, and Green Biologics are the main corporate leaders in the butanol market, according to Technavio's market research analysis (Global bio-based butanol market 2017–2021).

Butalco (Switzerland) is improving industrial yeast strains for biobutanol production from lignocellulose. BP Biofuels and DuPont (UK) have undertaken a joint project (Butamax) to produce an innovative low-cost bio-isobutanol from various raw materials, including corn and sugar cane. GEVO (USA) has built a demonstration plant by adapting a bioethanol plant from corn and sorghum, able to produce 105 g/L of iso-butanol. Green Biologics (UK) has developed new butanol-producing microbial strains to improve the yields of the fermentation process.

The market of bio-based butanol is expected to increase steadily at a CAGR of more than 9% by 2021, according to the Technavio report. The market is geographically segmented (USA and Canada, Europe, Middle East, Africa and Asia-Pacific). North America is the main contributor to the biobutanol market due to the appropriate climatic conditions for growing feedstocks and government incentives for butanol production using bio-synthetic routes.

1.3 Biobutanol Sustainability Considerations

Several studies, based on Life Cycle Assessment (LCA), sustainability parameters and techno-economic analysis, show the potential environmental, economic and societal advantages of using fuels obtained from lignocellulosic residues compared to conventional fuels (gasoline, diesel, natural gas, etc.). It has been shown that NOx emissions in diesel or biodiesel combustion are significantly reduced by adding 20% by volume of butanol (Liu et al. 2014). The literature indicates that biobutanol produced from lignocellulosic residues (wheat straw, sugarcane bagasse) has a much lower environmental impact than that produced from crops (corn, sugarcane) (Brito and Martins 2017). LCA shows substantial decreases in environmental impacts for climate change, the quality of the ecosystem and resources in comparison to natural gas (Levasseur et al. 2017), as well as reductions in GHG emissions of 50% compared to gasoline (Pereira et al. 2015). The sustainable production of biofuels offers a range of benefits to society. The major benefits include the reduction of dependency on fossil fuel, fuel price stability, the reduction of contaminated gases, employment creation and rural development, the conversion of wastes to fuels, among others (The European Technology and Innovation Platform Bioenergy).

2. Biobutanol Production from Agro-Industrial Residues

2.1 General Description of the Process

The transformation of agro-industrial waste into biofuels and value-added products has gained enormous relevance in the last few years. Vegetable waste from agricultural activities and agro-food industries are a renewable, inedible, low-cost, and sustainable source of resources available the entire year. Its use does not interfere with food production and can contribute to increasing the profitability of agriculture in rural areas. The valorization of agro-industrial biomass therefore entails an alternative to reduce the dependency on fossil fuels.

Butanol-producing clostridia are not able to directly ferment agro-industrial biomass; a pretreatment step is required to improve the saccharification of complex carbohydrates. The biochemical conversion of agro-industrial waste to biobutanol involves four main stages (1) pretreatment, (2) enzymatic saccharification, (3) ABE fermentation, and (4) product recovery (Fig. 1A).

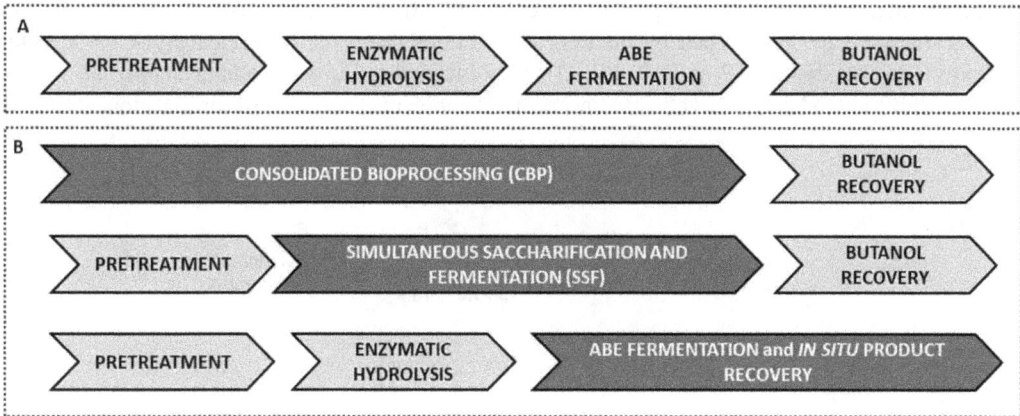

Fig. 1. Production of biobutanol from lignocellulosic residues (A) conventional process, (B) process intensification alternatives.

The objective of the pretreatment is to modify the recalcitrant structure of lignocellulose, reduce the crystallinity of the cellulose, increase porosity and favor the access of the hydrolytic enzymes. Intensive pretreatment could increase the price of butanol manyfold, limiting its commercial production (Jiang et al. 2015, Ibrahim et al. 2018). A wide range of physical, chemical and biological methods have been investigated for biobutanol production from agro-industrial residues (Baral et al. 2016, Amiri and Karimi 2018, Huzir et al. 2018, Amiri 2020, Veza et al. 2021). The main criteria for an effective pretreatment for biobutanol production have been critically discussed (Amiri and Kamiri 2018). For ABE fermentation, the formation of several compounds, mainly phenolics derived from lignin degradation in the pretreatment, play a critical role; so they exert a strong inhibitory effect in *Clostridium* species at concentrations as low as 0.5 g/L (Maiti et al. 2016, Bellido et al. 2018). For efficient ABE production, the pretreatment method and conditions have to be optimized considering the recovery of fermentable sugars and also the formation of inhibitors (López-Linares et al. 2020).

Solvent producing clostridia have no cellulolytic activity, so butanol is produced after enzymatic hydrolysis of such complex polymeric carbohydrates as cellulose. Enzymes are one of the main costs in the production of biofuels from lignocellulosic feedstocks. The optimization of enzyme loadings for maximizing the sugar released with the minimum enzyme consumption is vital to the viability of the process. On the other hand, the use of high solid loadings (> 15% w/w) in enzymatic hydrolysis is a requisite for the economic viability of biofuel production (Plaza et al. 2020). A detoxification step before fermentation is usually required to remove potential ABE inhibitors, mainly phenolic acids, from the plant or those formed during the pretreatment from lignin degradation.

ABE fermentation is carried out by anaerobic bacteria of the genus *Clostridium*, which can utilize a wide variety of substrates, including hexoses, pentoses and even more complex sugars such as starch, cellobiose and other oligomers obtained from the hydrolysis of lignocellulose (Amiri and

Karimi 2018). *C. acetobutylicum* is the most widely used on an industrial scale, while *C. beijerinkii* is pointed out as the most suitable candidate for ABE fermentation from lignocellulose derivatives (Lee et al. 2015, Gottumukkala et al. 2017). *C. saccharobutylicum* and *C. saccharoperbutylacetonicum* are other wild-type strains that have also been used to produce butanol (Zetty-Arenas et al. 2019).

Anaerobic metabolism by *Clostridium* has two phases: acidogenesis and solventogenesis. Acidogenesis occurs during the exponential growth phase, with the formation of acetic and butyric acids as soluble metabolites. The acidification of the environment (pH 4–5) and the high levels of NAD(P)H/NAD(P)$^+$ lead to the beginning of the stationary phase of cell growth. At the same time, it triggers solventogenesis (Li et al. 2020a). In solvent-*Clostridium* strains, acetic and butyric acids are reabsorbed by the cell, in reactions catalyzed by acetoacetyl-CoA acetate/butyrate-CoA transferases (9 and 10, Fig. 2), thus increasing the intracellular supply of acetyl-CoA and butyryl-CoA, respectively. Two acetyl-CoA molecules combine to form acetoacetyl-CoA

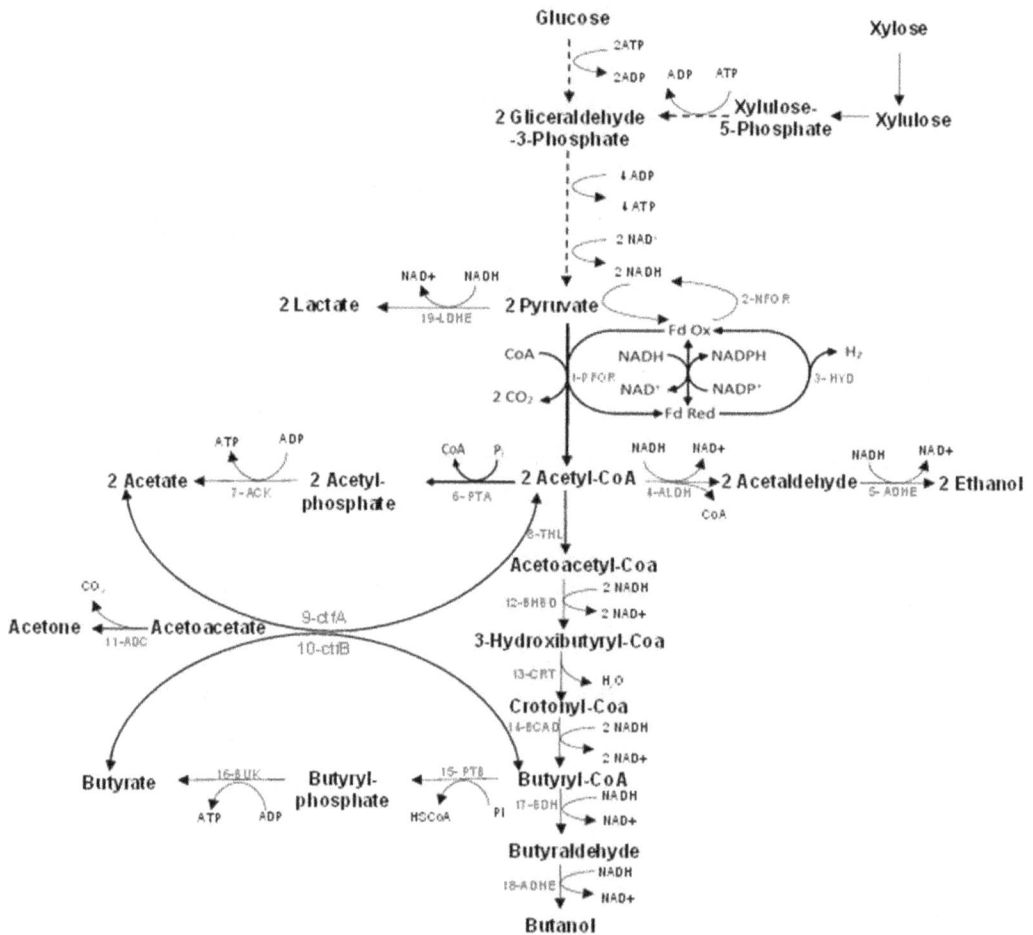

Fig. 2. Carbohydrate metabolism in solventogenic *Clostridium* (dotted lines indicate several reactions). Enzymes of acidogenesis and solventogenesis in *C. acetobutylicum*: 1-Pyruvate-ferredoxin oxidoreductase (PFOR); 2-NAD(P)H ferredoxin oxidoreductase (NFOR); 3-Hydrogenase (HYD); 4-Aldehyde dehydrogenase (ALDH); 5-Ethanol dehydrogenase (ADHE); 6-Phosphate acetyltransferase (PTA); 7-Acetate kinase (ACK); 8-Acetyl-CoA acetyltransferase (thiolase-TLH); 9-Acetoacetyl-CoA-acetate CoA-transferase (ctfA); 10-Acetoacetyl-CoA-butyrate CoA-transferase (ctfB); 11-Acetoacetate decarboxilase (ADC); 12-β-hydroxy-butyryl-CoA dehydrogenase; 13-hydroxybutyryl dehydratase (crotonase); 14-butyryl-CoA dehydrogenase; 15-Phosphate butyryltransferase (PTB); 16-Butyrate Kinase (BUK); 17-Butyryl-CoA dehydrogenase (BDH); 18-Butanol dehydrogenase (ADHE); 19-Lactate dehydrogenase (LDH). Adapted from Lee et al. (2008).

by the thiolase (8, Fig. 2), which is then converted into 3-hydroxybutyryl-CoA, crotonyl-CoA, butyryl-CoA, butyraldehyde and finally into butanol, by the enzymes, β-hydroxy-butyryl CoA dehydrogenase (12); 3-hydroxybutyryl dehydratase (crotonase) (13); butyryl CoA dehydrogenase (14); butyraldehyde dehydrogenase (17), Butanol dehydrogenase (18) (Fig. 2), respectively. As products of the solventogenic phase, an ABE mass ratio of 3:6:1 is usually observed, i.e., butanol is a major product of the carbohydrate metabolism (Li et al. 2020a).

According to the stoichiometry, 1 mol of glucose can generate 1 mol of butanol (0.41 g/g). However, the actual butanol yield is much lower due to the production of ethanol and acetone as by-products of ABE fermentation, the non-reabsorbed acetic and butyric acids, and the formation of other intermediates. The low product titer in the fermentation broth is mainly caused by butanol toxicity. Butanol is an amphipathic molecule which significantly affects the clostridial metabolism due to membrane transport impairment, dissolution of membrane lipids, and finally, low culture viability (Patakova et al. 2018). Butanol toxicity depends on the clostridial strain, but generally bacteria can rarely tolerate a concentration higher than 1.2–1.6% v/v (Amiri 2020).

Despite the extensive knowledge of the past few years, the economic viability of biobutanol production by *Clostridium* still presents some technological challenges. The major constraints that have impaired the biobutanol production process are the cost of the raw material, the low yield and productivity of butanol, the low butanol titer and the high separation and purification costs. Several papers have reviewed the strategies followed to overcome these drawbacks (Xue et al. 2017, Abo et al. 2019, Birgen et al. 2019). To be competitive with the petrochemical route, the selection of the feedstock, strain improvement and the integration of process steps are still challenging issues.

2.2 Agro-Industrial Residues as Feedstocks

The substrate selection is a critical decision for the profitability of ABE fermentation plants as feedstock accounts for up 30–55% of total butanol production costs (Jiang et al. 2015, Baral et al. 2016, Qureshi et al. 2020). In the last few years, substantial efforts have been made to use agro-industrial residues as substrates, and hence reduce the cost of butanol production.

Agro-industrial waste can be classified into agricultural and agro-food residues. Agricultural residues are generated in the cultivation and processing of crops (stems, stalks, straw, leaves, pulp, bagasse, husks). The content in complex carbohydrates (hemicellulose and cellulose) ranges from 50–80% w/w and approximately 10–25% w/w is composed of lignin (Sadh et al. 2018). Agro-food waste is organic waste produced in the food processing industries, such as fruit and vegetable processing (peel, pulp, seeds, hides). The content of lignin is usually lower, which could facilitate the saccharification of the carbohydrates (Sadh et al. 2018).

Many studies have been published on biobutanol production from agricultural lignocellulosic residues. Birgen et al. (2019) reported a dataset of butanol batch fermentation of 77 lignocellulosic hydrolysates by clostridial strains. The medians of butanol and total ABE solvents concentrations and yields were 6.95 g/L and 9.33 g/L, and 0.25 g/g and 0.34 g/g, respectively. The highest reported value of butanol was 14.5 g/L, produced by *C. beijerinckii* P260 from a corn stover hydrolysate after overliming (26.27 g ABE/L) (Qureshi et al. 2010a). The same authors reached a butanol concentration of 18 g/L from barley straw hydrolysates after overliming by *C. beijerinckii* P260 fermentation (Qureshi et al. 2010b).

By contrast, the studies that consider agro-food waste as raw material are scarcer. The literature reports butanol concentrations of between 1.1 and 12.6 g/L (Table 1). Butanol yields and concentrations are greatly affected by the feedstock, pretreatment and clostridial strain. An appropriate selection of the pretreatment method, operating conditions and clostridial strains could make detoxification unnecessary before the ABE fermentation of hydrolysates from potato peel,

Table 1. Biobutanol production from agro-food residues by batch fermentation.

Substrate	Microorganism	Pretreatment	Detoxification	S_0 (g/L)	Butanol (ABE) (g/L)	Butanol yield (g/g)	P_{BUT} (g/(L·h))	Reference
Potato peel waste	C. acetobutylicum PTCC 1492	Ethanol extraction + dilute acid	YES	36	6.22 (11.64)	NR	NR	Abedini et al. 2020
Potato peel waste	C. acetobutylicum PTCC 1492	Ethanol organosolv	YES	38	12.6 (24.8)	NR	NR	Abedini et al. 2020
Potato peel waste	C. saccharoperbutylacetonicum DSM 2152	Autohydrolysis	NO	38	8.1 (10.9)	0.203	0.068	Hijosa-Valsero et al. 2018a
Apple pomace	C. beijerinckii CECT 508	Surfactant (PEG 6000)	NO	42	9.11 (12.92)	0.276	0.095	Hijosa-Valsero et al. 2017
Apple pomace	C. beijerinckii P260	Dilute sulfuric acid	NO	40.3	5.4 (9.6)	0.32	0.1 (ABE)	Jin et al. 2019
Apple pomace	C. beijerinckii P260	Alkaline pretreatment	NO	36.6	4.8 (9.4)	0.34	0.13 (ABE)	Jin et al. 2019
Tomato pomace	C. beijerinckii DSM 6423	Hydrothermal	NO	44.1	6.52 (14.48 ABEI)	0.184	0.068	Hijosa-Valsero et al. 2019
Pea pod waste	C. acetobutylicum B527	Dilute sulfuric acid	YES	60	3.59 (5.64)	0.18 (ABE)	NR	Nimbalkar et al. 2018
Lettuce residues	C. acetobutylicum DSMZ 792	Alkaline pretreatment	NO	19.5	1.1 (1.44)	0.07	0.03 (ABE)	Procentese et al. 2017
Cauliflower waste	C. acetobutylicum NRRL B527	Dilute sulfuric acid	YES	60	2.99 (5.29)	0.17 (ABE)	NR	Khedkar et al. 2017
Coffee silverskin	C. beijerinckii CECT 508	Autohydrolysis	NO	34.4	7.02 (11.4)	0.269	0.073	Hijosa-Valsero et al. 2018b
Spent coffee grounds	C. beijerinckii DSM 6422	Microwave dilute sulfuric acid	NO	37.5	6.7 (10.4)	0.21	0.14	López-Linares et al. 2021

S_0: initial sugar concentration of the hydrolysate before fermentation.

P_{BUT}: productivity of butanol (or ABE).

Acetone–butanol–ethanol–isopropanol (ABEI).

NR: not reported.

tomato waste and apple pomace (Hijosa-Valsero et al. 2017, 2018a, 2019). On the contrary, Abedini et al. (2020) reported that glycoalkaloids in potato peel are severe inhibitors for clostridial bacteria and they were extracted by organosolv pretreatment with ethanol before enzymatic hydrolysis. Jin et al. (2019) reported that inhibitors generated during alkaline and dilute acid pretreatments of apple pomace negatively affect sugar consumption, ABE fermentation titers and rates. This negative effect was alleviated by mixing hydrolysates from structural carbohydrates with water soluble sugars extracted before pretreatment. The ABE fermentation of hydrolysates from lettuce, pea pod and cauliflower waste produced low butanol concentrations (1.1–3.6 g/L), probably due to the presence of inhibitors, mainly phenolic acids (Khedkar et al. 2017, Procentese et al. 2017, Nimbalkar et al. 2018). It should be highlighted that the ABE fermentation of coffee industry by-products could produce 7 g/L butanol without detoxification (Hijosa-Valsero et al. 2018b, López-Linares et al. 2021).

2.3 Strain Improvement

2.3.1 Solvent toxicity

The application of genetic tools using metabolic engineering strategies can improve butanol tolerance, yield and concentration. The development of a non-sporulating strain would allow more robust fermentation in long-term processes (Du et al. 2021a, b). Several approaches are being developed to obtain a suitable strain for industrial application. Concentrations of butanol as high as 20.9 g/L have been reported from pure glucose using the hyper-butanol producer *C. beijerinckii* BA101, a mutant of the *C. beijerinckii* NCIMB 8052 by N-methyl-N'-nitro-N-nitroso-guanidine (NTG) treatment. Genome analysis of *C. beijerinckii* BA101 revealed mutations on the genes encoding the transcriptional regulator, sensor kinase and phosphatase in comparison with the wild strain (Seo et al. 2021). A concentration of up to 21 g/L butanol from glucose was reported for *C. acetobutylicum* JB200, a mutant strain evolved from *C. acetobutylicum* ATCC 55025 by cultivation in a fibrous bed bioreactor in the presence of butanol. The increased butanol tolerance was due to the disruption of the gen *cac3319*, which regulates solventogenesis in *C. acetobutylicum* (Xu et al. 2015).

Genetic engineering is based on the insertion of heterogenetic genes and overexpression of endogenous genes. The disruption of the *hydA* gene in *C. acetobutylicum* ATCC 55025 with a plasmid pSY6 led to a strain able to produce 18.3% more butanol and 31.2% less acetone compared to the wild-type strain (Du et al. 2021a). The further addition of methyl viologen reduced acetone formation, reaching a maximum butanol yield of 0.28 g/g from corn stover hydrolysate as substrate (10 g/L butanol, 14.9 g/L ABE). Another work reported butanol concentrations of 18.2 g/L (0.24 g/g butanol yield, 0.41 g butanol/(L·h)) from the batch fermentation of glucose by a mutant strain of *C. acetobutylicum* with double HK knockouts (*cac3319/cac0323*) (Du et al. 2021b). The overexpression of *C. saccharoperbutylacetonicum* N1-4 (*adhE* genes) increased butanol production by 13.7% in comparison with the parental strain, reporting butanol concentrations of 17.4 g/L from glucose (23.6 g/L total ABE) (Wang et al. 2017a). The strain produced nearly 19 g/L of butanol by a mutant of *C. saccharoperbutylacetonicum* N1-4 obtained after the deletion of *pta* and *buk* genes (Wang et al. 2017b).

2.3.2 Tolerance to lignocellulose derivative inhibitors

Many *Clostridium* strains are inhibited by compounds derived from hemicellulose degradation, mainly organic acids (acetic and formic acids), furans (furfural and 5-hydroxymethylfurfural (HMF), and phenolics (ferulic and p-coumaric acids, syringaldehyde, vanillin) from lignin degradation (Jönsson and Martín 2016).

High levels of weak acid and a pH lower than pKa favor acid non-dissociated forms, which may enter the cell. The near neutral pH inside the cell promotes acid dissociation in the proton and related ion, resulting in intracellular acidity and metabolic disorder. In general, acetate addition induced butanol production during the early clostridial growth phase, once it becomes an intermediate for the solvent metabolism. Weak acid inhibition has been reported when the total undissociated (acetic and butyric) acid reached concentrations of 13 mmol/L and more (Fonseca et al. 2018). As for formic acid, the enhancement of the proline biosynthesis through overexpressing the proA, proB and proC by the *C. acetobutylicum* strain 824(proABC) improved the multiple tolerance to formic acid and phenolic compounds (Liao et al. 2019). The butanol production by the engineered strain proABC was 3.4-fold higher compared to the wild strain from undetoxified lignocellulosic hydrolysates.

The intracellular detoxification of HMF and furfural relies on their NADH-dependent reduction to furfuryl alcohol, which is less harmful to the cell. Thus, the chromosomal integration of an aldo-keto reductase in *C. beijerinckii* NCIMB 8052 improved 2- and 1.2-fold the furfural tolerance and butanol production relative to the wild type, respectively (Okonkwo et al. 2019). A highly furfural tolerant *C. acetobutylicum* strain was also successfully attained after the heterologous expression of the heat shock proteins GroESL and DnaK (Liao et al. 2017).

Transcriptional analysis revealed the broad effect of phenolics on *C. acetobutylicum* by altering the gene expression of the membrane transporters, glycolysis, and the heat shock proteins (Luo et al. 2020). Phenolic compounds can denature essential enzymes, or they can mutate genetic material by the generation of reactive oxygen species (ROS). Thus, metabolic engineering approaches that aim to enhance the tolerance of *Clostridium* to phenolic inhibitors frequently improve its scavenging ability of ROS. For example, *C. beijerinckii* has its tolerance to ferulic acid improved by genes encoding alkyl hydroperoxide reductase and a hypothetical NADPH-dependent FMN reductase (Liu et al. 2017).

Furthermore, efflux and heat shock protein genes were also related to ferulic acid tolerance. *C. beijerinckii*, expressing the groESL operon, has improved its tolerance to ferulic acid and the bioconversion of glucose into solvents (Lee et al. 2015). The overexpression of the efflux pump gene srpB from *Pseudomonas putida* in *C. saccharoperbutylacetonicum* has its tolerance towards furfural and ferulic acid increased by 17% and 50%, respectively (Jiménez-Bonilla et al. 2020).

3. Process Intensification in Biobutanol Production

Though various efforts have been made in genetic engineering, maximum butanol concentrations as high as 21–25 g/L have been reported by batch fermentation (Qureshi et al. 2014b). As the boiling point of butanol is 118°C, solvent recovery by conventional distillation is too energy-intensive, which hinders competitiveness with petrochemical synthesis. Process intensification aims to combine the different process steps, i.e., pretreatment, enzymatic hydrolysis, fermentation and recovery, to reduce capital and operation costs and enhance the competitiveness of butanol production by the biochemical route (Qureshi et al. 2014a, 2020). New developments on process intensification are described below.

3.1 Consolidated Bioprocessing

One important disadvantage of solventogenic clostridia, such as *C. beijerinckii*, *C. acetobutylicum* or *C. saccharoperbutylacetonicum*, is the limitation for secreting the enzymes (cellulases and xylanases) required to saccharify lignocellulose. Consolidated bioprocessing can accomplish hydrolytic enzyme production, polysaccharide hydrolysis and the ABE fermentation of pentoses and hexoses in the same reactor, using one microorganism or consortia. Therefore, CBP is considered to be a process intensification strategy (Fig. 1B). However, no natural microbial consortia with the ability to produce butanol are currently available.

In the last few years, a great effort has been made to achieve advances in the CBP of lignocellulosic biomass through the design of cellulolytic clostridia or consortium. Recent progress is mainly based on metabolic engineering tools and omics analysis (genome, transcriptome, proteome and metabolome) to optimize the structure of the consortium. Three strategies can be used for CBP in biobutanol production: (1) monocultures of engineered cellulolytic clostridia with heterologous solventogenesis pathways, (2) monocultures of engineered solventogenic clostridia to overexpress heterologous cellulases, and (3) co-cultures of cellulolytic and butanol producing clostridia (Jiang et al. 2018b, Wen et al. 2020a).

Several clostridia can grow on lignocellulose producing cellulases such as *C. thermocellum*, *C. cellulolyticum*, and *C. cellulovorans*. However, no wild-type cellulolytic/hemicellulolytic *Clostridium* sp. can produce butanol directly from lignocellulose (Jiang et al. 2018a). Microorganisms able to degrade lignocellulose have been engineered to produce biofuels. The combination of both cellulolytic and butanol-producing phenotypes through metabolic engineering provides an opportunity for biobutanol production through CBP. *C. thermocellum* can grow at temperatures higher than 60°C, showing high saccharification rates. Engineering a modification of *C. thermocellum* to extend the 2-keto acid metabolic pathway for isobutanol production has provided 5.4 g/L from cellulose within 75 h (Lin et al. 2015). Tiang et al. (2019) reported maximum concentrations of 357 mg/L of n-butanol from cellulose within 120 h using *C. thermocellum*. The engineered strain contained the enzymes thiolase–hydroxybutyryl-CoA dehydrogenase–crotonase (Thl-Hbd-Crt), trans-enoyl-CoA reductase (Ter) and butyraldehyde dehydrogenase and alcohol dehydrogenase (Bad-Bdh). The key enzymes were further optimized through protein engineering by introducing homologous mutations identified in *C. acetobutylicum*. The low butanol final titer is due to the net carbon flux conversion, mainly to ethanol. *C. cellulovorans* is quite a promising strain for CBP, since it can be cultivated on lignocellulose producing butyrate, a precursor of butanol. Cellulose was converted to 1.42 g/L butanol within 252 h by the recombinant *C. cellulovorans* through introducing an aldehyde/alcohol dehydrogenase (adhE2) from *C. acetobutylicum*, which converts butyryl-CoA to butanol (Yang et al. 2015). Metabolic and evolutionary engineering techniques can be applied to direct carbon flux from butyrate to butanol, producing 3.47–4.96 g/L of butanol by *C. cellulovorans* (Bao et al. 2019, Wen et al. 2019, 2020b). Low titer, productivity and yields are mainly related to inefficient lignocellulose hydrolysis and insufficient carbon flux supply (Wen et al. 2020a). Considering these results (Table 2), further improvements through strain engineering are needed.

The recombinant strategy implies the engineering of butanol-producing clostridia to degrade lignocellulose through the expression of cellulolytic enzymes. This challenging technology is reliant on the development of heterologous expression. Allele-coupled exchange (ACE) technology has been applied to integrate a hybrid cellulosome operon into the genome. The mini-cellulosome was anchored to the cell wall of *C. acetobutylicum* via the native sortase system (Willson et al. 2016). However, the modified strains produced low butanol concentrations (0.82–1.1 g/L) after 164 h (Willson et al. 2016). The heterologous expression levels of cellulase in *Clostridium* sp. are insufficient due to the complexity of cellulase systems, leading to low concentrations of butanol.

The use of consortia of cellulolytic and clostridia to produce butanol from lignocellulose without the addition of enzymes has also been explored. The production of butanol depends on the interaction between strains, which can be improved by genetic engineering and coordinated growth conditions. The consortium formed by the engineered *C. cellulovorans* DSM 713B and *C. beijerinckii* NCIMB 8052 produced 3.94 g/L butanol without pH control after 83 h (Wen et al. 2020c). Adaptative laboratory evolution and genetic engineering enhanced *C. cellulovorans* tolerance to the low pH required for biobutanol production. A previous study from the same authors

Table 2. Consolidated bioprocessing for biobutanol production.

Microorganism	Substrate	Butanol (g/L)	P_{BUT} (g/(L·h))	Reference
Monocultures				
C. thermocellum Thl-Hbd-Crt	Avicel	0.357	0.003	Tian et al. 2019
C. cellulovorans adhE2	Avicel	1.42	0.006	Yang et al. 2015
C. cellulovorans adhE1-ctfAB-adc	Corn cobs (alkali extracted)	3.47	0.041	Wen et al. 2019
C. cellulovorans adhE2	Cellulose	4.0	0.014	Bao et al. 2019
C. cellulovorans (engineered by pull-push modular metabolic)	Corn cobs (alkali extracted)	4.96	0.051	Wen et al. 2020b
Co-cultures				
C. cellulovorans 743B and C. beijerinckii NCIMB8052	Mandarin orange waste	0.28	0.0006	Tomita et al. 2019
C. cellulovorans DSM 743B + C. beijerinckii NCIMB 8052 (mesophilic co-culture and modular metabolic engineering)	Corn cobs (alkali extracted)	11.5	0.096	Wen et al. 2017
C. cellulovorans DSM 743B + C. beijerinckii NCIMB 8052 (adaptative laboratory evolution, genetic engineering)	Corn cobs (alkali extracted)	3.94	0.048	Wen et al. 2020c
White-rot fungus Phlebia sp. MG-60-P2 + C. saccharoperbutylacetonicum NBRC 109357	Unbleached hardwood kraft pulp	3.2	0.012	Tri et al. 2020
Thermoanaerobacterium sp. M5 + C. acetobutylicum NJ4	Xylan	8.34	0.05	Jiang et al. 2018a

P_{BUT}: butanol productivity.

reported butanol concentrations as high as 11.5 g/L from alkali extracted corncob, after a two-stage pH control strategy and multivariate modular metabolic engineering of *C. cellulovorans* and *C. beijerinckii* to strengthen their feeding-detoxification relationship (Wen et al. 2017).

To date, sequential co-culturing has provided a more feasible alternative than monocultures (Table 2). Adequate sugar rate production is a critical factor to reach high butanol concentration from lignocellulose. In this context, the co-culture of thermophilic cellulolytic/hemicellulolytic strains with solventogenic clostridia would be a better alternative, as they usually show higher hydrolysis rates (Jiang et al. 2020). Jiang et al. (2018a) co-cultured *Thermoanaerobacterium* sp. M5 and *C. acetobutylicum* NJ4. Although the isolated wild-type *Thermoanaerobacterium* sp. M5 can ferment xylan to butanol at 55°C, the co-culture improved the butanol titer. M5 secreted xylanase and xylosidase and produced butanol (0.78 g/L) from 60 g/L of xylan within 72 h. Butanol synthesis was attributed to the bifunctional alcohol/aldehyde dehydrogenase (AdhE) enzyme. The addition of *C. acetobutylicum* NJ4 to the *Thermoanaerobacterium* sp. M5 culture drives the carbon flux toward butanol, improving the concentration to 8.34 g/L, the highest titer reported from xylan through CBP. These two strains were also co-cultivated for butanol production from corncob (Jiang et al. 2020). Butyrate produced by the thermophilic strain M5 triggers solventogenesis of the strain NJ4. Butyrate could also be reutilized by CoA-transferase for butanol production. The positive interaction between the two strains produced 7.6 g/L of butanol from corncob through CBP after 168 h (productivity of 0.045 g/(L·h)). Co-culture between white-rot fungus *Phlebia* sp. MG-60-P2 and *C. saccharoperbutylacetonicum* for butanol from hardwood has been reported by Tri and Kamei (2020), who demonstrated that knocking out the pyruvate decarboxylase gen from

MG-60-P2 enhanced butanol production. Although co-culturing systems can achieve adequate butanol titers from lignocellulose, productivity is still low (Table 2). In addition, CBP challenges due to the different culture conditions of the strains (pH and temperature) need to be overcome by advances in biology techniques (Wen et al. 2020a).

3.2 *Simultaneous Saccharification and Fermentation*

The use of integrated processes for biobutanol production is an attractive alternative to diminish end-product inhibition or to shorten the whole processing time. In that sense, simultaneous saccharification and fermentation (SSF) is an alternative to separate hydrolysis and fermentation (SHF) (Fig. 1). Valles et al. (2020) reports the main advantages of SSF, in comparison to SHF, as it:

- Reduces the process steps, since only one reactor is necessary to carry out both the saccharification and fermentation stages.
- Reduces the inhibition of enzymes due to the immediate consumption of sugars as they are released.
- Diminishes potential contamination due to the presence of solvents.
- Decreases the overall production costs.

Several studies have reported the use of SSF for biobutanol production from agro-industrial residues (Table 3). SSF is mostly employed to obtain cellulosic butanol, the main operating parameters being the enzyme and the solid loadings. The lower enzyme loading was used by Li et al. (2016) (10 FPU/g), who reached 10.4 g butanol/L (0.07 g butanol/g substrate), whereas the higher enzyme load was 40 FPU/g (He et al. 2017), attaining 9.5 g butanol/L (0.1 g butanol/g substrate). The solid loading in SSF processes is usually around 5–6% w/v, although Dong et al. (2016) reported up to 12.9 g butanol/L working with 7% w/v solid loading and *C. saccharobutylicum* DSM 13864. The benefits of a pre-hydrolysis step for 12–24 h has been investigated by He et al. (2017) and Wu et al. (2021), who demonstrated that the fermentation process could be enhanced since only a small amount of monosaccharides is found in the fermentation medium before the inoculation of the microorganism. Although the butanol concentration obtained from SSF processes is 9–12% lower when compared to SHF processes (Sasaki et al. 2014), the detoxification of the pretreated solid prior to SSF allows butanol concentrations to increase by 6.6% (Qureshi et al. 2014a).

The major challenge in the SSF process is the higher saccharification temperature (45–50°C) in comparison with the optimal for ABE fermentation (37°C). Wu et al. (2021) obtained up to 10.8 g butanol/L from pretreated corn stover after 12 h of prehydrolysis working with a thermotolerant strain of *C. acetobutylicum* L7 (GlcG) at 42°C.

Khalili and Amiri (2020) suggested different alternatives for the conversion of sweet sorghum bagasse (SBB) to butanol based on the integration of autohydrolysis, enzymatic hydrolysis and ABE fermentation. The best results for hemicellulosic butanol (14 g butanol/kg SBB; 32 kg ABE/kg SBB) were obtained after two autohydrolysis steps (8% solid loading, 150°C, 30 min and 120°C, 60 min, respectively), enzymatic post-hydrolysis (250 mg protein/g Cellic Htec2, 48°C, 72 h) and the fermentation of the resulting liquid, whereas the application of SSF (5% solid loading, 25 FPU/g Cellic CTec2, 37°C, 96 h) led to 37 g BuOH/kg SBB (75 g ABE/kg SBB). Seifollahi and Amiri (2020) analyzed the effect of pretreatment conditions on the simultaneous co-saccharification and fermentation (SCSF) of the pretreated solid and liquid hydrolysates (5% (w/v) solid loading, 25 FPU/g cellulase, 37°C, 72 h). The best results reported 16.1 g butanol/L (24.1 g ABE/L) when dilute acid pretreatment (120°C, 60 min, 1% acid) was employed. This alternative allows the simultaneous fermentation of the oligomeric sugars present in the hydrolysates, pentoses and

Table 3. Comparison of butanol production from agro-industrial residues by simultaneous saccharification and fermentation.

Substrate	Microorganism	Pretreatment	Enzymatic Hydrolysis Conditions	Butanol (g/L)	Butanol (ABE) yield (g/g)	P_{BUT} (P_{ABE}) (g/(L·h))	Reference
Rice straw (RS)	*C. beijerinkii* DSM 6422	Microwave hydrothermal	12 FPU Cellic CTec2/g RS, 9% w/v solid loading, 37°C, 120 h	5.2	0.217 (0.341)	0.109 (0.172)	Valles et al. 2020
Wood chips (WC)	*C. acetobutylicum* NBRC 13948	Steam explosion	7.8 mg protein/g WC cellulase, 5% w/v solid loading, 37°C, 144 h	7.8	0.170 (0.300)	0.054 (0.093)	Sasaki et al. 2014
Napier grass (NG)	*C. acetobutylicum* ATCC 82	Alkaline (NaOH)	40 FPU/g NG cellulase, 24 h prehydrolysis, 10% w/v solid loading, 37°C, 129 h	9.5	0.100 (0.170)	0.080 (0.130)	He et al. 2017
Corn stover (CS)	*C. acetobutylicum* ATCC 824	Steam explosion	10 FPU cellulase/g CS; 17.5% w/v solid loading, 37°C, 84 h	10.4	0.070 (0.110)	0.120 (0.200)	Li et al. 2016
CS	*C. acetobutylicum* L7 (GlcG)	Dilute sulfuric acid + alkaline (NH_4OH)	15 FPU cellulase/g CS, 12 h prehydrolysis at pH 4.8, 4.5% w/v solid loading, 10 g/L $CaCO_3$, 42°C, 60 h	10.8	0.180 (0.310)	0.180 (0.300)	Wu et al. 2021
CS	*C. saccharobutylicum* DSM 13864	Alkaline (NaOH)	20 FPU Accellerase 1500/g CS, 7% w/v pretreated CS loading, 37°C, 50 h	12.9	0.175 (0.284) based on pretreated CS	0.257 (0.398)	Dong et al. 2016
Wheat straw (WS)	*C. acetobutylicum* ATCC 82	Ammonium sulfite	20 FPU/g WS cellulase; 40 IU xylanase, 6% w/v pretreated WS loading, 37°C, 108 h	11.3	0.148 (0.231) based on raw WS	0.105 (0.164)	Qi et al. 2019

P_{BUT}: butanol productivity (P_{ABE}: ABE productivity).
Yield: g product/g total sugar consumed in fermentation.

hexoses from both hydrolysates and pretreated solid, enhancing the butanol production process by up to 182 g ABE/kg cellulose.

3.3 ABE Fermentation with in situ Butanol Recovery

One of the main bottlenecks of the ABE fermentation process is the low concentration of butanol in the fermentation broth due to product inhibition. The separation of butanol by conventional distillation to obtain a commercial-grade product (concentration of butanol from 0.5% w/v to 99.5%) requires 79 MJ/kg of butanol, almost two-fold higher than the energy content of butanol (36 MJ/kg). Increasing the concentration of butanol in the broth to 4–5% w/v can reduce the energy demand to less than 10 MJ/kg (Lu et al. 2013). As genetic modification is still far from achieving these concentrations, the development of more energy-efficient recovery systems is essential.

The integration of fermentation with *in situ* product recovery (ISPR) (Fig. 1B) is an engineering approach to mitigate product toxicity and thereby improve the fermentation performance. The objective is to simultaneously remove the solvents produced in the bioreactor, reducing the butanol toxicity. ISPR can facilitate semi-continuous and continuous operation, which improves substrate consumption and productivity. Product removal techniques include *in situ* processes such as gas-stripping, liquid-liquid extraction, adsorption, pervaporation and perstraction. The principles of *in situ* recovery techniques are summarized in Table 4. Table 5 compares the performance of integrated recovery processes.

Table 4. Processes for *in situ* butanol recovery in ABE fermentation.

Process	Description	Advantages	Disadvantages
Gas-stripping	Gas-stripping of solvents by recirculation of the fermentation gases and further condensation of the vapors at temperatures below 10°C	Easy operation Simple scale-up No fouling Non-toxic to cells Continuous operation is possible Use of fermentation off-gas	Low selectivity towards butanol Foaming
Liquid-liquid extraction	Extractant is mixed with the fermentation broth. Butanol is selectively concentrated in the organic phase Subsequent recovery by distillation	High selectivity Simple	Loss of extractant due to the formation of emulsions Possible cellular inhibition by the extractant
Perstraction	Similar to L-L extraction, but a membrane is used to separate the broth and the extractant	Selective Separation of the organic and aqueous phase Less toxicity to cells Emulsions are avoided	Mass transfer limitations Fouling of membrane Low productivity
Pervaporation	Selective diffusion of solvents through a membrane in contact with the fermentation broth. The driving force is the chemical potential gradient of the permeating component from the broth to the vapor phase Further condensation of the permeate	Simple High selectivity Energy-efficiency Not harmful for microorganisms Simplicity in scaling-up	Fouling and clogging
Adsorption	Fermentation broth is circulated through an adsorbent where butanol is preferentially attached. Desorption of butanol previously adsorbed.	Energy efficient Simple scale-up Continuous operation is possible High selectivity	Regeneration is required Adsorbent fouling by cells Loss of nutrients by adsorption

Table 5. Performance of technologies for integrated fermentation and *in situ* butanol recovery.

Substrate	Microorganism	*In situ* Separation Technology	P$_{BUT}$ (P$_{ABE}$) (g/(L·h))	Butanol (ABE) in Concentrated Stream (g/L)	Reference
Wood pulp hydrolysate	*C. beijerinckii* CC101	GS	0.19 (0.25)	65–110 (80–130)	Lu et al. 2013
Barley straw	*C. beijerinckii* P260	GS	0.6 (ABE)	67.1–88.1 (100–135)	Qureshi et al. 2014b
Corn stover	*C. beijerinckii* P260	GS	0.7 (ABE)	57–79.9 (88.5–127.6)	Qureshi et al. 2014b
Corn stover	*C. acetobutylicum* ABE-P 1201	GS	0.09 (0.13)	77–136 (115–220)	Cai et al. 2017
Lactose (whey permeate)	*C. saccharobutylicum* P262	PV	0.43 (ABE)	72.39 (79.01)	Qureshi et al. 2014c
Glucose	*C. acetobutylicum* ABE 1201	PV	0.26 (0.45)	199.09 (346.45)	Cai et al. 2017
Glucose	*C. acetobutylicum* ABE 1201	PV	0.12 (0.22)	199.5 (344.2)	Wen et al. 2018
Glucose	*C. acetobutylicum* JB200	AD (activated carbon)	0.45	54.6 g/L butanol production (167 g/L butanol after desorption)	Xue et al. 2016b
Glucose	*C. acetobutylicum* JB200	GS + GS	0.4	420.3 (532.3)	Xue et al. 2013
Glucose	*C. acetobutylicum* JB200	GS + PV	NR	521.3 (622.9)	Xue et al. 2015
Glucose	*C. acetobutylicum* ABE 1401	GS + PV	NR	482.5 (706.7)	Cai et al. 2016
Glucose	*C. acetobutylicum* ABE 1201	PV + PV	NR	451.98 (782.5)	Cai et al. 2018

GS: gas-stripping; PV: pervaporation; AD: adsorption.
NR: not reported.

3.3.1 Gas-stripping

Gas-stripping can be easily coupled with the fermentation stage to alleviate the butanol toxicity and to obtain a condensate enriched in butanol. The process is simple to operate and scale up, can be integrated in fed-batch and continuous bioreactors, does not cause cell damage and does not require any chemicals or membrane (Rochón et al. 2017). The fermentation off-gas produced by the microorganism can be bubbled into the fermenter to aid stripping, which also enhances the agitation of the system (Fig. 3A). Butanol is removed by the off-gas and further condensed. The gases can be recycled to the fermenter. Butanol selectivity is lower in comparison to other separation methods because acetone, ethanol and water are also removed from the fermenter (Xue et al. 2012). Compared to other *in situ* separation strategies, gas-stripping could recover a cleaner butanol solution without cells, salts, and other macromolecules. As both acetic and butyric acids remained in the fermenter, they could be assimilated by the microorganism (Xue et al. 2014).

The gas flow rate, cooling temperature and butanol titer in the broth are important parameters that influence butanol concentration in the condensate. Xue et al. (2014) reported that gas-stripping should be started when the butanol concentration in the fermenter is higher than 8 g/L. In this case, a condensate with a butanol concentration higher than its solubility in water (7.7% w/w at 20°C) could be obtained, which would result in phase separation. The organic phase could have a butanol concentration of 80% v/v, which would further reduce the distillation costs. However, a butanol concentration of 8 g/L is often inhibitory to the bacteria. Lu et al. (2013) reported that

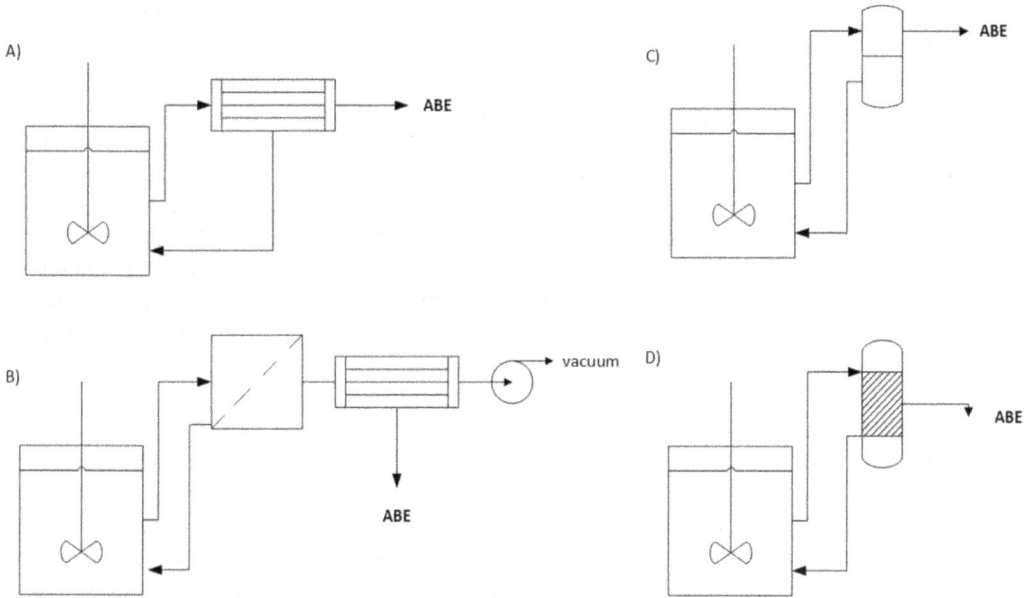

Fig. 3. Integrated fermentation-separation techniques for ABE fermentation. (A) gas-stripping, (B) pervaporation, (C) perstraction, (D) adsorption.

selective removal of butanol by gas-stripping improves butanol concentration and productivity from 9.14 g/L and 0.13 g/(L·h) to 13.46 g/L and 0.19 g/(L·h) compared to batch fermentation. Cumulative butanol concentrations from 13.5 to 34.7 g/L have been reported using lignocellulosic hydrolysates and fermentation by mutant strains (Lu et al. 2013, Qureshi et al. 2014b, Cai et al. 2017). Two-stage gas-stripping for butanol recovery have also been applied (Xue et al. 2013). The first stage reduces inhibition, while the second improves the concentration of butanol in the condensate (420.3 g/L butanol, 532.3 g/L ABE). It has been estimated that the conventional distillation of a solution with a butanol concentration of 1–1.5% w/v requires 24–36 MJ/kg butanol (Xue et al. 2013), which is similar to the energy content of butanol (36 MJ/kg). The energy requirement for butanol recovery by gas-stripping followed by distillation has been estimated at 14–31 MJ/kg butanol, considering a concentration of butanol in the condensate lower than 70 g/L. If the gas-stripping operates under optimized conditions for spontaneous phase separation in the condensate (about 80 g/L butanol), then the energy demand for purification could be decreased to 7–15 MJ/kg (Xue et al. 2013).

3.3.2 Pervaporation

Pervaporation combines membrane permeation and evaporation. It is usually used in many industrial applications due to its high selectivity and energy efficiency, higher than conventional separation techniques. Pervaporation involves the diffusion of two or more components through a membrane. A vacuum applied to the permeate side is coupled with the condensation of the permeated vapors, producing a condensed liquid (Fig. 3B). The retentate can be recycled to the fermenter (Huang et al. 2014). Hydrophobic membranes are used for the preferential permeation of organic compounds from a dilute aqueous solution, which is desirable for biobutanol recovery. Selectivity through the membrane and the flux of the permeate determine the effectiveness of pervaporation. Both parameters depend on the membrane type, the vacuum pressure, the feed temperature and composition, and the biomass concentration. The vacuum pressure applied on the permeate side is the driving force of the transference across the membrane. Values of 15 mbar and lower are the optimal values for this process (Rom and Friedl 2016, Gao et al. 2017, Rdzanek et al. 2018).

Hydrophobic polymers have been investigated for organophilic pervaporation, including polypropylene (PP), polydimethylsiloxane (PDMS), poly(octylmethyl siloxane) (POMS), polyvinylidene fluoride (PVDF), polyether block polyamide (PEBA), polytetrafluoroethylene (PTFE), poly(1-trimethylsilyl-1-propyne) (PTMSP) and the polymer of intrinsic microporosity PIM-1 (Huang et al. 2014, Lee et al. 2019). Inorganic membrane materials, supported ionic liquid membranes and composite membranes have also been employed (Gao et al. 2017, Rdzanek et al. 2018, Li et al. 2020b). The material used for the preparation of pervaporation membranes should accomplish such properties as long term stability, high permeability, stability at high temperatures and high selectivity for the target compound (Rdzanek et al. 2018). Model solutions of butanol are usually prepared to determine the performance of the membranes fabricated with new materials for *ex situ* butanol recovery. However, the performance of the membranes has to be tested in conjunction with fermentation.

Pervaporation with a silicone membrane has been applied to recover butanol produced from concentrated whey permeate by fermentation with *C. saccharobutylicum* P262, reaching a butanol concentration of 72.39 g/L in the permeate (Qureshi et al. 2014c) and an improved productivity (0.43 g/(L·h)) in comparison with perstraction and gas-stripping. Sequential *in situ* pervaporation (PDMS/PVDF membrane) and salting out (K_3PO_4) has proven its effectiveness for ABE removal by the fed-batch fermentation of a model solution of glucose (Wen et al. 2018). The permeate contained 199.5 g/L butanol and 344.2 g/L ABE and was further concentrated by salting out to 486.74 g butanol/kg (805.52 g ABE/kg). Two-stage pervaporation, based on a PDMS/PVDF membrane, resulted in a permeate with 199.09 g/L butanol (346.45 g/L ABE) after the first stage *in situ* pervaporation and 451.98 g/L butanol and 782.5 g/L ABE after the second stage. Fermentation coupled with two-stage pervaporation could reduce the energy demand to 13.2 MJ/kg butanol (Cai et al. 2017). Other authors reported that the hybrid process pervaporation/distillation involves an energy cost of about 4–8.2 MJ/kg butanol (Friedl 2016).

The main operational problems are related to fouling and clogging, which can occur due to the adsorption of cell and macromolecules when the membrane module is placed inside the fermenter. To solve this problem, a two-stage gas-stripping-pervaporation has been proposed as an alternative to take advantage of both technologies. The clean condensate obtained by *in situ* gas-stripping can be concentrated in the pervaporation stage (Cai et al. 2016, Xue et al. 2016a). The energy requirement for butanol recovery of this multiple stage approach (gas-stripping-pervaporation-distillation) has been estimated at 20–23 MJ/kg butanol. The application of hybrid systems offers the possibility of optimizing the operation conditions of the second unit without interfering with the fermentation stage.

3.3.3 Solvent-based processes

In extractive fermentation or fermenters integrated with liquid–liquid extraction, a water insoluble organic extractant selectively recovers the butanol from the broth. The fermentation broth is put in contact with a non-miscible solvent in which butanol has preferential partition. Operating conditions, such as the agitation rate, temperature, or pH should respect fermentation operability. The extractant must fulfill proper characteristics, such as biocompatibility, non-emulsion forming with the broth, high selectivity, low cost, high partition coefficient, and density significantly different from that of the broth (Huang et al. 2014). The extracted biobutanol can be further purified through distillation and the solvent can be returned to the extractive process.

A common extractant is oleyl alcohol because of its low toxicity and proper selectivity as well as its partition coefficient through butanol. The advantages of ionic liquids as solvent include negligible vapor pressure, which produce low solvent loss and less energy consumption. Its non-volatility makes it easily separable from the butanol after extraction by evaporation or flash distillation (Huang et al. 2014). The use of mixed extractants, such as decanol in oleyl alcohol and sodium hydroxide, improved the distribution coefficient and the extraction efficiency (up to 97.7%)

with lowered cell toxicity (Khedkar et al. 2020). Other solvents (2-butyl-1-octanol, vegetable oils) have been compared in terms of biocompatibility towards *C. acetobutylicum* and butanol extraction capacity in batch extractive fermentations. The use of 2-butyl-1-octanol improves the substrate consumption, butanol yield and the butanol:acetone ratio in comparison to oleyl alcohol (González-Peñas et al. 2020a). Batch and fed-batch extractive fermentation can decrease the production cost, lowering the minimum butanol selling price by 29% in comparison to conventional fermentation (González-Peñas et al. 2020b). Other authors reported an energy consumption for butanol production as low as 4–6 MJ/kg for the concentration of butanol from 0.8% w/w to 99.5% w/w (Huang et al. 2014). The scaling up of extractive fermentation is challenging because the organic and aqueous phases cannot be mixed vigorously to prevent emulsion formation. Hence, the recovery rates are relatively low (Kim et al. 2020).

Problems associated with extractive fermentation, mainly low productivity and solvent losses, can be solved by membrane solvent extraction or perstraction. In this case, the extractive solvent and the broth are separated by a membrane (Fig. 3C). Therefore, there is no direct contact between the phases and butanol diffuses selectively through the polymeric membrane from the aqueous phase to the organic phase. Toxic extractants with higher partition coefficients can be used as the membrane protects microorganisms. The selection of the membrane materials and dimensions are important design considerations, because the membrane module involves an additional transport resistance.

The application of perstraction coupled with ABE fermentation by *C. saccharoperbutylacetonicum* N1-4, with recovery based on 1-dodecanol and a PTFE membrane, reached a butanol concentration of 20.1 g/L and butanol productivity 0.394 g/(L·h) (Tanaka et al. 2012). Oleyl alcohol and 2-ethyl-1-hexanol have partition coefficients of 3.6 and 9.3, respectively. The latter has better partitioning properties that can be exploited using a membrane to protect the microorganisms. In this sense, a continuous perstraction system, based on a spray-coated thin-film composite membrane using 2-ethyl-1-hexanol as an extractant, has demonstrated its efficiency to reduce energy consumption to 3.9 MJ/kg butanol (Kim et al. 2020). Outram et al. (2016) have compared different ISPR techniques based on simulations of the ABE fermentation process and concluded that perstraction had the higher profit margin compared to a batch process.

3.3.4 Adsorption

Adsorption can remove butanol from the fermenter. In an integrated fermentation-adsorption process, the fermentation broth can first be fed to an ultrafiltration membrane where cells are separated and recycled to the bioreactor. In this way, fouling of adsorbent and cell loss is avoided. Then, the particle-free permeate enters the adsorption column, where butanol is adsorbed. Adsorption is followed by absorbent regeneration, usually by thermal treatment or displacement, to obtain a concentrated butanol solution (Fig. 3D).

Butanol is a hydrophobic compound, so hydrophobic adsorbents including activated carbon, zeolites and polymeric resins, can be applied for selective separation (Huang et al. 2014). Adsorbents should have such properties as a high adsorption capacity, selectivity, biocompatibility for *in situ* recovery, a low price and easy regeneration. The performance of activated carbon (Norit ROW 0.8), zeolite (CBV901) and polymeric resins (Dowex Optipore L-493 and SD-2) were compared for butanol recovery by integrating adsorption with fed-batch immobilized-cell ABE fermentation (Xue et al. 2016b). Activated carbon turned out to be more biocompatible and effective for *in situ* butanol recovery. Butanol production and productivity increased by 230% and 32%, respectively, compared to the control without *in situ* adsorption recovery. Three external packed columns of activated carbon were used for butanol adsorption to maintain the butanol concentration in the bioreactor below the toxicity limits. The culture broth was circulated through the adsorption bed until saturation. Limitations arise through acetone accumulation and inhibition, as the activated carbon has relatively low adsorbent properties towards acetone. Xue et al. (2016b) reported that activated carbon adsorption can reduce the energy requirement from a dilute butanol solution

(1% w/v) to 14.1 MJ/kg. Other authors reported values as low as 3.4 MJ/kg of butanol for a silicate solvent and a butanol concentration from 2% w/w to 98% w/w (Huang et al. 2014).

4. Concluding Remarks

Butanol is a commodity chemical also considered as a drop-in fuel with important advantages compared to ethanol. During the last few years, great efforts have been made to solve the bottlenecks of industrial-scale biobutanol production, which limit its competitiveness with the petrochemical route. The combination of robust engineered strains, the use of low-cost and abundant feedstocks, and technical advances in process intensification can all address the challenges of biobutanol production. The application of genetic engineering tools would improve the tolerance, yield and viability of the strains. Agro-industrial waste is an abundant and non-edible resource for sustainable biobutanol production. Process integration is a promising approach to reduce capital and operation costs. In addition to the sale of the by-products, acetone and ethanol, the valorization of the waste streams generated in the process in a biorefinery approach is also important. Cell biomass can be used as cattle feed, as well as the lignin in the spent solid, and hydrogen can be used to produce energy.

Fermentation integrated with *in situ* recovery processes can alleviate butanol toxicity, improving concentration and productivity. Among *in situ* recovery alternatives, perstraction, adsorption and pervaporation are considered the most energy efficient to remove butanol from the bioreactor. Integrated fermentation-separation techniques should satisfy such prerequisites as biocompatibility, scalability, high selectivity for butanol, robustness, fouling resistance, enhanced butanol production and cost effectiveness. All alternatives need further proofs on a larger scale to be successfully integrated into fed-batch or continuous bioreactors fermenting real agro-industrial waste hydrolysates in order to prove competitive capital and operation costs, maximum butanol recovery and assure long term stability. The combination of assessment methods (LCA, sustainability parameters, techno-economic analysis) is also necessary to analyze the potential of biobutanol as an emerging fuel, as well as to design more sustainable integrated biorefinery processes.

Acknowledgments

The authors would like to thank Ana Mª Rodriguez Rodríguez (Chemical Engineering Department, University of Vigo, Spain) and María Eugenia Guazzaroni (Biology Department, FFCLRP, Universidade de São Paulo, Brazil) for reviewing this paper and providing constructive comments that improved its quality. The authors acknowledge the financial support from the Spanish Ministry of Science and Innovation (PID2020-115110RB-I00) and the Regional Government of Castilla y Leon (VA028G19, CLU 2017-09, UIC 320).

References

Abedini, A., Amiri, H. and Karimi, K. 2020. Efficient biobutanol production from potato peel wastes by separate and simultaneous inhibitors removal and pretreatment. Renew. Energy 160: 269–277.

Abo, B.O., Gao, M., Wang, Y., Wu, C., Wang, Q. and Ma, H. 2019. Production of butanol from biomass: Recent advances and future prospects. Environ. Sci. Pollut. Res. 26: 20164–20182.

Amiri, H. and Karimi, K. 2018. Pretreatment and hydrolysis of lignocellulosic wastes for butanol production: Challenges and perspectives. Bioresour. Technol. 270: 702–721.

Amiri, H. 2020. Recent innovations for reviving the ABE fermentation for production of butanol as a drop-in liquid biofuel. Biofuel Res. J. 28: 1256–1266.

Bao, T., Zhao, J., Li, J., Liu, X. and Yang, S.T. 2019. n-Butanol and ethanol production from cellulose by *Clostridium cellulovorans* overexpressing heterologous aldehyde/alcohol dehydrogenases. Bioresour. Technol. 285: 121316.

Baral, N.R., Slutzky, L., Shah, A., Ezeji, T.C., Cornish, K. and Christy, A. 2016. Acetone-butanol-ethanol fermentation of corn stover: Current production methods, economic viability and commercial use. FEMS Microbiol. Lett. 363: fnw033.

Bellido, C., Lucas, S., González-Benito, G., García-Cubero, M.T. and Coca, M. 2018. Synergistic positive effect of organic acids on the inhibitory effect of phenolic compounds on Acetone-Butanol-Ethanol (ABE). Food Bioprod. Process. 108: 117–125.

Birgen, C., Dürre, P., Preisig, H.A. and Wentzel, A. 2019. Butanol production from lignocellulosic biomass: Revisiting fermentation performance indicators with exploratory data analysis. Biotechnol. Biofuels 12: 167.

Birgen, C., Degnes, K.F., Markussen, S., Wentzel, A. and Sletta, H. 2021. Butanol production from lignocellulosic sugars by *Clostridium beijerinckii* in microbioreactors. Biotechnol. Biofuels 14: 34.

Brito, M. and Martins, F. 2017. Life cycle assessment of butanol production. Fuel 208: 476–482.

Cai, D., Chen, H., Chen, C., Hu, S., Wang, Y., Chang, Z. et al. 2016. Gas stripping-pervaporation hybrid process for energy saving product recovery from acetone-butanol-ethanol (ABE) fermentation broth. Chem. Eng. J. 287: 1–10.

Cai, D., Hu, S., Miao, Q., Chen, C., Chen, H., Zhang, C. et al. 2017. Two-stage pervaporation process for effective *in situ* removal acetone-butanol-ethanol from fermentation broth. Bioresour. Technol. 224: 380–388.

Dong, J.J., Ding, J.C., Zhang, Y., Ma, L., Xu, G.C., Han, R.Z. et al. 2016. Simultaneous saccharification and fermentation of dilute alkaline-pretreated corn stover for enhanced butanol production by *Clostridium saccharobutylicum* DSM 13864. FEMS Microbiol. Lett. 363.

Du, G., Che, J., Wu, Y., Wang, Z., Jiang, Z., Ji, F. and Xue, C. 2021a. Disruption of hydrogenase gene for enhancing butanol selectivity and production in *Clostridium acetobutylicum*. Biochem. Eng. J. 171: 108014.

Du, G., Zhu, C., Xu, M., Wang, L., Yang, S.T. and Xue, C. 2021b. Energy-efficient butanol production by *Clostridium acetobutylicum* with histidine kinase knockouts to improve strain tolerance and process robustness. Green Chem. 23: 2155.

Fonseca, B.C., Schmidell, W. and Reginatto, V. 2018. Impact of glucose concentration on productivity and yield of hydrogen production by the new isolate *Clostridium beijerinckii* Br21. Can. J. Chem. Eng. 97: 1092–1099.

Friedl, A. 2016. Downstream process options for the ABE fermentation. FEMS Microbiol. Lett. 363.

Gao, L., Alberto, M., Gorgojo, P., Szekely, G. and Budd, P.M. 2017. High-flux PIM-1/PVDF thin film composite membranes for 1-butanol/water pervaporation. J. Membr. Sci. 529: 207–214.

González-Peñas, H., Eibes, G., Lu-Chau, T.A., Moreira, M.T. and Lema, J.M. 2020a. Altered Clostridia response in extractive ABE fermentation with solvents of different nature. Biochem. Eng. J. 154: 107455.

González-Peñas, H., Lu-Chau, T.A., Eibes, G. and Lema, J.M. 2020b. Energy requirements and economics of acetone–butanol–ethanol (ABE) extractive fermentation: a solvent-based comparative assessment. Bioprocess. Biosyst. Eng. 43: 2269–2281.

Gottumukkala, L.D., Haigh, K. and Görgens, J. 2017. Trends and advances in conversion of lignocellulosic biomass to biobutanol: Microbes, bioprocesses and industrial viability. Renew. Sust. Energ. Rev. 76: 963–973.

He, C.R., Kuo, Y.Y. and Li, S.Y. 2017. Lignocellulosic butanol production from Napier grass using semi-simultaneous saccharification fermentation. Bioresour. Technol. 231: 101–108.

Hijosa-Valsero, M., Paniagua-García, A.I. and Díez-Antolínez, R. 2017. Biobutanol production from apple pomace: The importance of pretreatment methods on the fermentability of lignocellulosic agro-food wastes. Appl. Microbiol. Biotechnol. 101: 8041–8052.

Hijosa-Valsero, M., Paniagua-García, A.I. and Díez-Antolínez, R. 2018a. Industrial potato peel as a feedstock for biobutanol production. New Biotechnol. 46: 54–60.

Hijosa-Valsero, M., Garita-Cambronero, J., Paniagua-García, A.I. and Díez-Antolínez, R. 2018b. Biobutanol production from coffee silverskin. Microb. Cell. Fact. 17: 154.

Hijosa-Valsero, M., Garita-Cambronero, J., Paniagua-García, A.I. and Diez-Antolínez, R. 2019. Tomato waste from processing industries as a feedstock for biofuel production. Bioenerg. Res. 12: 1000–1011.

Huang, H.J., Ramaswamy, S. and Liu, Y. 2014. Separation and purification of biobutanol during bioconversion of biomass. Sep. Purif. Technol. 132: 513–540.

Huzir, N.M., Maniruzzaman, M., Aziz, A., Ismail, S.B., Abdullah, B., Mahmood, N.A.N. et al. 2018. Agro-industrial waste to biobutanol production: Eco-friendly biofuels for next generation. Renew. Sust. Energ. Rev. 94: 476–485.

Ibrahim, M.F., Kima, S.W. and Abd-Aziz, S. 2018. Advanced bioprocessing strategies for biobutanol production from biomass. Renew. Sust. Energ. Rev. 91: 1192–1204.

Jiang, Y., Liu, J., Jiang, W., Yang, Y. and Yang, S. 2015. Current status and prospects of industrial bio-production of n-butanol in China. Biotechnol. Adv. 33: 1493–1501.

Jiang, Y., Guo, D., Lu, J., Dürre, P., Dong, W., Yan, W. et al. 2018a. Consolidated bioprocessing of butanol production from xylan by a thermophilic and butanologenic *Thermoanaerobacterium* sp. M5. Biotechnol. Biofuels 11: 89.

Jiang, Y., Zhang, T., Lu, J., Dürre, P., Zhang, W., Dong, W. et al. 2018b. Microbial co-culturing systems: Butanol production from organic wastes through consolidated bioprocessing. Appl. Microbiol. Biotechnol. 102: 5419–5425.

Jiang, Y., Lv, Y., Lu, J., Dong, W., Zhou, J., Zhang, W., Xin, F. and Jiang, M. 2020. Consolidated bioprocessing performance of a two-species microbial consortium for butanol production from lignocellulosic biomass. Biotechnol. Bioeng. 117: 2985–2995.

Jiménez-Bonilla, P., Zhang, J., Wang, Y., Blersch, D., de-Bashan, L.E., Guo, L. and Wang, Y. 2020. Enhancing the tolerance of *Clostridium saccharoperbutylacetonicum* to lignocellulosic-biomass-derived inhibitors for efficient biobutanol production by overexpressing efflux pumps genes from Pseudomonas putida. Bioresour. Technol. 312: 123532.

Jin, Q., Qureshi, N., Wang, H. and Huang, H. 2019. Acetone-butanol-ethanol (ABE) fermentation of soluble and hydrolyzed sugars in apple pomace by *Clostridium beijerinckii* P260. Fuel 244: 536–544.

Jönsson, L.J. and Martín, C. 2016. Pretreatment of lignocellulose: Formation of inhibitory by-products and strategies for minimizing their effects. Bioresource Technol. 199: 103–112.

Kattela, S.P., Vysyaraju, R.K.R., Surapaneni, S.R. and Ganji, P.R. 2019. Effect of n-butanol/diesel blends and piston bowl geometry on combustion and emission characteristics of CI engine. Environ. Sci. Pollut. Res. 26: 1661–1674.

Khalili, F. and Amiri, H. 2020. Integrated processes for production of cellulosic and hemicellulosic biobutanol from sweet sorghum bagasse using autohydrolysis. Ind. Crop. Prod. 145: 111918.

Khedkar, M.A., Nimbalkar, P.R., Chavan, P.V., Chendake, Y.J. and Bankar, S.P. 2017. Cauliflower waste utilization for sustainable biobutanol production: Revelation of drying kinetics and bioprocess development. Bioprocess Biosyst. Eng. 40: 1493–1506.

Khedkar, M.A., Nimbalkar, P.R., Gaikwad, S.G., Chavan, P.V. and Bankar, S.P. 2020. Solvent extraction of butanol from synthetic solution and fermentation broth: Batch and continuous studies. Sep. Purif. Technol. 249: 117058.

Kim, J.H., Cook, M., Peeva, L., Yeo, J., Bolton, L.W., Lee, Y.M. and Livingston, A.G. 2020. Low energy intensity production of fuel-grade bio-butanol enabled by membrane-based extraction. Energy Environ. Sci. 13: 4862.

Lamani, V.T., Yadav, A.K. and Gottekere, K.N. 2017. Performance, emission, and combustion characteristics of twin-cylinder common rail diesel engine fuelled with butanol-diesel blends. Environ. Sci. Pollut. Res. 24: 23351–23362.

Lee, S.Y., Park, J.H., Jang, S.H., Nielsen, L.K., Kim, J. and Jung, K.S. 2008. Fermentative butanol production by Clostridia. Biotechnol. Bioeng. 101: 209–228.

Lee, S., Lee, J.H. and Mitchell, R.J. 2015. Analysis of *Clostridium beijerinckii* NCIMB 8052's transcriptional response to ferulic acid and its application to enhance the strain tolerance. Biotechnol. Biofuels 8: 68.

Lee, J.Y., Hwanga, S.O., Kimb, H.J., Hong, D.Y., Lee, J.S. and Lee, J.H. 2019. Hydrosilylation-based UV-curable polydimethylsiloxane pervaporation membranes for n-butanol recovery. Sep. Purif. Technol. 209: 383–391.

Levasseur, A., Bahn, O., Beloin-Saint-Pierre, D., Marinova, M. and Vaillancourt, K. 2017. Assessing butanol from integrated forest biorefinery: A combine techno-economic and life cycle approach. Appl. Energy 198: 440–452.

Li, J.W., Wang, L. and Chen, H.Z. 2016. Periodic peristalsis increasing acetone-butanol-ethanol productivity during simultaneous saccharification and fermentation of steam-exploded corn straw. J. Biosci. Bioeng. 122: 620–626.

Li, S., Huang, L., Ke, C., Pang, Z. and Liu, L. 2020a. Pathway dissection, regulation, engineering and application: Lessons learned from biobutanol production by solventogenic clostridia. Biotechnol. Biofuels 13: 39.

Li, W., Li, J., Wang, N., Li, X., Zhang, Y., Ye, Q. et al. 2020b. Recovery of bio-butanol from aqueous solution with ZIF-8 modified graphene oxide composite membrane. J. Membr. Sci. 598: 117671.

Liao, Z., Zhanga, Y., Luoa, S., Suoa, Y., Zhanga, S. and Wan, J. 2017. Improving cellular robustness and butanol titers of *Clostridium acetobutylicum* ATCC824 by introducing heat shock proteins from an extremophilic bacterium. J. Biotechnol. 252: 1–10.

Liao, Z., Guo, X., Hu, J., Suo, Y., Fu, H. and Wang, J. 2019. The significance of proline on lignocellulose-derived inhibitors tolerance in *Clostridium acetobutylicum* ATCC. Bioresour. Technol. 272: 561–569.

Lin, P.P., Mi, L., Morioka, A.H., Yoshino, K.M., Konishi, S., Xu, S.C. et al. 2015. Consolidated bioprocessing of cellulose to isobutanol using *Clostridium thermocellum*. Metab. Eng. 31: 44–52.

Liu, H., Wang, X., Zheng, Z., Gu, J., Wang, H. and Yao, M. 2014. Experimental and simulation investigation of the combustion characteristics and emissions using n-butanol/biodiesel dual-fuel injection on a diesel engine. Energy 74: 741–752.

Liu, J., Guo, T., Yang, T., Xu, J., Tang, C., Liu, D. and Ying, H. 2017. Transcriptome analysis of *Clostridium beijerinckii* adaptation mechanisms in response to ferulic acid. Int. J. Biochem. Cell. B 86: 14–21.

López-Linares, J.C., García-Cubero, M.T., Coca, M. and Lucas, S. 2020. Integral valorization of cellulosic and hemicellulosic sugars for biobutanol production: ABE fermentation of the whole slurry from microwave pretreated brewer's spent grain. Biomass Bioenerg. 135: 105524.

López-Linares, J.C., García-Cubero, M.T., Coca, M. and Lucas, S. 2021. Efficient biobutanol production by acetone-butanol-ethanol fermentation from spent coffee grounds with microwave assisted dilute sulfuric acid pretreatment. Bioresour. Technol. 320: 124348.

Lu, C., Dong, J. and Yang, S.T. 2013. Butanol production from wood pulping hydrolysate in an integrated fermentation–gas stripping process. Bioresour. Technol. 143: 467–475.

Luo, H., Zheng, P., Bilal, M., Xie, F., Zeng, Q., Zhu, C. et al. 2020. Efficient bio-butanol production from lignocellulosic waste by elucidating the mechanisms of *Clostridium acetobutylicum* response to phenolic inhibitors. Sci. Total Environ. 710: 136399.

Maiti, S., Gallastegui, G., Sarma, S.K., Brar, S.K., Bihan, Y.L., Drogui, P. et al. 2016. A re-look at the biochemical strategies to enhance butanol production. Biomass Bioenerg. 94: 187–200.

Nimbalkar, P.R., Khedkar, M.A., Chavan, P.V. and Bankar, S.P. 2018. Biobutanol production using pea pod waste as substrate: Impact of drying on saccharification and fermentation. Renew. Energy 117: 520–529.

Okonkwo, C.C., Ujor, V. and Ezeji, T.C. 2019. Chromosomal integration of aldo-keto-reductase and short-chain dehydrogenase/reductase genes in *Clostridium beijerinckii* NCIMB 8052 enhanced tolerance to lignocellulose-derived microbial inhibitory compounds. Sci. Rep. 9: 7634.

Outram, V., Lalander, C.A., Lee, J.G.M., Davis, E.T. and Harvey, A.P. 2016. A comparison of the energy use of *in situ* product recovery techniques for the Acetone Butanol Ethanol fermentation. Bioresour. Technol. 220: 590–600.

Patakova, P., Koleka, J., Sedlarb, K., Koscovab, P., Branskaa, B., Kupkovab, K. et al. 2018. Comparative analysis of high butanol tolerance and production in clostridia. Biotechnol. Adv. 36: 721–738.

Pereira, L.G., Dias, M.O.S., Amriano, A.P., Maciel Filho, R. and Bonomi, A. 2015. Economic and environmental assessment of n-butanol production in an integrated first and second generation sugarcane biorefinery: Fermentative versus catalytic routes. Appl. Energy 160: 120–131.

Plaza, P.E., Coca, M., Lucas, S., Fernández-Delgado, M., López-Linares, J.C. and García-Cubero, M.T. 2020. Efficient use of brewer's spent grain hydrolysates in ABE fermentation by *Clostridium beijerinkii*. Effect of high solid loads in the enzymatic hydrolysis. J. Chem. Technol. Biotechnol. 95: 2393–2402.

Procentese, A., Raganati, F., Olivieri, G., Russo, M.E. and Marzocchella, A. 2017. Pre-treatment and enzymatic hydrolysis of lettuce residues as feedstock for bio-butanol production. Biomass Bioenerg. 96: 172–179.

Qi, G., Huang, D., Wang, J., Shen, Y. and Gao, X. 2019. Enhanced butanol production from ammonium sulfite pretreated wheat straw by separate hydrolysis and fermentation and simultaneous saccharification and fermentation. Sustain. Energy Technol. Assess. 36: 100549.

Qureshi, N., Saha, B.C., Hector, R.E., Dien, B., Hughes, S., Liu, S. et al. 2010a. Production of butanol (a biofuel) from agricultural residues: Part II—Use of corn stover and switchgrass hydrolysates. Biomass Bioenerg. 34: 566–571.

Qureshi, N., Saha, B.C., Dien, B., Hector, R.E. and Cotta, M.A. 2010b. Production of butanol (a biofuel) from agricultural residues: Part I—Use of barley straw hydrolysate. Biomass Bioenerg. 34: 559–565.

Qureshi, N., Singh, V., Liu, S., Ezeji, T.C., Saha, B.C. and Cotta, M.A. 2014a. Process integration for simultaneous saccharification, fermentation, and recovery (SSFR): Production of butanol from corn stover using Clostridium beijerinckii P260. Bioresour. Technol. 154: 222–228.

Qureshi, N., Cotta, M.A. and Saha, S.C. 2014b. Bioconversion of barley straw and corn stover to butanol (a biofuel) in integrated fermentation and simultaneous product recovery bioreactors. Food Bioprod. Process. 92: 295–308.

Qureshi, N., Friedl, A. and Maddox, I.S. 2014c. Butanol production from concentrated lactose/whey permeate: Use of pervaporation membrane to recover and concentrate product. Appl. Microbiol. Biotechnol. 98: 9859–9867.

Qureshi, N., Lin, X., Liu, S., Saha, B.C., Mariano, A.P., Polaina, J. et al. 2020. Global view of biofuel butanol and economics of its production by fermentation from sweet sorghum bagasse, food waste, and yellow top presscake: Application of novel technologies. Fermentation 6: 58.

Rdzanek, P., Marszałek, J. and Kamiński, W. 2018. Biobutanol concentration by pervaporation using supported ionic liquid membranes. Sep. Purif. Technol. 196: 124–131.

Rochón, E., Ferrari, M.D. and Lareo, C. 2017. Integrated ABE fermentation-gas stripping process for enhanced butanol production from sugarcane-sweet sorghum juices. Biomass Bioenerg. 98: 153–160.

Rom, A. and Friedl, A. 2016. Investigation of pervaporation performance of POMS membrane during separation of butanol from water and the effect of added acetone and ethanol. Sep. Purif. Technol. 170: 40–48.

Sadh, P.K., Duhan, S. and Duhan, J.S. 2018. Agro-industrial wastes and their utilization using solid state fermentation: A review. Bioresour. Bioprocess. 5: 1.

Sasaki, C., Kushiki, Y., Asada, C. and Nakamura, Y. 2014. Acetone–butanol–ethanol production by separate hydrolysis and fermentation (SHF) and simultaneous saccharification and fermentation (SSF) methods using acorns and wood chips of *Quercus acutissima* as a carbon source. Ind. Crops Prod. 62: 286–292.

Seifollahi, M. and Amiri, H. 2020. Enhanced production of cellulosic butanol by simultaneous co-saccharification and fermentation of water-soluble cellulose oligomers obtained by chemical analysis. Fuel 263: 116759.

Seo, S.O., Lu, T., Jin, Y.S. and Blaschek, H.P. 2021. A comparative phenotypic and genomic analysis of *Clostridium beijerinckii* mutant with enhanced solvent production. J. Biotechnol. 329: 49–55.

Tanaka, S., Tashiro, Y., Kobayashi, G., Ikegami, T., Negishi, H. and Sakaki, K. 2012. Membrane-assisted extractive butanol fermentation by *Clostridium saccharoperbutylacetonicum* N1-4 with 1 dodecanol as the extractant. Bioresour. Technol. 116: 448–452.

Technavio's market research analysis "Global Bio-based Butanol Market 2017–2021". www.technavio.com/report/global-bio-based-butanol-market. Last accessed 11 August 2021.

The European Technology and Innovation Platform Bioenergy "Societal benefits of biofuels". www.etipbioenergy.eu. Last accessed 18 May 2021.

Tiang, L., Conway, P.M., Cervenka, N.D., Cui, J., Maloney, M., Olson, D.G. and Lynd, L.R. 2019. Metabolic engineering of *Clostridium thermocellum* for *n*-butanol production from cellulose. Biotechnol. Biofuels 12: 186.

Tomita, H., Okazaki, F. and Tamaru, Y. 2019. Direct IBE fermentation from mandarin orange wastes by combination of *Clostridium cellulovorans* and *Clostridium beijerinckii*. AMB Express 9: 1.

Tri, C.L. and Kamei, I. 2020. Butanol production from cellulosic material by anaerobic co-culture of white-rot fungus *Phlebia* and bacterium *Clostridium* in consolidated bioprocessing. Bioresour. Technol. 305: 123065.

Valles, A., Álvarez-Hornos, F.J., Martínez-Soria, V., Marzal, P. and Gabaldón, C. 2020. Comparison of simultaneous saccharification and fermentation and separate hydrolysis and fermentation processes for butanol production from rice straw. Fuel 282: 118831.

Veza, I., Said, M.F.M. and Latiff, Z.A. 2021. Recent advances in butanol production by acetone-butanol-ethanol (ABE) fermentation. Biomass Bioenerg. 144: 105919.

Wang, S., Donga, S. and Wang, Y. 2017a. Enhancement of solvent production by overexpressing key genes of the acetone-butanol-ethanol fermentation pathway in *Clostridium saccharoperbutylacetonicum* N1-4. Bioresour. Technol. 245: 426–433.

Wang, S., Dong, S., Wang, P., Tao, Y. and Wan, Y. 2017b. Genome Editing in *Clostridium saccharoperbutylacetonicum* N1-4 with the CRISPR-Cas9 System. Appl. Environ. Microbiol. 83: e00233–17.

Wen, Z., Minton, N.P., Zhang, Y., Li, Q., Liu, J., Jiang, Y. and Yang, S. 2017. Enhanced solvent production by metabolic engineering of a twin-clostridial consortium. Metab. Eng. 39: 38–48.

Wen, H., Gao, H., Zhang, T., Wu, Z., Gong, P., Li, Z., Chen, H. et al. 2018. Hybrid pervaporation and salting-out for effective acetone-butanol-ethanol separation from fermentation broth. Bioresour. Technol. Rep. 2: 45–52.

Wen, Z., Ledesma-Amaro, R., Lin, J., Jiang, Y. and Yang, S. 2019. Improved *n*-butanol production from *Clostridium cellulovorans* by integrated metabolic and evolutionary engineering. Appl. Environ. Microbiol. 85: e02560–18.

Wen, Z., Li, Q., Liu, J., Jin, M. and Yang, S. 2020a. Consolidated bioprocessing for butanol production of cellulolytic Clostridia: development and optimization. Microb. Biotechnol. 13: 410–422.

Wen, Z., Ledesma-Amaro, R., Lu, M., Jin, M. and Yang, S. 2020b. Metabolic engineering of *Clostridium cellulovorans* to improve butanol production by consolidated bioprocessing. ACS Synth. Biol. 9: 304–315.

Wen, Z., Ledesma-Amaro, R., Lu, M., Jiang, Y., Gao, S., Jin, M. and Yang, S. 2020c. Combined evolutionary engineering and genetic manipulation improve low pH tolerance and butanol production in a synthetic microbial *Clostridium* community. Biotechnol. Bioeng. 117: 2008–2022.

Willson, B.J., Kovacs, K., Wilding-Steele, T., Markus, R., Winzer, K. and Minton, N.P. 2016. Production of a functional cell wall-anchored minicellulosome by recombinant *Clostridium acetobutylicum* ATCC 824. Biotechnol. Biofuels 9: 109.

Wu, Y., Wang, Z., Ma, X. and Xue, C. 2021. High temperature simultaneous saccharification and fermentation of corn stover for efficient butanol production by a thermotolerant *Clostridium acetobutylicum*. Process Biochem. 100: 20–25.

Xu, M., Zhao, J.B., Yu, L., Tang, I.C., Xue, C. and Yang, S.T. 2015. Engineering *Clostridium acetobutylicum* with a histidine kinase knockout for enhanced n-butanol tolerance and production. Appl. Microbiol. Biotechnol. 99: 1011–1022.

Xue, C., Zhao, J., Lu, C., Yang, S.T., Bai, F. and Tang, I.C. 2012. High-Titer n-butanol production by *Clostridium acetobutylicum* JB200 in fed-batch fermentation with intermittent gas stripping. Biotechnol. Bioeng. 109: 2746–2756.

Xue, C., Zhao, J., Liu, F., Lu, C., Yang, S.T. and Bai, F. 2013. Two-stage in situ gas stripping for enhanced butanol fermentation and energy-saving product recovery. Bioresour. Technol. 135: 396–402.

Xue, C., Du, G.Q., Sun, J.X., Chen, L.J., Gao, S.S., Yu, M.L. et al. 2014. Characterization of gas stripping and its integration with acetone–butanol–ethanol fermentation for high-efficient butanol production and recovery. Biochem. Eng. J. 83: 55–61.

Xue, C., Liu, F., Xu, M., Zhao, J., Chen, L., Ren, J. et al. 2016a. A novel *in situ* gas stripping-pervaporation process integrated with acetone-butanol-ethanol fermentation for hyper n-butanol production. Biotechnol. Bioeng. 133: 120–129.

Xue, C., Liu, F., Xu, M., Tang, I.C., Zhao, J., Bai, F. and Yang, S.T. 2016b. Butanol production in acetone-butanol-ethanol fermentation with *in situ* product recovery by adsorption. Bioresour. Technol. 219: 158–168.

Xue, C., Zhao, J., Chen, L., Yang, S.T. and Fengwu, B. 2017. Recent advances and state-of-the-art strategies in strain and process engineering for biobutanol production by *Clostridium acetobutylicum*. Biotechnol. Adv. 35: 310–322.

Yang, X., Xu, M. and Yang, S.T. 2015. Metabolic and process engineering of *Clostridium cellulovorans* for biofuel production from cellulose. Metab. Eng. 32: 39–48.

Zetty-Arenas, A.M., Ferraz Alves, R., Freixo Portela, C.A., Pinto Mariano, A., Olitta Basso, T., Plazas Tovar, L. et al. 2019. Towards enhanced n-butanol production from sugarcane bagasse hemicellulosic hydrolysate: Strain screening, and the effects of sugar concentration and butanol tolerance. Biomass Bioenerg. 126: 190–198.

CHAPTER 10

Microbiology of Biodiesel Production

Juan Jáuregui Rincón,[1,]* *David Chaos Hernández,*[2]
Hilda Elizabeth Reynel Avila[2] and *Didilia Ileana Mendoza Castillo*[2]

1. Introduction

Currently, petroleum hydrocarbons continue to be the main source of energy worldwide, even though renewable sources have increased in the last 20 years. The production and consumption of biofuels has increased in many developed and developing countries, but the main limitation is the type and cost of the raw material (oils and fats). Therefore, it is necessary to look for other alternative sources of raw feedstock for their production (Patel et al. 2020, Younes et al. 2020).

Among the most widely produced liquid biofuels are bioethanol and biodiesel, biodiesel being produced from vegetable oils and animal fats. Annual global biodiesel production in 2019 was: Indonesia 7.9, United States 6.5, Brazil 5.9, Germany 3.8, France 3, Argentina 2.5, Netherlands 2.1, Spain 2.0, Thailand 1.7, Malaysia 1.6, Italy 1, Poland 1 billion liters (Statista, Global biodiesel production by country, 2019). During the last years, new sources of triglycerides (TG) have been sought that are inexpensive and easy to obtain, since the use of vegetable oils for food consumption depends on climatic and environmental conditions, as well as the availability of soil for cultivation.

An alternative raw material is oleaginous microorganisms. Oleaginous microorganisms comprise bacteria, fungi, yeasts, microalgae, and cyanobacteria, which can produce lipid concentrations above 20% (w/w); most of them require inexpensive culture media and easy-to-establish production conditions. Microbiologically, lipid-producing oleaginous microorganisms can be classified into three different groups: microalgae, fungi (molds and yeasts) and bacteria (Ma et al. 2018). Among them, bacteria are less capable of producing lipids as they can only synthesize specific lipids and polyunsaturated fatty acids (PUFA). Therefore, microalgae and fungi have been considered the main lipid producers. While yeasts, fungi and microalgae can synthesize TG, which are similar in composition to vegetable oils, prokaryotic bacteria can synthesize specific lipids.

---

[1] Universidad Autónoma de Aguascalientes, Av. Universidad # 940 Fracc. Ciudad Universitaria C.P. 20100 Aguascalientes, Ags. México.
[2] Instituto Tecnológico de Aguascalientes, Av. Adolfo López Mateos Ote. 1801, Bona Gens, 20256 Aguascalientes, Ags., Mexico.
* Corresponding author: jjaureg@correo.uaa.mx

2. Oleaginous Bacteria

Microbial oil, also known as single cell oil (i.e., yeast, fungi, algae, and bacteria) has attracted great attention in the last years because these organisms can produce and store significant amounts of lipophilic compounds and the fatty acids show a profile similar to that of vegetable oils commonly used as feedstock in biodiesel production (Kumar et al. 2015, 2017). A recent interest in the less explored bacterial lipids has emerged (Behera et al. 2019). Only few bacteria belonging to the actinomycetes group such as *Streptomyces*, *Nocardia*, *Rhodococcus*, *Mycobacterium*, *Dietzia* and *Gordonia* have been identified to accumulate lipids and TG under conditions limited of nitrogen (Behera et al. 2019). Fig. 1 shows the metabolic pathway of bacteria to produce lipids. As can be seen, lipidic production in biomass cells involves complex biochemical reactions (Qadeer et al. 2017). This lipidic production is carried out when a nutrient deficiency exists, specifically of nitrogen, and then the metabolism shifts to the production of lipids for the storage of energy (Quadeer et al. 2017). The main process to produce lipids is carried out via acetyl Co-A formation. These oleaginous compounds are accumulated in the cytosol as small droplets (Quadeer et al. 2017) and/or in the membrane (extracellular) (Silva et al. 2021).

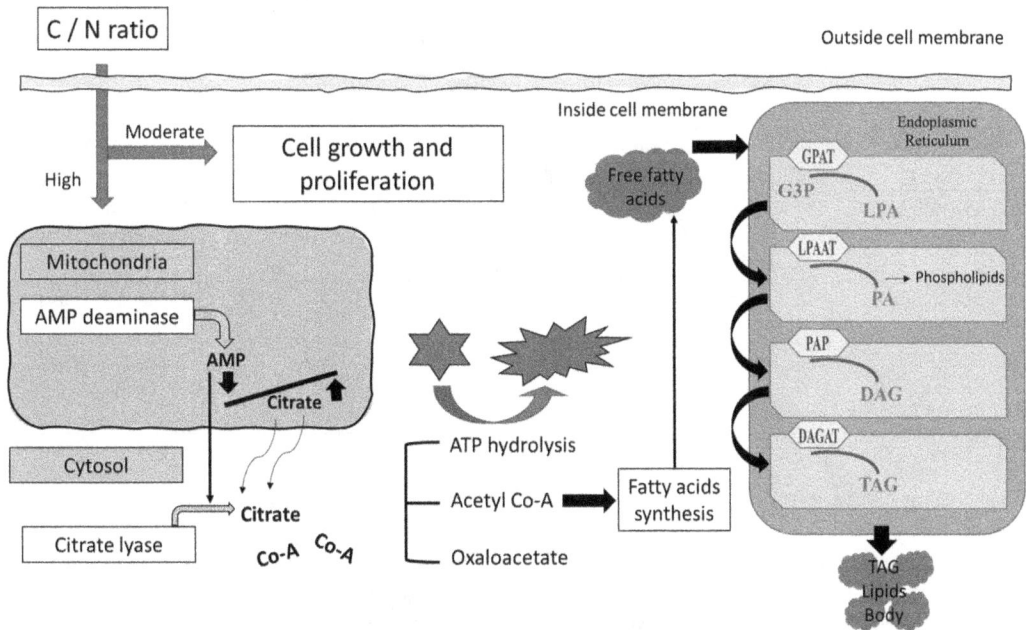

Fig. 1. Mechanism of lipid synthesis in bacteria. AMP: adenosine monophosphate, ATP: adenosine 5'-triphosphate, G3P: Glycerol-3-phosphate, GPAT: Glycerol-3-phosphate O-acyltransferase, LPA: Lysophosphatidic acid, LPAAT: Lysophosphatidic acid acyltransferase, PA: Phosphatidic acid, PAP: Phosphatidate phosphatase, DAG: Diacylglyceride, DAGAT: Diacylglyceryl hydroxy-methyl-trimethyl-β-alanine, TAG: Triacylglyceride (Adapted from De Bhowmick et al. 2015, Qadeer et al. 2017).

Several carbon sources have been tested as culture of bacteria such as glucose, dextrose, and lactose (Kumar et al. 2015, Behera et al. 2019). However, it is well known that the cost of this carbon source represents about 60% of the total cost of conventional fermentation (Fei et al. 2015). In this sense, waste lignocellulosic biomass can be transformed into fermentable sugars, becoming abundant and low-cost carbon sources that microorganisms can utilize (Fei et al. 2015). Also, the use of wastewaters such as urban (Kumar and Takur 2018) and milk processing (Cea et al. 2015, Behera et al. 2019) has received increasing attention in order to explore new alternatives to minimize costs for biofuel production in relation to carbon sources for bacteria. Table 1 shows some of the

Table 1. Oleaginous bacteria.

Bacteria	Sustrate	Lipid Concentration	Reference
Serratia sp. ISTD04	Sodium bicarbonate	0.647 mg/g	Bharti et al. (2014a)
Serratia sp. ISTD04	CO_2 contained in a solution of sodium bicarbonate with 0.2% of glucose and 0.05% of yeast extract	466 mg/L	Bharti et al. (2014b)
Acinetobacter sp. V4, *Pseudomonas* sp. T2 and sp. T15, *Bacillus* sp. V10	Milk processing wastewater supplemented with glucose	*Acinetobacter* sp. V4 – 3.0% *Pseudomonas* sp. T2 – 3.6% *Pseudomonas* sp. T15 – 4.8% *Bacillus* sp. V10 – 7.4%	Cea et al. (2015)
Rhodococcus opacus	Dextrose Dairy wastewater Mineral-enhanced dairy wastewater	71% 14.28% 31.5%	Kumar et al. (2015)
Serratia sp. ISTD04	1% Glucose 1% Glycerol 1% Mollasse Municipal sewage sludge (25 g/L)	66.7% 56.4% 53.7% 12.1%	Kumar and Thakur (2018)
Bacillus sp. SS105	$NaHCO_3$ + 1% Glucose CO_2	38.8% 29.7%	Maheshwari et al. (2018)
DS-7 isolate	Glucose Lactose Dairy wastewater	83.3% 90.0% 72.0%	Behera et al. (2019)
Rhodococcus opacus	Refinery wastewater containing hydrocarbon	18.0	Paul et al. (2019)
Lentibacillus salarius NS12IITR	Total reducing sugar of all food-waste hydrolysate as glucose substitute	0.70 ± 0.03 g/L	Singh and Choudhury (2019)
Serratia sp. ISTD04	$NaHCO_3$ 0.1% Glucose	----	Kumar and Thakur (2020)
Firmicutes Bacteroidia Bacilli	Waste food and seed sludge	0.91 g/g	Zhang et al. (2020b)

studies about lipid extraction from bacteria with potential to be used as feedstock in biodiesel production.

As example, Bharti et al. (2014a) reported the use of *Serratia* sp. ISTD04 bacteria for the extraction of lipids and hydrocarbons. For this purpose, rocks of palaeoproterozoic metasediments dissolved in distillated water were employed as inoculum. To enrich the bacteria, sodium bicarbonate was used as a carbon source. For the extraction of lipids and hydrocarbons, bacteria sample was mixed with a chloroform/methanol solution and shaken. The organic phase was separated from lipid phase, which was concentrated on a rotary evaporator. In addition, the obtained lipids were employed to obtain fatty acid methyl esters (FAMEs). The results showed 0.647 and 0.487 mg/g content of lipids and hydrocarbons, respectively. They also showed a conversion of bacterial lipid to FAMEs of 96%.

In other research reported by Bharti et al. (2014b), they obtained the *Serratia* sp. ISTD04 bacteria from marble rocks of the palaeoproterozoic metasediments. In order to enrich the bacteria, a solution of sodium bicarbonate with 0.2% of glucose and 0.05% of yeast extract was employed. Also, to extract the extracellular lipids, the bacterial samples were mixed with chloroform/methanol solution and shaken vigorously during 30 min. Organic phase was separated from lipid phase. Lipids extracted were used to obtain FAMEs via transesterification under the next conditions: 40°C,

1% of catalyst concentration, lipid/methanol molar ratio of 1:6 and 350 rpm during 3 h. The results showed a lipid content of 466 mg/L and a maximum conversion of 94% of FAMEs.

Kumar et al. (2015) valorized the dairy wastewater in the production of lipids by *Rhodococcus opacus*. First, a synthetic media of dextrose was used as carbon source achieving a lipid percentage of 71% (w/w). As alternative carbon source, the raw dairy wastewater was tested, reaching a 14.28% (w/w) of lipids. However, when this media was enriched with mineral salts, the lipid percentage increased to 33%. Nuclear magnetic resonance spectroscopy showed that bacteria accumulated lipids containing more saturated fatty acids than unsaturated fatty acids, with a biodiesel profile of methyl palmitate (34.90%), methyl stearate (35.48%), methyl myristate (29.79%), methyl linoleate (27.87%), and methyl palmitate (25.85) as the main esters.

In a study reported by Cea et al. (2015), the use of sewage sludge as a feedstock of microbial oils to obtain transesterifiable lipids was studied. The main bacteria detected were *Acinetobacter* sp. V4, *Pseudomonas* sp. T2 and sp. T15 and *bacillus* sp. V10. As carbon source, they employed milk processing wastewater enriched with glucose. In order to extract the lipids from the sewage sludge, chloroform and methanol were used. In addition, to identify the transesterifiable lipids, the extracted lipids were converted in methyl esters via esterification. According to the results, *Bacillus* sp. V10 bacteria obtained the highest transesterifiable lipid content, which corresponded to 7.4%. Finally, authors concluded that sewage sludge from wastewater plants can be used as a feedstock of oils for biodiesel production.

The use of *Bacillus* sp. SS105 to produce microbial lipids and their conversion to biodiesel was tested (Maheshwari et al. 2018). For this purpose, *Bacillus* sp. SS105 was cultivated and supplemented with CO_2 and $NHCO_3$ with glucose, as carbon sources. For lipid extraction, the samples were centrifugated at 10,000 rpm during 10 min, then the biomass obtained was lysed using sonication. After, the biomasses obtained were mixed with a solution of chloroform/methanol. With the obtained lipids, a transesterification reaction in presence of NaOH as catalyst was studied to produce fatty acid alkyl esters. The results showed a lipid content of 38.80 and 29.65% for $NHCO_3$ with glucose and CO_2, respectively, and a production of 120 mg/L of biodiesel. Authors concluded that *Bacillus* sp. SS105 can be used to obtain lipids and to produce biodiesel.

Singh and Choudhury (2019) reported the employment of food-wastes such as rice, wheat bran, orange, and mango peels to obtain *Lentibacillus salarius* NS12IITR bacteria through fermentation. This halophilic bacterium was cultured in a medium that contained sodium chloride, yeast extract, potassium chloride, magnesium sulfate, potassium dihydrogen phosphate and ferrous sulfate. Total reducing sugar of all food-waste hydrolysate was used as glucose substitute for carbon source. For lipid extraction, the food-waste hydrolysates were centrifugated and separated from solvent. In addition, the obtained microbial lipids were employed in transesterification reaction to obtain FAMEs. Results showed a lipid content of 0.70 ± 0.029 g/L and 81 ± 4.72% of biodiesel produced using these microbial lipids.

On the other hand, the production of bio-oil from oleaginous microbial biomass using hydrothermal liquefaction (HTL) has been less studied (Paul et al. 2019). In this study, *Rhodococcus opacus* bacteria was used to treat refinery wastewater (containing hydrocarbons) due to its capability to degrade complex compounds. Among several reactor configurations, the continuous cell recycle tubular membrane was the most efficient, reducing the chemical oxygen demand in 99% with a lipid production of 86% (w/w). Then, the residual bacteria were treated through HTL, the produced oil was extracted with dichloromethane, achieving 18% of oil with suitable biofuel properties.

Kumar and Thakur (2020) studied the use of *Serratia* sp. ISTD04 to extract microbial lipids using the methodology reported by Bharti et al. (2014) where the bacterial samples were mixed with chloroform/methanol solution, shaken for 30 min and the organic phase was separated from lipid phase. To culture the bacteria, a carbon source consisting of a solution of $NaHCO_3$ with glucose was used for feeding. In addition, to determine the FAMEs obtained from the transesterification of bacterial oils, an immobilized lipase was used as heterogeneous catalyst. The reaction conditions were oil/methanol molar ratio of 1:6 and 100 mg of catalyst. The results indicated that a 97.41% of FAMEs was obtained. Authors concluded that the use of this bacteria is suitable to obtain lipids and to produce biodiesel.

As can be seen, the study of bacterial oil in biofuels' productions is an interesting and wide topic with a clear progress where efforts must be addressed in order to achieve an industrial and efficient process using these microorganisms.

3. Oleaginous Yeasts

Currently, the production of biodiesel from oil obtained from oleaginous yeasts is still under development and the production process needs to be optimized to make it competitive when compared to other oil sources. Of the total population of yeasts, only 5% can accumulate more than 25% lipids (Ageitos et al. 2011).

In the last 20 years, many works have been reported. Table 2 summarizes the microorganisms that in recent years have been isolated and identified, as well as the concentration of lipids they produce when cultivated in different culture media. These yeasts are characterized for being cheap, which reduces production costs, but are also capable of using new sources of raw material such as crude glycerol obtained from biodiesel production, wastewater, hydrolyzed ligninolytic residues (sugarcane bagasse and corn residues), hydrolyzed banana peel, among others, to produce biodiesel.

Within the group of fungi, there are six genus of yeasts that are the most studied: genus *Rhodotorula* (Ayadi et al. 2018, Maza et al. 2020, Miao et al. 2020, Zhang et al. 2020, Chmielarz et al. 2021, Maza et al. 2021, Ngamsirisomsakul et al. 2021); genus *Cryptococcus* (Han et al. 2019, Annamalai et al. 2020, Kamal et al. 2020, Younes et al. 2020, Qian et al. 2021); genus *Rhodosporidium* (Dai et al. 2019, Li et al. 2020, Younes et al. 2020, Carmona-Cabello et al. 2021); genus *Trichosporon* (Li et al. 2017, Hu et al. 2018, Qian et al. 2021); genus *Cutaneotrichosporon* (Tang et al. 2020, Younes et al. 2020) and genus *Lipomyces* (Wild et al. 2010, Sutanto et al. 2018, Takaku et al. 2020).

Some species of the genus *Rhodotorula* have been identified as oleaginous yeasts, and among those that produce higher percentages of lipids are: *Rhodotorula glutinis*, which uses hydrolysates from sugarcane bagasse as a source of carbon and energy, which are only used as fuel. The percentage of lipids obtained from *R. glutinis* cells was 37.8%. Maza et al. (2020) published the isolation of *R. glutinis* R4 strain from Antarctic soil. The yeast was grown on glucose, malt and yeast extract medium (GMY) with a glucose concentration of 40 g/L and showed a cell concentration of 14 g/L, a μ_{max} of 0.092 1/h and a lipid content of 47% w/w. Furthermore, it was found that the oil produced by *R. glutinis* R4 was similar to vegetable oil, with 61% oleic acid, which makes it suitable for producing biodiesel. Also, Maza et al. (2021) published that *R. glutinis* T13 yeast grown in GMY culture medium, which contained glucose as carbon and energy source and low level of nitrogen source, produced cells with 40% lipids and a concentration of 6.3 g/L after 144 h of culture and the lipid profile showed that only 85% were TAG.

Recently, mixed culture of microalgae and yeast with the aim of improving the economics of microbial lipid production has shown attention. Liu et al. (2018) reported the mixed culture of *Chlorella pyrenoidosa* and *R. glutinis*, where the microalgae: yeast ratio of 3:1 had the highest biomass concentration and lipid productivity $395 \pm 12.5 \times 10^5$ cell/mL and 90 ± 3 mg/L/day, respectively.

Another well-studied yeast genus is *Cryptococcus*, among which *C. curvatus* ATCC 20509 stands out, showing that it can grow and produce high lipid concentrations of 29.0 to 48.8% when grown in culture media consisting of mixtures of amino acids from wastewater from the meat industry; these types of substrates are economical and high lipid percentages can be obtained (Kamal et al. 2020).

Table 2. Oleaginous yeast/fungi.

Yeast/fungi	Substrate	Lipid Concentration	Reference
Kazachstania unispora	Deproteinized potato wastewater	6.3 g/L	Gientka et al. (2017)
Trichosporon fermentans *T. cutaneum*	Crude glycerol	32.29%	Liu et al. (2017)
Rhodotorula babjevae YS-L7	Wheat bran hydrolysate	39.17%	Ayadi et al. (2018)
Trichosporon cutaneum	Phenolic aldehydes	0.85 g/L	Hu et al. (2018)
Chlorella pyrenoidosa and *Rhodotorula glutanis*	Medium GMY (glucose)	21% y 2.4 g/L	Lu et al. (2018)
Lipomyces starkeyi	Glucose, molasses, ethanol, sweet whey permeates, sewage sludge, oleic, myristic, palmitic acid	60–80%	Sutanto et al. (2018)
Rhodosporidium toruloides	Corn stover	6.2 g/L	Dai et al. (2019)
Cryptococcus sp	Banana peel hydrolysates	34%	Han et al. (2019)
Cryptococcus curvatus	Volatile fatty acids derived from wastepaper	38.6%	Annamalai et al. (2020)
Aspergillus ochraceus	Acid hydrolyzate from pongamia seeds	28.93%	Jathanna et al. (2020)
Cryptococcus curvatus ATCC 20509	Amino acid-rich wastes	48.8–44.5%	Kamal et al. (2020)
Rhodosporidium toruloides CGMCC 2.1389	Amino acid wastes	28.7%	Li et al. (2020)
Yarrowia lipolytica and *Bacillus subtilis*	Palm oil industrial wastes	54.21%	Louhasakul et al. (2020)
Rhodotorula glutinis R4	Glucose	47%	Maza et al. (2020)
Rhodotorula taiwanewnsis AM2352	Corn cob hydrolyzate	55.8 g/Kg corn cob	Miao et al. (2020)
Cryptococcus curvatus	Waste office paper	4.95 g/L	Nair et al. (2020)
Naganishia albida	Agricultural waste (onion waste) and industrial waste (crude glycerol)	34%	Sathiyamoorthi et al. (2020)
Cutaneotrichosporon oleaginosum	Deproteinized potato wastewater	10.1 g/L	Tang et al. (2020)
Cutaneotrichosporon oleaginosus, Cryptococcus curvatus and *Rhodosporidium toruloides*	*Scenedesmus obtusiusculus* hydrolyzate	35%	Younes et al. (2020)
Yarrowia lipolytica modificada genéticamente y *S. cervisieae*	Complex synthetic medium (2% glucose)	4.8 mg/L	Yu et al. (2020)
Rhodotorula glutinis	Wastewater from ethanol production	2.18 g/L	Zhang et al. (2020b)
Pichia cactophila, P. fermentans, P. anomala, Rhodotorula mucilaginosa, R. dairenensis	Crude glycerol	61%	Berikten et al. (2021)
Rhodosporidium toruloides Y-27012	Good waste discarded	26.7 g/L	Carmona-Cabello et al. (2021)
Rhodotorula toruloides CB514	Crude glycerol and hemicellulose hydrolyzate	10.6 g/L	Chmielarz et al. (2021)
Rhodotorula glutinis	Glucose	40%	Maza et al. (2021)
Rhodotorula glutinis	Cane bagasse hydrolyzate	37.8%	Ngamsirisomsakul et al. (2021)
Apiotrichum porosum DSM27194	Volatile fatty acids from organic waste	36.2%	Qian et al. (2021)

Regarding the genus *Rhodosporidium*, Xu et al. (2017) reported that yeasts of this genus have been an important candidate in the last two decades, since some strains present a high capacity to accumulate lipids using diverse substrates, in addition to the fact that they are capable of producing some carotenoids.

In 2019, Dai et al. (2019) reported that it is possible to improve lipid production from corn stover using a novel process called pre-hydrolysis followed by simultaneous saccharification and lipid production. Results showed that pre-hydrolysis at 50°C and then lipid production process at 30°C improved lipid yield by more than 17.0%.

Wen et al. (2020) have reported that the red yeast *Rhodosporidium toruloides* naturally produces lipids and carotenoids using very diverse substrates such as carbohydrates: pentoses, hexoses, sucrose, maltose, cellobiose, trehalose, raffinose and melezitose; alcohols: ethanol, glycerol, mannitol and sorbitol; acids such as acetate, lactate, succinate, succinate, citrate and long-chain fatty acids, as well as D-galacturonic acid. It can grow in a wide range of pH and temperatures, can be genetically manipulated and the metabolic pathways of TAG synthesis are well known.

The use of different carbon sources for microbial lipid biosynthesis has been reported. Li et al. (2020) demonstrated the use of waste amino acids that can be obtained from meat waste or other protein source using the yeast *Rhodosporidium toruloides* CGMCC 2.1389, isolated in China. This work studied the effect of each of the twenty amino acids that conform the proteins, finding that the presence of the amino acid L-proline favored more lipid accumulation inside the yeast cells, reaching 37.3%. The lipids obtained showed great similarity with those obtained from vegetable oils.

Recently, Carmona-Cabello et al. (2021) have studied the *Rhodosporidium toruloides* strain Y-27012 to produce microbial lipids from hospital food waste hydrolysates and use them to produce biodiesel. In this process, hydrolysates from hospital food waste were used as culture medium and 32.9 g/L of dry biomass and 36.4% of lipids were obtained. When cultured in a bioreactor, 53.9 g/L of biomass and a lipid concentration of 26.7 g/L were obtained.

Another yeast genus that has attracted attention due to its potential for lipid production from crude glycerol is *Trichosporon*. Within this genus, we find some yeast species with good prospects: *Trichosporon fermentans* CICC 1368 and *Trichosporon cutaneum* AS 2.0571. These yeasts can accumulate lipids when cultivated in media based on lignocellulosic biomass hydrolysates, but Liu et al. (2017) demonstrated the capacity to use crude glycerol obtained from biodiesel. The strains showed good lipid biosynthesis capacity, despite the presence of residual methanol and potassium ions. The authors were able to obtain for *T. fermentans* an optimum biomass production of 14.5 g/L and a lipid content of 29% when using 50 g/L of crude glycerol, while for *T. cutaneum*, the optimum biomass production was 12. 5 g/L with a lipid content of 27% when using 70 g/L crude glycerol. In all cases, the optimal inoculum concentration was 10%, temperatures of 28–30°C and a pH of 6.0. It is noteworthy that the presence of residual methanol in the culture medium presented a low effect.

The oleaginous yeast *Trichosporon cutaneum* possesses high tolerance to lignocellulosic inhibitors and can not accumalate high concentrations of lipids using phenolic aldehydes as a substrate as a carbon and energy source. Hu et al. (2018) reported that *T. cutaneum* ACCC 20271 grew on a medium with three different phenolic aldehydes at different concentrations, in a bioreactor operating in fed-batch mode and the best result was with 4-hydroxybenzaldehyde, accumulating 0.85 g/L lipid which were 3.7 times higher than the control without the substrate.

In the search for new substrates for oleaginous yeast culture, Younes et al. (2020) used biomass hydrolysates of *Scenedesmus obtusiusculus*, which is used as a carbon source for the yeast *Cutaneotrichosporon oleaginosus* growing in minimal nitrogen medium, producing a lipid content of 61%.

Naganishia albida is an oleaginous yeast characterized by using different carbon and energy sources including lignocellulosic hydrolyzed complex sugars and fatty acids. Sathiyamoorthi et al. (2020) reported a co-fermentation of onion waste hydrolysate and crude glycerol, producing a higher biomass concentration 21.1 g/L and 34% lipids after 168 h of processing with intermittent feeding of culture medium.

Another genus of oleaginous yeasts is the *Lipomyces*, where specifically the yeast *Lipomyces starkeyi* is an excellent lipid producer and accumulates TAG up to 70% in dry base weight (Takaku et al. 2020) and is of great interest at industrial level to produce biodiesel (Wild et al. 2010). This yeast is characterized by a low capacity to degrade its own lipids and metabolize different substrates such as glucose, molasses, ethanol, sweet whey permeates, sewage sludge, D-xylose, glucose, fructose, sucrose, soluble potato starch, myristic acid, palmitic acid, oleic acid, among others (Sutanto et al. 2018). Furthermore, Sutanto et al. (2018) reported that 14% of the oil production can be obtained with different *Lipomyces starkeyi* strains and different types of substrates. Moreover, the lipid content was higher and in the range 60–80% using bioreactors operating in batch or fed batch mode. Wild et al. (2010) studied the effect of C:N molar ratio on the growth of *L. starkeyi* yeast and its lipid content in media with glucose or potato starch as a source of C. With a C:N molar ratio of 61.2 in glucose medium, the lipid content was 30% on a dry basis. They established that cell lipid content increased with increasing C:N molar ratio, but cell yield decreased.

Fatty acid biosynthesis in yeasts takes place in the cytosol in a series of reactions that convert the biosynthetic precursor acetyl-CoA to long-chain fatty acids and the subsequent synthesis of TAG as shown in Fig. 2. TAG are the major storage lipids in yeast and accumulate in the stationary phase of growth within organelles called lipid bodies (Sandager et al. 2002, Wang et al. 2016). There are several papers that report lipid synthesis in oleaginous yeasts such as *L. starkeyi* (Sutanto et al. 2018, Takaku et al. 2020), *Yarrowia lipolytica* (Ledesma-Amaro and Nicaud 2016, Fakas 2017) and *Rhodotorula toruloides* (Li et al. 2020, Maza et al. 2020, Carmona-Cabello et al. 2021).

The study of new strains isolated from nature is an important contribution to the knowledge of oleaginous yeasts and shows a good number of potentially interesting organisms for biodiesel

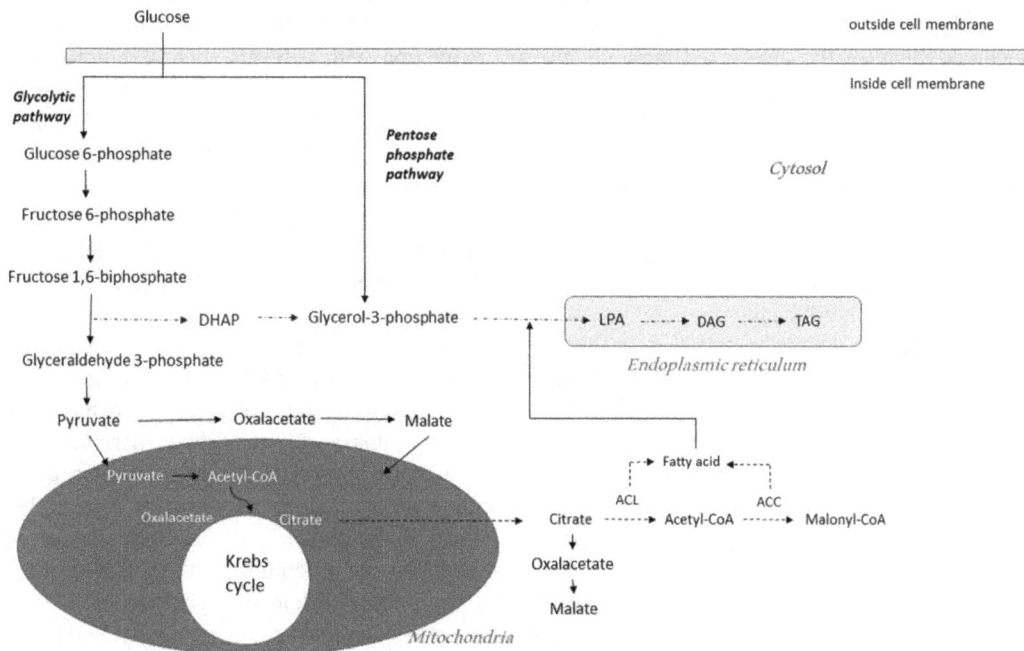

Fig. 2. Mechanism of lipid synthesis of yeast. ACL: ATP Citrate lyase, ACC: acetyl-coA carboxylase, LPA: lyosophosphatidic acid, DAG: diacylglyceride, DHAP: dihydroxyacetone phosphate, TAG: triacylglyceride (adapted from Sutanto et al. 2018, Takakun et al. 2020).

production. There are several reports on the use of genetically modified microorganisms, whose objective is to have oleo microorganisms with higher lipid production capacity.

4. Oleaginous Microalgae

Biofuels from microalgae are considered a promising sustainable substitute from petroleum-based fuels. Specifically, microalgae have become an increasingly competitive raw material for biodiesel production due to their high lipid content, growth rate, photosynthetic efficiency, and less water consumption (Al-Ameri and Al-Zuhair 2019, Cheng et al. 2020, Figueroa-Torres et al. 2020). Biofuels obtained from this raw biomass are known as third generation biofuels with great benefits over the first and second generation due to their lower environmental impact (Al-Ameri and Al-Zuhair 2019, Atmanli 2020, Cheng et al. 2020). They overcome the dilemma of using arable lands for cultivation or interfere in food production, although microalgal-biofuel production is limited by the high costs and energy requirements (Branco-Vieria et al. 2020, Figueroa-Torres et al. 2020). It is noticeable that microalgae not only contain lipids which can transform into biodiesel, but they are also rich in proteins, sugars, carotenoids, and vitamins, useful to produce food additives, supplements for animal feed, cosmetics, or pharmaceuticals (Mata et al. 2016, Branco-Vieria et al. 2020, Figueroa-Torres et al. 2020).

The production of biodiesel from microalgae involves several stages such as cultivation, harvesting, lipid extraction and transesterification of the recovered oils (Al-Ameri and Al-Zuhair 2019). In cultivation stage, it is possible to produce more oil per land area in comparison to conventional oil crops. This process is renewable and can help reducing the emissions of CO_2 in the atmosphere due to the photosynthetic process (Velazquez-Lucio et al. 2018, Shahi et al. 2020). Microalgae can grow in almost all media such as fresh, salty water or even wastewater; moreover, they can grow in the presence of sunlight or in dark ecosystems (Mata et al. 2016, Branco-Vieria et al. 2020). In the lipid extraction stage, around 20–50% (dry w/w) of lipids can be recovered from microalgae and the residual solids can be used as soil fertilizers (Wahidin et al. 2018). This stage is very important because its feasibility and success depend on the extraction method. Some of the most used methods are: (a) extraction with solvents such as chloroform/methanol mixture in the Folch method (El-Sheekh et al. 2018), methanol/chloroform/water in the Bligh and Dyer method (Shahi et al. 2020), hexane (Atmanli 2020), green solvents such as 2-methyltetrahydrofuran or cyclopentylmethyl ether (de Jesus et al. 2020), switchable solvents such as N,N-dimethylcyclohexylamine (DMCHA), N-ethylbutylamine (EBA), dipropylamine (Al-Ameri and Al-Zuhair 2019), and N,N,N',N'-tetraethyl-1,3-propanediamine (TEPDA) (Cheng et al. 2020), (b) extraction with supercritical fluids such as CO_2 (Jafari et al. 2021), (c) ultrasound (Onay 2020), (d) mechanical methods such as bead milling (Shafiei-Alavijeh et al. 2020), (e) the use of enzymes (He et al. 2020), among others.

Understanding lipid metabolism in algae is important. Fig. 3 shows that lipids are mainly obtained from the plasma membrane, endomembranes, chloroplast and lipid bodies (TAG and free fatty acids) (De Bhowmick et al. 2015). Due to the availability of Acetyl-Co, malonyl-coA (first product of lipid biosynthesis) is then formed. This acetyl-coA set is derived from cytosolic/plastid glycolysis or directly from dihydroxyacetone phosphate during the Calvin cycle (Radakovits et al. 2010). Subsequently, the fatty acids from acyl carrier protein (ACP) complex are directly transferred to glycerol-3-phosphate (G3P) by chloroplast-resident acyltransferases or transported to cytosol for sequential acylation in the endoplasmic reticulum (E.R) by E.R-resident acyltransferases (Banerjee et al. 2020). The first acylation occurs at the sn-1 position of G3P by glycerol-3-phosphate acyltransferases (GPAT). After, lysophosphatidic acid (LPA) is acylated to the sn-2 position resulting in the formation of phosphatidic acid (PA) by lysophosphatidate acyltranferase (LPAAT). PA is dephosphorylated to produce diacylglycerol (DAG) which can be used as a primary precursor for the synthesis of structural lipids in the chloroplast to produce membrane and storage

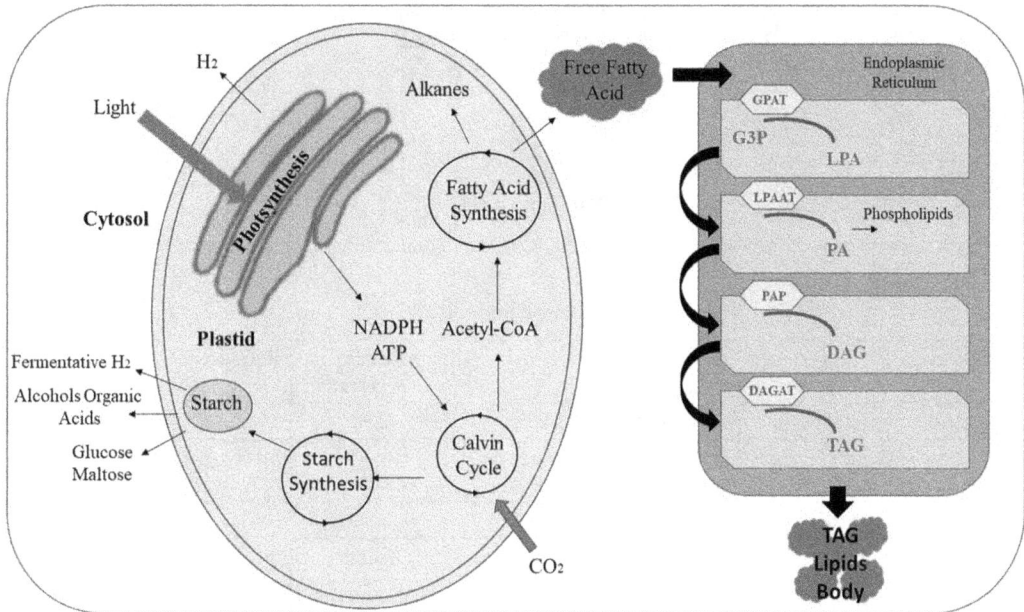

Fig. 3. Mechanism of lipid synthesis in microalgae. ATP: Adenosine 5'-triphosphate, G3P: Glycerol-3-phosphate, GPAT: Glycerol-3-phosphate O-acyltransferase, LPA: Lysophosphatidic acid, LPAAT: Lysophosphatidic acid acyltransferase, PA: Phosphatidic acid, PAP: Phosphatidate phosphatase, DAG: Diacylglyceride, DAGAT: Diacylglyceryl hydroxy-methyl-trimethyl-β-alanine, TAG: Triacylglyceride, NADPH: Nicotinamide adenine dinucleotide phosphate (Adapted from De Bhowmick et al. 2015).

lipids. The TAG is then deposited into the cytosol as ER-derived lipid droplets (De Bhowmick et al. 2015).

On the other hand, within the wide variety of microalgae species, the biofuel production of green algae has been demonstrated to be high in comparison to other genus of algae (red, blood red, yellow green), because they have autotrophic species with rich nutrition and a high degree of photosynthetic utilization (Atmanli 2020). In Fig. 4, the structure of three most used green microalgae species as raw biomass for biodiesel is shown. The genus *Chlorella* is identified as spherical unicellular green microalgae with an approximate size of 2–10 μm in diameter; this is one of the most cultivated microalgae due to its main characteristics such as high photosynthetic efficiency, nutritional value, and fast growth (Spain et al. 2021).

The genus *Nannochloropsis* is one of the microalgae with wide diversity of species and can be easily found in any aquatic ecosystem. They have a high reproductive capacity since their cells asexually divide to produce two daughter cells that later shed their mother cell wall. This type of microalgae has been widely studied as a promising alternative feedstock in large-scale biofuel installations due to its resistant growth profiles in the open air and high lipid yields (Scholz et al. 2014, Ma et al. 2016).

The genus *Scenedesmu* is one of the most common green microalgae in freshwater and can be found in coenobia of four or eight cells inside a common wall. This thick wall enables *Scenedesmus* species to be resistant to digestion and predation (Pignolet et al. 2013). This species is normally harvested for oil production.

Considering all the stages of the process for obtaining microalgae derived biodiesel, the selection of the most efficient species to produce oil is one of the main factors recently studied (Al-Ameri and Al-Zuhair 2019, Felix et al. 2019, Ma et al. 2019, de Jesus et al. 2020, Shafiei Alavijeh et al. 2020, Tejada Carbajal et al. 2020, Morais et al. 2021). Table 3 shows relevant studies using several microalgae species for lipid extraction.

Fig. 4. Microalgae species (adapted from Subramani et al. 2015, Pham and Bui 2020, Zanella and Vianello 2020).

Table 3. Oleaginous microalgae.

Microalgae	Substrate	Lipid Concentration	Reference
Scenedesmus obliquus	Brewery effluent (maltose)	27%	Mata et al. (2013)
Arthrospira platensis *Chlorella vulgaris*	1.00 g/L of glucosa and salinity stress	15.4% 23.0%	Mata et al. (2016)
Nannochloropsis gaditana	-----	31%	Navarro et al. (2016)
Scenedesmus sp.	Chu13 medium and biogas $(CO_2:CH_4)$ (40:60)	34.1%	Srinuanpan et al. (2018)
Chlorella sp.	Bold's Basal Medium and CO_2	13.6%	Al-Ameri and Al-Zuhair (2019)
Chlorella vulgaris	BG-11 culture medium and CO_2	7.5%	Ma et al. (2019)
Scenedesmus abundans	Bold's Basal Medium and CO_2	24.3%	Nayak and Ghosh (2019)
Shizochitrium limacinum	-----	4.34%	Rathnam and Madras (2019)
Chlorella vulgaris *Chlorella pyrenoidosa*	-----	28.7% 7.3%	Sadvakasova et al. (2019)
Chlorella pyrenoidosa	-----	12.6%	De Jesus et al. (2020)
Nannochloropsis oculata	f/2 medium and CO_2	35.39%	He et al. (2020)
Chlorella sp.	Bold's Basal Medium and CO_2	8.56%	Ismail and Al-Zuhair (2020)
Chlorella variabilis	-----	24.67%	Nirmala and Dawn (2020)
Chlorella sp. + *Scenedesmus* sp.	Domestic wastewater	34.83%	Silambarasan et al. (2020)
Tetradesmus obliquus SGM19	BG-11 medium and CO_2	28.5%	Singh et al. (2020)
Scenedesmus dimorphus	-----	44%	Tejada et al. (2020)
Nannochloropsis oculata	f/2 medium in artificial sea water and CO_2	24.2%	Jafari et al. (2021)

Al-Ameri and Al-Zuhair (2019) carried out the cultivation of the microalgae *Chlorella* sp. in a 150 L pipeline system under permanent lighting and aeration. A Bold´s Basal Medium (BBM) was used at room temperature. Harvesting of the microalgae was achieved through centrifugation at 6000 rpm for 5 min. For the extraction of the lipids contained in the microalgae, three switchable solvents N,N-dimethylcyclohexylamine (DMCHA), N-ethylbutylamine (EBA) and dipropylamine were used, obtaining yields of 13.6, 12.3 and 7.0% (dry w/w), respectively. The efficiency of these solvents was also assessed in extraction-reaction systems to transform the oils into biodiesel, using enzymes as a catalyst. Using DMCHA and enzyme load of 30%, a biodiesel yield of 47.5% was reached at 35°C with a methanol: oil molar ratio of 6:1.

Also, Ma et al. (2019) used directly wet *Chlorella vulgaris* microalgae for biodiesel production, avoiding the steps of dehydration and lipid extraction. Wet microalgae were heated with radiofrequency to break cell walls, then the esterification/transesterification reaction was assisted with radiofrequency at 55°C during 20 min, obtaining FAMEs formation of 79.5%.

Isolated strains from thermal springs contained *Chlorella vulgaris* sp-1, *Ankistrodesmus* sp-21, *Scenedesmus obliquus* sp-21, *Chlorella pyrenoidosa* sp-13 and *Chlamydomonas* sp-22, which are lipid producers (Sadvakasova et al. 2019). Among these species, *Chlorella vulgaris* sp-1 and *Scenedesmus obliquus* sp-21 presented the highest lipid content of 28.7 and 29.8% (dry w), respectively.

Nannochloropsis is another microalga highly studied due to its high lipid content (Brennan and Regan 2020). As example, Navarro López et al. (2016) used the marine microalga *Nannochloropsis gaditana* to extract saponifiable lipids and convert into biodiesel through methanolysis catalyzed with the *Rhizopus oryzae* lipase. The lipids were extracted with ethanol, obtaining 31% of the dry biomass. It was possible to convert 83% of these lipids to biodiesel.

He et al. (2020) proposed a novel process from cultivation to extract lipids contained in three marine *Nannochloropsis* microalgae species (*N. oculata*, *Nannochloropsis* sp. and *N. oceanica*) for its conversion to biodiesel. Bubble column photobioreactors were implemented to carry out the cultivation of microalgae (BC-PBR, 40 mm internal diameter, 1000 mm height) containing a culture medium f/2 with sea salt enriched with $NaNO_3$ and NaH_2PO_4. Microalgae were cultured with air enriched with 5% of CO_2 at 23 ± 1°C. The rupture cell of microalgae was performed with four hydrolytic enzymes (cellulase, papain, hemicellulase and pectinase) and three phase partitioning (TPP). The TPP method was more efficient for lipid extraction. After 12 h of hydrolysis, the disrupted microalgal cells were collected at 5000 rpm for 5 min. Among these three species studied, *Nannochloropsis oculata* proved to be the most efficient with a total fatty acids of 35.39% wt and a conversion to biodiesel of 90.24%.

On the other hand, Onay (2020) reported the use of several effective extraction methods for the lipids contained in the *Nannochloropsis gaditana* microalgae incorporating osmotic shock (0–20 KCl%), ultrasound (0–50 kHz) and lysozyme (U/mL), to achieve maximum yields and minimize the cost of the method, using the Central Composite Design (CCD) of the Response Surface Methodology (RSM). After being examined, the maximum percentage of lipids (37% wt) was obtained at 10% KCl of osmotic shock, 30 kHz ultrasound and 10 U/mL of lysozyme.

Recently, Jafari et al. (2021) used wet biomass of the *Nannochloropsis oculata* microalgae to evaluate an ecological method of direct conversion to biodiesel in a single step, taking advantage of the supercritical state of CO_2 to extract the lipids from the microalgae. The yield of biodiesel production reached a maximum of 24.2% dry weight biomass.

Atmanli (2020) studied the freshwater green microalgae *Scenedesmus dimorphus* and the sea bream *Isochrysis aff. galbana* to know its potential as a source of lipids in the production of biodiesel. The culture was prepared under the conditions of the Bristol and Erdschreiber's medium in presence of 1.5% CO_2 to produce a mixing effect with large and slow bubbles to take advantage of the CO_2 injection. The results showed that the microalgae *Scenedesmus dimorphus* and *Isochrysis aff* can be used as lipid source since it was possible to extract total lipids of 15.87 and 42.65% in dry weight of biomass, respectively.

Tejada Carbajal et al. (2020) carried out several processes to produce biodiesel, dihydroxyacetone (DHA), glycerol and vegetable oil substituent from the microalgae species *Scenedesmus dimorphus*. Open channel reactors were implemented to perform the cultivation of biomass with total surfaces of 1.2 and 6.3 m^2. The cultivation area was designed circularly to achieve mixing through a paddle wheel system applying a surface flow rate of 0.30 m/s. Single-stage centrifugation was used as harvesting technique due to its high yields in biomass concentrations of up to 15–20% of dry mass. The biomass showed a high oil yield under the growth conditions used in the cultivation stage, achieving 44% in the lipid extraction process. A biodiesel conversion percentage of 98.34 was obtained.

5. Conclusion

This chapter reviewed and analyzed the current and advanced technology for obtaining lipids contained in microorganisms such as bacteria, yeasts, fungi, and microalgae as emerging alternative to produce biodiesel. This chapter demonstrates the enormous potential of biodiesel derived from microbial lipids from an ecological, sustainable, and profitable point of view. Several species of microalgae, bacteria, fungi, and yeasts can synthesize more than 20% w/w of lipids on a dry basis based on their cellular metabolism and are known as oleaginous microorganisms. Some species can synthesize lipids up to 70% w/w on a cell dry weight basis depending on culture conditions, such as under high C/N ratio. Of the species reviewed, yeasts are considered to have the greatest potential, specifically *Lipomyces starkeyi*, for being able to use a wide variety of residues as substrates and being cultivated under aerobic conditions and temperatures around 30°C with high oil contents between 60 and 80% on a dry basis and good yields, in addition to being able to be genetically modified.

Nowadays, the industrialization of biodiesel production derived from microorganism lipids has not been achieved. There are several technical drawbacks which need to be addressed to consider microbial biodiesel production a mature and cost-effective technology. The main challenge for the biodiesel industry is the availability of low-cost feedstock which directly impact the production cost. The use of refined vegetable oils for biofuel production increases the total cost of production and generates the food vs. fuel debate. Therefore, to reduce the high feedstock cost and social impact, microbial oil can be an alternative resource for both food and fuel applications.

However, the high costs associated with culture media for oleaginous microorganisms can be solved by using renewable carbon and energy sources, mostly obtained from waste. In addition, the integration of biofuels from oleaginous microorganisms with various value-added products helps to reduce the overall cost of production, since in some cases other compounds such as carotenoids can be separated in addition to oil. Another great advantage is that these organisms can be genetically modified and achieve higher yields or produce certain types of fatty acids to improve the properties of the biodiesel to be produced, and the synergistic use of oleaginous yeasts and microalgae can contribute to this objective.

On the other hand, the optimization of fermentation process, and reactor type and parameters to control incubation, quality of the substrate, temperature, among others, are paramount to achieve high lipid accumulation in microorganisms. Another important aspect is the extraction step which can be complicated. The selection of the extraction methodology (solvents, operational parameters, extra steps such as pre-drying of microorganisms, cell-disruption, among others) depends on the lipidic content in cells and the microorganism conditions, requiring additional steps to the extraction process. Also, it is important to consider the use of green solvents, their recovery and reuse, as well as the minimal use of extra energy to reduce the environmental impacts associated. The topics must be studied to consolidate and offer an integrated and efficient biodiesel production system.

Acknowledgments

We are grateful for the support of Autonomous University of Aguascalientes, Technological Institute of Aguascalientes and to the co-authors of this chapter for their support and dedication.

References

Al-Ameri, M. and Al-Zuhair, S. 2019. Using switchable solvents for enhanced, simultaneous microalgae oil extraction-reaction for biodiesel production. Biochem. Eng. J. 141: 217–224.

Annamalai, N., Sivakumar, N., Fernandez-Castane, A. and Oleskowicz-Popiel, P. 2020. Production of microbial lipids utilizing volatile fatty acids derived from wastepaper: a biorefinery approach for biodiesel production. Fuel 276: 118087.

Atmanli, A. 2020. Experimental comparison of biodiesel production performance of two different microalgae. Fuel 278: 118311.

Ayadi, I., Belghith, H., Gargouri, A. and Guerfali, M. 2018. Screening of new oleaginous yeasts for single cell oil production, hydrolytic potential exploitation, and agro-industrial by-products valorization. Process Saf. Environ. 119: 104–114.

Banerjee, S., Banerjee, S., Ghosh, A.K. and Das, D. 2020. Maneuvering the genetic and metabolic pathway for improving biofuel production in algae: Present status and future prospective. Renew. Sustain. Energy Rev. 133: 110155.

Behera, A.R., Dutta, K., Verma, P. Daverey, A. and Sahoo, D.K. 2019. High lipid accumulating bacteria isolated from dairy effluent scum grown on dairy wastewater as potential biodiesel feedstock. J. Environ. Manage. 252: 109686.

Berikten, D., Hoşgün, E.Z., Gökdal Otuzbiroğlu, A., Bozan, B. and Kıvanç, M. 2021. Lipid production from crude glycerol by newly isolated oleaginous yeasts: strain selection, molecular identification and fatty acid analysis. Waste Biomass Valorization 2021: 1–10.

Bharti, R.K., Srivastava, S. and Thakur, I.S. 2014a. Production and characterization of biodiesel from carbon dioxide concentrating chemolithotrophic bacteria, *Serratia* sp. ISTD04. Bioresour. Technol. 153: 189–197.

Bharti, R.K., Srivastava, S. and Thakur, I.S. 2014b. Extraction of extracellular lipids from chemoautotrophic bacteria *Serratia* sp. ISTD04 for production of biodiesel. Bioresour. Technol. 165: 201–204.

Branco-Vieira, M., Costa, D.M.B., Mata, T.M., Martins, A.A., Freitas, M.A.V. and Caetano, N.S. 2020. Environmental assessment of industrial production of microalgal biodiesel in central-south Chile. J. Clean. Prod. 266: 121756.

Brennan, B. and Regan, F. 2020. *In situ* lipid and fatty acid extraction methods to recover viable products from *Nannochloropsis* sp. Sci. Total Environ. 748: 142464.

Carmona-Cabello, M., García, I.L., Papadaki, A., Tsouko, E., Koutinas, A. and Dorado, M.P. 2021. Biodiesel production using microbial lipids derived from food waste discarded by catering services. Bioresour. Technol. 323: 124597.

Cea, M., Sangaletti-Gerhard, N., Acuña, P., Fuentes, I., Jorquera, M., Godoy, K. and Osses, F.N. 2015. Screening transesterifiable lipid accumulating bacteria from sewage. Biotechnol. Rep. 8: 116–123.

Cheng, J., Guo, H., Qiu, Y., Zhang, Z., Mao, Y. and Qian, L. 2020. Switchable solvent N, N, N′, N′-tetraethyl-1, 3-propanediamine was dissociated into cationic surfactant to promote cell disruption and lipid extraction from wet microalgae for biodiesel production. Bioresour. Technol. 312: 123607.

Chmielarz, M., Blomqvist, J., Sampels, S., Sandgren, M. and Passoth, V. 2021. Microbial lipid production from crude glycerol and hemicellulosic hydrolysate with oleaginous yeasts. Biotechnol. Biofuels 14: 1–11.

Dai, X., Shen, H., Li, Q., Rasool, K., Wang, Q., Yu, X. and Zhao, Z.K. 2019. Microbial lipid production from corn stover by the oleaginous yeast *Rhodosporidium toruloides* using the preSSLP process. Energies 12: 1053.

De Bhowmick, G., Koduru, L. and Sen, R. 2015. Metabolic pathway engineering towards enhancing microalgal lipid biosynthesis for biofuel application—A review Renew. Sustain. Energy Rev. 50: 1239–1253.

de Jesus, S.S., Ferreira, G.F., Moreira, L.S. and Filho, R.M. 2020. Biodiesel production from microalgae by direct transesterification using green solvents. Renew. Energy 160: 1283–1294.

El-Sheekh, M., Abomohra, A.E., Eladel, H., Battah, M. and Mohammed, S. 2018. Screening of different species of *Scenedesmus* isolated from Egyptian freshwater habitats for biodiesel production. Renew. Energy 129: 114–120.

Fei, Q., Wewetzer, S.J., Kurosawa, K., Rha, C. and Sinskey, A.J. 2015. High-cell-density cultivation of an engineered *Rhodococcus opacus* strain for lipid production via co-fermentation of glucose and xylose. Process Biochem. 50: 500–506.

Felix, C., Ubando, A., Madrazo, C., Sutanto, S., Tran-Nguyen, P.L. and Go, A.W. 2019. Investigation of direct biodiesel production from wet microalgae using definitive screening design. Energy Procedia 158: 1149–1154.

Figueroa-Torres, G.M., Mahmood, W.M.A., Pittman, J.K. and Theodoropoulos, C. 2020. Microalgal biomass as biorefinery platform for biobutanol and biodiesel production. Biochem. Eng. J. 153: 107396.

Gientka, I., Kieliszek, M., Jermacz, K. and Błazejak, S. 2017. Identification and characterization of oleaginous yeast isolated from kefir and its ability to accumulate intracellular fats in deproteinated potato wastewater with different carbon sources. Biomed. Res. Int., 1–19.

Han, S., Kim, G.Y. and Han, J.I. 2019. Biodiesel production from oleaginous yeast, *Cryptococcus* sp. by using banana peel as carbon source. Energy Rep. 5: 1077–1081.

He, Y., Zhang, B., Guo, S., Guo, Z., Chen, B. and Wang, M. 2020. Sustainable biodiesel production from the green microalgae *Nannochloropsis*: Novel integrated processes from cultivation to enzyme-assisted extraction and ethanolysis of lipids. Energy Convers. Manag. 209: 112618.

Hu, M., Wang, J., Gao, Q. and Bao, J. 2018. Converting lignin derived phenolic aldehydes into microbial lipid by *Trichosporon cutaneum*. J. Biotechnol. 281: 81–86.

Ismail, M. and Al-Zuhair, S. 2020. Thermo-responsive switchable solvents for simultaneous microalgae cell disruption, oil extraction-reaction, and product separation for biodiesel production. Biocatal. Agric. Biotechnol. 26: 101667.

Jafari, A., Esmaeilzadeh, F., Mowla, D., Sadatshojaei, E., Heidari, S. and Wood, D.A. 2021. New insights to direct conversion of wet microalgae impregnated with ethanol to biodiesel exploiting extraction with supercritical carbon dioxide. Fuel 285: 119199.

Jathanna, H.M., Rao, C.V. and Goveas, L.C. 2020. Exploring pongamia seed cake hydrolysate as a medium for microbial lipid production by *Aspergillus ochraceus*. Biocatal. Agric. Biotechnol. 24: 101503.

Kamal, R., Shen, H., Li, Q., Wang, Q., Yu, X. and Zhao, Z.K. 2020. Utilization of amino acid-rich wastes for microbial lipid production. Appl. Biochem. Biotechnol. 191: 1594–1604.

Kumar, D., Singh, B. and Korstad, J. 2017. Utilization of lignocellulosic biomass by oleaginous yeast and bacteria for production of biodiesel and renewable diesel. Renew. Sust. Energy Rev. 73: 654–671.

Kumar, M. and Thakur, I.S. 2018. Municipal secondary sludge as carbon source for production and characterization of biodiesel from oleaginous bacteria. Bioresour. Technol. Rep. 4: 106–113.

Kumar, S., Gupta, N. and Paksjirajan, K. 2015. Simultaneous lipid production and dairy wastewater treatment using *Rhodococcus opacus* in a batch bioreactor for potential biodiesel application. J. Environ. Chem. Eng. 3: 1630–1636.

Kumar, V. and Thakur, I.S. 2020. Biodiesel production from transesterification of *Serratia* sp. ISTD04 lipids using immobilised lipase on biocomposite materials of biomineralized products of carbon dioxide sequestrating bacterium. Bioresour. Technol. 307: 123193.

Li, Q., Kamal, R., Wang, Q., Yu, X. and Zhao, Z.K. 2020. Lipid production from amino acid wastes by the oleaginous yeast *Rhodosporidium toruloides*. Energies 13: 1–9.

Liu, L., Chen, J., Lim, P.E. and Wei, D. 2018. Dual-species cultivation of microalgae and yeast for enhanced biomass and microbial lipid production. J. Appl. Psychol. 30: 2997–3007.

Liu, L., Hu, Y., Lou, W., Li, N., Wu, H. and Zong, M.H. 2017. Use of crude glycerol as sole carbon source for microbial lipid production by oleaginous yeasts. Appl. Biochem. Biotechnol. 182: 495–510.

Louhasakul, Y., Cheirsilp, B., Intasit, R., Maneerat, S. and Saimmai, A. 2020. Enhanced valorization of industrial wastes for biodiesel feedstocks and biocatalyst by lipolytic oleaginous yeast and biosurfactant-producing bacteria. Int. Biodeterior. Biodegradation 148: 1–7.

Ma, X.N., Chen, T.P., Yang, B., Liu, J. and Chen, F. 2016. Lipid Production from *Nannochloropsis*. Mar. Drugs 14: 61.

Ma, Y., Gao, Z., Wang, Q. and Liu, Y. 2018. Biodiesels from microbial oils: Opportunity and challenges. Bioresour. Technol. 263: 631–641.

Ma, Y., Liu, S., Wang, Y., Adhikari, S., Dempster, T.A. and Wang, Y. 2019. Direct biodiesel production from wet microalgae assisted by radio frequency heating. Fuel 256: 115994.

Maheshwari, N., Kumar, M., Thakur, I.S. and Srivastava, S. 2018. Carbon dioxide biofixation by free air CO_2 enriched (FACE) bacterium for biodiesel production. J. CO_2 Util. 27: 423–432.

Martinez-Silveira, A., Villarreal, R., Garmendia, G., Rufo, C. and Vero, S. 2019. Process conditions for a rapid *in situ* transesterification for biodiesel production from oleaginous yeasts. Electron. J. Biotechnol. 38: 1–9.

Mata, T., Martins, A., Oliveira, O., Oliveira, S., Mendes, A. and Caetano, N. 2016. Lipid content and productivity of *Arthrospira platensis* and *Chlorella vulgaris* under mixotrophic conditions and salinity stress. Chem. Eng. Trans. 49: 187–192.

Mata, T.M., Meloa, A.C., Meirelesb, S., Mendesa, A.M., Martinsc, A.A. and Caetano, N.S. 2013. Potential of microalgae *Scenedesmus obliquus* grown in brewery wastewater for biodiesel production. Chem. Eng. Trans. 32: 901–906.

Maza, D.D., Viñarta, S.C., Su, Y., Guillamón, J.M. and Aybar, M.J. 2020. Growth and lipid production of *Rhodotorula glutinis* R4, in comparison to other oleaginous yeasts. J. Biotechnol. 310: 21–31.

Maza, D.D., Viñarta, S.C., García-Ríos, E., Guillamón, J.M. and Aybar, M.J. 2021. *Rhodotorula glutinis* T13 as a potential source of microbial lipids for biodiesel generation. J. Biotechnol. 331: 14–18.

Miao, Z., Tian, X., Liang, W., He, Y. and Wang, G. 2020. Bioconversion of corncob hydrolysate into microbial lipid by an oleaginous yeast *Rhodotorula taiwanensis* AM2352 for biodiesel production. Renew. Energy 161: 91–97.

Morais, K.C.C., Conceição, D., Vargas, J.V.C., Mitchell, D.A., Mariano, A.B. and Ordonez, J.C. 2021. Enhanced microalgae biomass and lipid output for increased biodiesel productivity. Renew. Energy 163: 138–145.

Nair, A.S., Al-Bahry, S., Gathergood, N., Tripathi, B.M. and Sivakumar, N. 2020. Production of microbial lipids from optimized waste office paper hydrolysate, lipid profiling and prediction of biodiesel properties. Renew. Energy 14: 124–134.

Navarro López, E., Robles Medina, A., González Moreno, P.A., Esteban Cerdán, L., Martín Valverde, L. and Molina Grima, E. 2016. Biodiesel production from *Nannochloropsis gaditana* lipids through transesterification catalyzed by *Rhizopus oryzae* lipase. Bioresour. Technol. 203: 236–44.

Nayak, J.K. and Ghosh, U.K. 2019. Post treatment of microalgae treated pharmaceutical wastewater in photosynthetic microbial fuel cell (PMFC) and biodiesel production. Biomass Bioenerg. 131: 105415.

Ngamsirisomsakul, M., Reungsang, A. and Kongkeitkajorn, M.B. 2021. Assessing oleaginous yeasts for their potentials on microbial lipid production from sugarcane bagasse and the effects of physical changes on lipid production. Bioresour. Technol. Report. 14: 100650.

Nirmala, N. and Dawn, S.S. 2020. Phylogenetic analysis for identification of lipid enriched microalgae and optimization of extraction conditions for biodiesel production using response surface methodology tool. Biocatal. Agric. Biotechnol. 25: 101603.

Onay, A. 2020. Optimization of lipid content of *Nannochloropsis gaditana* via quadratic models using Matlab Simulink. Energy Rep. 6: 128–133.

Patel, A., Karageorgou, D., Rova, E., Katapodis, P., Rova, U., Christakopoulos, P. and Matsakas, L. 2020. An overview of potential oleaginous microorganisms and their role in biodiesel and omega-3 fatty acid-based industries. Microorganisms 8: 434.

Pham, T.L. and Bui, M.H. 2020. Removal of nutrients from fertilizer plant wastewater using *Scenedesmus* sp.: Formation of bioflocculation and enhancement of removal efficiency. J. Chem. 2020: 1–9.

Pignolet, O., Jubeau, S., Vaca-Garcia, C. and Michaud, P. 2013. Highly valuable microalgae: Biochemical and topological aspects. J. Ind. Microbiol. Biotechnol. 40: 781–796.

Qadeer, S., Khalid, A., Mahmood, S., Anjum, M. and Ahmad, Z. 2017. Utilizing oleaginous bacteria and fungi for cleaner energy production. J. Clean. Prod. 168: 917–928.

Qian, X., Zhou, X., Chen, L., Zhang, X., Xin, F., Dong, W. and Jiang, M. 2021. Bioconversion of volatile fatty acids into lipids by the oleaginous yeast *Apiotrichum porosum* DSM27194. Fuel 290: 119811.

Radakovits, R., Jinkerson, R., Darzins, A. and Posewitz, M. 2010. Genetic engineering of algae for enhanced biofuel production. Eukaryot. Cell 9: 486–501.

Rathnam, V.M. and Madras, G. 2019. Conversion of *Shizochitrium limacinum* microalgae to biodiesel by non-catalytic transesterification using various supercritical fluids. Bioresour. Technol. 288: 121538.

Sadvakasova, A.K., Akmukhanova, N.R., Bolatkhan, K., Zayadan, B.K., Usserbayeva, A.A. and Bauenova, M.O. 2019. Search for new strains of microalgae-producers of lipids from natural sources for biodiesel production. Int. J. Hydrog. Energy 44: 5844–5853.

Sandager, L., Gustavsson, M.H., Stahl, U., Dahlqvist, A., Wiberg, E., Banas, A., Lenman, M., Ronne, H. and Stymne, S. 2002. Storage lipid synthesis is non-essential in yeast. Journal of Biological Chemistry 277(8): 6478–6482.

Sathiyamoorthi, E., Dikshit, P.K., Kumar, P. and Kim, B.S. 2020. Co-fermentation of agricultural and industrial waste by *Naganishia albida* for microbial lipid production in fed-batch fermentation. J. Chem. Technol. Biotechnol. 95: 813–821.

Scholz, M.J., Weiss, T.L., Jinkerson, R.E., Jing, J., Roth, R. and Goodenough, U. 2014. Ultrastructure and composition of the *Nannochloropsis gaditana* cell wall. Eukaryot. Cell 13: 1450–64.

Shafiei Alavijeh, R., Karimi, K., Wijffels, R.H., Berg, C. and Eppink, M. 2020. Combined bead milling and enzymatic hydrolysis for efficient fractionation of lipids, proteins, and carbohydrates of *Chlorella vulgaris* microalgae. Bioresour. Technol. 309: 123321.

Shahi, T., Beheshti, B., Zenouzi, A. and Almasi, M. 2020. Bio-oil production from residual biomass of microalgae after lipid extraction: The case of *Dunaliella* sp. Biocatal. Agric. Biotechnol. 23: 101494.

Silambarasan, S., Logeswari, P., Sivaramakrishnan, R., Incharoensakdi, A., Cornejo, P. and Kamaraj, B. 2020. Removal of nutrients from domestic wastewater by microalgae coupled to lipid augmentation for biodiesel production and influence of deoiled algal biomass as biofertilizer for *Solanum lycopersicum* cultivation. Chemosphere 268: 129323.

Silva, R.M., Castro, A.R., Machado, R. and Alcina, M. 2021. Dissolved oxygen concentration as a strategy to select type and composition of bacterial storage lipids produced during oilfield produced water treatment. Environ. Technol. Innov. 23: 101693.

Singh, N., Batghare, A.H., Choudhury, B.J., Goyal, A. and Moholkar, V.S. 2020. Microalgae based biorefinery: Assessment of wild fresh water microalgal isolate for simultaneous biodiesel and β-carotene production. Bioresour. Technol. Rep. 11: 100440.

Singh, N. and Choudhury, B. 2019. Valorization of food-waste hydrolysate by *Lentibacillus salarius* NS12IITR for the production of branched chain fatty acid enriched lipid with potential application as a feedstock for improved biodiesel. J. Waste Manag. 94: 1–9.

Sutantoa, S., Zullaikahb, S., Tran-Nguyenc, P.L., Ismadjid, S. and Ju, Y.-Hsu. 2018. *Lipomyces starkeyi*: Its current status as a potential oil producer. Fuel Process. Technol. 177: 39–55.

Spain, O., Plöhn, M. and Funk, C. 2021. The cell wall of green microalgae and its role in heavy metal removal. Physiol. Plant, 1–10.

Srinuanpan, S., Cheirsilp, B. and Prasertsan, P. 2018. Effective biogas upgrading and production of biodiesel feedstocks by strategic cultivation of oleaginous microalgae. Energy 148: 766–774.

Statistics Leading biodiesel producers worldwide in 2019, by country (in billion liters). https://www.statista.com/statistics/271472/biodiesel-production-in-selected-countries/.

Subramani, K., Sengottian, M. and Venkatachalam, C.D. 2015. Extraction of algal oil from *chlorella* sp. using combined osmotic shock and ultra-sonication. https://www.researchgate.net/publication/305411697.

Takaku H., Tomohiko, M., Yaoi, K. and Yamazaki, H. 2020. Lipid metabolism of the oleaginous yeast *Lipomyces starkeyi*. Appl. Microbiol. Biotechnol. 104: 6141–6148.

Tang, M., Wang, Y., Zhou, W., Yang, M., Liu, Y. and Gong, Z. 2020. Efficient conversion of chitin-derived carbon sources into microbial lipid by the oleaginous yeast *Cutaneotrichosporon oleaginosum*. Bioresour. Technol. 315: 123987.

Tejada Carbajal, E.M., Martínez Hernández, E., Fernández Linares, L., Novelo Maldonado, E. and Limas Ballesteros, R. 2020. Techno-economic analysis of *Scenedesmus dimorphus* microalgae biorefinery scenarios for biodiesel production and glycerol valorization. Bioresour. Technol. Rep. 12: 100605.

Velazquez-Lucio, J., Rodríguez-Jasso, R.M., Colla, L.M., Sáenz-Galindo, A., Cervantes-Cisneros, D.E. and Aguilar, C.N. 2018. Microalgal biomass pretreatment for bioethanol production: A review. Biofuel Res. J. 5: 780–791.

Wahidin, S., Idris, A., Yusof, N.M., Kamis, N.H.H. and Shaleh, S.R.M.. 2018. Optimization of the ionic liquid-microwave assisted one-step biodiesel production process from wet microalgal biomass. Energy Convers. Manag. 171: 1397–1404.

Wang, R., Wang, J., Xu, R., Fang, Z. and Liu, A. 2014. Oil production by the oleaginous yeast *Lipomyces starkeyi* using diverse carbon sources. BioResources 9(4): 7027–7040.

Wang Wei, Wei Hui, Knoshaug Eric, Van Wychen Stefanie, Xu Qi, Himmel Michael E., Zhang Min. 2016. Fatty alcohol production in Lipomyces starkeyi and Yarrowia lipolytica. Biotechnology for Biofuels 9: 227.

Wild, R., Patil, S., Popovi, M., Zappi, M., Dufreche, S. and Bajpai, R. 2010. Lipids from *Lipomyces starkeyi*. Food Technol. Biotechnol. 48(3): 329–335.

Younes, S., Bracharz, F., Awad, D., Qoura, F., Mehlmer, N. and Brueck, T. 2020. Microbial lipid production by oleaginous yeasts grown on *Scenedesmus obtusiusculus* microalgae biomass hydrolysate. Bioproc. Biosystems Eng. 43: 1629–1638.

Yu, A., Zhao, Y., Li, J., Li, S., Pang, Y., Zhao, Y. and Xiao, D. 2020. Sustainable production of FAEE biodiesel using the oleaginous yeast *Yarrowia lipolytica*. Microbiology Open 9: 1–14.

Zanella, L. and Vianello, F. 2020. Microalgae of the genus *Nannochloropsis*: Chemical composition and functional implications for human nutrition. J. Funct. Foods 68: 103919.

Zhang, L., Chao, B. and Zhang, X. 2020a. Modeling and optimization of microbial lipid fermentation from cellulosic ethanol wastewater by *Rhodotorula glutinis* based on the support vector machine. Bioresour. Technol. 301: 122781.

Zhang, L., Loh, K., Dai, Y. and Tong, Y.W. 2020b. Acidogenic fermentation of food waste for production of volatile fatty acids: Bacterial community analysis and semi-continuous operation. Waste Manag. 109: 75–84.

CHAPTER 11

Harvesting and Lipid Extraction of Microalgal Biomass
Sustainable Routes to Biodiesel

Natasha Nabila Ibrahim,[1] Imran Ahmad,[1,]* Norhayati Abdullah,[1] Laila Amalia Dina Purba,[1] Anas Al-Dailami,[1] Iwamoto Koji,[1] Shaza Eva Mohamad[1] and Ali Yuzir[2]

1. Introduction

The current situation of planet Earth is unsettling with environmental pollution and resources depletion. To mention some of the critical points, it includes atmospheric carbon dioxide accumulation that has reached up to 36 billion tonnes and diminishing fossil fuel reserves, yet the demand is hiking as we speak. In contrast to fossil fuel, biodiesel emits lesser carbon dioxide making it eco-friendly because the cycle of carbon dioxide release and consumption is balanced (Sadaf et al. 2018). The alarming state of our current fossil fuel sources that keeps on depleting has compelled the researchers and industrialists to look for a plausible renewable substitute to fossil fuels. This topic has been widely and hotly argued over the years, and up until now, there are three different feedstocks of biodiesel. Edible oil from plants like soybean, corn, brassica, jatropha and palm are the first-generation of biofuels.

This plan was probable at first, but it is not future proofed in the long run because food and fuel issues have risen. The need for agricultural land and possible pollution has led to the search for second-generation biofuel. This type of biofuel is sourced from non-edible parts of the previous resources, such as the stems. However, the synthesis of second-generation biofuel is problematic because the oil yield is relatively low, pollution potential and are unsustainable (Aslam et al. 2018, Nisar et al. 2018, Obi et al. 2020). Hossain et al. (2008) expressed that biodiesel is a type of renewable and biologically degradable fuel that can scale down sulphur and crude matter emissions while functioning in transportation engines like petroleum-based fuels (Hossain et al. 2008).

[1] Algae and Biomass, Research Laboratory, Malaysia-Japan International Institute of Technology (MJIIT), Universiti Teknologi Malaysia (UTM), Jalan Sultan Yahya Petra, 54100, Kuala Lumpur, Malaysia.
[2] Department of Chemical and Environmental Engineering (ChEE), Malaysia-Japan International Institute of Technology (MJIIT), Universiti Teknologi Malaysia (UTM), Jalan Sultan Yahya Petra, 54100, Kuala Lumpur, Malaysia.
* Corresponding author: mustafwibinqamar@gmail.com

Consequently, microalgae were envisioned as the next sustainable biodiesel feedstock and have been studied ever since (Bagul et al. 2018). This candidacy is no longer a surprise because this microorganism has promising characteristics that are viable for biofuel synthesis: high growth rates, large lipid, and carbohydrates content, and immense biomass production without the need to be grown on land, unlike crops. Microalgae are also advantageous in terms of its robustness because they can grow well under various conditions, even when introduced to wastewater and flue gas (Ahmad et al. 2021). Furthermore, the environmental stressors resulted in a change of physiology, making biomass configuration rich and their nutritional value increase greatly (Hussain et al. 2021). Some of the significant characteristics and advantages of microalgae are incorporated in Table 1.

Table 1. Salient characteristics and advantages of microalgae (Ahmad et al. 2021).

Characteristics of Microalgae	Advantages of Microalgae
They are unicellular (eukaryotic or prokaryotic) microorganisms, having sizes ranging from 5–50 μm	Capability of CO_2 sequestration and wastewater remediation
Microalgae can be green, blue-green, red, brown, and golden taxonomically	Can be cultivated in different types of habitats (salinity, pH, temperature)
300,000 species explored and about 30,000 archived	Faster growth rate (10–50 times other terrestrial plants) and high photosynthetic efficiency (about 10%)
Well defined cell wall and nucleus	Microalgal biomass can be utilized to produce biofuels and other products
Photosynthetic organisms rich in chlorophyll and produce adenosine triphosphate (ATP) energy currency of life	Requires little or no agriculture land, can be cultivated in closed photobioreactors

Microalgae possess versatile metabolism in terms of growth, such as autotrophic, heterotrophic and mixotrophic cultivation. In autotrophic cultivation, CO_2 was photosynthetically converted to saccharides. The growth rate and biomass yield of microalgae depend on the rate of photosynthesis (Zhou et al. 2020). Therefore, CO_2 and light intensity should be specifically controlled. Initially, this is to assure that by virtue of microalgae cells, the emission of CO_2 is less than the fixation of CO_2, and to achieve that, the intensity of light is kept above the compensation point (Wahidin et al. 2013). But if the light intensity is too high, it will limit the microalgal growth due to the oxidation stresses induced by light irradiation. At the same time, the concentration of CO_2 in the medium should be kept above the CO_2 compensation point.

In heterotrophic cultivation microalgae, species grow without illumination by utilizing organic carbon such as glycerol, glucose, acetate, volatile fatty acids (VFAs) etc. (Perez-Garcia et al. 2011). The biomass yield is better than autotrophic cultivation, but the cost of carbon sources is more than CO_2. To overcome this cost issue, the large-scale cultivation of microalgae carbon can be obtained from wastewater streams (Kim et al. 2013). Mixotrophic cultivation can improvise the technical deficiencies of both types of cultivation by contributing in the following way. Microalgae cells grow in the heterotrophic mode by assimilating organic carbon and thus increasing biomass productivity and lipid synthesis (Ho et al. 2011). Microalgae cells also utilize inorganic carbon under autotrophic mode and photosynthetically produce oxygen, thereby controlling CO_2 emissions. Therefore, the mixotrophic cultivation mode is having a positive impact on the cellular compositional synthesis and microalgal growth (Lowrey et al. 2015). The growth modes are illustrated in Fig. 1.

Over recent years, microalgae that accumulate high lipid content have been the limelight of researchers and industrialists in search of potential feedstock for biodiesel synthesis. The most acknowledged sources of biodiesel, owing to its high growth rates as well as high accumulation of lipids, are these two groups: green algae or Chlorophyta, and diatoms, also known as *Bacillariophyta*. Microalgae is unique in metabolism as they can utilise both types of carbon for the

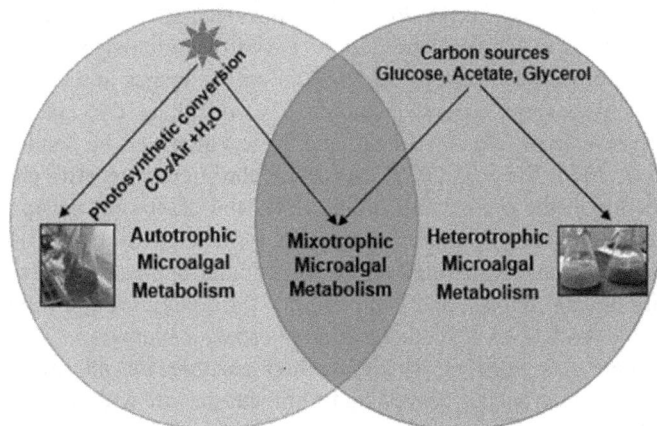

Fig. 1. Versatility of microalgae metabolism.

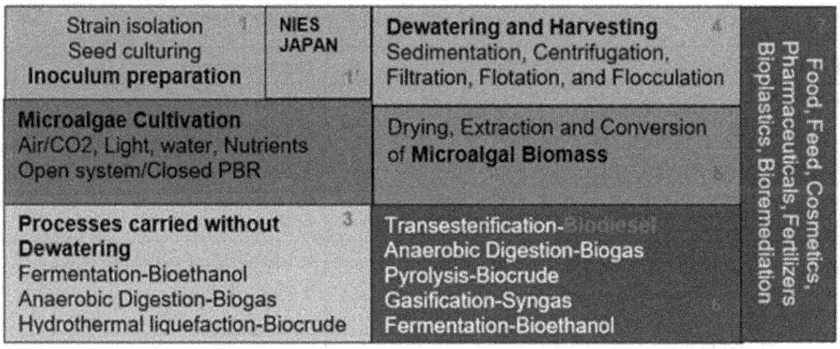

Fig. 2. Routes for obtaining biofuels from microalgae.

biosynthesis of lipids which are organic and inorganic carbon. Some organic carbon examples are acetate and glucose, while atmospheric carbon dioxide (CO_2) is an inorganic carbon. Lipids that are accumulated in microalgae can be divided into two types of lipids which are polar and neutral lipids. Phospholipids and triglycerides are examples of polar and neutral lipids, respectively. According to Bagnato et al., in 2017 (Bagnato et al. 2017), to produce biodiesel, the essential raw material for biosynthesis is neutral lipids. The routes to obtain biofuels (biodiesel) and other myriad products from microalgae are illustrated in Fig. 2. The whole process involves the cultivation of microalgae in closed or open systems depending upon the optimum biotic conditions for the specific microalgae species. The cultivation plays a vital role in the productivity and yield of biomass, which is later harvested for further processing (Ahmad et al. 2021).

Essentially, biofuel exploitation from microalgae encompasses several processes that include cultivation of microalgae, harvesting, dewatering, or drying of biomass, extraction of microalgal lipids, and, lastly, transesterification. The chapter provides insight into different methods of harvesting microalgae, drying and dewatering of microalgal biomass, and lipid extraction techniques. The chapter will also elaborate on different transesterification processes adopted to attain biodiesel.

2. Harvesting of Microalgae

The biodiesel production from microalgae biomass begins with microalgae cultivation, followed by microalgae harvesting and further downstream processes, such as dewatering, extraction, and transesterification steps (Tan et al. 2021). Previous studies reported that these downstream processes contribute more than 80% to biodiesel production cost from microalgae biomass, including the microalgae harvesting process (Vasistha et al. 2020). The challenges in the microalgae harvesting process are affected by several factors, including diluted culture concentration (0.2–5 g/L), a relatively small cell size (typically 2–10 μm), and large volume to be handled (Hønsvall et al. 2016, Laamanen et al. 2016, Xia et al. 2017). Moreover, the slightly negative charge on the surface of microalgae cells due to the presence of amine (NH) and carboxylic group (COO–) cause the microalgae cells to be relatively stable in suspension form (Barros et al. 2015). Therefore, it is crucial to select a suitable harvesting method with maximum biomass recovery, cost and energy-efficient, low environmental impact, and compatibility for a large group of microalgae strains to support the sustainable production of biodiesel from microalgae biomass.

Nevertheless, there has been no single method suitable for all microalgae strains and cultivation types. Among microalgae, harvesting techniques that are widely used are sedimentation, centrifugation, filtration, flotation, and flocculation. The advantages and disadvantages of each method and its recent application in biodiesel production are summarized in Table 2.

Table 2. Advantages and disadvantages of different harvesting methods.

Method	Mechanism	Advantages	Disadvantages
Sedimentation	Gravity settling	• Zero energy • Simple method	• Low efficiency • Time-consuming
Centrifugation	Rotation	• High efficiency • Rapid method • Universal	• Energy intensive • High capital and maintenance cost • May damage cell
Filtration	Membrane separation	• High efficiency • No contamination • Rapid method	• Membrane fouling/clogging • High cost
Flotation	Bubble attachment	• Low-cost • Less energy intensive	• Contamination from additional chemicals • Less efficient
Flocculation	Sedimentation by flocculant	• Rapid method • High efficiency • Cost-effective	• Contamination • Hard to upscale

2.1 Sedimentation

In the sedimentation process, the separation of microalgae cells from a liquid medium is mainly due to the gravitational force. This method is feasible to be applied in small scale and large scale microalgae cultivation with low energy requirements since separation is mainly due to the gravitational force (Laamanen et al. 2021). However, due to the negative surface charge of microalgae cells, this method is usually time-consuming, which is not preferable in commercialization (Laamanen et al. 2016). Often, despite the low energy requirement, the biomass recovery of this method is considerably low (10–50%), which subsequently leads to high biomass loss and low lipid productivity (Fasaei et al. 2018). In most cases, additional chemicals are added to the microalgae culture to destabilize microalgae cells and induce sedimentation (Xia et al. 2017). This method, namely flocculation, will be described in the upcoming section.

2.2 Centrifugation

Centrifugation is the most commonly used method for microalgae harvesting because of its high efficiency of biomass recovery (> 90%), simple process and can be applied to a wide range of microalgae strains (Najjar and Abu-Shamleh 2020). Moreover, the centrifugation method can be used to separate highly diluted microalgae culture, achieving concentration biomass harvested in a short period of time (Brennan and Owende 2010). This was proven by previous literature, which demonstrated high biomass recovery (up to 90%) within 2–5 minutes centrifugation time (Javed et al. 2019). Another advantage of using the centrifugation method is that the harvested biomass can be preserved and stored for a long time without damaging microalgae cells' quality (Najjar and Abu-Shamleh 2020).

Despite its numerous advantages, the application of centrifugation as microalgae harvesting is argued to be not sustainable in the long term, as it involves high capital and operational cost (Zhang and Zhang 2019). The centrifugation process is energy-intensive, especially in large-scale production (Wang et al. 2015). In addition to that, the labour cost should also be considered when using the centrifugation method. Although this issue can be solved with automation, this field is still under research and requires extensive studies. The centrifugation process involves high gravitational and shear forces, leading to cell breakage and subsequently, loss of bio-products. Many studies investigated the correlation between microalgae species, such as *Tetraselmis* sp., *Isochrysis* sp. and *Dunaliella* sp., and their tolerance towards such forces (Japar et al. 2017). It was suggested that optimising rotational speed is almost always needed to maintain microalgae cell structure and achieve high product yield.

2.3 Filtration

The filtration method involves the separation of microalgae biomass (solid phase) from the culture medium (liquid phase) through a filter or membrane (Junior et al. 2020). Conventional filtration methods include dead-end filtration, vacuum filtration, pressure filtration and microfiltration (Dragone et al. 2010). Meanwhile, the advanced filtration process includes counter-current filtration that has been proven to reduce clogging and fouling (Singh and Patidar 2018). Microalgae harvesting through the filtration method is highly dependent on the strain of microalgae and the size of their cells. For example, separation of larger microalgae cells (> 70 μm) or colony-forming microalgae can be achieved with conventional filtration method, such as dead-end filtration (Barros et al. 2015).

However, the same system cannot be utilized for harvesting smaller microalgae cells due to rapid clogging of the filters. Microalgae strain with smaller cell size (2–10 μm) is typically separated with microfiltration and ultrafiltration (Hapońska et al. 2018). It is important to find the most suitable filtration method and the filter types for the respective microalgae culture in order to maximize the biomass recovery and minimize the risk of membrane fouling and clogging (Deconinck et al. 2018). The cost of the filtration method is highly associated with the replacement of filtration membranes and the energy for the pumping system (Junior et al. 2020). However, when compared with other harvesting methods, filtration process is considered as cost-effective. Moreover, by using the filtration method for microalgae harvesting, the risk of contamination by other chemicals is eliminated, which is useful for the recovery of other valuable compounds (Singh and Patidar 2018).

2.4 Flotation

Flotation, commonly referred to as inverted sedimentation, is a separation process whereby biomass is attached to small rising bubbles that rise or float to the surface (Laamanen et al. 2021). This method is usually combined with the flocculation process by adding coagulants to destabilize microalgae cells in suspension form (Amaro et al. 2011, Coward et al. 2013). The flotation method possesses numerous advantages, including providing a high overflow rate, low detention period, and smaller footprints (Laamanen et al. 2016, Rubio et al. 2002). Moreover, microalgae cells sometimes have the tendency to float rather than get sedimented.

Flotation process is divided into three main steps: (1) bubble generation (2) microalgae adhesion to rising bubbles and (3) layer formation at the surface of the system (Zhang and Zhang 2019). There are many options for the operational system of flotation method, which can be divided based on the bubble generation method and types of pre-treatments used. The most commonly used bubble generation method is dissolved air flotation by using pressurized (400–650 kPa) saturated water to generate air bubbles between 10–100 µm size (Laamanen et al. 2016). A less energy-intensive method is dispersed air flotation, which introduces air through a porous medium (Rubio et al. 2002). However, this system is reported to be less efficient in trapping microalgae cells since the bubble produced has a larger size within 700–1500 µm (Garg et al. 2014). The flotation harvesting method is typically useful for bulk harvesting. Currently, the flotation system is still advancing to achieve higher biomass recovery. For example, a bubble encapsulation system, where the microalgae cells are encapsulated with the air bubble, creates a more stable foam and enhances separation efficiency (Zhang et al. 2017).

3. Dewatering and Drying of Microalgal Biomass

Microalgae harvesting aims to remove water content which is mainly the culture media from microalgal biomass. This procedure is important as it precedes lipid extraction, whereby it could affect the whole lipid yield later. Nonetheless, to advance to energy-dynamic and economic practice for obtaining microalgal biodiesel, these downstream processes (harvest, dewatering and lipid extraction) are often consequential bottlenecks (Patel et al. 2019) as it makes up to about 84.9% of overall energy usage in the total assembly line (Culaba et al. 2020, Lardon et al. 2009). Basically, most algal drying methods are designed from traditional dehydration of sewage slurry.

Factors of drying have an impact on the chemical properties of the biomass (Sahoo et al. 2017). If the microalgal biomass content is high in water, the efficacy of biodiesel synthesis, later on, will be reduced, thus reducing the outcome (Tan et al. 2018). However, as reported by Atadashi et al. in 2018, even if the water presence is as low as 0.1%, the yield of oil extracted is still affected and lessen the methyl esters that are about to be transesterified. Nevertheless, the processing of biodiesel from the totally dried microalgae sample is energy-intensive.

In the comparison of lipid extraction from a totally dried sample to the sample that is newly dehydrated, the former is less complicated. Sample of *Schizochytrium limacinum* strain that has been dewatered yielded 45% of lipid (dry base) through Soxhlet extraction coupled with n-hexane (Tang et al. 2011). In 2011, Sheng et al. acquired 48% dry-based lipid from a totally dried *Synechocystis* PCC 6803 sample by applying ethanol during extraction. Supercritical extraction using CO_2 and ethanol has been used to obtain a 34% (dry base) yield of lipid from dried *Schizochytrium limacinum* powder (Tang et al. 2011). Apart from that, via mixed solvent extraction whereby methanol-ethyl acetate was used in the ratio of 2:1, an 18.1% dry-based lipid yield was extracted from Chlorella spp. powder (Wu et al. 2017). Despite the seemingly successful extraction, microalgal sample drying consumes intense energy and power.

Dewatering step via a mechanical process by using thermal drying necessitates 3560 kJ of energy input, only to separate 1 kg of water. This exhibits the negative net energy balance and, consequently, reveals that the output of energy from extracted lipid is lower than the energy required to accumulate biodiesel (Ansari et al. 2017). Meanwhile, thermally dried microalgae that is used to produce biodiesel uses intense energy up to 4000 times higher than that of biodiesel synthesis from mechanically dehydrated microalgae with approximately 20% of biomass from microalgae slurry (Lardon et al. 2009). Therefore, thermal drying should be dismissed for a successful, positive net energy balance (Ansari et al. 2017).

It is required to dehydrate microalgae due to its easily spoiled attributes after harvesting to avoid decomposition of the sample. Upon harvesting microalgae from the growth medium, the microalgae slurry is further dewatered for next processing and to ensure stability. Dewatered and dehydrated microalgae subjected to a press is beneficial in biodiesel production since it can notably aid in accumulating algal oil or lipids (Show et al. 2013). The most plausible drying method should be summarised so that there is no microalgae disintegration resulting from the drying process.

Dehydrating microalgae is essential for the higher oil yield to be collected, but the process incurs so much damage on production capital cost as it uses around 75% of the overall cost. Some examples of drying and dewatering method include rotary drying, solar drying, vacuum-shelf drying, spray drying, crossflow air drying and freeze-drying (Show et al. 2013). Sahoo et al. reported in 2017 that these procedures are costly and unprofitable, and seasonal. Hence, several researchers have been focusing on the optimisation of lipid extraction from microalgae slurry or wet microalgae, which will be discussed further in the lipid extraction subtitle.

Thus far, there are a few different methods of drying microalgae from its culture medium suspension: rotary drying, spray drying, vacuum drying, flash drying, vacuum shelf drying, and crossflow drying (Culaba et al. 2020).

An investigation in determining microalgal lipid yield was conducted by Viswanathan et al. in (2012) that utilised convective drying with numerous temperatures as well as the velocity of wind. The use of the solar device in the solar drying method targeted specific microalgae species, which are *Scenedesmus* and *Spirulina* species (Prakash et al. 1997). The solar drying technique is presumed to be old-fashioned. Nevertheless, this method is at a disadvantage owing to its weather-dependent characteristics. Countries that are not blessed with ample sunlight would not be able to implement this method. In addition, the dehydrated samples will no longer possess their nutritional value; therefore, other types of drying methods came to light, like microwave drying (Al Rey et al. 2016, Show et al. 2015). Assessment on each drying method, particularly for human consumption purposes, needs to be conducted. In the following sections, these methods are discussed further.

3.1 Rotary Drying

In this drying system, the drying apparatus is a sloped cylinder that rotates by gravity and moves the algae from one end to the other to be dried. It is also known as a rotary dryer. An experiment by Soeder and Pabst (1957) had resulted in a successful product from drying of *Scenedesmus* sample via thin layer drum dryer. The bio-product also had been tested. This process has an advantage on sample sterilisation and cell wall disruption. In another report, a drum dryer with a surface area of 2.5 m^2 could aid in thickening microalgae sludge by up to 25% of dry samples (Mohn and Soeder 1978). In addition, a pilot study intended for testing electric drum-dryer was investigated by Becker and Venkataraman (1982). The wet sludge consisted of 30% of microalgae *Scenedesmus* sample and was run for 10 seconds at 120°C. This study was conducted with an energy usage of 52 kWh.

Meanwhile, instead of using electric drum dryer, a steam dryer could help lower the cost of processing by 6.8 times. This is possible because at a mere 8 ATM of steam pressure, up to 50 kg of water per every m^2 of the drum surface could be eliminated (Mohn and Soeder 1978). Moreover, the energy-wise cost could certainly be further lessened if waste steam is supplied. Soeder and Pabst had also reported their finding in 1975 regarding their evaluation of energy requirements. Dehydration of algae that has 4% of water bodies would necessitate heat energy that reached 15.7 Mcal. The energy dispersed as much as 18.2kg of water from every 1kg of dried algae. It is worth mentioning that an auxiliary electrical power input of 1.4 kW h was required for the dryer to be operating.

The suggestion to this method is that, since the necessity of electrical energy massively depends on water bodies of the ultimately dried microalgae, with larger amount of water in order to save energy costs.

3.2 Spray Drying

Shelef et al. (1984) observed that spray drying employs atomisation of liquid, mixing gas or droplets and liquid droplets drying. The typical process flow of vapourised water droplets is downward spraying into an upright tunnel. The water droplets will go into hot gases once they enter the tunnel, and in a few seconds, the drying process is started and completed. Finally, the dried sample is collected at the bottom of the vertical tunnel, and waste gas is dissipated via a rapid dust separator.

As the reference to research by Soeder (1980), spray drying is a suitable method for microalgal end-product that is intended for human consumption. Although this method is very adept, unbroken cells could be disrupted due to its atomisation process with high pressure. This would also affect the quality of the product, and this is rendered to be undesirable deterioration. Prevalently, the major disadvantage of spray drying method is its high capital cost and low digestibility of the dehydrated algae sample. Mohn and Soeder (1978) discovered that between drum drying and spray drying, the former is proposed to be better in terms of production cost, lower power consumption and ability of digesting.

3.3 Crossflow Air Drying

The fourth method of biomass drying is crossflow air drying. Wet slurry sample of microalgae *Spirulina* that still has about 55–66% of water content was processed through crossflow air drying. This drying method ran in a compartment dryer at 14 hours long with a temperature of 62°C. The resulting product is of high-quality dried algal cake with 2–3 mm thickness. This dehydrated microalgal cake still has around 4–8% of moisture (Becker and Venkataraman 1982). Mayol et al. stated in their study in 2015 that this drying method does not rupture cell walls, and it is relatively cost-saving (Mayol et al. 2015). Detailed evaluation on comparison of this method with solar drying and drum drying disclosed that crossflow air drying is much cheaper among the two (Show et al. 2021).

3.4 Vacuum-shelf Drying

This type of method was conducted at a temperature of 50–65°C with pressure at 0.06 ATM upon microalgae *Spirulina* slurry as the subject. The sludge was dehydrated until the remaining moisture reached only 4% in a vacuum-shelf dryer (Becker and Venkataraman 1982). These two researchers discovered that the dried algae sample possesses a porous structure of biomass and hygroscopic feature. Hygroscopic means it absorbs moisture from the surrounding ambience. However, vacuum-shelf drying exhibits a non-cost-effective strategy as the overall production cost, including capital and running, are comparatively higher than other methods, as highlighted by Becker and Venkataraman in the same study in 1982.

3.5 Flash Drying

This method of flash drying is derived from drying wastewater slurry that was designed in the 30s. To rapidly get rid of water content, the sample must have a mixture of dehydrated and wet algae and is subjected to a hot gas steam (Shelef et al. 1984). The turbulent hot gases serve as a carrier for the mass transfer of moisture from algae slurry to the gases. The source of the hot vapour portrays a strong influence on drying process cost and the microalgal end-product. Other than that, it would be of great help if the waste steam was uncontaminated. It would greatly aid in making sure that the final product is of good quality and lessening the cost.

4. Microalgal Lipids

4.1 Overview of Microalgal Lipids

Commercially valuable treasures offered by microalgae in tremendous amounts encourage scientific research through this route. Moreover, the universal characteristics of its lipid really aid in exploiting every potential application like health supplements, aquaculture feed, nutraceuticals, biofuels, and biopolymers.

Biosynthesis and metabolic pathways of most lipid levels are associated with fatty acids and their constitutions greatly defines the features and practicality. Maltsev and Maltseva (2021) state that the overall number of microalgal fatty acids derived from various habitats is formed by 135 fatty acids (Maltsev and Maltseva 2021). Considering the hydrocarbon chain length, the framework with the availability of substituents is categorised into a few groups. The total number of even carbon number groups are 81 (2 short chain, 14 medium chain, 28 long chain, 37 very long chain). In addition, there are 33 odd numbered carbon atoms and 21 branched hydrocarbon chains equipped with additional functional groups (Maltsev and Maltseva 2021). A graphical summary of microalgal lipid is illustrated below in Fig. 3.

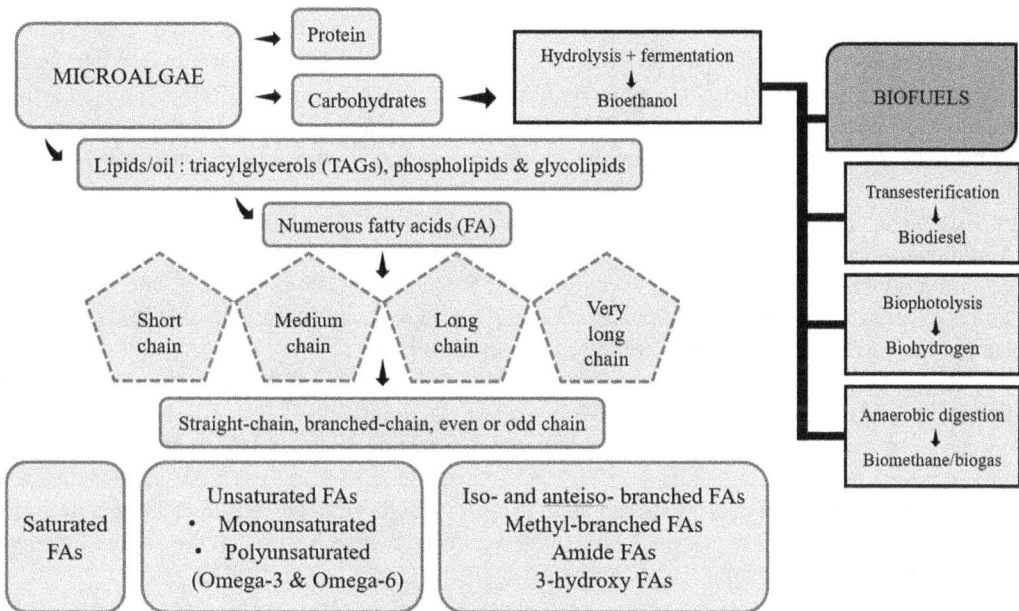

Fig. 3. Transformation of microalgal lipids.

In microalgae, lipid capacity makes up about 5–77% of its biomass, and the fatty acid configuration is not necessarily similar in all species. This amount, however, is species-specific (Bernaerts et al. 2019). It depends on the species of microalgae, the life cycle of microalgae and the condition of its cultivation. These condition parameters are growth medium, light intensity and period, temperature, airflow rate and others (Gour et al. 2018). Higher polar lipids content had been detected during the growth phase as compared to those during the stationary phase. In certain microalgae, oxygen depletion might promote lipid biosynthesis as the lipid yield had potentially increased from 10 wt % to approximately 20 wt % out of their biomass (Dunstan et al. 1993).

This statement is supported by a similar report by (Srinuanpan et al. 2018) portraying that microalga is highly affected by nutrient deprivation and responding by enhancing their metabolisms towards the accumulation of neutral lipids. Jónasdóttir (2019) stated that his study on microalgal fatty acids resulted in thousands of fatty acid profiles. His analytical finding has become the guidance for researchers from a wide variety of work and background to learn about numerous issues concerning microalgal fatty acids (Galloway and Winder 2015, Jónasdóttir 2019). The investigations conducted by these people are vital in the sense of generalisation and implementation of an up-to-date approach in feedstock candidacy of microalgae.

Each species of microalgae carries its own unique constitution of fatty acids and lipids. Microalgae that create a large volume of oil is acknowledged as a plausible feedstock of third-generation biofuels. Generally, these fatty acids that make up the microalgal lipid govern the synthesised lipid's purpose (Maltsev and Maltseva 2021). However, each distinct series of fatty acid do not have the same properties. A few important factors are considered regarding the commercialisation of fatty acids, including hydrocarbon chain length, fatty acids ratio, position, presence, and the volume of double bonds.

Basically, lipids are comprised of saturated and unsaturated fatty acids that carry at least 12 carbon atoms, up until 24 carbon atoms. Essential fatty acids are described as hydrocarbon chains that are accompanied by a carboxyl group and a methyl group on either end of the particle (COOH and CH, respectively). Saturated fatty acids (SFAs) are fatty acids that do not carry double bonds alongside the chain. In contrast, unsaturated fatty acids are fatty acids that carry double bonds within hydrocarbon chains, and they are divided into 2 types: monounsaturated fatty acids (MUFAs) and polyunsaturated fatty acids (PUFAs). To understand better, MUFAs are the ones that possess only one double bond, and PUFAs are fatty acids that are occupied with more than one double bond. PUFAs are also further divided into several fatty acids that include Omega-3 (N-3) and Omega-6 (N-6) fatty acids (Remize et al. 2021). Essential fatty acids that originate from Omega-3 are α-linolenic acid (ALA), docosahexaenoic acid (DHA), as well as eicosapentaenoic acid (Abdimomynova et al. 2019). Living up to its name, these fatty acids carry their first double bond at the third terminal-methyl carbon of the chain. As for N-6 acids or Omega-6 PUFAs, the first double bond is positioned at the sixth and seventh carbon. Arachidonic acid (ARA), linoleic acid (LA), and Gamma linoleic acid (GLA) are several types of Omega-6 PUFAs.

The prohibition capability against diverse diseases, including arthritis, thrombosis, and atherosclerosis, makes DHA and EPA widely known. Traditionally, EPA and DHA are primarily sourced from marine fish oil. It remained the major feedstock until studies on the efficiency of microalgae in accumulating omega-3 fatty acids became extensive. Essential fatty acids are associated with the health of humans and animals. Still, they need to be consumed by diet because our body can only synthesise carbon double bonds upon carbon number nine of the terminal-methyl carbon (Jones and Papamandjaris 2012).

Several researchers studied seven microalgae species (*Ankistrodesmus* sp., *Botryococcus braunii, Dunaliella bardawil, Dunaliella salina, Isochrysis* sp., *Nannochloropsis* sp., and *Nitzschia* sp.) regarding their fatty acid composition. From the study, all microalgae have similarities in the type of fatty acid that they generate, which are C14:0, C16:0, and C18:1 until C18:3, while for other fatty acids and their concentration, it is subject to the species themselves. For instance, *Nannochloropsis* sp. possesses C22:6, C16:2, C16:3, C20:5, C16:2 and C16:3. As for *Isochrysis* sp.

and *Nitzschia* sp., these two strains have a fatty acid chain of C18:4 and C20:5, respectively. C16:4 and C18:4 belonged to lipid bodies in *Ankistrodesmus* sp. (Mata et al. 2010, Thomas et al. 1984). In the prospect of quality, fatty acid chains could be as short as 10 carbons to 24 carbons long. The chains are usually analogous between each species of the same class or phyla, yet highly vary amongst classes and phyla from cyanobacteria to oleaginous species of eukaryotic alga.

Nevertheless, it is an exception for cyanobacteria as there is an oil level hike upon starvation of nitrogen (Moazami et al. 2011). Quantitatively, the total lipid productivity among species differs from each other with very little accumulation of merely 4.5% to up until 80% out of its dry biomass weight (Hu et al. 2008). It has been observed that lipid accumulation could differ throughout its lifespan depending on which growth phase the microalgae are in. More precisely, the lowest yield of lipid is often found at the log phase, and as it approaches lag and stationary phase, the oil accumulation could go higher or stable after it has reached the maximum amount according to its capability. Some species could also even exhibit higher lipid accumulation if the culture time is longer (Hu et al. 2008, Xu et al. 2008).

Cell configuration of microalgae is protected by membrane lipids that are based on glycerol, which is highly produced during logarithmic growth (Hu et al. 2008). This is contradicting to triacylglycerides (TAGs) as they are neutral lipids that do not have a structural function, and they are used for the depository. The differences in qualitative and quantitative aspects of oil level in microalgae certainly affect the oil's target. In conventional methods, microalgal oils are employed as the food source for the growth and metabolism health of aquaculture animals. Microalgal oil usage in this sector depends on fatty acid composition and content obtained from the microalgae because it promotes better productivity of the cultured animals. Therefore, the selection of microalgal species revolves around hyper-accumulating microalgae that is better in lipid production.

Examples of microalgae with lipid yield (% dry weight) are recorded in Table 3. As listed in the table, lipid capacity in microalgae can be achieved as high as 75% of its dry biomass weight. Nearly all microalgae listed possess oil content exceeding 10%, but optimum and suitable growth conditions and appropriate stress would promote higher oil productivity.

Table 3. Lipid content of several microalgae species.

Microalgae	Lipid Content (% dry wt.)	References
Botryococcus braunii	25–75	(Chisti 2007)
Coelastrum sp.	31.0	(Bhuyar et al. 2021)
Chlorella sp.	21.3	(Mathimani et al. 2021)
Scenedesmus sp.	26.5	
Monoraphidium contortum	22.2	(Bogen et al. 2013)
Nannochloropsis sp.	52	(Moazami et al. 2011)
Neochloris sp.	46	
Nannochloropsis oculata	22.7–29.7	
Chlorella sorokiniana	19–22	
Isochrysis galbana	7–40	(Ramaraj et al. 2015)
Pavlova lutheri	35.5	
Scenedesmus dimorphus	16–40	
Nannochloropsis sp.	12.53	
Dunaliella salina	6–25	(Mata et al. 2010)
Isochrysis sp.	7.1–33	

4.2 Lipid Extraction of Microalgal Biomass

Neutral lipids are the lipid with low unsaturation degree, and this type of lipid is substantial in producing biodiesel. Nearly all microalgae synthesise this type of oil. To productively extract components that are from inside the wall of cells, cell breakage of disruption is a general prerequisite. Be that as it may, some researchers opted for breaking of cells first then proceeded with lipid extraction, some other experimented with lipid extraction without cell disruption. Over the past decades, extraction techniques have been plentiful,the selection of extraction techniques relies on the characteristics of algal cell wall and end-product feature. The technique should be focused in obtaining the maximum value of the materials; thus, it must be fast and accurate. In an industrial scale, an adequate disruption method is chosen based on the sturdiness of the cell walls, downstream processing size, the risk of lower extraction rate, safety issues, and capital expense (Show et al. 2015).

The cell disruption process is essential as it demonstrates the value of the extracted products. Additionally, appropriate selection of cell disruption method and equipment is also substantial. Munir et al. (2013) mentioned that almost all extraction studies were conducted on thick biomass cake or paste weighing around 50–200 kg/m^3 revolving the required energy to be used and the production cost (Munir et al. 2013). Lipid extraction is then followed by identification and quantification of the lipid sample via analytical methods like gas-chromatography mass-spectrometry (GCMS) (Bhuyar et al. 2021), gas-chromatography flame-ionisation detection (GS-FID) and thin layer chromatography (TLC).

The extraction method selection highly relies upon competence and performance, simplicity, adequateness, robustness, and system throughput value. A wide variety of cell disruptions are derived to break the sturdy cell walls to remove the cell contents. These methods have their own advantage and disadvantages. This part of downstream processing can be sub-categorised into series, namely, mechanical and non-mechanical methods. Fig. 4 shows several other possible options for cell disruption methods.

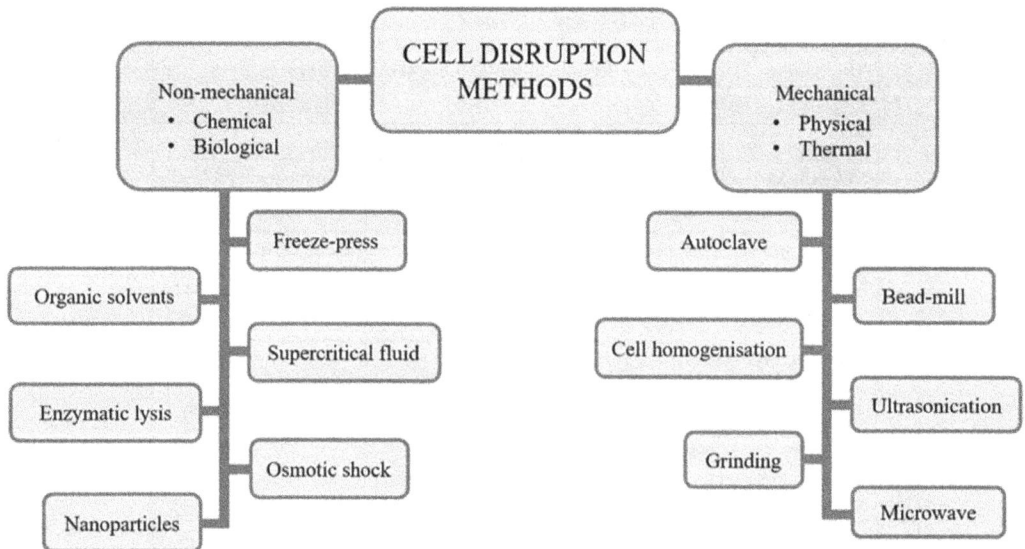

Fig. 4. Alternate techniques of lipid extraction.

4.2.1 Organic solvent extraction

Lipid extraction using solvent as the extractor is used quite frequently by scientists. There are numerous solvent types: chloroform, ethanol, methanol, n-hexane, acetone, acetonitrile, and benzene.

The benefit of applying solvent extraction techniques is that they are cost and time saving as well as very adept in extracting microalgal oil (Ramaraj et al. 2015). It is highly imperative that solvents to be used should not dissolve in water, be easily acquired, reused, and possess a low boiling point (Show et al. 2015). Arabian (2021) used the mixed solvents of chloroform-methanol for their lipid experiments to achieve up to 20.39% of lipid from its dry biomass (Arabian 2021). It has been observed by Halim et al. in 2011 that n-hexane is commonly used to extract lipid due to its lesser toxic traits and its non-polar attraction towards non-lipid contaminants (Halim et al. 2011).

(Bhuyar et al. 2021) had also observed a high extraction rate that yielded 31% of dry lipid from the biomass by using hexane as the solvent. Back in 1997, Yaguchi et al. utilised chloroform-methanol for oil extraction and resulted in almost 83% of dry lipid weight (Yaguchi et al. 1997). Hexane was used by Miao and Wu (2006) for oil extraction from *Chlorella protothecoides* that accumulated lipid amounts at the maximum of 55% per dry weight (Miao and Wu 2006). In addition, different ratios of mixed solvent would also affect the yield of extracted lipid, as investigated by (Ryckebosch et al. 2012), who obtained better oil yield when a comparative study was conducted. It resulted in better yield productivity from four different microalgae by using the ratio of 1:1 chloroform: methanol. However, solvent extraction has been discovered by Kapoore et al. (2018) to be not friendly to the environment and is not entirely cost saving. It also poses threat to human health and the original integrity of the end-yield is also not high quality.

4.2.2 Supercritical fluid extraction (SFE)

Extracting lipid from supercritical fluid extraction was one of the most productive techniques, as suggested by (Pagels et al. 2021). This method utilises the properties of liquid and gases that are subjected to high pressure and temperature. The most well-known supercritical fluid applied is carbon dioxide (CO_2), and every now and then, it is co-used with other solvents like methanol or ethanol (Cooney et al. 2009). Carbon dioxide is initially heated and condensed until it precedes the critical point or the liquid-gas mode is attained. This state allows the liquid to be the solvent, and it is returned to atmospheric pressure; there will be no excess matter left (Mercer and Armenta 2011). SFE is recommended as a rapid and eco-friendly method as no toxic solvent is involved and is generally safe for sensitive samples. Andrich et al. (2005) also shared that this method is easily recovered and usable at low temperatures. The swiftness of SFE is owing to carbon dioxide solvent that becomes less viscous and highly diffusive after achieving liquid-gas state (Andrich et al. 2005).

Besides that, the success of SFE technique is highly impacted by four major aspects, viz. extraction time, the flow rate of CO_2, pressure, and temperature. Mendes et al. (2006) discussed that employing ethanol as the co-solvent with 10–15% concentration yielded similar amount of extracted oil by using Bligh and Dyer method.

The oil sample was obtained from *Arthrospira maxima* and *Spirulina platensis*. Lipid extraction from *Chlorococcum* sp. that was conducted by Halim et al. in 2011 resulted in greater lipid concentration of 5.8% g lipid from dry biomass. SFE efficacy can be heightened with water presence in wet microalgae biomass (Ramaraj et al. 2015). SFE is an ideal method since it is non-toxic and easily recovered. Although SFE is an effective method of extraction, the main drawback with this technique is that the capital expense involved is very costly because of the electricity and pressure involved and high maintenance of the system (Show et al. 2021).

4.2.3 Enzymatic extraction

There has been an alternative method to prevent the toxicity of chemical solvents which is by using biological extraction. The benefit of using enzymes is that they enhance the disruption of cell walls through hydrolysis to levitate microalgal lipid to go into the chosen solvent medium. The enzymatic reaction can also be coupled with chemical extraction or physical methods to improve extracted oil concentration and reduce the processing time (Mercer and Armenta 2011). Unfortunately, there are limited studies on the biochemical extraction of microalgal oil. For instance, Fu et al. (2010) investigated enzyme pre-treatment to extract lipid from *Chlorella sp.* This method applied cellulase hydrolysis pre-treatment for 72 hours and yielded 70% of sugar, while as for lipid, 56% g of lipid per dry mass was attained. This value is higher than the extracted lipid that was not treated with enzyme prior to extraction, which was merely 32% g of lipid per dry biomass.

Besides that, enzymes are added as catalysts via biological extraction. There are few examples of such catalysts, i.e., cellulose, lysozyme, pectinase, and xylanase (Gong and Bassi 2016) and to complete the procedure, solvent washing is needed (Zuorro et al. 2016). Selective way of work by enzymatic reaction for cell wall breakage or lysis is what stands this method out from others. Cell wall disruption is operated with the help of cellulase because hemicellulose and cellulose are the major constituents of nearly all cell wall (Kumar et al. 2017). Sierra et al. (2017) discussed that enzymatic extraction causes only a little reaction state regarding temperature and pH and zero corrosion. This method does not need drying as a prerequisite, and the outcome is in higher concentration (Sierra et al. 2017).

Biological pre-treatment does not necessarily imply a single enzyme or microbe type but rather better with a mixture of enzymes, as shown by scientific studies. Combined enzyme extraction facilitated the process because each technology benefits the other and eventually demonstrated strong synergy between them (Barati et al. 2021). For example, protein from *Candida lipoytica* was extracted and the yield was proven to be in higher amount upon cell disruption in alkaline solution coupled with a high-pressure homogenizer (Munir et al. 2013). Combined enzymatic reaction with other disruption technology might also aid in better resistance towards disruption, whereby mechanical disruption alone might be expensive. It is proven to be more efficient when enzymatic extraction is combined with bead milling (Alavijeh et al. 2020). Bead milling process was followed by hydrolysis with few enzymes such as cellulase, lipase, protease, and phospholipase. For lipid yield, a total of 75% dry lipid was obtained without hydrolysis by enzymes, and critically, the recovery concentration had increased at a maximum of 88% lipid when lipase was introduced. The disintegration of microalgal cell walls is also critically affected by enzyme and bacterial strain selection.

Nevertheless, there is a certain limitation from this kind of procedure, for instance, the purity of the end-product is quite arguable, and the productivity is lesser. The resulted oil is considered to be tainted or 'unoriginal', hence leading to more downstream processing, which incurs larger cost and time (Kapoore et al. 2018). The expenditure for enzymes is also not cost-effective as there is requirement in ensuring the enzymes' stability and durability during extraction because they are thermal and pH sensitive (Vernès et al. 2019). Processing of extracted samples by enzymes is not favourable in an industrial scale due to the probably long hours of reaction (Gong and Bassi 2016).

In carotenoids (pigment) extraction, the enzymatic processing method is chosen by Tavanandi et al. (2019) to gather allophycocyanin from *Arthrospira platensis* (Tavanandi et al. 2019). Specifically, enzyme lysozyme was utilised at 37°C with pH 7.0 for 20 hours long. In comparison to surfactant-based extraction, the addition of enzymes promoted purer allophycocyanin. Moreover, the application of pre-treatment beforehand would support extraction productivity. There was also an increment of up to 30% of the collected phycocyanin when ultrasound pre-treatment was administered prior to enzymatic extraction. Even though the production of new enzymes and isolates are tedious, naturally existing enzymes and microbes from organisms' gut from cellulosic biomass feeding may still be found and used (Barati et al. 2021).

There has not been much study on microalgal lipid extraction via enzymatic approach or coupled technology. Hence, adequate and further research on bacterial strain and enzymes and the mixtures is required to enhance enzymatic extraction.

4.2.4 Freeze-press extraction

Another method of non-mechanical extraction is freeze-pressing. Hughes press is an example of a freeze-pressing method in which hard frozen microalgal samples are put through a slim opening and exposed to a disruptive temperature of around –25°C. The temperature could be varied from just slightly sub-zero or more than that. In 2011, Schwede et al., observed comparative research involving French press, freezing and microwave heating techniques on microalgae *Nannochloropsis salina* (Schwede et al. 2011). The freezing was done overnight at –15°C, and microwave heating was operated at 100°C for 8 hours. The biomass extracted was then further analysed by digesting it anaerobically for biogas collection in batch tests. The result concluded that cell wall deterioration is indeed the limitation of microalgal biomass anaerobic digestion for *N. salina*.

Disruption efficacy and its influence on anaerobic digestion depend on the method of cell lysis executed. There is a reasonable increase in particular production of biogas and decomposition rate via disruption methods of heating, French press, and microwave. Through pre-treatment, biogas generation had risen by 40% as a consequence of using the microwave method. Meanwhile, the biogas production rate increased up to 33% and 58% through the heating and French press approach. Freeze-drying is another interesting and functional approach for biomass drying in small scale application, but rather too costly and energy extensive to be implemented on the industrial scale of microalgal downstream processing (Munir et al. 2013).

4.2.5 Osmotic shock extraction

Next, the osmotic shock approach is associated with an abrupt change in the amount or movement of water that transverse through the microalgal cell membrane. This sudden change causes stress to the cells. A few examples of the change could be adding concentrated additives or other solutes such as dextran, salt, polyethylene glycol, salt, and substrates. The osmotic stress leads to cell lysis, delivering intracellular elements. Like extraction by enzymes, to date, there is still no upscale application due to the costly capital expense (Mercer and Armenta 2011). In comparison to sonication, bead-beating and microwaves technique, osmotic stress was found to be the most productive method in the disruption of cells. This is possible by using 10% sodium chloride solution combined with vortexing for 60 seconds and kept for 48 hours. Lee et al. (2010) reported their experiment of *Scenedesmus* sp. and *Chlorella vulgaris* that was conditioned to osmotic shock, which portrayed similar extraction efficacy compared to bead-beating. It is relatively a simple method. Nonetheless, this approach requires long treatment hours. Hence, it is likely unfavourable because of the time-consuming feature (Lee et al. 2010).

4.2.6 Bead-milling extraction

Extraction of microalgal biomass by using the bead-milling process is considered as one of the highly useful methods. In general, this technique is correlated with the usage of solvent to draw out lipid. This method is truly cost-effective and productive if the biomass is largely concentrated and the targeted end-product is easily removed after cell lysis. Greenwell et al. (2010) stated that this method would stand out over other methods if the biomass weight of 100–200 g/L is used, making this technique energy-saving and highly efficient (Greenwell et al. 2010).

The bead mill equipment is designed to have either horizontal or vertical cylinders supported by many discs or other beating elements that act as the motor. High-speed glass beads that have a diameter of 0.1 mm are filled into the compartment that acts as the beater and spins at high speed to promote cell lysis (Show et al. 2015). Microalgal cells are mechanically sheared from the vigorous beating and spinning at 2800 rpm that has been applied in down-scale and up-scale purposes (Geciova et al. 2002). The effect of the disruption basically relies upon a collision between the cells' sample and the beads.

Apart from that, the beads' shape, size, and structure and the robustness of the algal cell wall play a vital role in disruption efficacy (Doucha and Lívanský 2008). Other parameters like the design of both agitator and chamber and biomass concentration are also crucial towards disruption productivity (Postma et al. 2017). Lee et al., suggested in 1998 that microwaving and bead-milling are the two most significant approaches compared to other methods. This is because the lipid concentration extracted from Botryococcus sp. was 28.6% and 28.1%, respectively. Bead-milling is also preferred over other methods like French press, sonication, and French press because extracted lipid amount from *Botryococcus braunii* is depicted to be larger. In the absence of hydrolysis by enzymes, bead milled biomass recovered a lesser lipid amount by 44% (Alavijeh et al. 2020). Even so, bead-beating is not adequate for industrial-scale application owing to its energy extensive requirement.

5. Conclusion

Biodiesel from microalgae can become a renewable alternative to the diminishing resources of fossil fuels. It is expected to suffice the global energy demand in an economically and environmentally sustainable manner. But the uphill tasks to attain biodiesel in an economically viable way are the high cost of downstream processing of microalgal biomass (harvesting, dewatering, and thermochemical processes). Therefore, researchers are giving a lot of stress to improvise the processes and technologies involved in the downstream processing of microalgae. This chapter gives a better perception and understanding about different processes of harvesting, dewatering and lipid extraction to obtain biodiesel in an environmentally sustainable manner.

Acknowledgement

The authors are thankful to the administration of Malaysia-Japan International Institute of Technology to provide the facilities on the auspices of Algae and Biomass Research Laboratory.

Conflict of Interest

The authors find no conflict of interest.

References

Abdimomynova, A., Kolpak, E., Doskaliyeva, B., Stepanova, D. and Prasolov, V. 2019. Agricultural diversification in low-and middle-income countries: Impact on food security. Montenegrin Journal of Economics 15(3): 167–178.

Ahmad, I., Abdullah, N., Koji, I., Yuzir, A. and Muhammad, S.E. 2021. Evolution of photobioreactors: A Review based on microalgal perspective. Paper Presented at the IOP Conference Series: Materials Science and Engineering.

Al Rey, C.V., Mayol, A.P., Ubando, A.T., Biona, J.B.M.M., Arboleda, N.B., David, M.Y., Tumlos, R.B., Lee, H., Lin, O.H., Espiritu, R.A., Culaba, A.B. and Kasai, H. 2016. 2016. Microwave drying characteristics of microalgae (*Chlorella vulgaris*) for biofuel production. Clean Technologies and Environmental Policy 18(8): 2441–2451.

Alam, M.A., Vandamme, D., Chun, W., Zhao, X., Foubert, I., Wang Z., Muylaert, K. and Yuan, Z. 2016. Bioflocculation as an innovative harvesting strategy for microalgae. Reviews in Environmental Science and Bio/Technology 15(4): 573–583.

Alavijeh, R.S., Karimi, K., Wijffels, R.H., van den Berg, C. and Eppink, M. 2020. Combined bead milling and enzymatic hydrolysis for efficient fractionation of lipids, proteins, and carbohydrates of Chlorella vulgaris microalgae. Bioresource Technology 309: 123321.

Amaro, H.M., Guedes, A.C. and Malcata, F.X. 2011. Advances and perspectives in using microalgae to produce biodiesel. Applied Energy 88(10): 3402–3410.

Andrich, G., Nesti, U., Venturi, F., Zinnai, A. and Fiorentini, R. 2005. Supercritical fluid extraction of bioactive lipids from the microalga Nannochloropsis sp. European Journal of Lipid Science and Technology 107(6): 381–386.

Angles, E., Jaouen, P., Pruvost, J. and Marchal, L. 2017. Wet lipid extraction from the microalga Nannochloropsis sp.: Disruption, physiological effects and solvent screening. Algal Research 21: 27–34.

Ansari, F.A., Gupta, S.K., Shriwastav, A., Guldhe, A., Rawat, I. and Bux, F. 2017. Evaluation of various solvent systems for lipid extraction from wet microalgal biomass and its effects on primary metabolites of lipid-extracted biomass. Environmental Science and Pollution Research 24(18): 15299–15307.

Arabian, D. 2021. Optimization of cell wall disruption and lipid extraction methods by combining different solvents from wet microalgae.

Aslam, A., Thomas-Hall, S.R., Manzoor, M., Jabeen, F., Iqbal, M., Uz Zaman, Q., Schenk, P.M. and Tahir, M.A. 2018. Mixed microalgae consortia growth under higher concentration of CO2 from unfiltered coal fired flue gas: Fatty acid profiling and biodiesel production. Journal of Photochemistry and Photobiology B: Biology 179: 126–133.

Bagnato, G., Iulianelli, A., Sanna, A. and Basile, A. 2017. Glycerol production and transformation: A critical review with particular emphasis on glycerol reforming reaction for producing hydrogen in conventional and membrane reactors. Membranes 7(2): 17.

Bagul, S.Y., Chakdar, H., Pandiyan, K. and Das, K. 2018. Conservation and application of microalgae for biofuel production. Microbial Resource Conservation, pp. 335–352. Springer.

Barati, B., Zafar, F.F., Rupani, P.F. and Wang, S. 2021. Bacterial pretreatment of microalgae and the potential of novel nature hydrolytic sources. Environmental Technology & Innovation 101362.

Barros, A.I., Gonçalves, A.L., Simões, M. and Pires, J.C. 2015. Harvesting techniques applied to microalgae: A review. Renewable and Sustainable Energy Reviews 41: 1489–1500.

Barrut, B., Blancheton, J.-P., Muller-Feuga, A., René, F., Narváez, C., Champagne, J.-Y. and Grasmick, A. 2013. Separation efficiency of a vacuum gas lift for microalgae harvesting. Bioresource Technology 128: 235–240.

Becker, E. and Venkataraman, L.V. 1982. Biotechnology and Exploitation of Algae: The Indian Approach: Deutsche Gesellschaft fur Technische Zusammenarbeit.

Bermejo-Barrera, P., Muñiz-Naveiro, Ó., Moreda-Piñeiro, A. and Bermejo-Barrera, A. 2001. The multivariate optimisation of ultrasonic bath-induced acid leaching for the determination of trace elements in seafood products by atomic absorption spectrometry. Analytica Chimica Acta 439(2): 211–227.

Bernaerts, T.M., Gheysen, L., Foubert, I., Hendrickx, M.E. and Van Loey, A.M. 2019. The potential of microalgae and their biopolymers as structuring ingredients in food: A review. Biotechnology Advances 37(8): 107419.

Bhuyar, P., Sundararaju, S., Rahim, M.H.A., Maniam, G.P. and Govindan, N. 2021. Enhanced productivity of lipid extraction by urea stress conditions on marine microalgae Coelastrum sp. for improved biodiesel production. Bioresource Technology Reports 15: 100696.

Bogen, C., Klassen, V., Wichmann, J., La Russa, M., Doebbe, A., Grundmann, M., Uronen, P., Kruse, O. and Mussnug, J.H. 2013. Identification of Monoraphidium contortum as a promising species for liquid biofuel production. Bioresource Technology 133: 622–626.

Bosma, R., Van Spronsen, W.A., Tramper, J. and Wijffels, R.H. 2003. Ultrasound, a new separation technique to harvest microalgae. Journal of Applied Phycology 15(2): 143–153.

Brennan, L. and Owende, P. 2010. Biofuels from microalgae—A review of technologies for production, processing, and extractions of biofuels and co-products. Renewable and Sustainable Energy Reviews 14(2): 557–577.

Buchmann, L., Frey, W., Gusbeth, C., Ravaynia, P.S. and Mathys, A. 2019. Effect of nanosecond pulsed electric field treatment on cell proliferation of microalgae. Bioresource Technology 271: 402–408.

Carter, M. and Shieh, J.C. 2015. Guide to Research Techniques in Neuroscience. Academic Press.

Chen, J., Leng, L., Ye, C., Lu, Q., Addy, M., Wang, J., Liu, J., Chen, P., Ruan, R. and Zhou, W. 2018. A comparative study between fungal pellet-and spore-assisted microalgae harvesting methods for algae bioflocculation. Bioresource Technology 259: 181–190.

Chen, J., Zhang, R., Xiao, J., Wang, L. and Guan, Z. 2011. Rectangular pulse sharpening of high voltage pulse transformer based on magnetic compression switch technology. IEEE Transactions on Dielectrics and Electrical Insulation 18(4): 1163–1170.

Chisti, Y. 2007. Biodiesel from microalgae. Biotechnology Advances 25(3): 294–306.

Cooney, M., Young, G. and Nagle, N. 2009. Extraction of bio-oils from microalgae. Separation & Purification Reviews 38(4): 291–325.

Coward, T., Lee, J.G. and Caldwell, G.S. 2013. Development of a foam flotation system for harvesting microalgae biomass. Algal Research 2(2): 135–144.

Cravotto, G., Boffa, L., Mantegna, S., Perego, P., Avogadro, M. and Cintas, P. 2008. Improved extraction of vegetable oils under high-intensity ultrasound and/or microwaves. Ultrasonics Sonochemistry 15(5): 898–902.

Culaba, A.B., Ubando, A.T., Ching, P.M.L., Chen, W.-H. and Chang, J.-S. 2020. Biofuel from microalgae: Sustainable pathways. Sustainability 12(19): 8009.

Deconinck, N., Muylaert, K., Ivens, W. and Vandamme, D. 2018. Innovative harvesting processes for microalgae biomass production: A perspective from patent literature. Algal Research 31: 469–477.

Dey, S. and Rathod, V.K. 2013. Ultrasound assisted extraction of β-carotene from Spirulina platensis. Ultrasonics Sonochemistry 20(1): 271–276.

Doucha, J. and Lívanský, K. 2008. Influence of processing parameters on disintegration of Chlorella cells in various types of homogenizers. Applied Microbiology and Biotechnology 81(3): 431–440.

Dragone, G., Fernandes, B.D., Vicente, A.A. and Teixeira, J.A. 2010. Third generation biofuels from microalgae. Formatex Research Center. https://hdl.handle.net/1822/16807.

Dunstan, G., Volkman, J., Barrett, S. and Garland, C. 1993. Changes in the lipid composition and maximisation of the polyunsaturated fatty acid content of three microalgae grown in mass culture. Journal of Applied Phycology 5(1): 71–83.

Fasaei, F., Bitter, J., Slegers, P. and van Boxtel, A. 2018. Techno-economic evaluation of microalgae harvesting and dewatering systems. Algal Res. 31: 347–362.

Fernandes, A.S., Nogara, G.P., Menezes, C.R., Cichoski, A.J., Mercadante, A.Z., Jacob-Lopes, E. and Zepka, L.Q. 2017. Identification of chlorophyll molecules with peroxyl radical scavenger capacity in microalgae Phormidium autumnale using ultrasound-assisted extraction. Food Research International 99: 1036–1041.

Galloway, A.W. and Winder, M. 2015. Partitioning the relative importance of phylogeny and environmental conditions on phytoplankton fatty acids. PLoS One 10(6): e0130053.

Gao, S., Yang, J., Tian, J., Ma, F., Tu, G. and Du, M. 2010. Electro-coagulation–flotation process for algae removal. Journal of Hazardous Materials 177(1-3): 336–343.

Garg, S., Wang, L. and Schenk, P.M. 2014. Effective harvesting of low surface-hydrophobicity microalgae by froth flotation. Bioresource Technology 159: 437–441.

Garzon-Sanabria, A.J., Davis, R.T. and Nikolov, Z.L. 2012. Harvesting Nannochloris oculata by inorganic electrolyte flocculation: Effect of initial cell density, ionic strength, coagulant dosage, and media pH. Bioresource Technology 118: 418–424.

Geciova, J., Bury, D. and Jelen, P. 2002. Methods for disruption of microbial cells for potential use in the dairy industry—A review. International Dairy Journal 12(6): 541–553.

Goettel, M., Eing, C., Gusbeth, C., Straessner, R. and Frey, W. 2013. Pulsed electric field assisted extraction of intracellular valuables from microalgae. Algal Research 2(4): 401–408.

Gong, M. and Bassi, A. 2016. Carotenoids from microalgae: A review of recent developments. Biotechnology Advances 34(8): 1396–1412.

Gour, G., Jennings, D., Buscemi, F., Duan, R. and Marvian, I. 2018. Quantum majorization and a complete set of entropic conditions for quantum thermodynamics. Nature Communications 9(1): 1–9.

Greenwell, H.C., Laurens, L., Shields, R., Lovitt, R. and Flynn, K. 2010. Placing microalgae on the biofuels priority list: A review of the technological challenges. Journal of the Royal Society Interface 7(46): 703–726.

Grima, E.M., Belarbi, E.-H., Fernández, F.A., Medina, A.R. and Chisti, Y. 2003. Recovery of microalgal biomass and metabolites: Process options and economics. Biotechnology Advances 20(7-8): 491–515.

Halim, R., Gladman, B., Danquah, M.K. and Webley, P.A. 2011. Oil extraction from microalgae for biodiesel production. Bioresource Technology 102(1): 178–185.

Halim, R., Harun, R., Danquah, M.K. and Webley, P.A. 2012. Microalgal cell disruption for biofuel development. Applied Energy 91(1): 116–121.

Haponska, M., Clavero, E., Salvadó, J. and Torras, C. 2018. Application of ABS membranes in dynamic filtration for Chlorella sorokiniana dewatering. Biomass and Bioenergy 111: 224–231.

Henderson, R.K., Parsons, S.A. and Jefferson, B. 2010. The impact of differing cell and algogenic organic matter (AOM) characteristics on the coagulation and flotation of algae. Water Research 44(12): 3617–3624.

Ho, S.-H., Chen, C.-Y., Lee, D.-J. and Chang, J.-S. 2011. Perspectives on microalgal CO2-emission mitigation systems—A review. Biotechnology Advances 29(2): 189–198.

Hønsvall, B.K., Altin, D. and Robertson, L.J. 2016. Continuous harvesting of microalgae by new microfluidic technology for particle separation. Bioresource Technology 200: 360–365.

Hossain, A.S., Salleh, A., Boyce, A.N., Chowdhury, P. and Naqiuddin, M. 2008. Biodiesel fuel production from algae as renewable energy. American Journal of Biochemistry and Biotechnology 4(3): 250–254.

Hu, Q., Sommerfeld, M., Jarvis, E., Ghirardi, M., Posewitz, M., Seibert, M. and Darzins, A. 2008. Microalgal triacylglycerols as feedstocks for biofuel production: Perspectives and advances. The Plant Journal 54(4): 621–639.

Hu, Y.-R., Guo, C., Xu, L., Wang, F., Wang, S.-K., Hu, Z. and Liu, C.-Z. 2014. A magnetic separator for efficient microalgae harvesting. Bioresource Technology 158: 388–391.

Hussain, F., Shah, S.Z., Ahmad, H., Abubshait, S.A., Abubshait, H.A., Laref, A., Manikandan, A., Kusuma, H.S. and Munawar, I. 2021. Microalgae an ecofriendly and sustainable wastewater treatment option: Biomass application in biofuel and bio-fertilizer production. A review. Renewable and Sustainable Energy Reviews 137: 110603.

Japar, A.S., Azis, N.M., Takriff, M.S. and Yasin, N.H.M. 2017. Application of different techniques to harvest microalgae. Trans. Sci. Technol. 4(2): 98–108.

Javed, F., Aslam, M., Rashid, N., Shamair, Z., Khan, A.L., Yasin, M., Fazal, T., Hafeez, A., Rehman, F., Ur Rehman, M.S., Khan, Z., Iqbal, J. and Bazmi, A.A. 2019. Microalgae-based biofuels, resource recovery and wastewater treatment: A pathway towards sustainable biorefinery. Fuel 255: 115826.

Jónasdóttir, S.H. 2019. Fatty acid profiles and production in marine phytoplankton. Marine Drugs 17(3): 151.

Jones, P.J. and Papamandjaris, A.A. 2012. Lipids: Cellular metabolism. Present Knowledge in Nutrition, 132–148.

Junior, W.G.M., Gorgich, M., Corrêa, P.S., Martins, A.A., Mata, T.M. and Caetano, N.S. 2020. Microalgae for biotechnological applications: Cultivation, harvesting and biomass processing. Aquaculture 528: 735562.

Kapoore, R.V., Butler, T.O., Pandhal, J. and Vaidyanathan, S. 2018. Microwave-assisted extraction for microalgae: From biofuels to biorefinery. Biology 7(1): 18.

Kim, S., Park, J.-e., Cho, Y.-B. and Hwang, S.-J. 2013. Growth rate, organic carbon and nutrient removal rates of Chlorella sorokiniana in autotrophic, heterotrophic and mixotrophic conditions. Bioresource Technology 144: 8–13.

Kumar, S.J., Kumar, G.V., Dash, A., Scholz, P. and Banerjee, R. 2017. Sustainable green solvents and techniques for lipid extraction from microalgae: A review. Algal Research 21: 138–147.

Laamanen, C., Desjardins, S., Senhorinho, G. and Scott, J. 2021. Harvesting microalgae for health beneficial dietary supplements. Algal Research 54: 102189.

Laamanen, C.A., Ross, G.M. and Scott, J.A. 2016. Flotation harvesting of microalgae. Renewable and Sustainable Energy Reviews 58: 75–86.

Lam, M.K. and Lee, K.T. 2012. Microalgae biofuels: A critical review of issues, problems and the way forward. Biotechnology Advances 30(3): 673–690.

Lardon, L., Hélias, A., Sialve, B., Steyer, J.-P. and Bernard, O. 2009. Life-cycle assessment of biodiesel production from microalgae: ACS Publications.

Lee, J.-Y., Yoo, C., Jun, S.-Y., Ahn, C.-Y. and Oh, H.-M. 2010. Comparison of several methods for effective lipid extraction from microalgae. Bioresource Technology 101(1): S75–S77.

Lowrey, J., Brooks, M.S. and McGinn, P.J. 2015. Heterotrophic and mixotrophic cultivation of microalgae for biodiesel production in agricultural wastewaters and associated challenges—A critical review. Journal of Applied Phycology 27(4): 1485–1498.

Maltsev, Y. and Maltseva, K. 2021. Fatty acids of microalgae: Diversity and applications. Reviews in Environmental Science and Bio/Technology, 1–33.

Mata, T.M., Martins, A.A. and Caetano, N.S. 2010. Microalgae for biodiesel production and other applications: A review. Renewable and Sustainable Energy Reviews 14(1): 217–232.

Mathimani, T., Sekar, M., Shanmugam, S., Sabir, J.S., Chi, N.T.L. and Pugazhendhi, A. 2021. Relative abundance of lipid types among Chlorella sp. and Scenedesmus sp. and ameliorating homogeneous acid catalytic conditions using central composite design (CCD) for maximizing fatty acid methyl ester yield. Science of the Total Environment 771: 144700.

Mayol, A.P., Ubando, A., Biona, J.B., Ong, H.L., Espiritu, R., Lee, H., Tumlos, R., Arboleda, N. and Culaba, A.B. 2015. Investigation of the drying characteristics of microalgae using microwave irradiation. Paper Presented at the 2015 International Conference on Humanoid, Nanotechnology, Information Technology, Communication and Control, Environment and Management (HNICEM).

Mercer, P. and Armenta, R.E. 2011. Developments in oil extraction from microalgae. European Journal of Lipid Science and Technology 113(5): 539–547.

Miao, X. and Wu, Q. 2006. Biodiesel production from heterotrophic microalgal oil. Bioresource Technology 97(6): 841–846.

Moazami, N., Ranjbar, R., Ashori, A., Tangestani, M. and Nejad, A.S. 2011. Biomass and lipid productivities of marine microalgae isolated from the Persian Gulf and the Qeshm Island. Biomass and Bioenergy 35(5): 1935–1939.

Moh, Y. 2017. Solid waste management transformation and future challenges of source separation and recycling practice in Malaysia. Resources, Conservation and Recycling 116: 1–14.

Mohn, F. and Soeder, C. 1978. Improved technologies for the harvesting and processing of microalgae and their impact on production costs. Archiv fur Hydrobiologie, Beihefte Ergebnisse der Limnologie 1: 228–253.

Munir, N., Sharif, N., Naz, S . and Manzoor, F. 2013. Algae: A potent antioxidant source. Sky J. Microbiol. Res. 1(3): 22–31.

Najjar, Y.S. and Abu-Shamleh, A. 2020. Harvesting of microalgae by centrifugation for biodiesel production: A review. Algal Research 51: 102046.

Natarajan, R., Ang, W.M.R., Chen, X., Voigtmann, M. and Lau, R. 2014. Lipid releasing characteristics of microalgae species through continuous ultrasonication. Bioresource Technology 158: 7–11.

Neis, U., Nickel, K. and Tiehm, A. 2000. Enhancement of anaerobic sludge digestion by ultrasonic disintegration. Water Science and Technology 42(9): 73–80.

Nisar, N., Mehmood, S., Nisar, H., Jamil, S., Ahmad, Z., Ghani, N., Oladipo, A.A., Qadri, R.W., Latif, A.A., Ahmad, S.R., Iqbal, M. and Abbas, M. 2018. Brassicaceae family oil methyl esters blended with ultra-low sulphur diesel fuel (ULSD): Comparison of fuel properties with fuel standards. Renewable Energy 117: 393–403.

Obi, C., Ibezim-Ezeani, M.U. and Nwagbo, E.J. 2020. Production of biodiesel using novel *C. lepodita* oil in the presence of heterogeneous solid catalyst. Chem. Int. 6(2): 91–97.

Pagels, F., Pereira, R.N., Vicente, A.A. and Guedes, A. 2021. Extraction of pigments from microalgae and cyanobacteria—A review on current methodologies. Applied Sciences 11(11): 5187.

Patel, A.K., Joun, J.M., Hong, M.E. and Sim, S.J. 2019. Effect of light conditions on mixotrophic cultivation of green microalgae. Bioresource Technology 282: 245–253.

Perez-Garcia, O., Escalante, F.M., De-Bashan, L.E. and Bashan, Y. 2011. Heterotrophic cultures of microalgae: Metabolism and potential products. Water Research 45(1): 11–36.

Postma, P., Suarez-Garcia, E., Safi, C., Yonathan, K., Olivieri, G., Barbosa, M., Wijffels, R.H. and Eppink, M.H.M. 2017. Energy efficient bead milling of microalgae: Effect of bead size on disintegration and release of proteins and carbohydrates. Bioresource Technology 224: 670–679.

Prakash, J., Pushparaj, B., Carlozzi, P., Torzillo, G., Montaini, E. and Materassi, R. 1997. Microalgal biomass drying by a simple solar device*. International Journal of Solar Energy 18(4): 303–311.

Ramaraj, S., Hemaiswarya, S., Raja, R., Ganesan, V., Anbazhagan, C., Carvalho, I.S. and Juntawong, N. 2015. Microalgae as an attractive source for biofuel production. Environmental Sustainability, pp. 129–157. Springer.

Remize, M., Brunel, Y., Silva, J.L., Berthon, J.-Y. and Filaire, E. 2021. Microalgae n-3 PUFAs production and use in food and feed industries. Marine Drugs 19(2): 113.

Roux, J.-M., Lamotte, H. and Achard, J.-L. 2017. An overview of microalgae lipid extraction in a biorefinery framework. Energy Procedia 112: 680–688.

Rubio, J., Souza, M. and Smith, R. 2002. Overview of flotation as a wastewater treatment technique. Minerals Engineering 15(3): 139–155.

Ryckebosch, E., Muylaert, K. and Foubert, I. 2012. Optimization of an analytical procedure for extraction of lipids from microalgae. Journal of the American Oil Chemists' Society 89(2): 189–198.

Sadaf, S., Iqbal, J., Ullah, I., Bhatti, H.N., Nouren, S., Nisar, J. and Iqbal, M. 2018. Biodiesel production from waste cooking oil: an efficient technique to convert waste into biodiesel. Sustainable Cities and Society 41: 220–226.

Sahoo, N.K., Gupta, S.K., Rawat, I., Ansari, F.A., Singh, P., Naik, S.N. and Bux, F. 2017. Sustainable dewatering and drying of self-flocculating microalgae and study of cake properties. Journal of Cleaner Production 159: 248–256.

Sati, H., Mitra, M., Mishra, S. and Baredar, P. 2019. Microalgal lipid extraction strategies for biodiesel production: A review. Algal Research 38: 101413.

Schwede, S., Kowalczyk, A., Gerber, M. and Span, R. 2011. Influence of different cell disruption techniques on mono digestion of algal biomass. Paper Presented at the World Renewable Energy Congress, Linkoping, Sweden.

Shao, W., Zhang, J., Wang, K., Liu, C. and Cui, S. 2018. Cocamidopropyl betaine-assisted foam separation of freshwater microalgae Desmodesmus brasiliensis. Biochemical Engineering Journal 140: 38–46.

Shelef, G.A., Sukenik, A. and Green, M. 1984. Microalgae harvesting and processing: A literature review. Report. Solar Energy Research Institute, Golden Colorado, SERI/STR-231-2396.

Sheng, J., Vannela, R. and Rittmann, B.E. 2011. Evaluation of methods to extract and quantify lipids from Synechocystis PCC 6803. Bioresource Technology 102(2): 1697–1703.

Shirley, S.A., Heller, R. and Heller, L.C. 2014. Electroporation gene therapy. Gene Therapy of Cancer, 93–106.

Show, K.-Y., Lee, D.-J. and Chang, J.-S. 2013. Algal biomass dehydration. Bioresource Technology 135: 720–729.

Show, K.-Y., Lee, D.-J., Tay, J.-H., Lee, T.-M. and Chang, J.-S. 2015. Microalgal drying and cell disruption–recent advances. Bioresource Technology 184: 258–266.

Show, K., Yan, Y. and Lee, D.J. 2021. Advances in drying and milling technologies for algae. Recent Advances in Micro and Macroalgal Processing: Food and Health Perspectives, 72–95.

Sierra, L.S., Dixon, C.K. and Wilken, L.R. 2017. Enzymatic cell disruption of the microalgae Chlamydomonas reinhardtii for lipid and protein extraction. Algal Research 25: 149–159.

Singh, G. and Patidar, S. 2018. Microalgae harvesting techniques: A review. Journal of Environmental Management 217: 499–508.

Tan, C.H., Nomanbhay, S., Shamsuddin, A.H. and Show, P.L. 2021. Recent progress in harvest and recovery techniques of mammalian and algae cells for industries. Indian Journal of Microbiology 61(3): 279–282.

Tan, X.B., Lam, M.K., Uemura, Y., Lim, J.W., Wong, C.Y. and Lee, K.T. 2018. Cultivation of microalgae for biodiesel production: A review on upstream and downstream processing. Chinese Journal of Chemical Engineering 26(1): 17–30.

Tang, S., Qin, C., Wang, H., Li, S. and Tian, S. 2011. Study on supercritical extraction of lipids and enrichment of DHA from oil-rich microalgae. The Journal of Supercritical Fluids 57(1): 44–49.

Tavanandi, H.A., Vanjari, P. and Raghavarao, K. 2019. Synergistic method for extraction of high purity Allophycocyanin from dry biomass of Arthrospira platensis and utilization of spent biomass for recovery of carotenoids. Separation and Purification Technology 225: 97–111.

Thomas, W.H., Tornabene, T.G. and Weissman, J. 1984. Screening for lipid yielding microalgae: Activities for 1983. Final subcontract report (No. SERI/STR-231-2207). Solar Energy Research Inst., Golden, CO (USA).

Toepfl, S., Mathys, A., Heinz, V. and Knorr, D. 2006. Potential of high hydrostatic pressure and pulsed electric fields for energy efficient and environmentally friendly food processing. Food Reviews International 22(4): 405–423.

Vasistha, S., Khanra, A., Clifford, M. and Rai, M. 2020. Current advances in microalgae harvesting and lipid extraction processes for improved biodiesel production: A review. Renewable and Sustainable Energy Reviews 110498.

Vernès, L., Li, Y., Chemat, F. and Abert-Vian, M. 2019. Biorefinery concept as a key for sustainable future to green chemistry—The case of microalgae. Plant Based Green Chemistry 2.0, pp. 15–50. Springer.

Viswanathan, T., Mani, S., Das, K., Chinnasamy, S., Bhatnagar, A., Singh, R. and Singh, M. 2012. Effect of cell rupturing methods on the drying characteristics and lipid compositions of microalgae. Bioresource Technology 126: 131–136.

Wahidin, S., Idris, A. and Shaleh, S.R.M. 2013. The influence of light intensity and photoperiod on the growth and lipid content of microalgae Nannochloropsis sp. Bioresource Technology 129: 7–11.

Wang, S.-K., Stiles, A.R., Guo, C. and Liu, C.-Z. 2015. Harvesting microalgae by magnetic separation: A review. Algal Research 9: 178–185.

Wu, J., Alam, M.A., Pan, Y., Huang, D., Wang, Z. and Wang, T. 2017. Enhanced extraction of lipids from microalgae with eco-friendly mixture of methanol and ethyl acetate for biodiesel production. Journal of the Taiwan Institute of Chemical Engineers 71: 323–329.

Xia, L., Li, Y., Huang, R. and Song, S. 2017. Effective harvesting of microalgae by coagulation–flotation. Royal Society Open Science 4(11): 170867.

Xu, L., Guo, C., Wang, F., Zheng, S. and Liu, C.-Z. 2011. A simple and rapid harvesting method for microalgae by in situ magnetic separation. Bioresource Technology 102(21): 10047–10051.

Xu, Z., Yan, X., Pei, L., Luo, Q. and Xu, J. 2008. Changes in fatty acids and sterols during batch growth of Pavlova viridis in photobioreactor. Journal of Applied Phycology 20(3): 237–243.

Yaguchi, T., Tanaka, S., Yokochi, T., Nakahara, T. and Higashihara, T. 1997. Production of high yields of docosahexaenoic acid by Schizochytrium sp. strain SR21. Journal of the American Oil Chemists' Society 74(11): 1431–1434.

Zhang, H., Lin, Z., Tan, D., Liu, C., Kuang, Y. and Li, Z. 2017. A novel method to harvest Chlorella sp. by co-flocculation/air flotation. Biotechnology Letters 39(1): 79–84.

Zhang, H. and Zhang, X. 2019. Microalgal harvesting using foam flotation: A critical review. Biomass and Bioenergy 120: 176–188.

Zhang, R., Fu, X. and Wan, M. 2016. Influence of high voltage pulsed electric fields on disrupture of chlorella. High Volt. Eng. 42: 152–155.

Zhang, R., Gu, X., Xu, G. and Fu, X. 2021. Improving the lipid extraction yield from Chlorella based on the controllable electroporation of cell membrane by pulsed electric field. Bioresource Technology 330: 124933.

Zhou, W., Lu, Q., Han, P. and Li, J. 2020. Microalgae cultivation and photobioreactor design. Microalgae Cultivation for Biofuels Production, pp. 31–50. Elsevier.

Zou, T.-B., Jia, Q., Li, H.-W., Wang, C.-X. and Wu, H.-F. 2013. Response surface methodology for ultrasound-assisted extraction of astaxanthin from Haematococcus pluvialis. Marine Drugs 11(5): 1644–1655.

Zuorro, A., Maffei, G. and Lavecchia, R. 2016. Optimization of enzyme-assisted lipid extraction from Nannochloropsis microalgae. Journal of the Taiwan Institute of Chemical Engineers 67: 106–114.

CHAPTER 12

Biodiesel from Microalgae
In-depth Extraction Processes and Transesterification Strategies

Natasha Nabila Ibrahim, Imran Ahmad,* Norhayati Abdullah,
Iwamoto Koji and *Shaza Eva Mohamad*

1. Introduction

Microalgae are microorganisms that can exist in unicellular or multicellular form. These photosynthetic microorganisms favour both terrestrial and aquatic environments, and inhabit either one of them. They could survive in such various places where there are enough water and sunlight like river, ocean, ponds, lakes, soils, moist rocks and in tree barks (Martins et al. 2020). The entire microalgae species can be divided into two categories: eukaryotic and prokaryotic microalgae. Examples of eukaryotic microalgae include Chlorophyta (green algae), Rhodophyta (red algae) and Bacillariophyta (diatoms).

Meanwhile, cyanobacteria (blue-green algae) are defined to be Chloroxybacteria in the prokaryotic microalgae category (Pignolet et al. 2013). Microalgae is so diverse that approximately 80,000 different species exist on Earth. According to Khan et al. (2018), 40,000 species of microalgae in total have been investigated for industrial purposes.

Both eukaryotic and prokaryotic microalgae have different metabolic mode, viz., autotrophic, heterotrophic, photoheterotrophic and mixotrophic (Arif et al. 2020). The size of microalgal cells, also known as phytoplanktons, range from 1 to 50 μm in diameter. The microalgae's propagation rate is efficient due to large surface to volume ratio (Yousuf 2020, Enamala et al. 2018, Yin et al. 2020). Nearly all species comprise of chlorophyll that is essential in converting solar energy, water, and carbon dioxide by photosynthesis into oxygen, adenosine triphosphate (ATP), glycerate 3-phosphate and reducing power. The resulting elements are utilized to support growth (make biomass) that ultimately leads to the production of biomolecules like carbohydrates, lipids, and proteins sustainably (Pignolet et al. 2013).

Algae and Biomass, Research Laboratory, Malaysia-Japan International Institute of Technology (MJIIT), Universiti Teknologi Malaysia (UTM), Jalan Sultan Yahya Petra, 54100, Kuala Lumpur, Malaysia.
* Corresponding author: mustafwibinqamar@gmail.com

Moreover, microalgae offer a significant advantage over other higher plants for exploitation of their lipid. Microalgae have the upper hand in terms of high growth rate, high biomass capacity and efficient photosynthesis. Microalgae is also better in taking up atmospheric carbon dioxide compared to terrestrial plants by 200-fold (Martins et al. 2020). A few examples of species that are widely studied for biodiesel feedstock are *Scenedesmus* sp., *Nannochloropsis* sp., and *Chlorella* sp. (Chhandama et al. 2021, Ahmad et al. 2021a). Furthermore, microalgae do not require agricultural land as they can be cultivated in ponds and photobioreactors (Ahmad et al. 2021b). This is also due to their versatility to intense ecosystem and environment. They are also robust since they can withstand a wide range of temperature, pH, and salinity (Chhandama et al. 2021).

Besides that, the food versus fuel debate can be prevented since microalgal biodiesel are more promising than first- and second-generation biodiesel. Being capable of continuous cultivation and not seasonally harvested is also one of the vital benefits of microalgae. Microalgae are renewable and sustainable as they can sequester harmful and excessive industrial as well as atmospheric carbon dioxide (Yousuf 2020). As a result, air pollution and greenhouse gases can be diminished, and global warming can also be decreased (Ahmad et al. 2021c).

Apart from that, microalgae can be exploited in the simultaneous process of wastewater treatment and biodiesel synthesis. The reason is that microalgae can fix organic pollutants that include nitrite, nitrate, ammonia, and orthophosphate that are found in wastewater (Chhandama et al. 2021). Furthermore, microalgae can be cultivated in a medium with different concentrations of nutrients and can adapt to change in growth aspects and the ability of nutrient input (Nascimento et al. 2013). This type of research is advantageous in reducing capital costs and will save time.

A research study by Arif et al. (2020) explained that the phosphorus and nitrogen removal rate by *Chlorella sorokiniana* that was isolated from wastewater treatment plants was high. The capacity of C18 and C16 fatty acids generated was also high, and the synthesised biodiesel met the requirement of international standards. Ryskamp et al. (2017) described that microalgal biodiesel portrayed a high number of cetane or high reactivity, and vehicle's engine alteration is not required. Biodiesel generated from microalgae eases the engine's start-up that is related to its high reactivity. Consequently, the engine's mechanical damage can be avoided, and initial ignition and combustion temperature can be reduced (Chhandama et al. 2021).

In addition, other valuable elements such as carotenoids, antioxidants and sugars can be extensively analyzed as essential value-added products (Ahmad et al. 2020). Next, the remaining biomass upon extraction of lipid is concentrated with phosphorus and nitrogen that can be further utilized as animal feed, bio-manure, and bioethanol feedstock. Moreover, biodiesel derived from microalgae symbolizes a sense of energy security and diversified energy supply, which will overcome the consequences of uncertain oil supply and price. This could also mean that oil imports' credence can be lessened (Chhandama et al. 2021).

Going back 2000 years ago, the Chinese pioneered human usage of microalgae during the famine outbreak. To date, various industrial and commercial uses of microalgae are being ventured in. Genetic engineering and modification of genes and metabolic pathways are meaningful in enhancing the function of existing useful compounds and achieving new products (Lu et al. 2021). Aside from biofuel production, microalgae are also comprehensively being studied for the synthesis of bioplastics or biopolymer, biofertilizer, carotenoids, carbohydrates, vitamins, animal and aqua feed, wastewater treatment, and bioelectricity (Bhatia et al. 2021, Madadi et al. 2021).

Over the past decades, scientists have utilised microalgae for recycling wastewater as nutrient sources, altogether with treating wastewater. Wastewater treatment could range from industrial, domestic, textile and palm oil mill effluent (POME). It is imminent that the global population will keep on growing, contributing to much more wastewater. It is implied that wastewater is harmful if it is not purified. Henceforth, the strategy of treating wastewater coupled with biodiesel production looks very promising (Liu et al. 2021).

Furthermore, producers might be able to discover optimized ways to generate raw feedstocks for biofuel production and gain extra profit. As a result, the quality and quantity of agricultural goods can be improved. This should lead to enhancing farmers' end-products and reducing poverty and potential disease. Besides that, this step should assist in cutting down the utilization of nitrogen fertilizers, emission of harmful carbon dioxide, wastage, and feed capital cost. Atmospheric oxygen capacity is also targeted to be boosted and the quality of food to be significantly improved. Therefore, human living standards and quality could become better. Exploiting microalgae on an industrial scale, i.e., biofuel production and wastewater treatment and biocycles, will be one of the keys to solving climate change and unsustainable technological advances. Soon, the issues can possibly be better resolved, contributing to the worldwide redirection of the sustainable living standard policy that is focused on the overall development of microalgae synthesis. This chapter provides insight into the integrated model of synthesizing and obtaining biodiesel from microalgae, incorporating different types of transesterification reactions.

In previous scientific literature, researchers have reported that there is no one true and specific conversion process of algal biomass transesterification to biofuels as we speak. Scientists are still studying on this matter regarding the transformation of biofuels. There are quite a handful of pathways that have been investigated, depending on practicality and cost of production of the whole upstream and downstream processes. Shin et al. (2018) and Marrone et al. (2018) have addressed those generative methods of lipid extraction which chiefly influence the expedience of biofuel transesterification and value-added product synthesis. One way of extraction method to be effective is that lipid extraction must be more particular to the targeted bio-products and concurrently ensure better purity because impurities produced are most likely to be lesser.

2. Achieving Biodiesel from Microalgae

Microalgae are well-exploited for their value-added secondary metabolites, including biofuels (Saad et al. 2019). Hydrocarbons that are attained from microalgae can be transformed into gasoline, kerosene, and diesel. Ranga et al. (2007) explained that *Botryococcus braunii* accumulates hydrocarbons that have great oil content extracellularly, which is advantageous for lipid extraction. Next, carbohydrates from microalgae can be converted into bioethanol through hydrolysis and fermentation, which is supported by reports from Markou et al. (2012) and Chen et al. (2013), mentioning that some microalgae can produce starch more than 50%. Finally, hemicellulose and cellulose from microalgae are the feedstocks that are converted to sugar and then ethanol (Hamelinck et al. 2005).

Meanwhile, extracted algal lipids undergo transesterification to produce biodiesel. Next, via bio-photolysis and anaerobic digestion, biohydrogen and biogas/biomethane are created by microalgae, respectively. Biohydrogen does not release greenhouse gases which would not harm our environment (Tiwari and Pandey 2012). As stated by Ward et al. (2014), the composition of biogas is 65–75% of methane, and the remaining 25–35% is made up of carbon dioxide. Other than that, Buxy et al. (2012) indicated that bio-syngas are generated via biomass gasification with oxygen and air or water vapour. Fig. 1 below depicts an outline of the integrated model of microalgal biofuel synthesis.

Successful biodiesel synthesis is highly dependent on what type of lipid is being converted into biodiesel. Neutral lipid is vital in biodiesel production and microalgae have been the most promising candidate. This is due to the neutral lipid that is produced by nearly all species of microalgae. Aside from that, the fatty acid constitution is also essential in characterising biodiesel (Ramaraj et al. 2015). Not to mention, the lipid composition and abundance are also distinct from each species, depending on cultivation conditions.

It has been proven in previous studies that the cultivation conditions could be varied in terms of type and composition of growth medium (Jayakumar et al. 2021, Valdez-Odeja et al. 2021), light

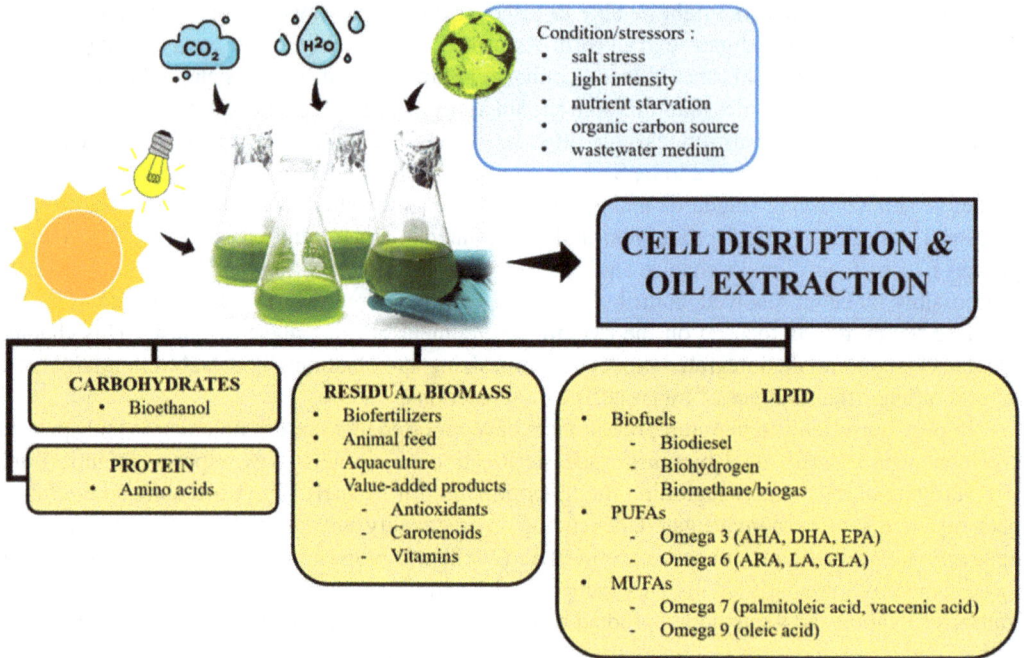

Fig. 1. Integrated model of microalgal biofuel synthesis.

intensity (Jayakumar et al. 2021), nutrient deprivation (Jayakumar et al. 2021), culture time (Valdez-Odeja et al. 2021) as well as temperature (Brindhadevi et al. 2021). In numerous microalgae species, when they are exposed to a nitrogen-starved environment, there has been an indisputable increment in their neutral lipid content and declined protein content. Therefore, choosing the suitable species for certain experiments, especially for biodiesel generation, involves considering all the possible factors.

Formally, bulk oil naturally manufactured by microalgae existed in triacylglycerides (TAGs) form (Yousuf 2019). TAGs are the accurate kind of oil in synthesizing biodiesel. Within the microalgal cells, fatty acids that are bound to the TAGs can be either long or short hydrocarbon chains. Ideally, the short fatty acids are the right kind of fatty acid for biodiesel composition, while the other type, the long chained fatty acids, could be useful for other exploitation. Freshwater and marine microalgae species have depicted typical yield of biomass and lipid. The decision to choose microalgae for cultivation is mostly dependent on the purpose of the experiment and database reported by preceding investigations (Ramaraj et al. 2015).

The good standard of a biodiesel that complies to ASTM D6751 standard (Howell 2012) relies on a few parameters which are heat of combustion, stability of oxidative, cold filter plugging point (CFPP), emission of exhaust, quality of ignition, viscosity, and fluidity. These qualities are regulated by every single FA alkyl ester that is majorly connected to the accumulated fatty acids constitution (Lee et al. 2010). The fatty acids' structure is determined by the number of double bonds or easily described as unsaturation, length of chain and chain branch. To manage the quality of the targeted biodiesel, it is possible with a suitable mixture of different fatty acids via selective microalgae species. In a way, this suggestion maximises oil production by enriching the microalgae with the inquired fatty acids or altering them through genetic modifications. Schenk et al. (2008) recommended ratios of optimal fatty acid property to be C14:0, C16:1, C16:4, C16:5, and C18:1.

3. Transesterification of Microalgal Lipids

Transesterification is a method of converting lipid or oil to biodiesel with the aid from a catalyst alongside alcohol. This process is comprised of several successive reactions that are reversible (Kumar et al. 2020). According to Guldhe et al. (2017), a catalyst is an advocate or component that mediates and is auxiliary to accelerate the chemical process. Another capability of a catalyst is that it improves the reaction rate towards condensation and proceeds to secondary response. In general, transesterification engages in the utilization of acid, alkali, and enzyme catalysts. Catalytic transesterification will be discussed further in the catalyst's subsection later.

The resulting product from this conversion process is 3 mols of biodiesel in fatty acid methyl esters (FAMEs) and 1 mol of glycerol (Kumar and Thakur 2018, Kumar et al. 2017). This transformation process can be operated at a temperature of approximately 140° F or 60°C with atmospheric pressure (Hussain et al. 2021). According to Munir et al. (2013), glycerol has bigger density compared to biodiesel; hence, the glycerol portion can be segregated periodically during the process. The resulting end-product of adding 90% lipid and 10% methanol is 90% and 10% of biodiesel and glycerine, respectively.

Amid transesterification, the reaction for end-products occurs in a sequential manner in the sense that triglycerides are firstly generated to diglycerides. Later, it is converted to monoglycerides and, lastly to glycerol. Transesterified biodiesel is pure oil derived from animal fats, vegetable oil, or microalgal oil, labelled as B100 biodiesel. Any mixture of biodiesel with petroleum-based fuel is labelled as biodiesel blend (BXX), depending on the percentage of the mixture of each oil represented by XX (Howell 2012).

The transesterification process can be categorized into two types: reactive and extractive transesterification. Reactive transesterification also goes by the name of direct and *in situ* transesterification. Reactive transesterification involves a direct transformation of wet microalgal biomass that does not conduct any prior lipid extraction, whereas the extractive transesterification method requires lipid extraction first and then is proceeded to biodiesel synthesis through transesterification. Extractive transesterification is also known as indirect and *ex situ* transesterification (Park et al. 2015).

The typical conventional extractive transesterification method involves dewatering and drying of microalgal biomass prior to lipid extraction. After lipid is successfully extracted, transesterification is carried, and this is called the three-level method. In contrast, the direct transesterification procedure does not require biomass to be dewatered or dried first. In recent years, *in situ* transesterification of wet biomass has been utilized to lessen the steps as well as the operational equipment and total capital expense of microalgal biodiesel.

Apart from that, it is essential to include the mention of catalysts in the transesterification process that is required in either direct or extractive transesterification. Previous studies discussed that catalysts used are acid, alkali, and enzymes (biological) (Guldhe et al. 2017, Bharathiraja et al. 2014). Fig. 2 shows the types of the transesterification process.

Typically, microalgal biodiesel that is transformed accounts for up to 80% of the extracted lipid (El-Shimi et al. 2013). Fundamentally, numerous alcohols have been used in the transesterification process that act as co-solvent, including propanol, ethanol, amyl alcohol, methanol, and butanol. However, in big-scale applications, ethanol and methanol are more favourable because of their benefits chemically and physically. They are also cost-saving organic solvents that are commonly used (Musa 2016). On the other hand, it is important to note the removal of any catalytic, methanolic and soap elements from the entire produced biodiesel because of choking and engine failure issues that would happen when biodiesel is being used as the power source.

The elimination step must be done after transesterification and through water washing and distillation. The following Fig. 3 demonstrates the biodiesel conversion process via transesterification.

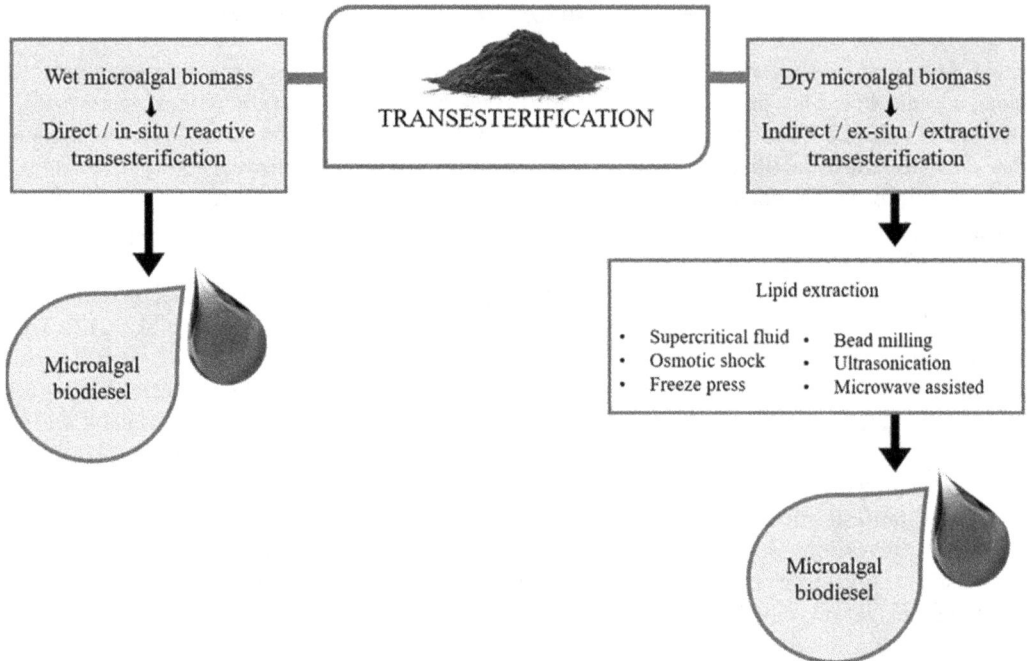

Fig. 2. Types of the transesterification process.

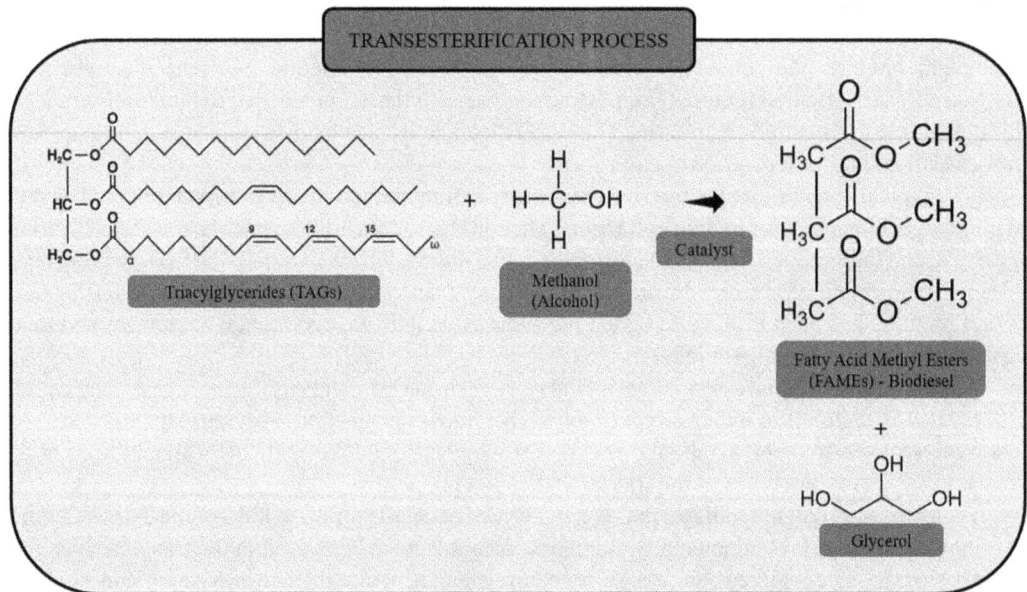

Fig. 3. Biodiesel synthesis via transesterification.

3.1 Acid-catalysed Transesterification

In acid-catalysed transesterification, several examples of utilized acid catalysts are hydrochloric acid (HCl), phosphoric acid (H_3PO_4), sulphuric acid (H_2SO_4), sulfonic acid (RSO_3H) and boron trifluoride (BF_3). Owing to their properties of higher transformation rates, sulfonic acid and sulphuric acid are largely preferred amongst others (Marchetti et al. 2007). Im et al. (2014) reported that obtained biodiesel from microalgae *Nannochloropsis oceania* was very high; 91% when it was trans-esterified using sulphuric acid. Microalgae *Nannochloropsis gaditana* was also subjected to acid-catalysed transesterification via sulphuric acid by Macías-Sánchez et al. (2018) and Kim et al. (2015). The resulting biodiesel was comparably high, which is 87% and 89%, respectively. Vasić et al. (2020) and Galadima and Muraza (2014) found increment of microalgal biodiesel upon acid-catalysed transesterification.

Recently, Jazie et al. (2020) applied dodecylbenzene sulfonic acid (DBSA) as an acid catalyst to transesterify *Chlorella* sp. lipid cultivated in a packed bed reactor. It was determined that DBSA is more reactive compared to sulphuric acid catalyst with merely 30 minutes of reaction time. In comparison to that, sulphuric acid-based transesterification needs more than 12 hours to yield FAME from *Chlorella* sp.

Macías-Sánchez et al. (2018) discussed that acid catalysts work by disrupting microalgal cell walls, letting methanol flow through and bringing out the lipid content. This type of reaction makes way to the production and extraction of fatty acid methyl esters from polar and non-polar lipids.

Although acid-catalysed transesterification could yield a high amount of biodiesel, the reaction rate during the process was sluggish, which contradicts when base catalysts are used. Ehimen et al. (2010) disclosed that usage of acid catalysts is advantageous towards *in situ* or direct transesterification of microbial biodiesel in the aspect of transformation rate. Even so, biodiesel from *Chlorella* via acid-based transesterification is prohibited by water presence. If the water volume is bigger than 115% w/w (on an oil weight basis), it was found out that this will affect the equilibrium of biodiesel yield. The explanation behind this is that acid catalyst deactivation is made probable due to the reaction by water molecules that can lead to the available protons' capability (Sathish et al. 2014). It is worth mentioning that inexpensive oil sources like animal oil or used cooking oil that are abundant with lots of fatty acids (more than 40%) require catalytic transesterification from acids to avoid saponification reaction. Saponification is the formation of soap when there is the presence of oil or lipid and aqueous alkali. Henceforth, in this type of oil, the acid catalyst is preferred to be used (Boruggada and Goud 2012).

Another main provocation when utilizing acid as the catalyst is the probability of corrosion in the fuel due to leachates of solid acid catalyst and downstream purification (Singh et al. 2012). Furthermore, *in situ* transesterification that involves acid catalysts and methanol also pose a corrosion risk for the reactors because the required working volume is larger. Therefore, to curb this condition, Ehimen et al. (2012) suggested using diethyl ether and n-pentane to reduce methanol amount by improving the reaction rate since these solvents can promote enhanced microalgal oil flow and diffusion across the cell wall.

Other than that, the resulting biodiesel would also be highly sensitive to any matter that is high in water content. Commercialization of acid-based biodiesel is also unlikeable because long reaction and recovery hours are a necessity. The risk of corrosion is also challenging, making it possible for industrial-scale application (Saifuddin et al. 2015). Moreover, the production expenditure is increased in acid-catalysed biodiesel processing due to more steps being required, and excessive alcohols are used (Brennan and Owende 2010).

3.2 Alkali-catalysed Transesterification

Alkali-based transesterification is broadly used to transform lipid to FAMEs (Razzak et al. 2013). There are several alkaline catalysts in transesterification; a few among them are potassium hydroxide, sodium hydroxide and sodium methoxide. The reasons that basic catalyst is widely used are that the conversion rate is relatively high, which is at the maximum of 97% in a period of not more than 30 minutes (Kumar et al. 2017). The reaction rate is about 4000-fold faster if compared to acid-catalysed transesterification (Bagul et al. 2018). It is also not costly because the working volume is only 0.5% to 1% by oil weight (Bagul et al. 2018, Yousuf 2019). The transesterification procedure that is alkali-based is conducted under atmospheric pressure at 60°C because methanol has a boiling point at 65°C. Under these parameters, the conversion time would take around 90 minutes faster than acid-based transesterification. Relatively, higher temperature and higher pressure can be applied, but this inflicts the cost, becoming more expensive because more energy and electricity are used (Fukuda et al. 2001).

Makarevičienė et al. (2020) investigated basic catalyst (sodium hydroxide) usage for transesterification of oil from microalgae *Chlorella protothecoides*. The yielded biodiesel makes up to 98% at maximum. In comparison, potassium hydroxide was chosen to become the basic catalyst for *Chlorella vulgaris* by both Cercado et al. (2018) and Rahman et al. (2017), with their chosen microalgae, *Spirulina maxima*. The resulting biodiesel was totalled in a maximum of up to 85% and 86.1%, respectively. The most reactive basic catalyst is the alkoxide sodium methoxide (CH_3ONa). This soluble metal oxide can convert biodiesel even with a low concentration at 0.5%, at a 498% transformation rate in merely 30 minutes.

Alcohols like methanol and ethanol are the two most typically used organic solvents for transesterification. Basic catalysts coupled with organic solvents tend to react better than acid catalysts. However, the major difficulties of the alkali-based transesterification method are eliminating alkaline catalysts during post-processing from the resulted FAMEs. These difficulties also include glycerol recovery, the necessity to treat alkaline wastewater to prevent environmental pollution when discarded, and moisture that reacts with fatty acids. According to Noiroj et al. (2007), the yielded biodiesel will be less fluid and more viscous when there is moisture in basic-catalyzed transesterification due to saponification. Alkali-based transesterification is significantly limited in saponification risk because basic catalysts will react to excessive fatty acids and form soap. Saponification must be avoided because it affects the richness of the biodiesel and makes the reaction mixture more adhesive (Veillette et al. 2017). To prevent this, a molar ratio of 6:1 is regarded to be favourable. However, this drawback would impose downstream processing expenditure, which will incur higher production costs (Kumar and Thakur 2018).

3.3 Enzyme-catalysed Transesterification

Biodiesel that is generated through enzyme-catalysed transesterification necessitates the enzyme lipase to be the mediator. Generally, lipase-based transesterification is preferred over acid and basic transesterification due to various factors such as the easiness in co-created products' elimination, uncomplicated glycerol retrieval, and universal biotransformation purposes (Bharathiraja et al. 2014). Studies from several articles have operated on lipase-based transesterification (Amoah et al. 2017, Guldhe et al. 2016, 2015). Teo et al. (2014) experimented on microalgal lipid from *Tetraselmis* sp. that was subjected to enzyme-catalysed transesterification. The researchers identified that lipase-based transesterification reacts better, and the yielded biodiesel amount was very much higher than basic-catalysed reaction by sevenfold.

As mentioned by Taher et al. (2011) and Guldhe et al. (2015), lipase enzymes are hydrolases capable of catalysing the transesterification process, assembling free fatty acids and triglycerides to produce esters. Thus, extracted microalgal lipid was subjected to extractive transesterification and coupled with lipase catalyst (Surendhiran et al. 2015, Tran et al. 2012). Lai et al. (2012) reported that Penicillium expansum lipase mediates enzyme-based transesterification towards lipids from *Chlorella pyrenoidosa* and resulted in a high amount of biodiesel around 90.7% in reaction medium ionic liquid solvent.

In executing biological transesterification, several parameters affect the process, namely, alcohol inhibition, moisture level, glycerol inhibition, pretreatment to enhance the stability of lipase, temperature as well solvent influence (Guldhe et al. 2015). Lipase enzyme can be divided into three levels categorized according to definitive position, definitive types of fatty acids and definitive of acylglycerol types, i.e., monoglyceride, diglycerides, or triglycerides. Typically, lipase enzymes are directly utilized in transesterification as catalyst to generate biodiesel from microalgal lipid, but lipase enzyme catalysts are apparently too expensive in contrast to other types of catalysts. Therefore, to curb this problem, an immobilization technique has been applied to the enzyme. The immobilization is executed on supportive materials so that reusability is assured, enhanced stability is achieved, and cell disruption is no longer needed (Borges et al. 2021). Immobilization of enzymes is constructed via distinct procedures like microwave assistance, magnetic nanoparticles, and enzyme cross-linkage (Yousuf 2019).

López et al. (2015) revealed that enzymatic transesterification of lipid from *Nannochloropsis gaditana* was accounted to more than 94%. Surendhiran et al. (2014) had used lipase from yeast *Rhodotorula mucilaginosa* immobilized to attain biodiesel from microalgal lipid. The immobilization has decreased the cost of production due to its reusability and the characteristics of being highly stable. This immobilized lipase can be used again up to ten runs with no decrement in productivity. This study is also relatively supported by another research proof by Guldhe et al. (2016) which utilizes immobilised lipase from *Aspergillus niger*. Transesterified biodiesel that was obtained had amounted to 80% when the lipase was applied, and it could be reused for a maximum of two times without loss of conversion relevance. According to Brennan and Owende (2010), the significance of biological transesterification by lipase enzyme is that neutralisation is not needed and the desired amount of alcohol to be used is lesser.

Enzyme pretreatments before the real transesterification have uncovered the enhanced lipase stability, leading to longer reusability of immobilised enzymes and reduced cost. Pretreatments using methanol, saline solution, and glutaraldehyde are among the methods applied (Guldhe et al. 2015). Moreover, microalgal lipids that have loads of free fatty acid and water can be introduced to the enzyme-base conversion process without saponification and emulsion risk (MacArio et al. 2007). However, despite that, this type of transesterification is still rather costly. Enzymes are found to be the protein elements in cells that execute as a catalyst in a chemical reaction.

These enzymes can be attained from various microorganisms such as bacteria, yeast and fungi, tissues, and plants. Noticeable characteristics of an enzyme are eco-friendliness, high efficacy, and selectivity. Lipase is favourable due to its high effectiveness even in a moderate state (MacArio et al. 2007). The enzymatic reaction can be administered into various applications and the main operations are tabulated in Table 1.

Table 1. Major enzyme applications in different industries.

Industry	Enzyme	Function	References
Leather making	Protease	Unhairing and bating	Yousuf 2019
Cooking	Glucose oxidase	Mixture enhancement	
	Amylase	Fluffy bread with volume	
Detergents	Amylase Proteinase Lipase	Discard stains of starch, proteins as well as fats	
Dairy	Renin	Cheddar planning	
	Lipase	Lactose removal	
Biodiesel	Lipase	Vegetable oil biodiesel synthesis	
Materials/textile	Amylase	Removing starch from woven surface	
Paper industry	Cellulase and xylanase	Dissolving pulps from cellulosic impurities	Kumar 2021
Rubber industry	Rubber oxygenase	Degradation of rubber, latex clearing	Basik et al. 2021

In terms of direct transesterification combined with lipase catalyst, cell wall disruption through pretreatment is necessary (Tran et al. 2013). Tran et al. (2012) revealed that up to 97.3 wt.% biodiesel was recovered upon immobilised lipase transesterification from wet *Chlorella vulgaris* ESP-31 biomass cake that was pre-sonicated. The lipase was achieved from *Burkholdeira* sp. C20. In comparison to this finding, only 72.1 wt.% of oil was gained from extracted lipid (extractive transesterification). The immobilised lipase was then reused for six consecutive cycles that accounted to 288 hours without relevant loss of original efficiency (Tran et al. 2012).

Tran et al. (2013) had also discussed direct transesterification by using lipase immobilised in alkyl grafted iron oxide-silicon dioxide (Fe_3O_4-SiO_2). This procedure bore more than 90% of biodiesel under optimised conditions. Meanwhile, in another study of biological transesterification by lipase enzyme, cross-linkage support medium between glutaraldehyde and nanozeolites that has been combined with 3-aminopropyl trimethoxysilane (APTMS) were utilised. The support chain was observed to improve the stability of lipase. de Vasconcellos et al. (2018) stated that higher lipase conversion activity was detected when combined with nanozeolites to become lipase-nanozeolites complex compared to free enzymes.

In addition, fatty acid ethyl esters (FAEE) yield was found to exceed 93% when lipase-nanozeolites complex was introduced. Transesterification with ethanol was able to operate with the same enzyme-nano zeolites for five cycles because the catalyst could be recovered. Scientific research was administered by Kazemifard et al. (2019) with the purpose of simple recovery for nano-catalyst in *in situ* transesterification of microalgae that was grown in wastewater. Magnetic KOH/Fe_2O_3-Al_2O_3 core was employed as the nano-catalyst for 6 hours and exhibited a positive result of 95.6% converted biodiesel. The enzyme was successfully retrieved and used repeatedly for as many times as six cycles without losing its original productivity.

3.4 *Direct or* in situ *Transesterification*

Direct or *in situ* or reactive transesterification represents transesterification method that utilises alcohol and a catalyst with no earlier cell disruption and extraction process. There is no one universal way of transesterification because methods could vary depending on the research goal, the microalgae species, the cost of production, and the uniqueness of the experiment itself. Several methods could be combined or co-executed at the same time to maximise microalgal biodiesel yield.

Several examples of *in situ* transesterification are the coupled reaction with sulphuric acid (H_2SO_4) and hydrochloric acid (HCl) as well as acetyl chloride (CH_3COCl). Amongst the three, CH_3COCl is found to be the best, producing 56% g of FAMEs per dry weight (Cooney et al. 2009). Microwave assistance could promote better conversion rate as proven by Koberg et al. (2011). Koberg et al. (2011) operated the experiment using strontium oxide (SrO) as a heterogenous catalyst and co-reacted with microwave heating and sonication. Dry *Nannochloropsis* sp. biomass was purchased from Seambiotic Company, so there was no lipid extraction involved. As a result, the yield of FAMEs rose from 7% to 37% g per dry weight.

Despite the common one-step approach of *in situ* transesterification (Ghosh et al. 2017, Kwon and Yeom 2015, Ma et al. 2015), transesterification can be conducted in double-step method whereby esterification of acids is employed to decrease the amount of free fatty acids (Dong et al. 2013, Suganya et al. 2013). Despite that, double-step transesterification method resulted in decreased amount of free fatty acid which is from 6.3% to 0.34%. The catalyst used were H_2SO_4 and methanol-oil mixture (Suganya et al. 2013). The reaction time was run for 90 minutes at 60ºC. Suganya et al. (2013) explained that alkaline transesterification was carried out in the second step by utilising NaOH and methanol-oil mixture for 70 minutes with the same temperature. From here, up to 90.6% of biodiesel was transformed.

Besides that, Dong et al. (2013) also carried out both single and double-step *in situ* microalgal lipid from *Chlorella sorokiniana* UTEX 1602. Total biodiesel attained from the two-step method is greater than the single-step approach, which is 94.87 ± 0.86% and 60.89%, respectively, after 1 hour and 10 minutes of conversion time at 90°C.

This study conducted acid-based transesterification earlier before the alkaline-base transesterification via acid catalyst, viz., Amberlyst-15. Moreover, Ma et al. (2015) had also disclosed that they applied the double-step method of *in situ* transesterification and gained biodiesel esters of 35.5 ± 1.27 mg/g biomass. The researchers exploited lipid from *Chlorella vulgaris* with the first transesterification with acid catalyst Amberlyst BD20 for 50 minutes, followed by a basic catalyst which is KOH for 40 minutes. Compared to single-step direct transesterification using alkali catalyst, this study yielded three-fold amount of biodiesel. All in all, *in situ* or direct transesterification is more favoured since it is advantageous in terms of low production expenditure and fewer processing steps, time, and energy. In addition, cost-effectiveness is possible in direct transesterification since prior biomass drying and cell disruption are unnecessary (Mandik et al. 2020, Ghosh et al. 2017, Ma et al. 2019).

Additionally, direct transesterification using acid as a catalyst for biodiesel from *Nannochloropsis gaditana*, has obtained a greater yield which is 64.98% (Torres et al. 2017). In addition, the maximum yield of biodiesel at 99.32% was successfully achieved by Shirazi et al. (2017) via direct transesterification method that includes usage of mixed solvents. There was another investigation in 2017, which was administered by Park et al., that yielded 80% of biodiesel through direct transesterification with ethanol-chloroform co-solvent setup. Other than that, there was an outstanding accumulation of biodiesel up to 97.1% resulting from the utilisation of ethyl acetate in direct transesterification. Moreover, it eradicated any extra co-solvent because it is unnecessary, thus lowering the entire production cost. As a result, direct transesterification coupled with optimum solvents is highly possible for microalgal cell wall disruption and converting the lipids into biodiesel in an *in situ* manner.

3.5 *Indirect or* ex situ *Transesterification*

Contrary to the *in situ* approach, *ex situ* transesterification offers more processing steps in regard to pre-treatment processes such as dewatering and drying of biomass, disruption of cells, lipid extraction, transesterification, and lastly, purifying of biodiesel (Saifuddin et al. 2015). However, due to extra steps involved in this approach, this method is less preferred and counterproductive. Apart from that, the high water content in microalgal biomass, especially in upscale production, hinders future commercialisation. In the cell disruption and lipid extraction step, organic solvents are usually used because those solvents can permeate the cell membrane and mediate dissolved lipid to be released into the reaction medium.

In the case of industrial biodiesel synthesis, solvents are conventionally applied in gaseous smoke form, but this technology is unsafe towards environment. Henceforth, this issue is needed to be addressed so that the production is lowered in terms of money, energy, time and is more favourable towards the environment (Patil et al. 2012).

During downstream processing of microalgae, biomass drying turned out to be the most energy draining that requires as high as 84% of the overall energy usage. Moreover, lipid extraction and the following lipid transesterification often requires up to 3 hours. Therefore, energy usage in the dewatering process needs to be deducted. A few researchers have suggested direct transesterification method as it integrates lipid extraction and transesterification altogether in a single step (Torres et al. 2017, Shirazi et al. 2017, Park et al. 2017, Wahidin et al. 2018). The efficiency of concurrent microwave-assisted extraction and transesterification was investigated, and it was determined that biodiesel yield resulting from this method is only 42.22% (Wahidin et al. 2018).

4. Conclusion

Albeit the numerous industrial applications, microalgae are mostly known for their up-and-coming potential as a biofuel source. Recent trend and advances in scientific knowledge and its industrial-scale production brings about possibilities in better living. Advances in technology lead to economic growth in developing countries and significantly aid the rich countries financially. The future of microalgal fuel is a bright and promising area that is still lacking feasible literature. Various possibilities could be implied in microalgal biofuel like genetic engineering and nanoparticles in achieving sustainable net zero lifestyle.

Acknowledgement

The authors are thankful to the staff of Albio ikohza for their support.

References

Ahmad, I., Abdullah, N., Yuzir, A., Koji, I. and Mohamad, S.E. 2020a. Efficacy of microalgae as a nutraceutical and sustainable food supplement. *In*: 3rd ICA Research Symposium (ICARS) 2020, p. 6.

Ahmad, I., Abdullah, N., Koji, I., Yuzir, A. and Muhammad, S.E. 2021b, April. Evolution of photobioreactors: A review based on microalgal perspective. In IOP Conference Series: Materials Science and Engineering (Vol. 1142, No. 1, p. 012004). IOP Publishing.

Ahmad, I., Abdullah, N., Koji, I., Yuzir, A. and Mohamad, S.E. 2021c. Potential of microalgae in bioremediation of wastewater. Bulletin of Chemical Reaction Engineering and Catalysis 16(2): 413–429.

Ahmad, I., Yuzir, A., Mohamad, S.E., Iwamoto, K. and Abdullah, N. 2021, February. Role of microalgae in sustainable energy and environment. In IOP Conference Series: Materials Science and Engineering (Vol. 1051, No. 1, p. 012059). IOP Publishing.

Amoah, J., Ho, S.H., Hama, S., Yoshida, A., Nakanishi, A., Hasunuma, T., Ogino, C. and Kondo, A. 2017. Conversion of Chlamydomonas sp. JSC4 lipids to biodiesel using Fusarium heterosporum lipase-expressing Aspergillus oryzae whole-cell as biocatalyst. Algal Research 28: 16–23.

Arif, M., Wang, L., Salama, E.S., Hussain, M.S., Li, X., Jalalah, M., Al-Assiri, M.S., Harrz, F.A., Ji, M-K. and Liu, P. 2020. Microalgae isolation for nutrient removal assessment and biodiesel production. BioEnergy Research 13: 1247–1259.

Arora, N., Patel, A., Sartaj, K., Pruthi, P.A. and Pruthi, V. 2016. Bioremediation of domestic and industrial wastewaters integrated with enhanced biodiesel production using novel oleaginous microalgae. Environmental Science and Pollution Research 23(20): 20997–21007.

Bagul, S.Y., Chakdar, H., Pandiyan, K. and Das, K. 2018. Conservation and application of microalgae for biofuel production. pp. 335–352. *In*: Sharma, S.K. and Varma, A. (eds.). Microbial Resource Conservation: Conventional to Modern Approaches. Springer International Publishing.

Basik, A.A., Sanglier, J.J., Yeo, C.T. and Sudesh, K. 2021. Microbial degradation of rubber: Actinobacteria. Polymers 13(12): 1989.

Bharathiraja, B., Chakravarthy, M., Kumar, R.R., Yuvaraj, D., Jayamuthunagai, J., Kumar, R.P. and Palani, S. 2014. Biodiesel production using chemical and biological methods–A review of process, catalyst, acyl acceptor, source and process variables. Renewable and Sustainable Energy Reviews 38: 368–382.

Bhatia, S.K., Mehariya, S., Bhatia, R.K., Kumar, M., Pugazhendhi, A., Awasthi, M.K., Atabani, A.E., Kumaar, G., Kim, W., Seo, S-O. and Yang, Y.H. 2021. Wastewater based microalgal biorefinery for bioenergy production: Progress and challenges. Science of the Total Environment 751: 141599.

Borges, J.P., Junior, J.C.Q., Ohe, T.H.K., Ferrarezi, A.L., Nunes, C.D.C.C., Boscolo, M., Gomes, E., Bocchini, D.A. and da Silva, R. 2021. Free and substrate-immobilised lipases from Fusarium verticillioides P24 as a biocatalyst for hydrolysis and transesterification reactions. Applied Biochemistry and Biotechnology 193(1): 33–51.

Borugadda, V.B. and Goud, V.V. 2012. Biodiesel production from renewable feedstocks: Status and opportunities. Renewable and Sustainable Energy Reviews 16(7): 4763–4784.

Brennan, L. and Owende, P. 2010. Biofuels from microalgae—A review of technologies for production, processing, and extractions of biofuels and co-products. Renewable and Sustainable Energy Reviews 14(2): 557–577.

Brindhadevi, K., Mathimani, T., Rene, E.R., Shanmugam, S., Chi, N.T.L. and Pugazhendhi, A. 2021. Impact of cultivation conditions on the biomass and lipid in microalgae with an emphasis on biodiesel. Fuel 284: 119058.

Buxy, S., Diltz, R. and Pullammanappallil, P. 2012. Biogasification of marine algae Nannochloropsis oculata. Materials Challenges in Alternative and Renewable Energy II: Ceramic Transactions, 59–67.

Cercado, A.P., Ballesteros, Jr, F. and Capareda, S. 2018. Ultrasound assisted transesterification of microalgae using synthesized novel catalyst. Sustainable Environment Research 28(5): 234–239.

Chen, C.Y., Zhao, X.Q., Yen, H.W., Ho, S.H., Cheng, C.L., Lee, D.J., Bai, F.W. and Chang, J.S. 2013. Microalgae-based carbohydrates for biofuel production. Biochemical Engineering Journal 78: 1–10.

Chhandama, M.V.L., Satyan, K.B., Changmai, B., Vanlalveni, C. and Rokhum, S.L. 2021. Microalgae as a feedstock for the production of biodiesel: A review. Bioresource Technology Reports 100771.

Cooney, M., Young, G. and Nagle, N. 2009. Extraction of bio-oils from microalgae. Separation and Purification Reviews 38(4): 291–325.

Dong, T., Wang, J., Miao, C., Zheng, Y. and Chen, S. 2013. Two-step in situ biodiesel production from microalgae with high free fatty acid content. Bioresource Technology 136: 8–15.

Ehimen, E.A., Sun, Z.F. and Carrington, C.G. 2010. Variables affecting the *in-situ* transesterification of microalgae lipids. Fuel 89(3): 677–684.

Ehimen, E.A., Sun, Z. and Carrington, G.C. 2012. Use of ultrasound and co-solvents to improve the *in-situ* transesterification of microalgae biomass. Procedia Environmental Sciences 15: 47–55.

El-Shimi, H., Attia, N.K., El-Sheltawy, S. and El-Diwani, G. 2013. Biodiesel production from Spirulina-platensis microalgae by *in-situ* transesterification process. Journal of Sustainable Bioenergy Systems 3(03): 224.

Enamala, M.K., Enamala, S., Chavali, M., Donepudi, J., Yadavalli, R., Kolapalli, B., Aradhyula, T.V., Velpuri, J. and Kuppam, C. 2018. Production of biofuels from microalgae—A review on cultivation, harvesting, lipid extraction, and numerous applications of microalgae. Renewable and Sustainable Energy Reviews 94: 49–68.

Fukuda, H., Kondo, A. and Noda, H. 2001. Biodiesel fuel production by transesterification of oils. Journal of Bioscience and Bioengineering 92(5): 405–416.

Galadima, A. and Muraza, O. 2014. Biodiesel production from algae by using heterogeneous catalysts: A critical review. Energy 78: 72–83.

Ghosh, S., Banerjee, S. and Das, D. 2017. Process intensification of biodiesel production from Chlorella sp. MJ 11/11 by single step transesterification. Algal Research 27: 12–20.

Griffiths, M.J., van Hille, R.P. and Harrison, S.T. 2012. Lipid productivity, settling potential and fatty acid profile of 11 microalgal species grown under nitrogen replete and limited conditions. Journal of Applied Phycology 24(5): 989–1001.

Guldhe, A., Singh, B., Rawat, I., Permaul, K. and Bux, F. 2015. Biocatalytic conversion of lipids from microalgae Scenedesmus obliquus to biodiesel using Pseudomonas fluorescens lipase. Fuel 147: 117–124.

Guldhe, A., Singh, P., Ansari, F.A., Singh, B. and Bux, F. 2017. Biodiesel synthesis from microalgal lipids using tungstated zirconia as a heterogeneous acid catalyst and its comparison with homogeneous acid and enzyme catalysts. Fuel 187: 180–188.

Guldhe, A., Singh, P., Kumari, S., Rawat, I., Permaul, K. and Bux, F. 2016. Biodiesel synthesis from microalgae using immobilized Aspergillus niger whole cell lipase biocatalyst. Renewable Energy 85: 1002–1010.

Hamelinck, C.N., Van Hooijdonk, G. and Faaij, A.P. 2005. Ethanol from lignocellulosic biomass: Techno-economic performance in short-, middle-and long-term. Biomass and Bioenergy 28(4): 384–410.

Howell, S. 2012. Biodiesel Industry and ASTM Standards for: Washington State DOT March 22, 2012. National Biodiesel Board.

Hussain, F., Shah, S.Z., Ahmad, H., Abubshait, S.A., Abubshait, H.A., Laref, A., Manikandan, A., Kusuma, H.S. and Iqbal, M. 2021. Microalgae an ecofriendly and sustainable wastewater treatment option: Biomass application in biofuel and bio-fertilizer production. A review. Renewable and Sustainable Energy Reviews 137: 110603.

Im, H., Lee, H., Park, M.S., Yang, J.-W. and Lee, J.W. 2014. Concurrent extraction and reaction for the production of biodiesel from wet microalgae. Bioresource Technology 152: 534–537.

Jayakumar, S., Bhuyar, P., Pugazhendhi, A., Rahim, M.H.A., Maniam, G.P. and Govindan, N. 2021. Effects of light intensity and nutrients on the lipid content of marine microalga (diatom) Amphiprora sp. for promising biodiesel production. Science of the Total Environment 768: 145471.

Jazie, A.A., Abed, S.A., Nuhma, M.J. and Mutar, M.A. 2020. Continuous biodiesel production in a packed bed reactor from microalgae Chlorella sp. using DBSA catalyst. Engineering Science and Technology, an International Journal 23(3): 642–649.

Kazemifard, S., Nayebzadeh, H., Saghatoleslami, N. and Safakish, E. 2019. Application of magnetic alumina-ferric oxide nano-catalyst supported by KOH for *in-situ* transesterification of microalgae cultivated in wastewater medium. Biomass and Bioenergy 129: 105338.

Khan, M.I., Shin, J.H. and Kim, J.D. 2018. The promising future of microalgae: Current status, challenges, and optimization of a sustainable and renewable industry for biofuels, feed, and other products. Microbial Cell Factories 17(1): 1–21.

Kim, B., Im, H. and Lee, J.W. 2015. *In situ* transesterification of highly wet microalgae using hydrochloric acid. Bioresource Technology 185: 421–425.

Koberg, M., Cohen, M., Ben-Amotz, A. and Gedanken, A. 2011. Bio-diesel production directly from the microalgae biomass of Nannochloropsis by microwave and ultrasound radiation. Bioresource Technology 102(5): 4265–4269.

Kumar, A. 2021. Dissolving pulp production: Cellulases and xylanases for the enhancement of cellulose accessibility and reactivity. Physical Sciences Reviews 6(5): 111–129.

Kumar, M. and Thakur, I.S. 2018. Municipal secondary sludge as carbon source for production and characterization of biodiesel from oleaginous bacteria. Bioresource Technology Reports 4: 106–113.

Kumar, S.J., Kumar, G.V., Dash, A., Scholz, P. and Banerjee, R. 2017. Sustainable green solvents and techniques for lipid extraction from microalgae: A review. Algal Research 21: 138–147.

Kwon, M.H. and Yeom, S.H. 2015. Optimization of one-step extraction and transesterification process for biodiesel production from the marine microalga Nannochloropsis sp. KMMCC 290 cultivated in a raceway pond. Biotechnology and Bioprocess Engineering 20(2): 276–283.

Lai, J.Q., Hu, Z.L., Wang, P.W. and Yang, Z. 2012. Enzymatic production of microalgal biodiesel in ionic liquid [BMIm][PF6]. Fuel 95: 329–333.

Lee, J.-Y., Yoo, C., Jun, S.-Y., Ahn, C.-Y. and Oh, H.-M. 2010. Comparison of several methods for effective lipid extraction from microalgae. Bioresource Technology 101(1): S75–S77.

Liu, C., Hu, B., Cheng, Y., Guo, Y., Yao, W. and Qian, H. 2021. Carotenoids from fungi and microalgae: A review on their recent production, extraction, and developments. Bioresource Technology 125398.

López, E.N., Medina, A.R., Moreno, P.A.G., Callejón, M.J.J., Cerdán, L.E., Valverde, L.M., López, B.C. and Grima, E.M. 2015. Enzymatic production of biodiesel from Nannochloropsis gaditana lipids: Influence of operational variables and polar lipid content. Bioresource Technology 187: 346–353.

Lu, Y., Gan, Q., Iwai, M., Alboresi, A., Burlacot, A., Dautermann, O., Takahashi, H., Crisanto, T., Peltier, G., Morosinotto, T., Melis, A. and Niyogi, K.K. 2021. Role of an ancient light-harvesting protein of PSI in light absorption and photoprotection. Nature Communications 12(1): 1–10.

Ma, G., Hu, W., Pei, H., Song, M. and Qi, F. 2015. *In situ* transesterification of microalgae with high free fatty acid using solid acid and alkali catalyst. Fresenius Environmental Bulletin 24(1): 90–95.

MacArio, A., Giordano, G., Setti, L., Parise, A., Campelo, J.M., Marinas, J.M. and Luna, D. 2007. Study of lipase immobilization on zeolitic support and transesterification reaction in a solvent free-system. Biocatalysis and Biotransformation 25(2-4): 328–335.

Macías-Sánchez, M., Robles-Medina, A., Jiménez-Callejón, M., Hita-Peña, E., Estéban-Cerdán, L., González-Moreno, P., Navarro-López, E. and Molina-Grima, E. 2018. Optimization of biodiesel production from wet microalgal biomass by direct transesterification using the surface response methodology. Renewable Energy 129: 141–149.

Madadi, R., Maljaee, H., Serafim, L.S. and Ventura, S.P. 2021. Microalgae as contributors to produce biopolymers. Marine Drugs 19(8): 466.

Makarevičienė, V., Lebedevas, S., Rapalis, P., Gumbyte, M., Skorupskaite, V. and Žaglinskis, J. 2014. Performance and emission characteristics of diesel fuel containing microalgae oil methyl esters. Fuel 120: 233–239.

Mandik, Y.I., Cheirsilp, B., Srinuanpan, S., Maneechote, W., Boonsawang, P., Prasertsan, P. and Sirisansaneeyakul, S. 2020. Zero-waste biorefinery of oleaginous microalgae as promising sources of biofuels and biochemicals through direct transesterification and acid hydrolysis. Process Biochemistry 95: 214–222.

Markou, G., Angelidaki, I. and Georgakakis, D. 2012. Microalgal carbohydrates: An overview of the factors influencing carbohydrates production, and of main bioconversion technologies for production of biofuels. Applied Microbiology and Biotechnology 96(3): 631–645.

Marrone, B.L., Lacey, R.E., Anderson, D.B., Bonner, J., Coons, J., Dale, T., Downes, C.M., Fernando, S., Fuller, C. and Goodall, B. 2018. Review of the harvesting and extraction program within the National Alliance for advanced biofuels and bioproducts. Algal Research 33: 470–485.

Martins, A.A., Mata, T.M., de Sá Caetano, N., Morais, W.G., Gorgich, M. and Corrêa, P.S. 2020. Microalgae for biotechnological applications: Cultivation, harvesting and biomass processing. Ciências Tecnológicas Technological Sciences. https://hdl.handle.net/10216/128023.

Medina, A.R., Grima, E.M., Giménez, A.G. and González, M.I. 1998. Downstream processing of algal polyunsaturated fatty acids. Biotechnology Advances 16(3): 517–580.

Munir, M., Sharif, N., Naz, S., Saleem, F. and Manzoor, F. 2013. Harvesting and processing of microalgae biomass fractions for biodiesel production. Science Technology and Development 32(3): 235–243.

Musa, I.A. 2016. The effects of alcohol to oil molar ratios and the type of alcohol on biodiesel production using transesterification process. Egyptian Journal of Petroleum 25(1): 21–31.

Nascimento, I.A., Marques, S.S.I., Cabanelas, I.T.D., Pereira, S.A., Druzian, J.I., de Souza, C.O., Vich, D.V., de Carvalho, G.C. and Nascimento, M.A. 2013. Screening microalgae strains for biodiesel production: Lipid productivity and estimation of fuel quality based on fatty acids profiles as selective criteria. Bioenergy Research 6(1): 1–13.

Noiroj, K., Intarapong, P., Luengnaruemitchai, A. and Jai-In, S. 2009. A comparative study of KOH/Al2O3 and KOH/NaY catalysts for biodiesel production via transesterification from palm oil. Renewable Energy 34(4): 1145–1150.

Park, J.Y., Park, M.S., Lee, Y.C. and Yang, J.W. 2015. Advances in direct transesterification of algal oils from wet biomass. Bioresource Technology 184: 267–275.

Park, J., Kim, B., Chang, Y.K. and Lee, J.W. 2017. Wet *in situ* transesterification of microalgae using ethyl acetate as a co-solvent and reactant. Bioresource Technology 230: 8–14.

Patil, P.D., Reddy, H., Muppaneni, T., Mannarswamy, A., Schuab, T., Holguin, F.O., Lammers, P., Nirmalakhandan, N., Cooke, P. and Deng, S. 2012. Power dissipation in microwave-enhanced *in situ* transesterification of algal biomass to biodiesel. Green Chemistry 14(3): 809–818.

Pignolet, O., Jubeau, S., Vaca-Garcia, C. and Michaud, P. 2013. Highly valuable microalgae: Biochemical and topological aspects. Journal of Industrial Microbiology and Biotechnology 40(8): 781–796.

Rahman, M., Aziz, M., Al-Khulaidi, R.A., Sakib, N. and Islam, M. 2017. Biodiesel production from microalgae *S. pirulina* maxima by two step process: Optimization of process variable. Journal of Radiation Research and Applied Sciences 10(2): 140–147.

Ramaraj, S., Hemaiswarya, S., Raja, R., Ganesan, V., Anbazhagan, C., Carvalho, I.S. and Juntawong, N. 2015. Microalgae as an attractive source for biofuel production. *In*: Environmental Sustainability, pp. 129–157. Springer.

Ranga, R., Sarada, R. and Ravishankar, G. 2007. Influence of CO_2 on growth and hydrocarbon production in Botryococcus braunii. Journal of Microbiology and Biotechnology 17(3): 414–419.

Razzak, S.A., Hossain, M.M., Lucky, R.A., Bassi, A.S. and De Lasa, H. 2013. Integrated CO2 capture, wastewater treatment and biofuel production by microalgae culturing—A review. Renewable and Sustainable Energy Reviews 27: 622–653.

Ryskamp, R., Thompson, G., Carder, D. and Nuszkowski, J. 2017. The Influence of High Reactivity Fuel Properties on Reactivity Controlled Compression Ignition Combustion (No. 2017-24-0080). SAE Technical Paper.

Saad, M.G., Dosoky, N.S., Zoromba, M.S. and Shafik, H.M. 2019. Algal biofuels: Current status and key challenges. Energies 12(10): 1920.

Saifuddin, N., Samiuddin, A. and Kumaran, P. 2015. A review on processing technology for biodiesel production. Trends in Applied Sciences Research 10(1): 1.

Sathish, A., Smith, B.R. and Sims, R.C. 2014. Effect of moisture on *in situ* transesterification of microalgae for biodiesel production. Journal of Chemical Technology and Biotechnology 89(1): 137–142.

Shin, Y.S., Choi, H.I., Choi, J.W., Lee, J.S., Sung, Y.J. and Sim, S.J. 2018. Multilateral approach on enhancing economic viability of lipid production from microalgae: A review. Bioresource Technology 258: 335–344.

Shirazi, H.M., Karimi-Sabet, J. and Ghotbi, C. 2017. Biodiesel production from Spirulina microalgae feedstock using direct transesterification near supercritical methanol condition. Bioresource Technology 239: 378–386.

Singh, B., Korstad, J. and Sharma, Y. 2012. A critical review on corrosion of compression ignition (CI) engine parts by biodiesel and biodiesel blends and its inhibition. Renewable and Sustainable Energy Reviews 16(5): 3401–3408.

Suganya, T., Gandhi, N.N. and Renganathan, S. 2013. Production of algal biodiesel from marine macroalgae Enteromorpha compressa by two step process: Optimization and kinetic study. Bioresource Technology 128: 392–400.

Surendhiran, D., Sirajunnisa, A.R. and Vijay, M. 2015. An alternative method for production of microalgal biodiesel using novel Bacillus lipase. 3 Biotech 5(5): 715–725.

Surendhiran, D., Vijay, M. and Sirajunnisa, A.R. 2014. Biodiesel production from marine microalga Chlorella salina using whole cell yeast immobilized on sugarcane bagasse. Journal of Environmental Chemical Engineering 2(3): 1294–1300.

Taher, H., Al-Zuhair, S., Al-Marzouqi, A.H., Haik, Y. and Farid, M.M. 2011. A review of enzymatic transesterification of microalgal oil-based biodiesel using supercritical technology. Enzyme Research 2011.

Tiwari, A. and Pandey, A. 2012. Cyanobacterial hydrogen production—A step towards clean environment. International Journal of Hydrogen Energy 37(1): 139–150.

Torres, S., Acien, G., García-Cuadra, F. and Navia, R. 2017. Direct transesterification of microalgae biomass and biodiesel refining with vacuum distillation. Algal Research 28: 30–38.

Tran, D.T., Chen, C.L. and Chang, J.S. 2013. Effect of solvents and oil content on direct transesterification of wet oil-bearing microalgal biomass of Chlorella vulgaris ESP-31 for biodiesel synthesis using immobilized lipase as the biocatalyst. Bioresource Technology 135: 213–22.

Tran, D.T., Yeh, K.L., Chen, C.L. and Chang, J.S. 2012. Enzymatic transesterification of microalgal oil from Chlorella vulgaris ESP-31 for biodiesel synthesis using immobilized Burkholderia lipase. Bioresource Technology 108: 119–127.

Valdez-Ojeda, R.A., del Rayo Serrano-Vázquez, M.G., Toledano-Thompson, T., Chavarria-Hernandez, J.C. and Barahona-Perez, L.F. 2021. Effect of media composition and culture time on the lipid profile of the green microalga Coelastrum sp. and its suitability for biofuel production. BioEnergy Research 14(1): 241–253.

Vasić, K., Hojnik Podrepšek, G., Knez, Ž. and Leitgeb, M. 2020. Biodiesel production using solid acid catalysts based on metal oxides. Catalysts 10(2): 237.

Veillette, M., Giroir-Fendler, A., Faucheux, N. and Heitz, M. 2017. A biodiesel production process catalyzed by the leaching of alkaline metal earths in methanol: from a model oil to microalgae lipids. Journal of Chemical Technology and Biotechnology 92(5): 1094–1103.

Wahidin, S., Idris, A., Yusof, N.M., Kamis, N.H.H. and Shaleh, S.R.M. 2018. Optimization of the ionic liquid-microwave assisted one-step biodiesel production process from wet microalgal biomass. Energy Conversion and Management 171: 1397–1404.

Ward, A., Lewis, D. and Green, F. 2014. Anaerobic digestion of algae biomass: A review. Algal Research 5: 204–214.

Yin, Z., Zhu, L., Li, S., Hu, T., Chu, R., Mo, F., Hu, D., Liu, C. and Li, B. 2020. A comprehensive review on cultivation and harvesting of microalgae for biodiesel production: Environmental pollution control and future directions. Bioresource Technology 301: 122804.

Yousuf, A. 2019. Microalgae Cultivation for Biofuels Production. Academic Press.

Yousuf, A. 2020. Fundamentals of microalgae cultivation. *In*: Microalgae Cultivation for Biofuels Production, pp. 1–9. Academic Press.

Zendejas, F.J., Benke, P.I., Lane, P.D., Simmons, B.A. and Lane, T.W. 2012. Characterization of the acylglycerols and resulting biodiesel derived from vegetable oil and microalgae (Thalassiosira pseudonana and Phaeodactylum tricornutum). Biotechnology and Bioengineering 109(5): 1146–1154.

CHAPTER 13

Biogas Production Enhancement Employing Bioelectrochemical Systems

Min-Hua Cui,[1],* Thangavel Sangeetha[2] and Wen-Zong Liu[3],*

1. Introduction

Conversion of organic wastes/wastewater to bioenergy is one practical strategy to reduce greenhouse gas emissions and reach carbon neutrality (Xu et al. 2021). For now, various organic wastes and wastewater, such as waste activated sludge (Calderon et al. 2021), food waste (Cui et al. 2021a), agricultural wastes (Li et al. 2021), beer wastewater (Sangeetha et al. 2020), dairy wastewater (Bella and Rao 2021), etc., have been utilized as substrates for the anaerobic digestion (AD) to produce biogas. It has attracted continuously increasing research attention due to the capabilities of AD in stabilizing organic wastes/wastewater and simultaneously generating biogas as an alternative energy source. However, the AD process has severe bottle necks like disadvantages due to low reaction rate and low quality of biogas, and more efforts are required to improve AD efficiency, especially in practical applications.

Up till now, multifarious investigation works have been made to improve the biogas yield and quality, pretreatment application (Atelge et al. 2020, Zhen et al. 2017), digestion condition optimization (Cui et al. 2021b, Li et al. 2019), conductive or nanomaterials (Ajay et al. 2020, Wu et al. 2020), co-digestion (Miryahyaei et al. 2020, Solé-Bundó et al. 2019), etc. Jiang et al., added granular activated carbon (GAC) in thermal pretreatment sludge AD and proved that the GAC obviously shortened the methanogenesis lag phases by 19.3%–30.6% via toxin disinhibition; however, the ultimate methane yields were reduced simultaneously (Jiang et al. 2021). The obstructions on the biogas production and quality improvement have not been completely solved yet.

[1] Jiangsu Key Laboratory of Anaerobic Biotechnology, Jiangnan University, Wuxi, 214122, P.R. China.
[2] Department of Energy and Refrigerating Air-Conditioning Engineering and Research Centre of Energy Conservation for New Generation of Residential, Commercial, and Industrial Sectors, National Taipei University of Technology, Taipei 10608, Taiwan.
[3] School of Civil and Environmental Engineering, Harbin Institute of Technology (Shenzhen), Shenzhen 518055, P. R. China.
* Corresponding authors: cuiminhua@jiangnan.edu.cn; liuwenzong@hit.edu.cn

Bioelectrochemical system (BES), a recently developed technology, has been widely applied to enhance AD performance (De Vrieze et al. 2018). According to the operational mode, BES was divided into microbial electrolysis cell (MEC) and microbial fuel cell (MFC), respectively. MEC requires an external applied voltage and MFC possesses net energy output capacity. MFCs and MECs are two examples of a rapidly developing BES that combine biological and electrochemical processes to generate useful by-products like electricity, hydrogen, biogas and also simultaneously removing the organic matters from the wastewater (Adekunle et al. 2019). The general working principle of the two significant BESs has been illustrated in Fig. 1.

(a) Microbial fuel cell (MFC)   **(b)** Microbial electrolysis cell (MEC)

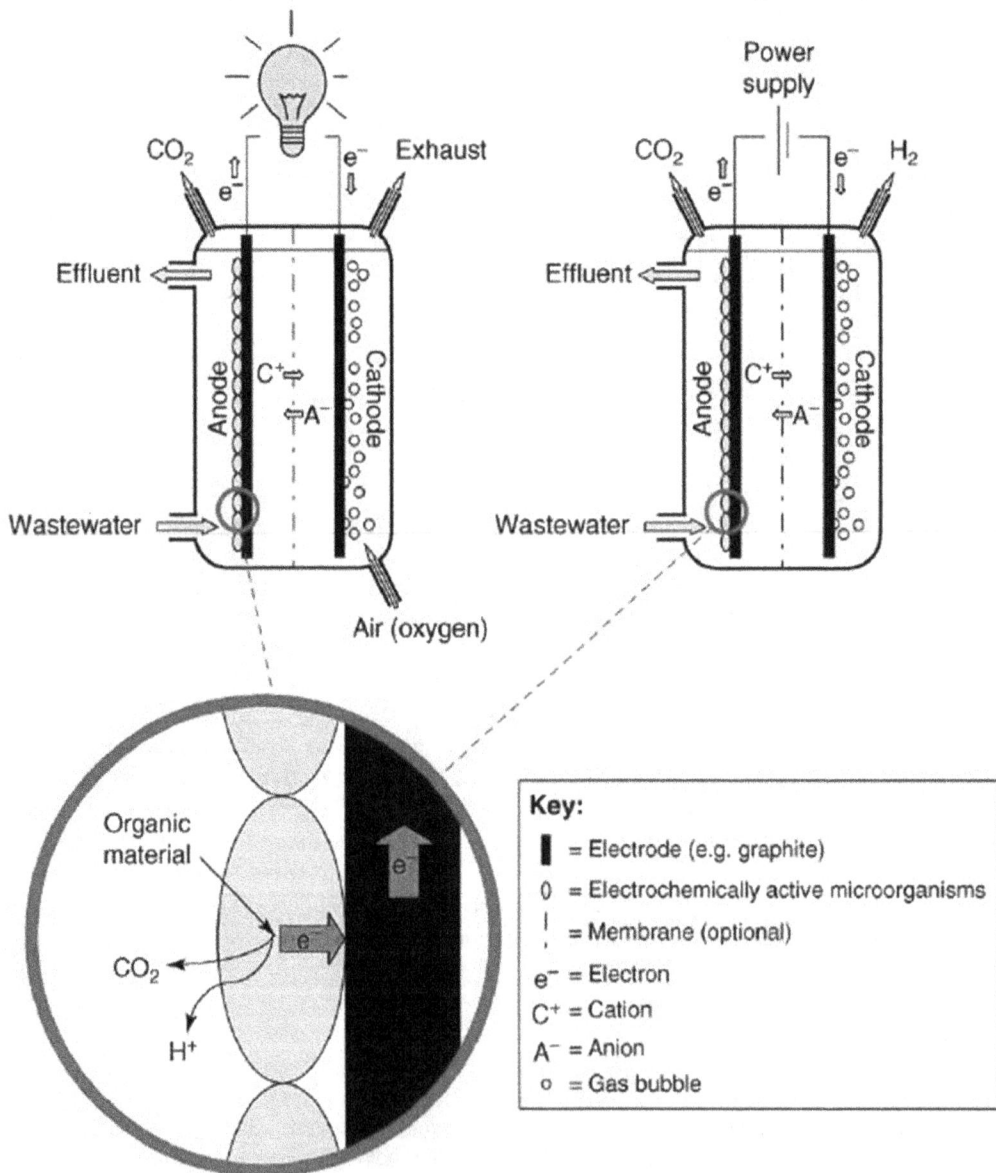

Fig. 1. Graphical representation of the working principle of MFC (a) and MEC (b) (European Commission 2013).

MFCs are prominent bio electrochemical reactors that are proficient in conversion of organic waste substances into beneficial energy such as bioelectricity, whereas MEC are almost similar to MFC, just varying in the voltage addition to the cathode (Sangeetha et al. 2021). Unlike traditional electrochemical technology, the microbial metabolic process plays an important role in substrate utilization and biogas generation in a BES. Specifically, the substrate degradation degree and reaction rate would be improved by the enhancement effect of anodic biofilm, which is obviously in favor of organic matters removal. The BES cathode provides a controllable reaction potential by applying an external voltage (normally in the range of 0.3 to 1.5 V) to facilitate biogas production. The proposed promoting mechanisms are summarized in Fig. 2. Basically, the cathode served as a constant and solid electron donor to drive direct and indirect cathodic reactions. (1) Specific methanogens directly accept electrons from the cathode to produce methane; (2) hydrogen that bioelectrochemically or electrochemically evolved from the cathode and further facilitate hydrogenotrophic methanogens; (3) specific microbes (e.g., homoacetogened) using cathodic electron, protons, and carbon dioxide to synthesize simple organic matters (e.g., acetate) serving as feedstock for methanogens to produce methane.

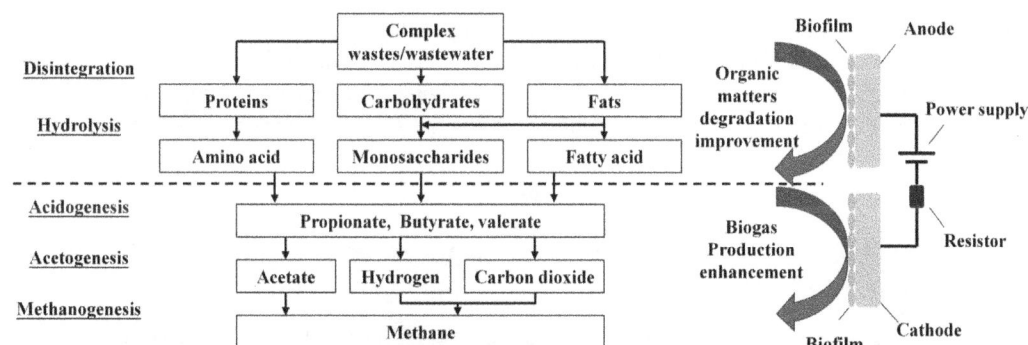

Fig. 2. The proposed mechanisms of BES promoting anaerobic digestion performance.

BES technology has been studied to improve each stage of the AD process as well as terminal biogas upgrading for more than a decade. However, those scattered reports require a systematic combing to illumine future investigations and potential engineering applications. In this chapter, the AD performance enhancement by introduction of BES was carefully surveyed in view of substrates degradation, biogas production, and biogas quality upgrading, especially by summarizing the effects of BES on the microbial community. Besides, the future perspectives of AD performance improvement by BES has also been proposed as well. These efforts will be beneficial to the development of BES technology in the AD and also potentially promote this technology closer to practical applications.

2. Bioelectrochemical Systems for Biogas Production

BES have been predominantly employed for various significant applications like wastes/wastewater treatment, and resource recovery, but the most essential is bioenergy generation in various forms like electricity, biogas and hydrogen. Almost all the BES have been implemented for biogas generation either individually or in an integrated form with AD as biogas. Biogas upgrading is an advanced process where impure biogas components are converted to bio methane (Thiruselvi et al. 2021). Dual chambered MFCs were fed with biogas slurry for biogas and electricity production (Wang et al. 2019a). This elegant technology was performed along with electricity production using an integrated MFC with bipolar membrane electrodialysis (BPMED) reactor by Chen et al.

(2013) with a mechanism called alkali CO_2 adsorption. MECs have always been employed for hydrogen generation as it is a pollution free energy source (Zhao et al. 2021a). But they have also been recently linked with biogas production which contains mainly methane as this has also been recognized as a significant renewable energy resource. Correspondingly, MECs have also been used for biogas production. Biogas constituted mainly of methane and hydrogen, and the biogas yield and methane to hydrogen mole ratio were mainly regulated by *Geobacter* and *Methanobacterium* (He et al. 2021). Wang et al. (2011) efficaciously integrated MFC and MEC for H_2 gas production from cellulose, where MFCs served as the power source for MECs. These reactors gained importance and further studies to upgrade their energy recovery as biogas was carried out by Almatouq et al. (2020) where the anode microbial communities were analyzed and the growth of methanogens was enhanced for further biogas production. Individual MECs achieved ammonia recovery along with biogas generation when operated with dark fermentation effluent (Zeppilli et al. 2021). Integration of MEC with AD has always proven to be fruitful for biogas production and wastewater treatment as combined reactors are always advantageous than the individual and conventional ones (Sangeetha et al. 2016, 2017). Coupling of two significant biological reactors such as MEC and AD can supply abundant hydrogen and carbon dioxide for enhanced methane generation. Electrochemical methanogenesis is more controllable and stable compared to conventional anaerobic methods (Sangeetha et al. 2020). Both anodic and cathodic reactions participated in the process of organic matters conversion to methane (Fig. 3a-d). The following sections will detail the techniques implemented in the BES for the exclusive ambition of enhancing the biogas production.

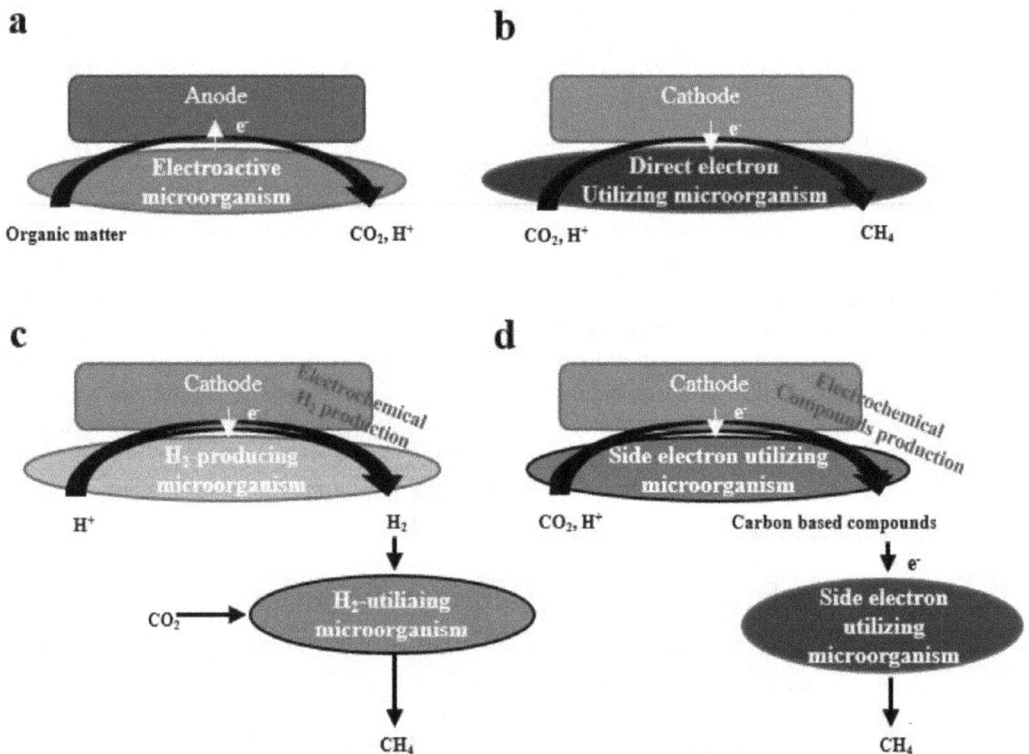

Fig. 3. Bioelectrochemical electron transfer pathways for methane production on electrodes. a, anodic oxidation of organic matters; b, direct bioelectrochemical methanogenesis, c, H_2-mediated methanogenesis; d, carbon compounds-mediated methanogenesis (Park et al. 2020).

3. Biogas Enhancement Strategies

Summarizing the stratagems for the augmentation of biogas production in BES is the exclusive and crucial objective of this chapter. The methods have been listed out in detail in the following sections.

3.1 Enhancement of Organic Matter Degradation

Traditional AD process is the groundwork technique for biogas generation, but it also suffers from a great bottleneck issue such as lower substrates degradation rate. The BES anode creates an oxidation environment and dredges the electron transfer pathway (Fig. 3a). BES technology has been verified to enhance organic matters conversion both of readily biodegradable and refractory substrates from food waste (Huang et al. 2020, Quashie et al. 2021), beer wastewater (Guo et al. 2016), sludge, and hydrolysate (Xiao et al. 2018, Zakaria et al. 2019, Zhang et al. 2020a), livestock wastewater (Ahn et al. 2017, Baek et al. 2021a), petrochemical wastewater (Arvin et al. 2019a, b), leachate (Hassan et al. 2018), blackwater (Huang et al. 2021, Liu et al. 2018), etc.

BES technology has been verified to enhance hydrolysis to facilitate methane production (e.g., waste sludge), and it made the MEC more applicable for carbohydrate-deficient substrates (Liu et al. 2016a). The presence of anode associated with EAB was considered as the major contributor to the enhancement of organic matters in AD. A combined AD-MEC reactor exhibited a 14.3% of chemical oxygen demand (COD) removal enhancement compared to the AD control in 23-day anaerobic digestion of food waste. It was also superior to the AD-MEC that only operated under closed-circuit mode for 5 days and was followed by no power supply for the subsequent 18 days (Hassanein et al. 2017). The AD-MEC enhanced the soluble carbohydrates to about 4 times compared to the AD reactor during food waste digestion. The degradation of VFAs, especially acetic acid, propionic acid, and butyric acid, was the major reason for the efficient utilization of organic matters in BES (Huang et al. 2020). Zhao et al. (2021b) reported that the applied voltage significantly enhanced the acidogenesis and methanogenesis processes from proteins in the AD. The methane production rate increased by 45.6% at the lower protein rate of 4 g/L, and 225.4% at higher rate of 20 g/L. Besides, the degradation of oil containing wastewater can also be improved by BES technology (Krishnan et al. 2019). Both MFC and MEC modes can sufficiently transform soybean edible oil refinery wastewater with maximum COD removal efficiencies of 96.4% and 95.8% in MFC and MEC, respectively. The methane yield of 45.4 ± 1.1 L/kg-COD was obtained under MEC mode with a constant applied voltage of 1.2 V (Yu et al. 2017).

The anodic oxidation rate was considered as the lagging issue in treating refractory substrates, thus more adjusting technologies were proposed. Peng et al. (2019) enhanced the anodic oxidation by adding nitrate into MEC. The anodic oxidation efficiency in the nitrate-added MEC increased by 55.9%, resulting in 21.9% of the volatile suspended solid removal efficiency being higher than that of control MEC. Although the cathodic electrons competition between nitrate and hydrogenotrophic methanogens deteriorated the initial cumulative methane production, the terminal cumulative methane production in 24 days was 8.9% higher. This indicated that the advantage of stimulating the anodic respiration was greater than the electrons lost in the nitrate reduction. Zhang et al. (2013) reported that dosing $Fe(OH)_3$ enhanced the degradation of reactive brilliant red X-3B dye and sucrose.

3.2 Biogas Production Enhancement

It is eminent that hydrolysis is the rate-limiting step during the anaerobic digestion of refractory substrates (e.g., wasted activated sludge) and further deteriorated the methane production. The BES technology has been widely employed for biogas production enhancement for various substrates, as shown in Table 1. The AD-MEC combined technology obviously increased hydrolysis and

Table 1. Summary of BES for biogas production enhancement from various substrates.

Substrate	Reactor Configuration	Electrode Material	Applied Voltage	Biogas Production	Energy Efficiency	Biogas Quality	References
Food waste	3.496 cm in diameter, 15.5 cm in height	Anode: graphite plate Cathode: stainless steel	0.9 V	0.59 m³/m³/d	Overall efficiency exceeded 400%	CH₄ of 71.9%	(Hassanein et al. 2017)
Acetate	5.0 cm in diameter, 9.2 cm in height	Anode: carbon felt Cathode: stainless steel	1.0 V	CH₄ yield increased 2.3 times	Overall energy efficiency of 66.7%	CH₄ content excess 98%	(Bo et al. 2014)
Pig slurry	Two-chamber MEC (0.5 L in each compartment)	Anode: carbon felt Cathode: granular graphite	Cathode potential poised at −800 mV vs. SHE	79 L/m³/d	–	–	(Cerrillo et al. 2018)
Acetate	10 cm in diameter, 7.6 cm in height	Anode: carbon felt Cathode: stainless steel	1.0 V	360.2 mL/g-COD	Overall energy efficiency of 74.6%	Carbon dioxide of 6.9%	(Yin et al. 2016)
Bovine serum albumin	7 cm in diameter, 25 cm in height	Anode: carbon brush Cathode: carbon cloth	0.6 V	Methanogenesis efficiency increased by 225.4%	–	–	(Zhao et al. 2021b)
Waste activated sludge	45 mm in diameter, 80 mm in length	Anode: graphite brush Cathode: carbon cloth	0.8 V	91.8 g CH₄/m³/d	–	–	(Liu et al. 2016b)
Thermal-alkaline pretreated sludge	6.5 cm in diameter, 9 cm in height	Ti/Ru alloy mesh plates	1.8 V	Methane productions increased by 79.3%	–	–	(Xiao et al. 2018)
Glucose	Working volume of 360 mL	Anode: carbon fibers integrated with a stainless-steel frame Cathode: stainless steel mesh	Applied potential turned off for 6 h/d	433 ± 7.9 L CH₄/m³	4.3 kJ per batch cycle	–	(Zakaria and Ranjan Dhar 2021)
Egeria densa	Working volume of 0.8 L	Ti/RuO₂	1.0 V	248.2 ± 21.0 mL/L/d	–	–	(Zhen et al. 2016)
Straw	100 mm in diameter, 150 mm in length	Anode: carbon brush Cathode: carbon cloth	1.0 V	116.18 mL/g VS	87.42% ~ 141.74%	87% ± 5%	(Yan et al. 2021)

acidogenesis of raw waste activated sludge, thus increasing methane productivity 7.8 times compared to the AD control. This enhancement not only occurred with 0.8 V external voltage, but also persisted to the open-circuit mode with a 6.2 times higher result. It was attributed to the enrichment of fermentative bacteria and syntrophic acetogenic bacteria by voltage application (Wang et al. 2021). Bioelectrolysis has been demonstrated to significantly regulate AD by pathways alteration, more acetate but less butyrate and propionate were produced in the acidogenesis fermentation of glucose. The average methane production with 0.8 V voltage supply reached 0.131 m³/m³/d, which was 1.4 times higher than that under open-circuit mode (0.055 m³/m³/d). Although the higher methane production was observed with a higher applied voltage (0.8 V), the preferable operation condition was 0.5 V external voltage due to the net energy profit increase by 14.2%. A systematic electron balance analysis was conducted to reveal the regulation of BES on electron transfer pathways. The bioelectrolysis reaction by employing BES created an additional pathway between acetate and hydrogen, which enhanced electron transfer to methane (see Fig. 4) (Guo et al. 2017).

Besides, another view deemed that the BES accelerated methane production and stabilization instead of improving ultimate methane yield. Park et al. (2018) suggested that the methane production rate and stabilization time of the AD-MEC reactor were approximately 1.7 and 4.0 times faster than those of the AD reactor; however, the final methane production yield was similar to the theoretical maximum.

The BES module characteristics, such as the electrode materials, electrode position, and electrode size, were seen as the major influencing factors related to the anaerobic digestion performance (Park et al. 2020, Zhao et al. 2021b). Sangeetha et al. (2017) declared that installing electrodes at the downside of an upflow-MEC was better than that at the upper side. A methane yield of 275.8 mL/g COD was obtained with the BES reactor at a hydraulic retention time (HRT) of 36 h. The superiority of bottom installed electrodes in an upflow MEC was verified by hydrodynamics analysis, in which a maximum COD removal efficiency of 92.1% and methane yield of 304.5 mL/L/d were achieved (Gao et al. 2019). Nickel seemed to be the preferable cathodic material versus stainless steel and copper in the upflow-MEC treating beer wastewater (Sangeetha et al. 2016). Gao et al. (2021) suggested that the ratio of cathode surface area to reaction region volume was a crucial factor to the bio electrochemical performance. Utilizing stacked nickel meshes with a cathode space ratio of 1.33 cm²/cm³ was recommended to increase the methane production rates to 332.0 and 334.8 mL/L/d at the HRT of 24 h and 36 h, respectively. Wang et al. (2019b) compared

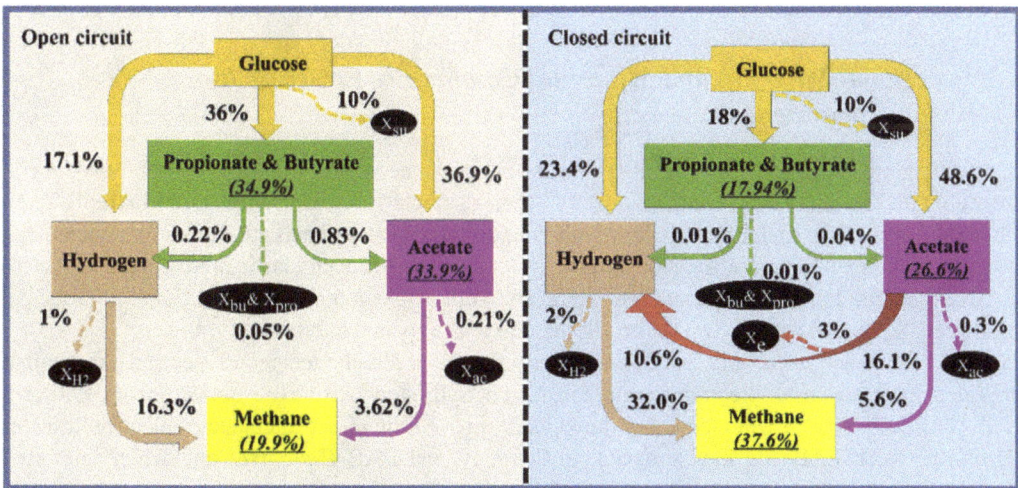

Fig. 4. The proposed electron balance analysis of anaerobic digestion with bioelectrochemical systems (left, without power supply; right, with power supply, viz., anaerobic digestion-bioelectrochemical system coupling mode) (Guo et al. 2017).

three different metal cathode materials, e.g., nickel, copper, and stainless steel, for bioelectrochemical methanogenesis and the MEC equipped with nickel cathode exhibited a maximum methane yield of 59.2 mL CH_4/gVSS with a higher current density of 9 A/m^2. It was inferred that the hydrogenotrophic methanogenic pathway was enhanced and related to the high hydrogen evolution reaction activity of nickel. The strategy of the applied voltage was optimized to improve the overall energy efficiency of AD-MEC. Biomethane production was not affected when the applied potential turned off for 6 h/d but was substantially deteriorated when the applied potential was turned off for 12 h/d (Zakaria and Ranjan Dhar 2021).

Recent research reported a new viewpoint that adding a high surface area carbon fiber brush was a more effective method for improving AD performance than using MEC electrodes with an applied voltage. Introducing a large brush provided a more specific surface area to support biomass retention, and further enhanced the AD performance by accelerating substrates consumption (Baek et al. 2021b). The conductive materials served as biomass carriers that promoted the direct interspecies electrons transfer between bacteria and archaea and this was attributed to the boost methanogenesis (Gahlot et al. 2020).

The BES could improve methane production by alleviating toxicity from hazardous matters. It has been widely reported that the high concentration of sulfate can be transformed to toxic sulfides by sulfate-reducing bacteria under anaerobic conditions and this will obviously inhibit the activity of methanogens and decrease methane production (Dai et al. 2017). Yuan et al. (2020) suggested that the cathode of MEC created alkaline conditions to convert unionized hydrogen sulfide (H_2S) to ionized sulfide (HS^-) that is less toxic to the methanogenesis. As a result, the methane production in AD-MEC increased by 303% compared to AD control.

In general, BES technology did accelerate methane production by enhancing electron transfer and regulating the methanogenic pathways. The controllable cathodic reduction environment facilitated the EAB to uptake the electrons from the cathode and the hydrogen evolution, thus improving the electrons' flow to the methanogens with both direct or indirect pathways. In terms of the methane yield, viewpoints have been suggested that it was probably dependent on the substrate properties and conversion rate. The easily degradable substrates could be converted with a high rate in AD and the BES can only adjust the methanogenic pathway instead of methane yield. BES could augment methane production by increasing the conversion rate of refractory substrates. Anodic EAB degraded more refractory substrates into small molecule matters that could feed methanogens. Thus, the methane increment was derived from the reinforcement of refractory substrates degradation.

3.3 Functional Microbes and Microbial Community Structure for Biogas Enhancement

Methanogens are the specific microorganisms that produce methane as a metabolic product in anaerobic digestion. Typical methanogens and their characteristics are summarized in Table 2. As elucidated above, the metabolic process is the major classification basis for the methanogens, and two dominated pathways (hydrogenotrophic methanogenesis and acetoclastic methanogenesis) were discussed herein. Different from the traditional anaerobic digestion process, functional microbes in BES or BES coupled systems would be altered due to the special habitats.

The electro-active bacteria (EAB) possessed a feature of direct interspecies electron transfer that provided methanogens an electron resource instead of hydrogen to drive the methanogenesis. Rotaru et al. (2014a, b) were the pioneers and they clearly proved the direct interspecies electron transfer (DIET) between EAB and methanogens. In 2014, typical EAB *Geobacter* and widely reported methanogens *Methanosaeta* species were found exchanging electrons via DIET, and in other work, *Geobacter metallireducens* and *Methanosarcina barkeri* were also capable of DIET. The indirect interspecies electron transfer was also in favor of methane production. By applying an external

Table 2. Typical methanogenesis (Fu et al. 2021).

Methanogens	Substrate	Reactions	Typical Methanogens	Typical Habitat
Hydrogenotrophic	H$_2$ and CO$_2$ Formate Methanol	$4H_2 + CO_2 \rightarrow CH_4 + 2H_2O$ $4HCOOH \rightarrow CH_4 + 3CO_2 + 2H_2O$ $4CH_3OH \rightarrow 3CH_4 + CO_2 + 2H_2O$	*Methanobacterium bryantii* *Methanobacterium formicicum* *Methanobacterium thermoalcaliphium* *Methanothermobacter thermoautotrophicum* *Methanothermobacter wolfeii* *Methanobrevibacter smithii* *Methanobrevibacter ruminantium* *Methanococcoides methylutens*	Deep marine sediments Termite hindguts Human gastrointestinal tracts Animal gastrointestinal tracts
Aceticlastic	Acetate	$CH_3COOH \rightarrow CH_4 + CO_2$	*Methanosaeta concilii (soehngenii)* *Methanosaeta thermophila*	Anaerobic digesters Rice fields Wetlands
Methylotrophic	Trimethylamine Dimethyl sulfate Methylated ethanolamines	$4(CH_3)_3N + 6H_2O \rightarrow 9CH_4 + 3CO_2 + 4NH_3$ $2(CH_3)_2NH + 2H_2O \rightarrow 3CH_4 + CO_2 + 2NH_3$ $4(CH_3)NH_2 + 2H_2O \rightarrow 3CH_4 + CO_2 + 4NH_3$ $2(CH_3)_2S + 2H_2O \rightarrow 3CH_4 + CO_2 + H_2S$	*Methanosarcina barkeri* *Methanosarcina mazei* *Methanosarcina thermophile*	Marine Hypersaline habitat Sulfate-rich sediments

voltage of 0.39 V, the methane production rate from BES was about 168% higher than the open circuit control. The microbial community analysis indicated that hydrogenotrophic methanogens (e.g., *Methanobacterium*) were enriched in the BES and methanogens utilizing carbon dioxide (e.g., *Methanosaeta* and *Methanosarcina*) were dominant in the control without applied voltage. The cathodic hydrogen evolution instead of DIET was speculated to be the major reason for the methane production enhancement (Lee et al. 2017). The abundance of fermentative bacteria at the anode was increased by 46.7% with external voltage applied and *Methanobacterium* at the cathode increased to 84.3%, indicating that the methanogenesis pathway transformed from acetoclastic to hydrogenotrophic. The electron balance analysis showed that only 10% of the produced methane was driven from direct interspecies electron transfer (Zhao et al. 2021b).

Even the same methanogens could produce methane by different pathways, for instance, Liu et al. (2019) revealed that the *Methanothrix* was found to be a significant contributor to the biogas upgrading in a BES. The transcriptomics revealed that two different pathways simultaneously occurred both on the electrode and in the bulk. *Methanothrix* on the cathode was using the carbon dioxide reduction pathway and in the bulk were using the acetate decarboxylation pathway for the production of methane.

3.4 Biogas Upgrading

The biogas from the traditional AD process consisted of about 60% methane, 40% carbon dioxide, and small amounts of impurity gases such as H_2, NH_3, H_2S, etc. (Demirel et al. 2010, Kougias et al. 2017). The high content of carbon dioxide in the biogas is the major limiting factor for the efficient high-value utilization. Thus, removing carbon dioxide and even converting it into methane simultaneously with BES technology is one promising strategy. Conventionally, the biogas upgrading technologies have been divided into *in-situ* upgrading, *ex-situ* upgrading, and hybrid upgrading methods according to the different scenes that upgrading technologies implement. Here, we have focused on the *in situ* biogas upgrading by BES technology.

The commonly used *in situ* biogas upgrading technology is introducing hydrogen into the AD system as the electron donor to drive the carbon dioxide conversion, such as directly producing methane utilizing hydrogen and carbon dioxide through hydrogenotrophic methanogenesis, and synthesizing simple organics (e.g., acetate, formate) through Wood-Ljungdahl pathway that indirectly facilitates the aceticlastic methanogenesis (Glueck et al. 2010). The cathode of BES served as a continuous solid electron donor which would potentially replace the hydrogen to improve methane content in biogas *in situ*. It was reported that the cathode poised at –500 mV vs. standard hydrogen electrode could efficiently upgrade biogas, methane content in biogas increased from 71% to 90% and carbon dioxide decreased by 8.2% (Liu et al. 2019). The proposed biogas upgrading mechanisms by BES technology have been depicted in Fig. 5.

Two possible electron transfer pathways were widely reported, namely direct electron transfer and indirect electron transfer. For the direct electron transfer, microbes can uptake electrons directly from the cathode and support the metabolic process. In 2009, for the first time, Cheng et al. (2009) reported that a pure methanogenic strain, namely *Methanobacterium palustre*, was able to directly uptake electrons from the cathode. Other methanogens *methanobacterium alcaliphilum* and *methanocorpusculum sinense* were found to produce methane using electrons from the electrode directly as well as hydrogen as the electron donor indirectly (Jiang et al. 2013). Both carbon brush and graphite plate can enrich functional cathodic biofilm to achieve carbon dioxide reduction, while the carbon brush was the preferable cathodic material that presented 22.7% of carbon dioxide into methane (602 mol/day/m³). The microbial community structure analysis indicated that *Methanothrix*, a genus of acetoclastic methanogen that can directly accept electrons for carbon dioxide reduction, was dominant in the archaeal community. Transcriptomic studies also revealed that *Methanothrix*

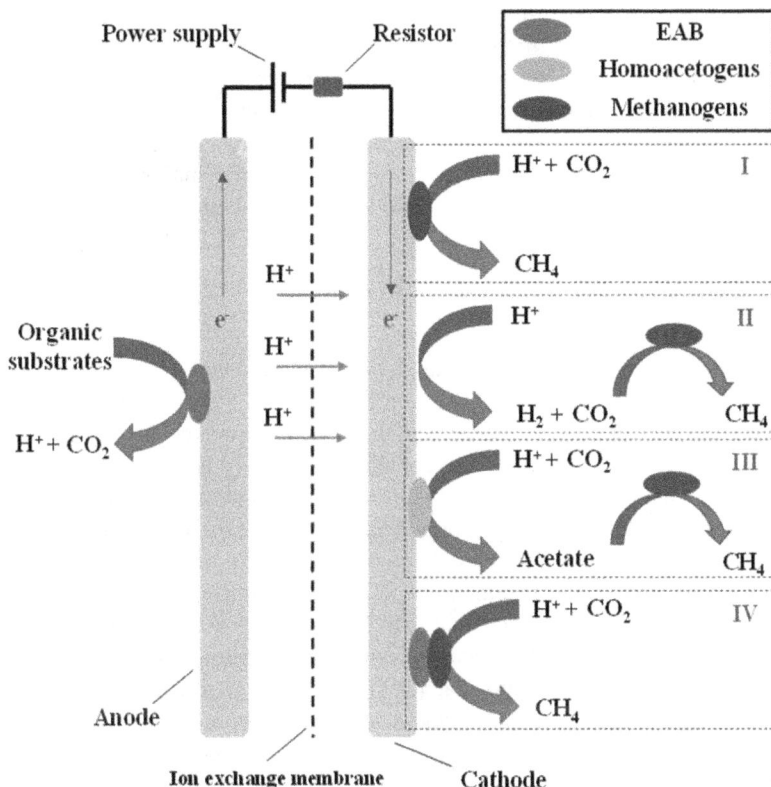

Fig. 5. Summary of biogas upgrading pathways by BES. I, methanogenesis through direct extracellular electron transfer; II, hydrogen mediated methanogenesis; III, homoacetogenesis coupled with acetoclastic methanogenesis; IV, methanogenesis through indirect extracellular electron transfer (direct interspecies electron transfer). EAB, electro-active bacteria.

species colonizing the carbon brush surface had actively expressing genes that coded for enzymes required for electromethanogenesis but not for acetoclastic methanogenesis (Liu et al. 2020a).

For the indirect electron transfer, some mediums (hydrogen, formate, acetate) played the role of electron carriers to drive the conversion of carbon dioxide. The cathode potential was considered as a key operational parameter to the bioelectrochemical methanogenesis. Methane was primarily produced indirectly via hydrogen and acetate when the cathode potential was fixed at –0.7 V or lower vs. normal hydrogen electrode (van Eerten-Jansen et al. 2015). However, in most cases that cultivated consortium in the cathodic biofilm, two electron transfer pathways were simultaneous in existence in the carbon dioxide reduction (Baek et al. 2017). Zhen et al. (2018) indicated that the cathode material seemed to be a critically influencing factor on the biofilm formation and further altered the carbon dioxide reduction pathway. Methane formation was hydrogen-concentration dependent (indirect electron transfer) with carbon cloth biocathode and non-hydrogen mediated (direct electron transfer) with graphite felt cathode, respectively.

Since the microbial community played an important role in the upgrading of biogas, Cerrillo et al. (2017) investigated the effect of two inoculum sources, the mixture of mature anode effluent with anaerobic granular sludge and biomass enriched in a methanol-fed UASB, on the BES performance. The results indicated there was no significant difference either in methane production or cathodic methane recovery efficiency. It was estimated that about 0.58 $m^3/m^3/d$ of biogas can be upgraded in this MEC to increase methane content to near 100%. This finding seems to suggest that the electrode biofilm formation and function were dependent on the habitat created by BES instead of the original microbial composition in the inoculum.

Besides, chemicals added to form insoluble carbonate can further improve the biogas upgrading in BES. Madeddu (2015) claimed that the alkaline-based extraction of $Mg(OH)_2$ from magnesium silicate minerals can efficiently be used for carbon dioxide capture and storage. Zhang et al. (2020a, b) employed wollastonite as a cost-efficient calcium source to sequester carbon dioxide in AD *in situ*. The carbon dioxide content in biogas decreased from 11.5% to 7.8%, and the methane content improved to > 90%. The calcite was identified as the major product during mineral carbonation, and the struvite precipitate was also found indicating the additional benefits of nutrients removal.

3.5 *Mathematical Modelling*

Modeling and simulation of reactors includes presenting the complex BES and related electrochemical mechanisms into mathematical forms and correlating them with computational softwares for a better understanding and performance prediction of the systems. Electromethanogenesis directed BES is an inevitable technology for power-to-gas (P2G) as well as wastes/wastewater treatment. Various simulation methods and mathematical models have been designed for predicting the exclusive performances of individual BES reactors like MFC, MEC, and AD. Simple form of ordinary differential equations and 3-D models using Ohm's law (Logan 2008), Monod kinetic equation (Deb et al. 2020), Nernst equation (Popat and Torres 2016), Butler-Volmer equation, Faraday's law (Rabaey et al. 2009) and other process-oriented equations are used to obtain the power output from MFCs under given set of operating conditions in MFCs. Pioneer reports on MEC models were from Pinto et al. (2011) where a dynamic model of MEC based on the microorganism biofilm growth was proposed based on a combination of MFC and AD models. Later on, many developments in MEC modeling were reported with models like Differential Algebraic Equation (DAE) Model (Dudley et al. 2019), proportional-integral-derivative control system (Yahya et al. 2015), parametric 2D dynamic mathematical model (Hernández-García et al. 2020), Artificial Neural Network (ANN) with Response Surface Methodology (RSM) (Sun et al. 2018), Dynamic Biofilm model (Flores-Estrella et al. 2019, 2020), Integrated mathematical modeling (Mardanpour et al. 2017), and NARX-BP hybrid neural network models (Xiao et al. 2021). AD mathematical models have been reported since 1970s (Andrews 1969). Since then, extensive investigations and reports have been published such as Simple mathematical model (Jeyaseelan 1997), Cyclic batch equations (Keshtkar et al. 2001), MATLAB Simulink models (Blumensaat and Keller 2005, Ramirez et al. 2009), Benchmark simulation models (Solon et al. 2015), Partial and ordinary differential equations (Belhachmi et al. 2021) and advanced ADMI models (Weinrich et al. 2021). Though numerous reports have been published in simulation and models in individual BES reactors, publications dealing with integrated reactors and their mathematical models are comparatively less. A combined mathematical model was assumed for MFC with AD for two approaches by Alavijeh et al. (2015). One was to validate MFC performance with a variety of substrates and the second model was to anticipate the AD for acetoclastic methanogen activity and thereby enhance the biogas production. A modelling approach based on equivalent circuit for BES-AD was performed by Shahparasti et al. (2020) and Ceballos-Escalera et al. (2020), for advantageous long-term operation and energy production from the integrated reactors. Models like 1D dynamic model integrated with Activated Sludge Models (ASM) were carried out in BES-AD and were used to predict the growth rate of methanogens for biogas production and COD degradation (Gadkari et al. 2018). Liu et al. (2021) have implemented innovative and novel Artificial Neural Network (ANN) models in BES-AD systems for predicting the enhanced conversion of methane from 50% to 97% in biogas, thereby reducing the CO_2 content from 50% to 3%. Modeling parameters like Akaike Information Criterion (AIC), Bayesian Information Criterion (BIC) and F-test were used to predict biogas production and performance of the bioelectrochemical digester. Exponential and Gompertz models were implemented for envisioning methane yield and production (Prajapati and Singh 2020). Previous publication from our research group have revealed a dynamic simulation model for prediction of methane production and microbial dynamics (Guo et al. 2017). The results revealed that bioelectrochemistry had a momentous impact on the profusion of microorganisms involved in acidogenesis and methanogenesis processes. To formulate the reaction rate, an extended ADM1 was established by incorporating the Nernst expression and Monod-type kinetic expression to achieve the control of electrical potential. The simulation results indicated that biogas methane content can be increased up to 85% under optimized settings (Samarakoon et al. 2019). Nevertheless, advanced models like First order, Logistic, Gompertz and Back-propagation Artificial Neural Network (BP-ANN) were applied in MEC-AD coupled reactor for predicting the cumulative biogas and methane yields (Zou et al. 2021). Elreedy et al. (2019) used cutting-edge models for techno-economic assessment and kinetic modelling for psychrophilic hydrogen production in integrated anaerobic sequencing batch reactors. Technologically advanced genetic algorithms and optimal operational models for economic and environmental impact assessment of power to hydrogen (P2H) and power to methane (P2M) conversion was testified by Liu et al. (2020b) in integrated energy systems. Modeling and simulation studies are not only advantageous in predicting the performance of the reactors but also help in limiting and deciding the future experimental reactions, thus saving time, energy and capital costs.

3.6 Viewpoints of Dynamics in BES-AD Reactors

Biological reactors have been continuously monitored for fluid flow regimes, flow parameters, motions, forces, heat and temperature and how these parameters influence their performance. Such dynamical approaches are of various kinds like hydrodynamics, fluid dynamics and thermodynamics (Blanco-Aguilera et al. 2020). Like the models and simulations described in the previous section, dynamical approaches have been conducted in both BES and AD systems individually. Fluid flow and mass transfer efficiencies of MFC and MEC have been well documented in publications (Fujii et al. 2021, Sangeetha et al. 2021, Wang and Sangeetha 2019). The impacts of flow and heat on biogas production in AD systems has been reported (Mallikarjuna and Dash 2020, Sharma et al. 2021). But dynamical research works published exclusively on BES-AD are still less in number. CFD models for simulating the fluid flow and mixing velocity between the electrodes in a microbial methanogenesis cell was reported by Park et al. (2017) and the influence of mixer design and operating conditions in an AD designed for biohydrogen production was carried out by Trad et al. (2017). Significant previous publications from our research group have considered imperative hydrodynamics perspectives in BES-AD reactors. The hydrodynamic role of electrode placement and spatial distribution inside a BES-AD reactor and its positive impacts on biogas generation and COD removal were observed by Gao et al. (2019). CFD simulation and Retention Time Distribution (RTD) evaluations suggested that flow patterns of the wastewater positively altered the mass transfer of ions and increased reactor performance. Cui et al. (2020) had also notified the electrode positioning in BES-AD, where hydrodynamic and fluid dynamics characteristics were analyzed using tracer experiments and RTD. The report specified that the fluid pattern in the reactors was modified to a consummate mixing ability compared to traditional individual reactors and CFD simulation showcased favorable ionic mass transfer due to electron position alteration. Dissimilar cathode spatial ratios (i.e., ratio of cathode surface area to reaction region volume) were experimentally investigated to optimize the flow regime and hydrodynamic character of the interior flow field in BES-AD (Gao et al. 2021). These variations were interestingly detected to augment the overall performance of the reactor with heightened biogas production. These above-mentioned interesting research studies have emphasized that more research works are required in the field of computational dynamics for predicting and efficiently heightening the overall performance of integrated BES reactors.

4. Future Perspectives

Acquiring sustainable energy in form of biogas from wastes/wastewater is a promising approach to extend energy sources and also an efficient path to achieving carbon neutrality. The AD was an artificially reinforced process and has been employed to produce biogas for several decades. However, the traditional AD process suffered from disadvantages such as low substrates conversion rate, large occupied area, and unsatisfactory biogas quality. BES technology provides an alternative to significantly enhance biogas yield as well as biogas quality upgrading. For now, numerous studies have been investigated to verify the feasibility of BES assisting AD for performance improvement, yet the BES is still in its infancy and will remain under development for some time before its full-scale application.

Primarily, the BES technology was well studied on the lab- and pilot-scale but less full-scale application. The major challenge is the BES module design and management, especially the electrode material selection and configuration. To develop a low-resistance, cost-efficient, light-weight, and tractable material and further propose a practical electrode configuration is a crucial and urgent requirement. Meanwhile, it is valuable to evaluate the fluid and mass transfer features determined by the electrode configuration with the assistance of mathematical models.

Furthermore, microbes are the key member for the function performing both in traditional AD and BES electrode biofilm. The pure strains could present higher performance in biogas yield and organic matter conversion, but mixed consortia microbes have more utility value as they are much more stress-resistant with enhanced adaptability and stability than individual ones. The comprehension of the regulating strategy to maintain microbial community structure is the important factor to operate a practical BES for biogas production.

Additionally, the BES technology seems to be an effective individual system replacing existing facilities, and coupling with traditional AD and other pretreatment or post treatment methods, thus making it more practical and acceptable to the industrial community. In spite of the differences between BES and traditional AD, it is necessary to develop the operation specification to promote the BES technology closer to practical applications.

Acknowledgments

This work was financially supported by the National Natural Science Foundation of China (No. 52000088), the Natural Science Foundation of Jiangsu Province (No. BK20180633), and the Open Project of Key Laboratory of Environmental Biotechnology, CAS (Grant No. kf2020010).

References

Adekunle, A., Raghavan, V. and Tartakovsky, B. 2019. A comparison of microbial fuel cell and microbial electrolysis cell biosensors for real-time environmental monitoring. Bioelectrochemistry 126: 105–112.

Ahn, Y., Im, S. and Chung, J.-W. 2017. Optimizing the operating temperature for microbial electrolysis cell treating sewage sludge. Int. J. Hydrogen Energy 42(45): 27784–27791.

Ajay, C.M., Mohan, S., Dinesha, P. and Rosen, M.A. 2020. Review of impact of nanoparticle additives on anaerobic digestion and methane generation. Fuel 277: 118234.

Alavijeh, M., Yaghmaei, S. and Mardanpour, M.M. 2015. A combined model for large scale batch culture MFC-digester with various wastewaters through different populations. Bioelectrochemistry 106: 298–307.

Almatouq, A., Babatunde, A.O., Khajah, M., Webster, G. and Alfodari, M. 2020. Microbial community structure of anode electrodes in microbial fuel cells and microbial electrolysis cells. J. Water Process. Eng. 34: 101140.

Andrews, J.F. 1969. Dynamic model of the anaerobic digestion process. J. Sanit. Engng. Div. 95(1): 95–116.

Arvin, A., Hosseini, M., Amin, M.M., Najafpour Darzi, G. and Ghasemi, Y. 2019a. A comparative study of the anaerobic baffled reactor and an integrated anaerobic baffled reactor and microbial electrolysis cell for treatment of petrochemical wastewater. Biochem. Eng. J. 144: 157–165.

Arvin, A., Hosseini, M., Amin, M.M., Najafpour Darzi, G. and Ghasemi, Y. 2019b. Efficient methane production from petrochemical wastewater in a single membrane-less microbial electrolysis cell: the effect of the operational parameters in batch and continuous mode on bioenergy recovery. J. Environ. Health Sci. Eng. 17(1): 305–317.

Atelge, M.R., Atabani, A.E., Banu, J.R., Krisa, D., Kaya, M., Eskicioglu, C., Kumar, G., Lee, C., Yildiz, Y.Ş., Unalan, S., Mohanasundaram, R. and Duman, F. 2020. A critical review of pretreatment technologies to enhance anaerobic digestion and energy recovery. Fuel 270: 117494.

Baek, G., Kim, J., Lee, S. and Lee, C. 2017. Development of biocathode during repeated cycles of bioelectrochemical conversion of carbon dioxide to methane. Bioresour. Technol. 241: 1201–1207.

Baek, G., Rossi, R. and Logan, B.E. 2021a. Changes in electrode resistances and limiting currents as a function of microbial electrolysis cell reactor configurations. Electrochim. Acta 388: 138590.

Baek, G., Saikaly, P.E. and Logan, B.E. 2021b. Addition of a carbon fiber brush improves anaerobic digestion compared to external voltage application. Water Res. 188: 116575.

Batstone, D.J., Keller, J., Angelidaki, I., Kalyuzhnyi, S.V., Pavlostathis, S.G., Rozzi, A., Sanders, W.T.M., Siegrist, H. and Vavilin, V.A. 2002. The IWA Anaerobic Digestion Model No 1 (ADM1). Water Sci. Technol. 45(10): 65–73.

Belhachmi, Z., Mghazli, Z. and Ouchtout, S. 2021. Mathematical modelling and numerical approximation of a leachate flow in the anaerobic biodegradation of waste in a landfill. Math. Comput. Simul. 185: 174–193.

Bella, K. and Rao, P.V. 2021. Anaerobic digestion of dairy wastewater: Effect of different parameters and co-digestion options—A review. Biomass Convers. Biorefinery, 1–26.

Blanco-Aguilera, R., Lara, J.L., Barajas, G., Tejero, I. and Díez-Montero, R. 2020. Hydrodynamic optimization of multi-environment reactors for biological nutrient removal: A methodology combining computational fluid dynamics and dimensionless indexes. Chem. Eng. Sci. 224: 115766.

Blumensaat, F. and Keller, J. 2005. Modelling of two-stage anaerobic digestion using the IWA Anaerobic Digestion Model No. 1 (ADM1). Water Res. 39(1): 171–183.

Bo, T., Zhu, X., Zhang, L., Tao, Y., He, X., Li, D. and Yan, Z. 2014. A new upgraded biogas production process: Coupling microbial electrolysis cell and anaerobic digestion in single-chamber, barrel-shape stainless steel reactor. Electrochem. Commun. 45: 67–70.

Calderon, A.G., Duan, H., Meng, J., Zhao, J., Song, Y., Yu, W., Hu, Z., Xu, K., Cheng, X., Hu, S., Yuan, Z. and Zheng, M. 2021. An integrated strategy to enhance performance of anaerobic digestion of waste activated sludge. Water Res. 195: 116977.

Casula, E., Molognoni, D., Borràs, E. and Mascia, M. 2021. 3D modelling of bioelectrochemical systems with brush anodes under fed-batch and flow conditions. J. Power Sources 487: 229432.

Ceballos-Escalera, A., Molognoni, D., Bosch-Jimenez, P., Shahparasti, M., Bouchakour, S., Luna, A., Guisasola, A., Borràs, E. and Della Pirriera, M. 2020. Bioelectrochemical systems for energy storage: A scaled-up power-to-gas approach. Appl. Energy 260: 114138.

Cerrillo, M., Viñas, M. and Bonmatí, A. 2017. Startup of electromethanogenic microbial electrolysis cells with two different biomass inocula for biogas upgrading. ACS Sustain. Chem. Eng. 5(10): 8852–8859.

Cerrillo, M., Viñas, M. and Bonmatí, A. 2018. Anaerobic digestion and electromethanogenic microbial electrolysis cell integrated system: Increased stability and recovery of ammonia and methane. Renew. Energy 120: 178–189.

Chen, M., Zhang, F., Zhang, Y. and Zeng, R.J. 2013. Alkali production from bipolar membrane electrodialysis powered by microbial fuel cell and application for biogas upgrading. Appl. Energy 103: 428–434.

Cheng, S., Xing, D., Call, D.F. and Logan, B.E. 2009. Direct biological conversion of electrical current into methane by electromethanogenesis. Environ. Sci. Technol. 43(10): 3953–3958.

Cui, M.H., Sangeetha, T., Gao, L. and Wang, A.J. 2020. Hydrodynamics of up-flow hybrid anaerobic digestion reactors with built-in bioelectrochemical system. J. Hazard. Mater. 382: 121046.

Cui, M.H., Zheng, Z.Y., Yang, M., Sangeetha, T., Zhang, Y., Liu, H.B., Fu, B., Liu, H. and Chen, C.J. 2021b. Revealing hydrodynamics and energy efficiency of mixing for high-solid anaerobic digestion of waste activated sludge. Waste Manag. 121: 1–10.

Cui, Y., Mao, F., Zhang, J., He, Y., Tong, Y.W. and Peng, Y. 2021a. Biochar enhanced high-solid mesophilic anaerobic digestion of food waste: Cell viability and methanogenic pathways. Chemosphere 272: 129863.

Dai, X., Hu, C., Zhang, D. and Chen, Y. 2017. A new method for the simultaneous enhancement of methane yield and reduction of hydrogen sulfide production in the anaerobic digestion of waste activated sludge. Bioresour. Technol. 243: 914–921.

De Vrieze, J., Arends, J.B.A., Verbeeck, K., Gildemyn, S. and Rabaey, K. 2018. Interfacing anaerobic digestion with (bio)electrochemical systems: Potentials and challenges. Water Res. 146: 244–255.

Demirel, B., Scherer, P., Yenigun, O. and Onay, T.T. 2010. Production of methane and hydrogen from biomass through conventional and high-rate anaerobic digestion processes. Crit. Rev. Environ. Sci. Technol. 40(2): 116–146.

Dudley, H.J., Ren, Z.J. and Bortz, D.M. 2019. Competitive exclusion in a DAE model for microbial electrolysis cells. arXiv preprint arXiv:1906.02086.

Elreedy, A., Fujii, M. and Tawfik, A. 2019. Psychrophilic hydrogen production from petrochemical wastewater via anaerobic sequencing batch reactor: Techno-economic assessment and kinetic modelling. Int. J. Hydrog. Energy 44(11): 5189–5202.

European Commission. 2013. Bioelectrochemical systems Wastewater Treatment, Bioenergy and Valuable Chemicals Delivered by Bacteria.

Flores-Estrella, R.A., de Jesús Garza-Rubalcava, U., Haarstrick, A. and Alcaraz-González, V. 2019. A dynamic biofilm model for a microbial electrolysis cell. Processes 7(4): 183.

Flores-Estrella, R.A., Rodríguez-Valenzuela, G., Ramírez-Landeros, J.R., Alcaraz-González, V. and González-Álvarez, V. 2020. A simple microbial electrochemical cell model and dynamic analysis towards control design. Chem. Eng. Commun. 207(4): 493–505.

Fu, S., Angelidaki, I. and Zhang, Y. 2021. *In situ* biogas upgrading by CO_2-to-CH_4 bioconversion. Trends Biotechnol. 39(4): 336–347.

Fujii, K., Yoshida, N. and Miyazaki, K. 2021. Michaelis-Menten equation considering flow velocity reveals how microbial fuel cell fluid design affects electricity recovery from sewage wastewater. Bioelectrochemistry 140: 107821.

Gadkari, S., Gu, S. and Sadhukhan, J. 2018. Towards automated design of bioelectrochemical systems: A comprehensive review of mathematical models. Chem. Eng. J. 343: 303–316.

Gahlot, P., Ahmed, B., Tiwari, S.B., Aryal, N., Khursheed, A., Kazmi, A.A. and Tyagi, V.K. 2020. Conductive material engineered direct interspecies electron transfer (DIET) in anaerobic digestion: Mechanism and application. Environ. Technol. Innov. 20: 101056.

Gao, L., Liu, W., Cui, M., Zhu, Y., Wang, L., Wang, A. and Huang, C. 2021. Enhanced methane production in an up-flow microbial electrolysis assisted reactors: Hydrodynamics characteristics and electron balance under different spatial distributions of bioelectrodes. Water Res. 191: 116813.

Gao, L., Thangavel, S., Guo, Z.-C., Cui, M.-H., Wang, L., Wang, A.-J. and Liu, W.-Z. 2019. Hydrodynamics analysis for an upflow integrated anaerobic digestion reactor with microbial electrolysis under different hydraulic retention times: Effect of bioelectrode spatial distribution on functional communities involved in methane production and organic removal. ACS Sustain. Chem. Eng. 8(1): 190–199.

Glueck, S.M., Gumus, S., Fabian, W.M. and Faber, K. 2010. Biocatalytic carboxylation. Chem. Soc. Rev. 39(1): 313–328.

Guo, Z., Liu, W., Yang, C., Gao, L., Thangavel, S., Wang, L., He, Z., Cai, W. and Wang, A. 2017. Computational and experimental analysis of organic degradation positively regulated by bioelectrochemistry in an anaerobic bioreactor system. Water Res. 125: 170–179.

Guo, Z., Thangavel, S., Wang, L., He, Z., Cai, W., Wang, A. and Liu, W. 2016. Efficient methane production from beer wastewater in a membraneless microbial electrolysis cell with a stacked cathode: The effect of the cathode/anode ratio on bioenergy recovery. Energy Fuels 31(1): 615–620.

Hassan, M., Fernandez, A.S., San Martin, I., Xie, B. and Moran, A. 2018. Hydrogen evolution in microbial electrolysis cells treating landfill leachate: Dynamics of anodic biofilm. Int. J. Hydrogen Energy 43(29): 13051–13063.

Hassanein, A., Witarsa, F., Guo, X., Yong, L., Lansing, S. and Qiu, L. 2017. Next generation digestion: Complementing anaerobic digestion (AD) with a novel microbial electrolysis cell (MEC) design. Int. J. Hydrogen Energy 42(48): 28681–28689.

He, C., Zhang, B., Jiang, Y., Liu, H. and Zhao, H.P. 2021. Microbial electrolysis cell produced biogas as sustainable electron donor for microbial chromate reduction. Chem. Eng. J. 403: 126429.

Hernández-García, K.M., Cercado, B., Rodríguez, F.A., Rivera, F.F. and Rivero, E.P. 2020. Modeling 3D current and potential distribution in a microbial electrolysis cell with augmented anode surface and non-ideal flow pattern. Biochem. Eng. J. 162: 107714.

Huang, J., Feng, H., Huang, L., Ying, X., Shen, D., Chen, T., Shen, X., Zhou, Y. and Xu, Y. 2020. Continuous hydrogen production from food waste by anaerobic digestion (AD) coupled single-chamber microbial electrolysis cell (MEC) under negative pressure. Waste Manag. 103: 61–66.

Huang, Q., Liu, Y. and Dhar, B.R. 2021. Pushing the organic loading rate in electrochemically assisted anaerobic digestion of blackwater at ambient temperature: Insights into microbial community dynamics. Sci. Total Environ. 781: 146694.

Hurtado, A., Arroyave, C. and Peláez, C. 2021. Effect of using effluent from anaerobic digestion of vinasse as water reuse on ethanol production from sugarcane-molasses. Environ. Technol. Innov. 23: 101677.

Jeyaseelan, S. 1997. A simple mathematical model for anaerobic digestion process. Water Sci. Technol. 35(8): 185–191.

Jiang, Q., Liu, H., Zhang, Y., Cui, M.H., Fu, B. and Liu, H.B. 2021. Insight into sludge anaerobic digestion with granular activated carbon addition: Methanogenic acceleration and methane reduction relief. Bioresour. Technol. 319: 124131.

Jiang, Y., Su, M., Zhang, Y., Tao, Y. and Li, D. 2013. Simultaneous production of methane and acetate from carbon dioxide with bioelectrochemical systems. Chinese Journal of Appplied Environmental Biology 19(5): 833–837.

Keshtkar, A., Ghaforian, H., Abolhamd, G. and Meyssami, B. 2001. Dynamic simulation of cyclic batch anaerobic digestion of cattle manure. Bioresour. Technol. 80(1): 9–17.

Kondaveeti, S., Patel, S.K.S., Pagolu, R., Li, J., Kalia, V.C., Choi, M.-S. and Lee, J.-K. 2019. Conversion of simulated biogas to electricity: Sequential operation of methanotrophic reactor effluents in microbial fuel cell. Energy 189: 116309.

Kougias, P.G., Treu, L., Benavente, D.P., Boe, K., Campanaro, S. and Angelidaki, I. 2017. *Ex-situ* biogas upgrading and enhancement in different reactor systems. Bioresour. Technol. 225: 429–437.

Krishnan, S., Md Din, M.F., Taib, S.M., Nasrullah, M., Sakinah, M., Wahid, Z.A., Kamyab, H., Chelliapan, S., Rezania, S. and Singh, L. 2019. Accelerated two-stage bioprocess for hydrogen and methane production from palm oil mill effluent using continuous stirred tank reactor and microbial electrolysis cell. J. Clean Prod. 229: 84–93.

Lee, J.Y., Park, J.H. and Park, H.D. 2017. Effects of an applied voltage on direct interspecies electron transfer via conductive materials for methane production. Waste Manag. 68: 165–172.

Li, Y., Chen, Y. and Wu, J. 2019. Enhancement of methane production in anaerobic digestion process: A review. Appl. Energy 240: 120–137.

Li, Y., Qi, C., Zhang, Y., Li, Y., Wang, Y., Li, G. and Luo, W. 2021. Anaerobic digestion of agricultural wastes from liquid to solid state: Performance and environ-economic comparison. Bioresour. Technol. 332: 125080.

Liu, C., Sun, D., Zhao, Z., Dang, Y. and Holmes, D.E. 2019. Methanothrix enhances biogas upgrading in microbial electrolysis cell via direct electron transfer. Bioresour. Technol. 291: 121877.

Liu, C., Xiao, J., Li, H., Chen, Q., Sun, D., Cheng, X., Li, P., Dang, Y., Smith, J.A. and Holmes, D.E. 2021. High efficiency *in-situ* biogas upgrading in a bioelectrochemical system with low energy input. Water Res. 197: 117055.

Liu, C., Yuan, X., Gu, Y., Chen, H., Sun, D., Li, P., Li, M., Dang, Y., Smith, J.A. and Holmes, D.E. 2020a. Enhancement of bioelectrochemical CO2 reduction with a carbon brush electrode via direct electron transfer. ACS Sustain. Chem. Eng. 8(30): 11368–11375.

Liu, J., Sun, W. and Harrison, G.P. 2020b. The economic and environmental impact of power to hydrogen/power to methane facilities on hybrid power-natural gas energy systems. Int. J. Hydrog. Energy 45(39): 20200–20209.

Liu, H., Xiong, Y., Guan, Y. and Xu, S. 2018. Decarbonization and denitrification characteristics of a coupling ABR-MFC-MEC process treating black water. Desalin. Water Treat. 129: 43–52.

Liu, Q., Ren, Z.J., Huang, C., Liu, B., Ren, N. and Xing, D. 2016a. Multiple syntrophic interactions drive biohythane production from waste sludge in microbial electrolysis cells. Biotechnol. Biofuels 9: 162.

Liu, W., Cai, W., Guo, Z., Wang, L., Yang, C., Varrone, C. and Wang, A. 2016b. Microbial electrolysis contribution to anaerobic digestion of waste activated sludge, leading to accelerated methane production. Renew. Energy 91: 334–339.

Madeddu, S. 2015. Alkaline-based extraction of Mg(OH)$_2$ from magnesium silicate minerals for CO$_2$ capture and storage, Ph.D. Thesis, The University of Sheffield, Sheffield, UK.

Mallikarjuna, C. and Dash, R.R. 2020. A review on hydrodynamic parameters and biofilm characteristics of inverse fluidized bed bioreactors for treating industrial wastewater. J. Environ. Chem. Eng. 8(5): 104233.

Mao, C., Feng, Y., Wang, X. and Ren, G. 2015. Review on research achievements of biogas from anaerobic digestion. Renewable Sustainable Energy Rev. 45: 540–555.

Mardanpour, M.M. and Yaghmaei, S. 2017. Dynamical analysis of microfluidic microbial electrolysis cell via integrated experimental investigation and mathematical modeling. Electrochim. Acta 227: 317–329.

Miryahyaei, S., Das, T., Othman, M., Batstone, D. and Eshtiaghi, N. 2020. Anaerobic co-digestion of sewage sludge with cellulose, protein, and lipids: Role of rheology and digestibility. Sci. Total Environ. 731: 139214.

Nakasugi, Y., Himeno, M., Kobayashi, H., Ikarashi, M., Maeda, H. and Sato, K. 2017. Experimental and mathematical analyses of bio-electrochemical conversion of carbon dioxide to methane. Energy Procedia 114: 7133–7140.

Park, J., Lee, B., Tian, D. and Jun, H. 2018. Bioelectrochemical enhancement of methane production from highly concentrated food waste in a combined anaerobic digester and microbial electrolysis cell. Bioresour. Technol. 247: 226–233.

Park, J.G., Jiang, D., Lee, B. and Jun, H.B. 2020. Towards the practical application of bioelectrochemical anaerobic digestion (BEAD): Insights into electrode materials, reactor configurations, and process designs. Water Res. 184: 116214.

Park, J.G., Lee, B., Shi, P., Kim, Y. and Jun, H.B. 2017. Effects of electrode distance and mixing velocity on current density and methane production in an anaerobic digester equipped with a microbial methanogenesis cell. Int. J. Hydrog. Energy 42(45): 27732–27740.

Peng, H., Zhao, Z., Xiao, H., Yang, Y., Zhao, H. and Zhang, Y. 2019. A strategy for enhancing anaerobic digestion of waste activated sludge: Driving anodic oxidation by adding nitrate into microbial electrolysis cell. J. Environ. Sci. (China) 81: 34–42.

Prajapati, K.B. and Singh, R. 2020. Enhancement of biogas production in bio-electrochemical digester from agricultural waste mixed with wastewater. Renewable Energy 146: 460–468.

Quashie, F.K., Feng, K., Fang, A., Agorinya, S., Antwi, P., Kabutey, F.T. and Xing, D. 2021. Efficiency and key functional genera responsible for simultaneous methanation and bioelectricity generation within a continuous stirred microbial electrolysis cell (CSMEC) treating food waste. Sci. Total Environ. 757: 143746.

Ramirez, I., Volcke, E.I., Rajinikanth, R. and Steyer, J.P. 2009. Modeling microbial diversity in anaerobic digestion through an extended ADM1 model. Water Res. 43(11): 2787–2800.

Robles, A., Ruano, M.V., Charfi, A., Lesage, G., Heran, M., Harmand, J., Seco, A., Steyer, J.P., Batstone, D.J., Kim, J. and Ferrer, J. 2018. A review on anaerobic membrane bioreactors (AnMBRs) focused on modelling and control aspects. Bioresour. Technol. 270: 612–626.

Rotaru, A.-E., Shrestha, P.M., Liu, F., Shrestha, M., Shrestha, D., Embree, M., Zengler, K., Wardman, C., Nevin, K.P. and Lovley, D.R. 2014a. A new model for electron flow during anaerobic digestion: Direct interspecies electron transfer to Methanosaeta for the reduction of carbon dioxide to methane. Energy Environ. Sci. 7(1): 408–415.

Rotaru, A.E., Shrestha, P.M., Liu, F., Markovaite, B., Chen, S., Nevin, K.P. and Lovley, D.R. 2014b. Direct interspecies electron transfer between Geobacter metallireducens and Methanosarcina barkeri. Appl. Environ. Microbiol. 80(15): 4599–4605.

Samarakoon, G., Dinamarca, C., Nelabhotla, A.B., Winkler, D. and Bakke, R. 2019. Modelling Bio-Electrochemical CO2 Reduction to Methane. SINTEF Academic Press, Oslo, Norway.

Sangeetha, T., Guo, Z., Liu, W., Cui, M., Yang, C., Wang, L. and Wang, A. 2016. Cathode material as an influencing factor on beer wastewater treatment and methane production in a novel integrated upflow microbial electrolysis cell (Upflow-MEC). Int. J. Hydrogen Energy 41(4): 2189–2196.

Sangeetha, T., Guo, Z., Liu, W., Gao, L., Wang, L., Cui, M., Chen, C. and Wang, A. 2017. Energy recovery evaluation in an up flow microbial electrolysis coupled anaerobic digestion (ME-AD) reactor: Role of electrode positions and hydraulic retention times. Appl. Energy 206: 1214–1224.

Sangeetha, T., Li, I.T., Lan, T.H., Wang, C.T. and Yan, W.M. 2021. A fluid dynamics perspective on the flow dependent performance of honey comb microbial fuel cells. Energy 214: 118928.

Sangeetha, T., Rajneesh, C.P. and Yan, W.M. 2020. Integration of microbial electrolysis cells with anaerobic digestion to treat beer industry wastewater. pp. 313–346. Integrated Microbial Fuel Cells for Wastewater Treatment. Oxford, UK.

Shahparasti, M., Bouchakour, S., Luna, A., Molognoni, D., Bosch-Jimenez, P. and Borràs, E. 2020. Simplified modelling of nonlinear electromethanogenesis stack for power-to-gas applications. J. Energy Storage 31: 101633.

Sharma, M., Mohapatra, T. and Ghosh, P. 2021. Hydrodynamics, mass and heat transfer study for emerging heterogeneous Fenton process in multiphase fluidized-bed reactor system for wastewater treatment—A review. Chem. Eng. Res. Des. 171: 48–62.

Singh, A. and Kumar, V. 2021. Recent developments in monitoring technology for anaerobic digesters: A focus on bio-electrochemical systems. Bioresour. Technol. 329: 124937.

Solé-Bundó, M., Passos, F., Romero-Güiza, M.S., Ferrer, I. and Astals, S. 2019. Co-digestion strategies to enhance microalgae anaerobic digestion: A review. Renewable Sustainable Energy Rev. 112: 471–482.

Solon, K., Flores-Alsina, X., Mbamba, C.K., Volcke, E.I., Tait, S., Batstone, D. et al. 2015. Effects of ionic strength and ion pairing on (plant-wide) modelling of anaerobic digestion. Water Res. 70: 235–245.

Sun, Y., Yang, G., Wen, C., Zhang, L. and Sun, Z. 2018. Artificial neural networks with response surface methodology for optimization of selective CO2 hydrogenation using K-promoted iron catalyst in a microchannel reactor. J. CO_2 Util. 24: 10–21.

Thiruselvi, D., Kumar, P.S., Kumar, M.A., Lay, C.-H., Aathika, S., Mani, Y. et al. 2021. A critical review on global trends in biogas scenario with its up-gradation techniques for fuel cell and future perspectives. Int. J. Hydrogen Energy 46(31): 16734–16750.

Trad, Z., Fontaine, J.P., Larroche, C. and Vial, C. 2017. Experimental and numerical investigation of hydrodynamics and mixing in a dual-impeller mechanically-stirred digester. Chem. Eng. J. 329: 142–155.

van Eerten-Jansen, M.C.A.A., Jansen, N.C., Plugge, C.M., de Wilde, V., Buisman, C.J.N. and ter Heijne, A. 2015. Analysis of the mechanisms of bioelectrochemical methane production by mixed cultures. J. Chem. Technol. Biotechnol. 90(5): 963–970.

Wang, A., Sun, D., Cao, G., Wang, H., Ren, N., Wu, W.M. and Logan, B.E. 2011. Integrated hydrogen production process from cellulose by combining dark fermentation, microbial fuel cells, and a microbial electrolysis cell. Bioresour. Technol. 102(5): 4137–4143.

Wang, C.T. and Sangeetha, T. 2019. Advances in Sustainable Polymers, pp. 309–334. Springer.

Wang, F., Zhang, D., Shen, X., Liu, W., Yi, W., Li, Z. and Liu, S. 2019a. Synchronously electricity generation and degradation of biogas slurry using microbial fuel cell. Renew. Energy 142: 158–166.

Wang, L., He, Z., Guo, Z., Sangeetha, T., Yang, C., Gao, L., Wang, A. and Liu, W. 2019b. Microbial community development on different cathode metals in a bioelectrolysis enhanced methane production system. J. Power Sources 444: 227306.

Wang, X.T., Zhao, L., Chen, C., Chen, K.Y., Yang, H. et al. 2021. Microbial electrolysis cells (MEC) accelerated methane production from the enhanced hydrolysis and acidogenesis of raw waste activated sludge. Chem. Eng. J. 413: 127472.

Weinrich, S., Mauky, E., Schmidt, T., Krebs, C., Liebetrau, J. and Nelles, M. 2021. Systematic simplification of the Anaerobic Digestion Model No. 1 (ADM1)-Laboratory experiments and model application. Bioresour. Technol. 333: 125104.

Wilberforce, T., Sayed, E.T., Abdelkareem, M.A., Elsaid, K. and Olabi, A.G. 2021. Value added products from wastewater using bioelectrochemical systems: Current trends and perspectives. J. Water Process. Eng. 39: 101737.

Wu, Y., Wang, S., Liang, D. and Li, N. 2020. Conductive materials in anaerobic digestion: From mechanism to application. Bioresour. Technol. 298: 122403.

Xiao, B., Chen, X., Han, Y., Liu, J. and Guo, X. 2018. Bioelectrochemical enhancement of the anaerobic digestion of thermal-alkaline pretreated sludge in microbial electrolysis cells. Renew. Energy 115: 1177–1183.

Xiao, J., Liu, C., Ju, B., Xu, H., Sun, D. and Dang, Y. 2021. Estimation of *in-situ* biogas upgrading in microbial electrolysis cells via direct electron transfer: Two-stage machine learning modeling based on a NARX-BP hybrid neural network. Bioresour. Technol. 330: 124965.

Xu, S., Qiao, Z., Luo, L., Sun, Y., Wong, J.W., Geng, X. and Ni, J. 2021. On-site CO_2 bio-sequestration in anaerobic digestion: Current status and prospects. Bioresour. Technol. 332: 125037.

Yahya, A.M., Hussain, M.A. and Abdul Wahab, A.K. 2015. Modeling, optimization, and control of microbial electrolysis cells in a fed-batch reactor for production of renewable biohydrogen gas. Int. J. Energy Res. 39(4): 557–572.

Yan, X., Wang, B., Liang, H., Yang, J., Zhao, J., Ndayisenga, F., Zhang, H., Yu, Z. and Qian, Z. 2021. Enhanced straw fermentation process based on microbial electrolysis cell coupled anaerobic digestion. Chin. J. Chem. Eng. (in press).

Yin, Q., Zhu, X., Zhan, G., Bo, T., Yang, Y., Tao, Y., He, X., Li, D. and Yan, Z. 2016. Enhanced methane production in an anaerobic digestion and microbial electrolysis cell coupled system with co-cultivation of Geobacter and Methanosarcina. J. Environ. Sci. (China) 42: 210–214.

Yu, N., Xing, D., Li, W., Yang, Y., Li, Z., Li, Y. and Ren, N. 2017. Electricity and methane production from soybean edible oil refinery wastewater using microbial electrochemical systems. Int. J. Hydrogen Energy 42(1): 96–102.

Yuan, Y., Cheng, H., Chen, F., Zhang, Y., Xu, X., Huang, C. et al. 2020. Enhanced methane production by alleviating sulfide inhibition with a microbial electrolysis coupled anaerobic digestion reactor. Environ. Int. 136: 105503.

Zakaria, B.S., Lin, L. and Dhar, B.R. 2019. Shift of biofilm and suspended bacterial communities with changes in anode potential in a microbial electrolysis cell treating primary sludge. Sci. Total Environ. 689: 691–699.

Zakaria, B.S. and Ranjan Dhar, B. 2021. An intermittent power supply scheme to minimize electrical energy input in a microbial electrolysis cell assisted anaerobic digester. Bioresour. Technol. 319: 124109.

Zeppilli, M., Cristiani, L., Dell'Armi, E. and Villano, M. 2021. Potentiostatic vs galvanostatic operation of a microbial electrolysis cell for ammonium recovery and biogas upgrading. Biochem. Eng. J. 167: 107886.

Zhang, J., Zhang, Y., Quan, X., Chen, S. and Afzal, S. 2013. Enhanced anaerobic digestion of organic contaminants containing diverse microbial population by combined microbial electrolysis cell (MEC) and anaerobic reactor under Fe(III) reducing conditions. Bioresour. Technol. 136: 273–280.

Zhang, Y., Jiang, Q., Gong, L., Liu, H., Cui, M. and Zhang, J. 2020a. *In-situ* mineral CO_2 sequestration in a methane producing microbial electrolysis cell treating sludge hydrolysate. J. Hazard. Mater. 394: 122519.

Zhang, Y., Gong, L.L., Jiang, Q.Q., Cui, M.H., Zhang, J. and Lie, H. 2020b. *In-situ* CO_2 sequestration and nutrients removal in an anaerobic digestion-microbial electrolysis cell by silicates application: Effect of dosage and biogas circulation. Chem. Eng. J. 399: 125680.

Zhao, N., Liang, D., Li, X., Meng, S. and Liu, H. 2021a. Hydrophilic porous materials provide efficient gas-liquid separation to advance hydrogen production in microbial electrolysis cells. Bioresour. Technol. 337: 125352.

Zhao, L., Wang, X.T., Chen, K.Y., Wang, Z.H., Xu, X.J., Zhou, X. et al. 2021b. The underlying mechanism of enhanced methane production using microbial electrolysis cell assisted anaerobic digestion (MEC-AD) of proteins. Water Res. 201: 117325.

Zhen, G., Kobayashi, T., Lu, X., Kumar, G. and Xu, K. 2016. Biomethane recovery from *Egeria densa* in a microbial electrolysis cell-assisted anaerobic system: Performance and stability assessment. Chemosphere 149: 121–129.

Zhen, G., Lu, X., Kato, H., Zhao, Y. and Li, Y.Y. 2017. Overview of pretreatment strategies for enhancing sewage sludge disintegration and subsequent anaerobic digestion: Current advances, full-scale application and future perspectives. Renewable Sustainable Energy Rev. 69: 559–577.

Zhen, G., Zheng, S., Lu, X., Zhu, X., Mei, J., Kobayashi, T. et al. 2018. A comprehensive comparison of five different carbon-based cathode materials in CO_2 electromethanogenesis: Long-term performance, cell-electrode contact behaviors and extracellular electron transfer pathways. Bioresour. Technol. 266: 382–388.

Zou, L., Wang, C., Zhao, X., Wu, K., Liang, C., Yin, F. et al. 2021. Enhanced anaerobic digestion of swine manure via a coupled microbial electrolysis cell. Bioresour. Technol. 340: 125619.

CHAPTER 14

Green Fuel from Sewage Sludge
Roles of Microorganism

Simona Di Fraia, Nicola Massarotti and *M. Rakib Uddin**

1. Introduction

Water is used in every activity in daily life including cooking, washing of human bodies and clothes, keeping house and communities clean, recreation, etc. The average freshwater consumption rate per capita varies from country to country or even in different regions of the same country, based on living standards, industrial and agricultural productivity. For instance in Europe in 2018, the lowest water consumption was in Malta (30 m³/person/year) and the highest in Greece (157 m³/person/year) (Eurostat 2021a). In literature, data on per capita freshwater consumption are not updated specially for the developing countries, with only a few exceptions (e.g., Brazil, Bangladesh, Indonesia, Nepal, etc.) (AQUASTAT 2014).

Wastewater is generated from the daily activity completed in the home and industry where water is used. The wastewater generation rate is increasing all over the world continuously due to the population and industrial growth and improvement of human living standards. Wastewater contains human waste, food scraps, oils, soaps, antibiotics, chemicals, tiny fractions of metals that elute out from industrial activities (Czatzkowska et al. 2021). According to the developed WaterGAP model, the estimated global wastewater generation from domestic and industrial activities was 450 km³ in 2010 (Mateo-Sagasta et al. 2015). According to the model developed by Sato et al. (2013), the average annual municipal wastewater generation rate is more than 330 km³ in the world. Wastewater generated from daily activity in the home, restaurant, and industry must be treated properly through WasteWater Treatment Plants (WWTPs) to remove all the pollutants before discharge to natural sinks, such as rivers or seas, to ensure environmental safety and security. The WWTPs generate two streams, treated water and Sewage Sludge (SS), as a by-product.

This chapter presents briefly the SS generation system and its quantity with existing management policies. Environmental problems associated with the current management practices of SS are also described in this chapter. Conversion of SS to green fuel by biological route of Anaerobic Digestion (AD) with focus on the role of microorganisms on the conversion process and the effect of operating parameters on efficiency is presented. The influence of antimicrobials in SS on green fuel generation efficiency through the AD process is illustrated. Finally, the properties and applications of the produced green fuel (biogas) are illustrated.

Dipartimento di Ingegneria, Università degli Studi di Napoli "Parthenope", Napoli–80143, Italy.
* Corresponding author: mohammadrakib.uddin001@studenti.uniparthenope.it

1.1 Sewage Sludge Generation System

Wastewater passes through primary treatment (to remove suspended solids, partially pathogens, nutrients, and inorganic substances) followed by secondary treatment (for the removal of organic substances, phosphorous, nitrogen, and other nutrients) and finally to the tertiary system (to complete separation of pathogens and nutrient elements) to complete the treatment cycle. Effluent from the tertiary treatment is sometimes subjected to advanced treatment systems, such as adsorption or ozonation if the quality does not match with the standard imposed by a country or region (Metcalf and Eddy 1991, Kwarciak-Kozlowska et al. 2016).

The connection of people to the sewage system (proper collection and treatment of wastewater) varies from country to country. According to Eurostat 2021b, among 27 countries in European Union (EU–27), in 2018, almost all people (99.95%) in Austria are served by the sewage system whereas the worst scenario found in Romania (52.90%). This scenario of the population connected to the sewage system is almost the same in other countries like the USA, the UK, Australia, Canada, Japan, China, Russia, Saudi Arabia, etc. (Mateo-Sagasta et al. 2015). According to Eurostat 2021c, the average per capita SS generation rate in the EU–27 was 17.59 kg/person/year within the range of 6.05 kg/person/year in Croatia to 33.23 kg/person/year in the Netherlands in 2015. The variation of SS generation rate per capita in the EU–27 from 2005 to 2015 is presented in Fig. 1.

The Annual SS generation rate in EU–27 fluctuates in the range of 6.88 to 10.67 Mt of Dry Solid (DS) from 2007 to 2015 (Eurostat 2021c). The annual quantity of SS generation in other countries of the world for a particular year is presented in Table 1 (Eurostat 2021c, Grobelak et al. 2019, LeBlanc et al. 2009).

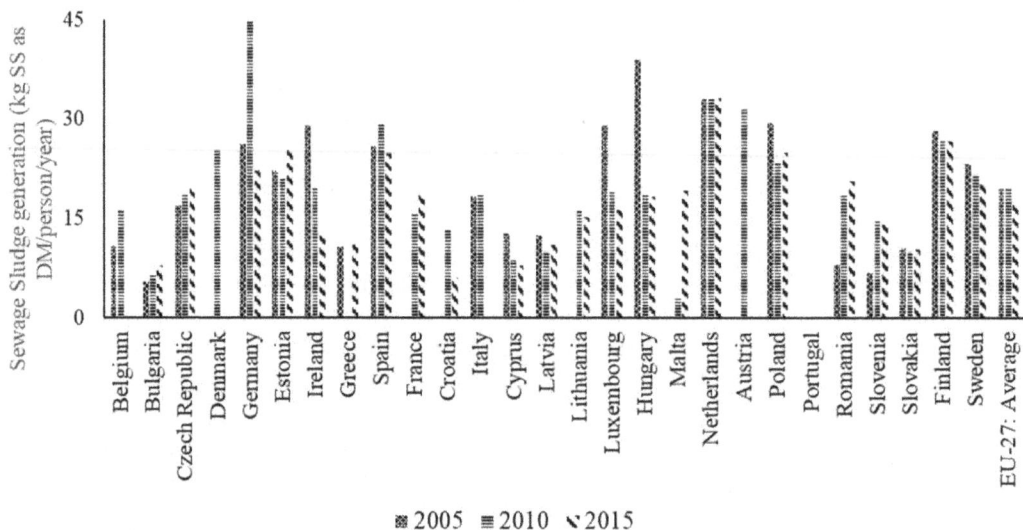

Fig. 1. Annual per capita SS generation rate in EU–27 countries (Eurostat 2021c).

1.2 Sewage Sludge Management

SS stream exiting from WWTPs is in suspension form with 98% impure water with a wide variety of solid particles that are presented as impurities in wastewater (Canziani and Spinosa 2019). Different kinds of organic (proteins, carbohydrates, oils, and fats), inorganic substances with wide varieties of microorganisms and antimicrobials, and tiny fractions of heavy metals are present as solid particles in SS (Czatzkowska et al. 2021, Magdziarz et al. 2016, Manara and Zabaniotou 2012, Harrison et al. 2006). Compounds present in SS are classified in six classes by Rulkens (2008) as follows:

Table 1. Annual SS production in different countries.

Country	Quantity of SS (Mt of DS)	Year	Reference
EU-27	9.00	2010	Eurostat 2021c
UK	1.09	2005	
Turkey	0.391	2004	
USA	6.514	2004	LeBlanc et al. 2009
Canada	0.550	2007	
Brazil	0.372	2001	
Australia and New Zealand	0.360	2006	
China	2.966	2006	Grobelak et al. 2019, LeBlanc et al. 2009
Japan	2.00	2006	
Korea	1.90	2006	LeBlanc et al. 2009

(i) organic carbonaceous compounds of nontoxic nature (about 60% on a DS basis), (ii) phosphorous, nitrogen, and sulfur-containing compounds, (iii) inorganic compounds of toxic nature due to the presence of heavy metals (e.g., Zn, Pb, Cu, Cr, Ni, Cd, Hg, and As) and organic pollutants (e.g., polychlorinated biphenyls, polycyclic aromatic hydrocarbons, dioxins, pesticides, linear-alkyl sulfonates, and nonyl-phenols), (iv) pathogens, (v) inorganic compounds (e.g., silicates, aluminates, and calcium and magnesium-containing compounds) and (vi) water.

SS can be used as fertilizer directly or after the conversion in the form of compost as it can supply organic compounds as well as nutrients, such as nitrogen, sulfur, and phosphorus to the plants (Mininni et al. 2019). Other common management practices of SS after agricultural applications are landfilling, incineration, pyrolysis, and further physical and chemical treatment before discharge (Spliethoff 2000). The current SS management practices in EU–27 are presented in Fig. 2 (Eurostat 2021).

In EU–27, the most common practice for SS management is agricultural reuse, followed by incineration or co-incineration, whereas the less diffused in landfilling. Less common treatments, designated as others in Fig. 2, include management by pyrolysis, temporary storage (e.g., Italy,

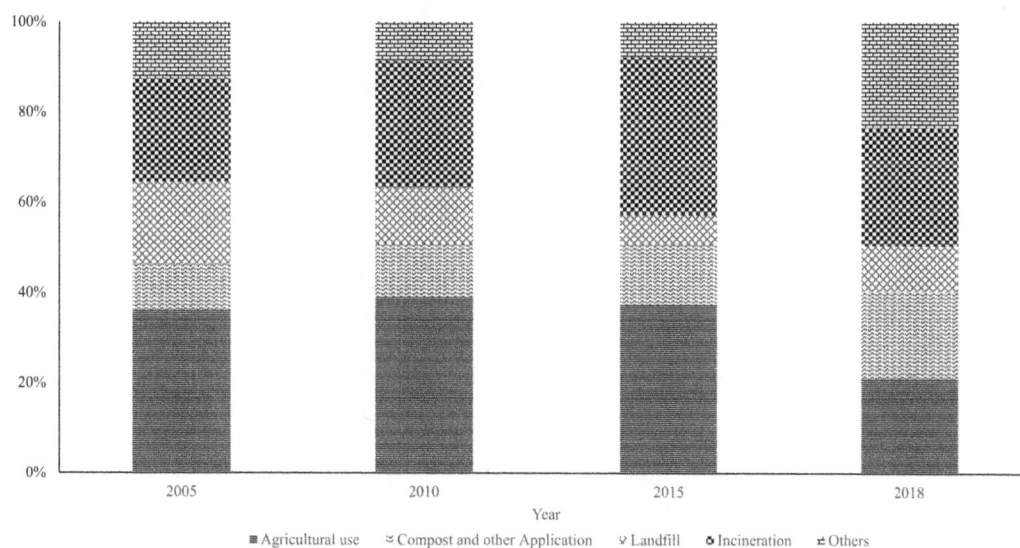

Fig. 2. SS management practiced in EU-27 countries (Eurostat 2021).

Greece), long time storage (e.g., Poland, Lithuania), reuse in forestry and green areas (e.g., Ireland, Slovakia), landfill cover (e.g., Sweden, Belgium), and export of SS to other countries (e.g., granulated SS for incineration from the Netherland to Germany, SS for composting or incineration from Luxembourg to Germany) (Kelessidis and Stasinakis 2012). The SS management in EU–27 is almost similar in other countries, such as USA, Japan, Australia, New Zealand, UK, and China (Christodoulou and Stamatelatou 2016, Wei et al. 2020).

1.3 Environmental Problems Associated with Current SS Management Strategy

1.3.1 Agricultural reuse and landfilling

Management of SS through reuse in agricultural land either directly or converted to compost is one of the most economic routes and offers long-term nutrient supply to the crops which reduces the dependency on chemical fertilizers in a country (Delibacak et al. 2020). SS used in agricultural land may cause the decrease or increase of pH of the soil; on the other hand, toxic substances, yeast population, pathogenic microorganisms, and aerobic bacteria in the soil may increase (Nielson et al. 1998, Kulling et al. 2001). Some organic toxic pollutants of Polycyclic Aromatic Hydrocarbons (PAHs) and PolyChlorinated Biphenyls (PCBs) are found in SS (Wyrwicka et al. 2014, Zhai et al. 2011). These PAHs and PCBs enter into the human body through the food chain and cause a serious problem on the human liver, skin, thyroid gland, cancer development, and immunological alteration (Vukasinovic et al. 2017).

Landfilling is the second most popular and cost-effective management method of SS especially in developing countries (Christodoulou and Stamatelatou 2016, Wei et al. 2020). Heavy metals (Pb, Cd, Cr, Zn, Cu, Hg, As) with other organic and inorganic pollutants present in SS elute out from landfill sites to the groundwater (Delibacak et al. 2020). In addition to this, pollutants and microorganisms reach the groundwater from SS landfill sites. Considering that more than 90% of the world's freshwater is abstracted from groundwater (Margat and Gun 2013, NGWA 2021), this may cause a serious threat to human life. According to the European Commission report (2009), SS usually contains different varieties of microorganisms such as *Salmonella, Faecal Streptococci, Enterovirus, Helminths Eggs, Escherichia Coli, Faecal Coli*, etc. These microorganisms are leached out to the groundwater from landfill sites and cause serious pollution which ultimately creates a potential risk to humans and aquatic life. Pollutants present in potable water cause diseases in the human body such as hepatitis, cholera, Blue Baby Syndrome (affecting the infants), neurological, kidney and liver dysfunction, and the risk of pregnancy (Al-Sudani 2019). Potable water pollution causes premature death of human life and this is a serious problem in developing or poor countries where the groundwater is directly used as potable water without further treatment.

Aerobic bacteria (e.g., *Salmonella, E. Coli, Proteus, Citrobacter*, etc.) continuously decompose the organic fractions present in SS when used in agricultural land (European Commission Report 2009) and form CO, CO_2, N_2O, and SO_2 that are continuously released to the environment. On the other hand, anaerobic bacteria (e.g., *Methanosarcina, Methanothrix, Methanococcus, Methanobacterium, Methanobacillus*, etc.) continuously decompose the organic, inorganic, and nutrients components present in SS in landfill sites, forming nitrate, methane, acetate, and sulfite that are released to the atmosphere (Nguyen et al. 2021). Gaseous compounds formed and released into the environment from agricultural land and landfill site are considered greenhouse gases and are responsible for global warming (Bunsow and Dobberstein 1987). According to the fifth assessment report of the Intergovernmental Panel on Climate Change (IPCC) (IPCC Report 2014), CH_4 and N_2O have 28 and 265 times, respectively, more potential as a greenhouse gas as compared to CO_2. Moreover, the odor is released continuously from the disposed site and stakeholders have strong objections against agricultural reuse and landfilling of SS.

It is mandatory to remove all the pollutants from SS before disposal to the agricultural land or landfill site to remove the potential threat of environmental pollution completely and ultimately reduce the threat to human life. The treatment process to remove these pollutants is more costly compared to the available advantages obtained through SS management in agricultural reuse or dumping to the landfill site. Application of SS to the agricultural land is prohibited in some EU countries (e.g., Belgium, Switzerland, and Romania) whereas restricted in other countries (e.g., Republic of Ireland) to overcome the environmental problem caused by the greenhouse gas release (Colón et al. 2017).

1.3.2 Incineration

Incineration or co-incineration of SS is gaining more attraction as a management strategy in the last decades over landfilling and agricultural reuse in EU, UK, USA, Japan, Australia, Canada, and China (Eurostat 2021, Wei et al. 2020, Spinosa 2007) since it can destroy the pathogenic microorganisms, both organic and inorganic pollutants. Incineration is conducted at a temperature higher than 925°C (Fericelli 2011). Large quantities of excess air are supplied to the incinerator during the combustion of SS to maximize energy extraction. However, the use of excess air drives the formation of CO_2, N_2O, SOx, and NOx that are emitted to the environment with particulate matter and metals as pollutants (Fericelli 2011). During incineration, about 90% of PAHs present in SS are released to the atmosphere at a temperature of 300–750°C being responsible for serious environmental pollution (Delibacak et al. 2020). Incineration allows reducing the volume of SS by around one-fourth but increases the heavy metals concentration in the ash. An advantage of incineration is that phosphorus content in the ash increases in the range of 4 to 9% compared to the parent SS which is comparable to the superphosphate (Colón et al. 2017). Therefore, application of SS incinerated ash to the agricultural land reduces the pressure on phosphorus extraction from rock. Phosphorus from SS incinerated ash can be recovered by applying advanced techniques (e.g., BioCon-Process, SEPHOS-Process, ASH DEC Umwelt AG, or RuePa-Process). However, currently, the operating cost involved to extract phosphorus from SS incinerated ash is not economically feasible (Dichtl et al. 2007).

1.3.3 Pyrolysis

Pyrolysis of SS generates three energy vectors: bio-gas, bio-oil, and bio-char. Biogas from pyrolysis of SS contains considerable fractions of H_2S (from 3.6 to 4.8 vol%) and SO_2 (in the range of 0.5 to 0.8 vol%) with other constituents (H_2, CO, CO_2, CH_4, and C_2–C_4 fractions) (Xue et al. 2021). H_2S and SO_2 are considered greenhouse gases responsible for the world temperature rising (IPCC Report 2014). These constituents need to be removed completely before the use of biogas in a further process to generate fuel or chemicals to ensure environmental safety.

Bio-oil mainly consists of organic compounds and is generated during pyrolysis usually at a temperature in the range of 300–600°C (Xue et al. 2021). In the bio-oil, around 94.8% is wastewater with pollutants (e.g., nitrous and phosphorous substances, heavy metals, etc.), and the remaining is oil (Dominguez et al. 2006). The oil obtained from the pyrolysis of SS is highly viscous and stuck with the combustion chamber surface during use in diesel engines (Xue et al. 2021). Kinematic viscosity of the available oil from pyrolysis process (81 x 10^{-6} m^2/s at 25°C) is much higher than the typical commercial diesel (3.88 x 10^{-6} m^2/s at 25°C) (Panchasara and Ashwath 2021). The moisture content is also higher (usually 25%) compared to the standard diesel (< 0.1%) and the oil is highly acidic (pH in the range of 2–3) (Panchasara and Ashwath 2021). Massive treatment is required to convert the bio-oil obtained in pyrolysis to biodiesel for use in an engine. In addition to this, generated wastewater has to be treated again to reduce the environmental pollution threat.

The heavy metals present in SS are transferred to the bio-char and their concentration is higher compared to the parent materials from where bio-char is formed due to the considerable reduction of volume. The generated char can be successfully used for industrial heat generation or production of construction materials or in agricultural reuse to supply nutrients to the plants (Hanif et al. 2020). Cleaning of the products generated through pyrolysis of SS for further applications may be not economically profitable.

2. Green Fuels

Fossil fuel (coal, oil, and natural gas) reserves have continuously declined due to the massive increase in demand causing the well-known world energy crisis (Bludowsky and Agar 2009, Shafiee and Topal 2009, Singh and Singh 2012). In addition, excessive use of fossil fuels increases CO_2 concentration in the atmosphere contributing to the continuous rise of world average temperature (Singh and Singh 2012). For these reasons, the research on renewable energy sources and sustainable conversion techniques to reduce the pressure on fossil fuel reserves and the environmental threat of global warming has become more and more attractive.

The alternative forms of energy that are generated from domestic resources and can supply energy services with zero or almost zero pollutants to the environment are considered renewable energy or green fuel. World primary energy demand has been increasing day by day with the rapid increase in population, urbanization, and industrial productivity from 10,031.53 Mtoe in 2000 to 13,757.52 Mtoe in 2016 (WBA Global Bioenergy Statistics 2018). According to the World Bioenergy Report (2018), in 2016, 81% of total primary energy demand was supplied by coal, natural gas, and oil, and the remaining comes from nuclear and renewable resources as shown in Fig. 3.

Biomass is the major contributor to the renewable energy sector with 1348.24 Mtoe followed by hydropower (346.69 Mtoe); solar, geothermal and wind have a similar contribution (each of 77.04 Mtoe) whereas the lowest input comes from tides, ocean, and other sources (0.10 Mtoe).

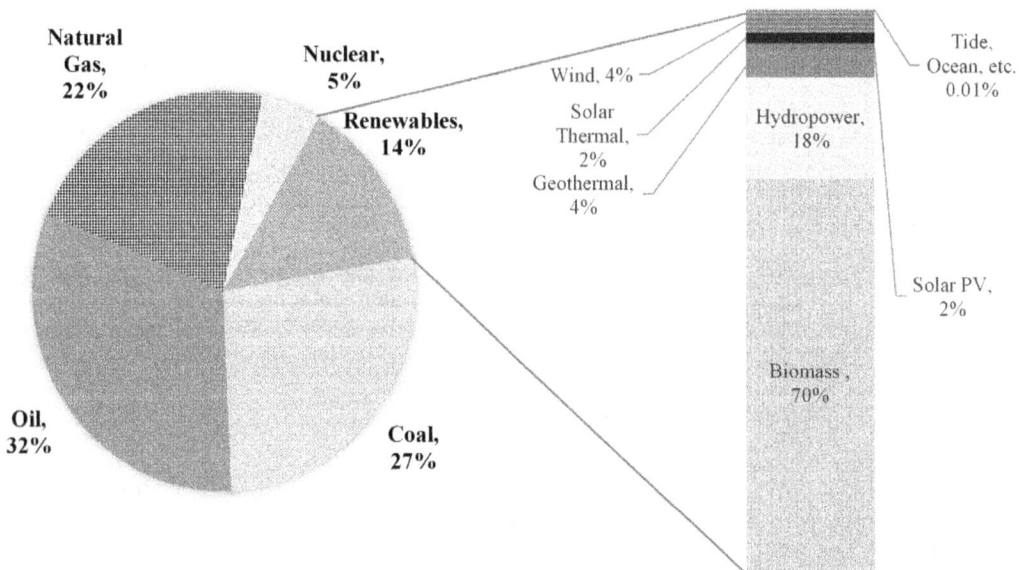

Fig. 3. Total global primary energy supply by source in 2016 (WBA Global Bioenergy Statistics 2018).

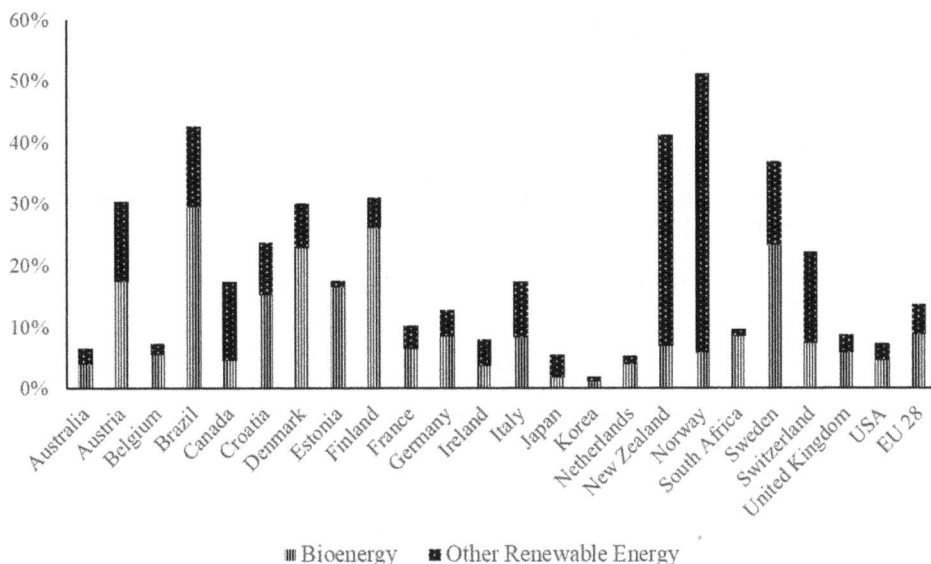

Fig. 4. Percentage of primary energy demand meet up from renewable energy sources (IEA Bioenergy Countries' Report 2018).

In 2016, more than 50% of primary energy demand had been supplied by renewable energy in Norway and around 40% in New Zealand, Sweden, and Brazil (IEA Bioenergy Countries' Report 2018). The contribution of renewable energy to the total primary energy demand and the share of bioenergy for some selected countries in different continents of the world are presented in Fig. 4.

In the available literature, there are no consistent data on the contribution of SS to bioenergy production to evaluate the renewable generation potentiality of SS despite energy recovery from SS through biological (Liu et al. 2019) or thermal treatment (gasification) (Abdelrahim et al. 2020, de Andrés et al. 2019) in lab scale.

2.1 Composition of Sewage Sludge

The composition of SS highly depends on the characteristics of pollutants present in wastewater and the techniques (mechanical, chemical, biological) applied during treatment. People's living standards and food habits are different from country to country at the same time and time to time (seasonal variation) in the same country, affecting the SS composition. The Lower Heating Value (LHV) of any biomass or SS depends on its constituents (Elias 2005). Proximate and ultimate analyses identify Moisture Content (MC), Volatile Matter (VM), Fixed Carbon (FC), and Ash content, and the elemental composition (C, H_2, N_2, S, Cl_2, F_2 and O_2), respectively. The LHV of SS can be calculated from the elemental composition by using the modified Dulong's formula presented in Equation (1) (Elias 2005).

$$LHV\left(\frac{kcal}{kg}\right) = 8.060 \cdot C + 33.910 \cdot \left(H - \frac{O}{8}\right) + 2.222 \cdot S + 556 \cdot N \tag{1}$$

where, C, H, O, S, and N represent the weight fraction of carbon, hydrogen, oxygen, sulfur, and nitrogen of the SS sample.

Some examples of proximate and ultimate analysis with LHV of SS generated in different counties at different times are presented in Table 2.

Table 2. Proximate and ultimate analysis results with LHV of SS.

	SS_1^{daf}	SS_2^{db}	SS_3^{daf}	SS_4^{daf}	SS_5^{db}	SS_6^{db}	SS_7^{daf}	SS_8^{db}	SS_9^{db}
Proximate Analysis (wt%)									
MC	20.0	14.1	10.0 ± 1.1	10.0 ± 0.6	7.0	4.43	n.r.	8.71	15.74 ± 0.23
VM	44.0	64.26	54.3 ± 0.8	60.9 ± 0.7	56.0	68.57	73.7	61.11	67.29 ± 0.16
FC	5.1	11.18	5.1 ± 0.1	4.8 ± 0.1	0	16.42	0.4	9.20	11.83 ± 0.38
Ash	30.9	24.56	30.6 ± 0.3	24.3 ± 0.1	44.0	15.01	25.9	26.89	20.88 ± 0.26
Ultimate Analysis (wt%)									
C	51.2	41.33	49.16 ± 0.06	51.75 ± 0.03	27.3	53.24	37.9	45.16	43.29 ± 0.55
H_2	8.2	5.82	8.50 ± 0.03	7.91 ± 0.02	4.8	7.39	5.5	7.20	6.13 ± 0.24
N_2	7.1	8.27	6.06 ± 0.09	6.70 ± 0.07	4.1	6.12	6.2	7.69	6.62 ± 0.13
S	1.7	n.r.	1.18 ± 0.14	1.37 ± 0.11	0.9	n.r.	n.r.	n.r.	0.51 ± 0.02
Cl_2	n.r.	n.r.	0.08 ± 0.40	5.36 ± 0.05	n.r.	n.r.	n.r.	n.r.	0.058 ± 0.01
O_2	31.8	20.02	35.02 ± 0.1	26.64 ± 0.09	18.9	33.25	50.4	27.50	22.52 ± 0.67

*db: Dry Basis; daf: Dry Ash Free; n.r.: Not Reported.

	SS_1^{daf}	SS_2^{db}	SS_3^{daf}	SS_4^{daf}	SS_5^{db}	SS_6^{db}	SS_7^{daf}	SS_8^{db}	SS_9^{db}
LHV (MJ/kg)	12.0	12.6	10.6 ± 0.1	14.8 ± 0.2	11.5	24.2	15.6	16.18	16.11 ± 0.20
Origin	Italy				USA		Greece	Taiwan	
Reference	Seggiani et al. 2012	Abdelrahim et al. 2020	Migliaccio et al. 2021		de Andrés et al. 2019	Xie et al. 2014	Agrafioti et al. 2013	Huang et al. 2015	Nguyen et al. 2021

*db: Dry Basis; daf: Dry Ash Free.

	SS_{10}^{db}	SS_{11}^{db}	SS_{12}^{db}	SS_{13}^{db}	SS_{14}	SS_{15}^{db}	SS_{16}^{db}	SS_{17}^{db}
Proximate Analysis (wt%)								
MC	7.0	8.7	5.6	5.6	7.0	10.8	5.30	5.30
VM	56.0	58.3	54.2	54.2	50.0	49.3	51.0	49.0
FC	n.r.	n.r.	8.6	8.6	3.0	0.5	7.20	1.50
Ash	44.0	41.7	37.2	37.2	40.0	50.2	36.50	44.20
Ultimate Analysis (wt%)								
C	27.3	29.5	25.5	40.6	27.9	29.0	31.79	27.72
H_2	4.8	4.9	4.5	7.1	4.7	3.8	4.36	3.81
N_2	4.1	4.1	4.9	7.7	4.5	3.8	4.88	3.59

*db: Dry Basis; daf: Dry Ash Free; n.r.: Not Reported.

	SS_{10}^{db}	SS_{11}^{db}	SS_{12}^{db}	SS_{13}^{db}	SS_{14}	SS_{15}^{db}	SS_{16}^{db}	SS_{17}^{db}
S	0.9	1.6	2.1	3.3	1.4	0.96	1.67	1.81
Cl_2	n.r.	n.r.	n.r.	n.r.	n.r.	0.05	0.22	0.03
F_2	n.r.	n.r.	n.r.	n.r.	n.r.	n.r.	0.013	0.003
O_2	18.9	15.0	25.8	41.2	34.6	12.2	20.57	18.84
LHV (MJ/kg)	11.5	13.1	11.1	11.1	12.50	10.20	12.96	10.76
Origin	Spain					Denmark	Poland	
Reference	de Andrés et al. 2011	Roche et al. 2014	Alvarez et al. 2015	Alvarez et al. 2016	Ruiz-Gómez et al. 2017	Ulusoy et al. 2021	Werle 2015	

*db: Dry Basis; daf: Dry Ash Free; n.r.: Not Reported.

Table 2 contd. ...

...Table 2 contd.

Proximate Analysis (wt%)

	SS_{18} db	SS_{19} db	SS_{20} db	SS_{21} db	SS_{22} db	SS_{23} daf	SS_{24} db
MC	7.18	n.r.	7.4	5.8	4.01	6.44 ± 0.50	7.27 ± 0.01
VM	38.05	52.10	63.1	54.1	73.10	56.00 ± 2.97	50.30 ± 0.02
FC	4.77	5.96	7.1	6.0	2.80	7.11 ± 2.30	7.73 ± 0.35
Ash	50.0	41.94	22.5	34.2	18.60	30.45 ± 1.24	34.70 ± 0.32

*db: Dry Basis; daf: Dry Ash Free.

Ultimate Analysis (wt%)

	SS_{18} db	SS_{19} db	SS_{20} db	SS_{21} db	SS_{22} db	SS_{23} daf	SS_{24} db
C	24.53	28.27	38.0	34.9	42.0	36.38 ± 0.06	29.88 ± 0.86
H_2	3.19	4.43	5.1	4.8	5.60	5.86 ± 0.17	4.61 ± 0.08
N_2	4.80	5.36	6.9	4.5	4.30	5.22 ± 0.16	4.34 ± 0.15
S	0.15	1.14	1.2	1.1	1.30	0.98 ± 0.05	1.06 ± 0.03
Cl_2	n.r.	n.r.	n.r.	n.r.	n.r.	0.74 ± 0.16	n.r.
O_2	10.4	18.86	19.0	14.8	28.30	47.98 ± 0.28	25.41 ± 1.06
LHV (MJ/kg)	10.4	11.3	17.7	12.1	11.72	15.32 ± 0.55	11.74
Origin	China					Thailand	Korea
Reference	Zuo et al. 2014	Liu et al. 2016	Fan and He 2016		Wang et al. 2017	Arjharn et al. 2013	Choi et al. 2017

*db: Dry Basis; daf: Dry Ash Free; n.r.: Not Reported.

2.2 Biological Treatment of Sewage Sludge: Energy Recovery

Energy content present in SS can be recovered by biological treatment of either aerobic or anaerobic digestion. Energy recovery from SS by Anaerobic Digestion (AD) is more advantageous compared to aerobic digestion since the latter one involves higher operating costs due to the supply of continuous aeration and the addition of water to ensure the constant Organic Loading Rate (OLR). In addition, AD has higher volatile substance degradation efficiency compared to the aerobic process for the same operating conditions of temperature, pH, OLR, and Hydraulic Retention Time (HRT). Treatment of SS through the aerobic process has a higher Humification Index (HIX) compared to AD which is the indication of resistance to biodegradability (Shao et al. 2013). The rising of HIX increases the processing time to complete the digestion process as well as reduces the digestion performance. AD is now considered a promising technology to recover green energy from SS as biogas with higher conversion efficiency (Amin et al. 2021). Biogas generated from SS through the AD process is an energy-rich product, a mixture of methane (CH_4) and carbon dioxide (CO_2) with a minor fraction of hydrogen sulfide (H_2S), ammonia (NH_3), and moisture (H_4O) (Rehman et al. 2019).

2.2.1 Anaerobic digestion of sewage sludge

Anaerobic Digestion (AD) of SS is completed in absence of oxygen by passing four consecutive stages: hydrolysis, acidogenesis, acetogenesis, and finally methanogenesis (Wang et al. 2018a). During hydrolysis, proteins, lipids, fats, etc., in SS are converted to fatty acids and sugars by the microbial activity of *hydrolytic bacteria* and amino acids by *fermentative bacteria*. Volatile Fatty Acids (VFAs), such as propionic acid and butyric acid, are formed in acidogenesis step due to the microbial activity of *acidogenic bacteria* on fatty acids, sugars, and amino acids. The products of the acidogenesis step are then converted to acetate, hydrogen, and carbon dioxide by the activity of *acetogenic bacteria* in the acetogenesis step. Finally, methanogenesis is completed by microbial action of three distinct *methanogenic archaea* of *Hydrogenotrophic, Acetoclastic* and *Methylotrophic methanogens* on acetate, formate, alcohols, amines, CO_2 and H_2, to form CH_4 (Wang et al. 2018a). AD process of SS to generate biogas is schematically presented in Fig. 5.

Different microorganisms have specific duties during AD to produce biogas from SS. The quality of biogas, as well as process performance, depends on the symbiotic relationship among the microbial activity involved in the four stages (Weiland 2010). Initially, the formation rate of *hydrolytic and fermentative bacteria* responsible for the hydrolysis steps of SS is very slow. Due to this limitation, hydrolysis of SS is considered a rate-limiting step (Chen et al. 2016). Around 75% of the CH_4 generated during the AD of SS through decarboxylation of acetate, formate, alcohols, and amines, and remaining are formed from H_2 and CO_2 (Perry et al. 1986). The growth rate of *methanogenic archaea* is very slow compared to the acetogenic and acidogenic steps and is sensitive to the concentration of H_2, NH_3, temperature, pH, and HRT (Chen et al. 2016). For these reasons, the final stage of biogas generation from SS through AD is the rate-limiting stage.

2.2.2 Role of microorganism

The literature related to microbial activity on the conversion of SS to biogas through the AD process is limited. *Hydrolytic* and *fermentative bacteria* complete the breakdown of complex protein, lipid and carbohydrates structure to the simpler amino acids, fatty acids, and soluble sugars, respectively. Hydrolysis of SS in the AD process is carried out by cooperative activities of phylogenetically diverse bacteria that mostly belong to phyla *Firmicutes* and *Bacteroidetes* (Li et al. 2019, Nguyen and Khanal 2018). Selected hydrolytic and fermentative bacteria with Family and Phylum as well as their functional activities in SS hydrolysis with optimum operating conditions (temperature and pH) are presented in Table 3.

Fatty acids, amino acids, and soluble sugars obtained after the hydrolysis of SS are converted to VFAs by the microbial action of different kinds of *acidogenic bacteria* in acidogenesis steps. The

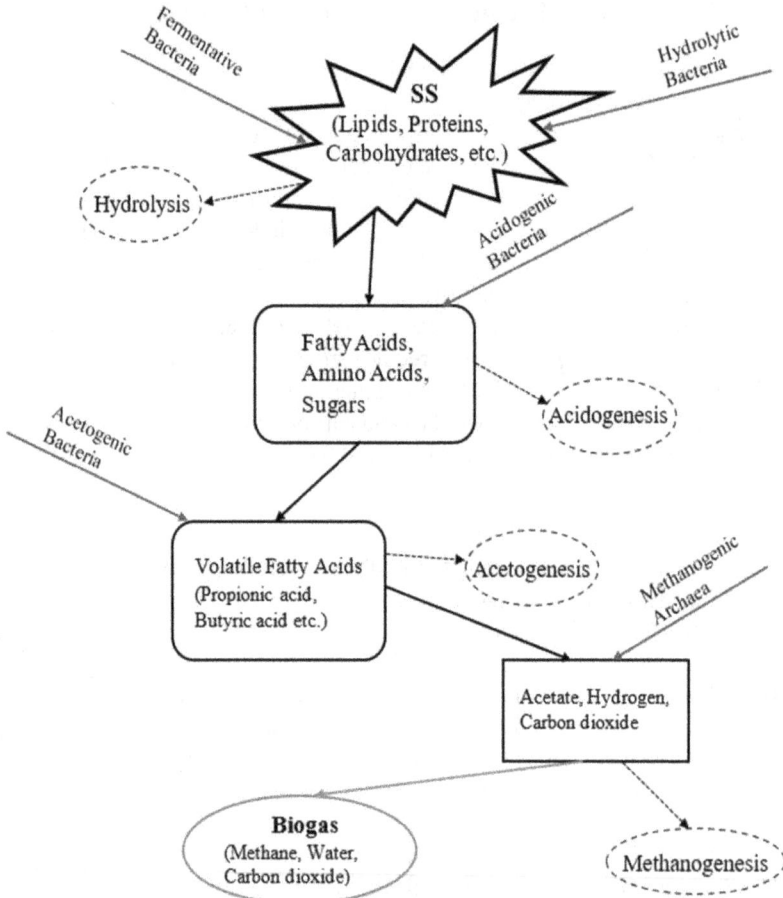

Fig. 5. Schematic diagram for biogas generation from SS through anaerobic digestion.

growth rate of *acidogenic bacteria* is 30 to 40 times higher compared to *methanogenic archaea* and has the ability to tolerate small changes in operational conditions especially temperature and pH (Li et al. 2019). Three typical *acidogenic bacteria* are selected from numerous kinds of *bacteria* involved in the AD process and presented in Table 4 with their functional activity.

Generated VFAs in acidogenic steps are transformed to acetate, formate, H_2, and CO_2 by the action of different kinds of *acetogenic bacteria* in acetogenesis steps. The growth rate of *acetogenic bacteria* is also higher than *methanogenic archaea* but lower than acidogenic microorganisms (Wang et al. 2018b). Selected *acetogenic bacteria* with their functions and products are shown in Table 5.

Acetate, formate, alcohols, H_2, and CO_2 obtained in the acetogenic steps are converted to CH_4 by the actions of *methanogenic archaea* in the final stage of the AD process. Depending on the substrate utilized to generate CH_4, *methanogenic archaea* are classified into three sub-class of hydrogenotrophic, acetoclastic, and methylotrophic methanogens. The growth rate of *hydrogenotrophic methanogens* is 18–24 times higher compared to *acetoclastic methanogens* (Amin et al. 2021). CH_4 is generated from formate, H_2, and CO_2 by the action of *Hydrogenotrophic methanogens* whereas acetate by the decomposition of *Acetoclastic methanogens* and from alcohol and amines by the activity of *Methylotrophic methanogens* (Amin et al. 2021). Three designated sub-classes of *methanogens* with their activity and products in AD of SS are presented in Table 6.

Table 3. Selected hydrolytic and fermentative bacteria for the hydrolysis of SS.

Name of Bacteria	Family; Phylum	Optimum Conditions		Function	Products	Reference
		pH	Temperature (°C)			
Bellilinea caldifistulae	Anaerolineaceae; Chloroflexi	7.0	55	Complete the hydrolysis of glucose, ribose, arabinose, galactose, mannose, xylose, raffinose, and sucrose	Acetate, formate, lactate, propionate, pyruvate, and H_2	Grégoire et al. 2011
Turicibacter sanguinis	Erysipelotrichaceae; Firmicutes	7.5	37	Hydrolyze: Carbohydrates, sorbitol, mannitol, xylose, mannose, glucose, fructose, arabinose, sucrose, and maltose	Acetate, ethanol, and lactate	Bosshard et al. 2002
Thermovirga lienii	Synergistaceae; Synergistetes	6.5–7.0	58	Fermentation of organic acids and amino acids (arginine, alanine, glutamic acid, leucine, cysteine)	Acetate, propionate, H_2, CO_2 and H_2S	Dahle and Birkeland 2006

Table 4. Functional activity of three selected acidogenic bacteria involved in the AD process of SS.

Name of Bacteria	Family, Phylum	Optimum Conditions		Function	Products	Reference
		pH	Temperature (°C)			
Advenella faeciporci	Alcaligenaceae; Proteobacteria	6.5	39	Activity on monomethyl ester, pyruvic acid, methyl ester, citric acid, α-and β-hydroxy-butyric acid, itaconic acid, α-ketobutyric acid, α-ketoglutaric acid, α-ketovaleric acid, L-threonine, L-proline, L-ornithine, L-leucine, L-glutamic acid, D- and L-alanine, alaninamide, glucuronamide, succinamide, succinic acid, bromo succinic acid, succinic acid, sebacic acid, and propionic acid	Volatile Fatty Acids (VFAs)	Xenofontos et al. 2016, Matsuoka et al. 2012
Cloacibacillus porcorum	Synergystaceae; Synergistetes	7.5	37	Decomposition of arginine, D-tryptophan, histidine, proline, serine, and threonine	Acetate, propionate, and formate, butyrate	Li et al. 2018, Looft et al. 2013
Dechloromonas denitrifican	Azonexaceae; Proteobacteria	7.0	30	Decomposition of acetate, butyrate, isobutyrate, propionate, lactate, isovalerate, succinate, pyruvate, glutamate, malate, and casamino. Responsible for the carrying out of denitrification and oxidation of organic acids	Esters	Chakraborty and Picardal 2013, Horn et al. 2005

Table 5. Functional activities and products of acetogenic bacteria involved in AD of SS.

Name of Bacteria	Family; Phylum	Optimum Conditions		Function	Products	Reference
		pH	**Temperature (°C)**			
Acetobacterium wieringae	Eubacteriaceae; Firmicutes	7.2–7.8	30	Complete the conversion of H_2 and CO_2	Acetic acid	Braun and Gottschalk 1982
Acetobacterium woodii		7.3–7.6	30–35	Decomposition of fructose, glycerate, glucose, and lactate. Oxidizes H_2 and reduces CO_2	Acetate; succinate	Bache and Pfennig 1981
Acetogenium kivui	Thermoanaerobacteraceae; Firmicutes.	6.4	66	Oxidizes H_2 and reduces CO_2	Acetate	Leigh et al. 1981
Hydrogenispora ethanolica	Clostridiaceae; Firmicutes	6.0–7.7	37–45	Fermentation of tryptone, fumarate, glycerol, starch, pectin, raffinose, mannose, galactose, sucrose, ribose, xylose, fructose, arabinose, maltose, and glucose	H_2, ethanol, and acetate	Liu et al. 2014
Levilinea saccharolytica	Anaerolineaceae; Chloroflexi	6.0–7.2	37	Complete the fermentation of sugars, pectin, pyruvate, tryptone, sucrose, ribose, raffinose, xylose, fructose, glucose, and amino acids	H_2, acetic, and lactic acids	Guo et al. 2015
Saccharofermentans acetigenes	Clostridiaceae; Firmicutes	6.5	37	Degradation of D-glucose, D-fructose, aesculin, starch, sucrose, adonitol, mannitol, dulcitol, and inositol	Acetate, lactate, and fumarate	Chen et al. 2010

Table 6. Functional activity and products formed of some selected methanogenic archaea in the AD process of SS.

Name of Bacteria	Family, Phylum	Optimum Conditions		Function	Products	Reference
		pH	Temperature (°C)			
Hydrogenotrophic methanogens						
Methanobacterium arcticum	Methanobacteriaceae; Euryarchaeota	6.8–7.2	37	Conversion of formate, H_2, and CO_2	CH_4	Shcherbakova et al. 2011
Methanoculleus thermophilicum	Methanomicrobiaceae; Euryarchaeota	7.0	55			Shimizu et al. 2013
Acetoclastic methanogens						
Methanothrix concilii		7.1–7.5	35–40			Patel and Sprott 1990
Methanothrix harundinacea	Methanosaetaceae; Euryarchaeota	7.2–7.6	34–37	Transform acetate	CH_4	Ma et al. 2006
Methanothrix soehngenii		7.4–.8	37			Jetten et al. 1992
Methanothrix thermophilla		7.0	50–60			Oren 2014
Methylotrophic methanogens						
Methanolobus chelungpuianus	Methanosarcinaceae; Euryarchaeota	7.0	37	Transformation of Methanol and Trimethylamine	CH_4	Wu and Lai 2011
Methanobacterium bryantii	Methanobacteriaceae; Euryarchaeota	6.9–7.0	37	Conversion of 2-Propanol, 2-butanol, Acetate		Shcherbakova et al. 2011

2.2.3 *Effect of antimicrobials on anaerobic digestion of sewage sludge and remediation*

Different types of drugs including antibiotics are frequently used for human and veterinary treatment in daily life and these are eluting out to the wastewater as pollutants. These substances are accumulated in SS during treatment of wastewater in WWTPs and considered as antimicrobials which act as an inhibitor for the archaea and bacteria involved to complete the AD process (Czatzkowska et al. 2020, Chen et al. 2014). Kumar et al. (2005) detected sulfonamides, quinolones, and macrolides-lincosamides-streptogramins drugs in SS as wastewater-borne antimicrobials. Antimicrobials retard the protein translation, cell division, and damage the membrane of archaea and bacteria (Schmidt et al., 2018). Consequently, the activity of methanogenic archaea and bacteria involved in the conversion of SS to green fuel through the AD process is disrupted and efficiency is decreased (Wang et al. 2019, Chen et al. 2008).

The presence of antimicrobials in SS decreases the biogas conversion efficiency and quality in terms of methane content. Czatzkowska et al. (2021) studied the effect of antimicrobials on degradation efficiency, biogas generation efficiency, and quality by adding eight different antimicrobial substances of metronidazole, amoxicillin, cefuroxime, oxytetracycline, doxycycline, sulfamethoxazole, ciprofloxacin, and nalidixic acid in SS to the concentration range of 512–1024 µg/g. The degradation was found between 10.6% (nalidixic acid) and 100% (metronidazole) but reduced the activity of acetogenesis bacteria, consequently increasing the VFAs' concentration during AD treatment of SS. Subsequently, decreases the biogas conversion efficiency in the range of 25.8 to 83.8% and methane content between 0.75% and 75.2%.

Removal of antimicrobials from SS during the AD process increases the metabolic activity of bacteria and archaea which improves the biogas quality and conversion efficiency (Zhang et al. 2018, Zhao et al. 2015). After the addition of granulated activated carbon and nano zero-valent iron mixture as a mediator in the AD process of SS, increases the total methanogens content from 74.7 to 81.74%, which ultimately increases the biogas production efficiency by 21.2% and methane content of 26.9% (Zhang et al. 2018). The utilization of graphite, biochar, and carbon cloth as conductive materials during the AD treatment of SS to generate green fuels improves the methane production rate in the range of 30–45% (Zhao et al. 2015).

2.3 *Influence of Operating Parameters*

The performance of microorganisms in the AD process to decompose SS for green fuel generation highly depends on operating parameters (temperature, pH, organic loading rate, hydraulic retention time, C/N ratio, and nutrients) and diversities of *bacteria* and *archaea*.

2.3.1 *Temperature*

Temperature has the highest influence on the growth rate of *bacteria* and *archaea* involved in the AD treatment of SS, especially *methanogens*. The growth rate of *methanogens* is highly sensitive to temperature (Liu et al. 2018). AD processes are categorized into three classes based on the operating temperature: psychrophilic AD (10–20°C), mesophilic AD (30–40°C), and thermophilic AD (50–60°C) (Liu et al. 2018, Hupfauf et al. 2018). As the temperature increases, the microbial activity of some *microorganisms* increases and some are washed out (Peces et al. 2018). The relative growth rate of *psychrophilic* and *mesophilic archaea* compared to *thermophilic methanogens* with temperature is presented in Fig. 6 based on the experimental data adopted from Wiegel (1990) on the activity of microbial strains between the temperature range of –14 to 110°C.

The growth rate of *methanogenic archaea* is reached only 23% compared to the growth rate of *archaea* in thermophilic conditions and this increases to 40% for mesophilic temperature. The temperature to maximize the conversion efficiency for biogas generation from SS is $\geq 30°C$ for the *mesophilic methanogens* and around 55°C for the thermophilic ones (Gonzalez et al. 2018).

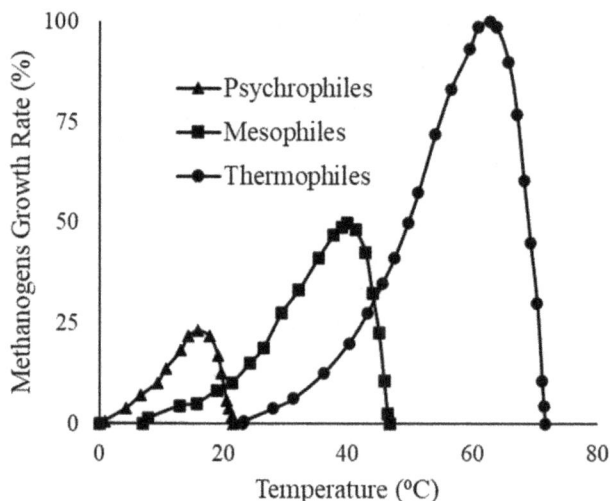

Fig. 6. Variation of the relative growth rate of psychrophilic, mesophilic, and thermophilic methanogens with temperature in AD process.

2.3.2 *pH*

pH is considered as the second most important operating parameter for biogas generation through AD treatment of SS (Amin et al. 2021). With the progress of AD, the concentration of VFAs, CO_2, acetate, amino acids increases and lowers the pH of digestate. Optimum pH for the *hydrolytic* and *acidogenic bacteria* is in the range of 5.0–6.0 to decompose the organic substance with the highest efficiency (Demirer and Chen 2004). With the decrease of pH, the ratio of undissociated to dissociated VFAs increases, being responsible for the cell destruction of *hydrolytic*, *acidogenic bacteria*, and *methanogenic archaea* (Kadam and Boone 1996). When the pH of the AD process drops down to less than 4.5, the fermentation process is predominant compared to the decomposition of protein, lipid, and carbohydrate, the concentration of ethanol increases, and that of VFAs decreases (Ren et al. 1997). Only methanogens of genera *Methanosarcina* can continue their metabolism at pH < 6.5 but all other *methanogenic archaea* can not work effectively when pH is lower than 6.7 (Strauber et al. 2018). Therefore, a pH of the system around 7.0 is suggested to generate biogas from SS through the AD process with the highest conversion efficiency.

2.3.3 *Organic loading rate*

Organic Loading Rate (OLR) is the fraction of degradable organics present in the biomass being processed through AD treatment. The portion of Volatile Substances (VS) in SS is the degradable organic portion and the remaining is considered as fixed solid. The concentration of VFAs in the AD process digestate increases with the increase of OLR and consequently decreases pH. The decrease in pH drives the instability of the process and consequently, drops down the conversion efficiency (Kadam and Boone 1996, Ren et al. 1997, Strauber et al. 2018).

2.3.4 *Hydraulic retention time*

Hydraulic Retention Time (HRT) is the amount of time that SS stays in the reactor during the AD treatment. As the HRT increases, the VFAs concentration increases, reducing the process efficiency. On the other hand, reducing the HRT causes the washed-out of slow-growing *methanogens* from the reactor (Chojnacka et al. 2015). Long or short HRT creates disturbances in digestion systems and as a consequence, the process performance decreases. It is recommended to maintain the HRT twofold greater than the generation time of slow-growing *methanogens* to achieve the highest performance.

The optimum value of HRT depends on the composition of feed, temperature, pH, and digester configuration (Kadier et al. 2018).

2.3.5 C/N ratio

Carbon present in the SS is used as an energy source whereas nitrogen is used for the synthesis of protein to build the cell structure of the *anaerobic bacteria* and *archaea* involved in the AD process. The C/N ratio of SS plays an important role in terms of AD process performance. Its optimum value is in the range of 20–30. A higher C/N ratio indicates that the SS has more readily degradable carbon and this causes an increase in VFAs' concentration during digestion. Consequently, the pH of digestate inside the reactor decreases, and the process performance decreases. On the other hand, a lower C/N ratio increases the ammonium (NH_4^+) in the digester and pH and creates a toxic environment for the microorganism especially for the *methanogens*, ultimately causing failure of the process (Braz et al. 2018).

2.3.6 Nutrients

Availability of inorganic trace elements such as Cobalt (Co), Nickel (Ni), Copper (Cu), and Iron (Fe) act as nutrients for the archaea involved in the methanogenesis steps. High concentrations of these metals boost up the growth rate of *methanogenic archaea* compared to the *fermentative bacteria* which increases the CH_4 formation rate (Hendriks et al. 2018). Iron act as an effective nutrient to stimulate the growth rate of *hydrogenotrophic methanogens* (Feng et al. 2014). The presence of Calcium (Ca) and Magnesium (Mg) in the SS increases the pH of the digestate; however, with a pH higher than 6.7, it is preferable to ensure the methanogenic metabolism of *archaea* (Thanh et al. 2016). The availability of inorganic nutrients depends on the origin and treatment process of wastewater from which SS is generated.

2.4 Application of Biogas

Biogas generated from SS through AD treatment contains CH_4 (55–75%), CO_2 (25–45%), H_2 (0–1%), H_2S (0–1%), O_2 (0–2%), N_2 (0–5%) and CO (0–1%) with a LHV in the range 21–24 MJ/Nm^3 (Kiselev et al. 2019, Demirbas et al. 2016, Andreoli et al. 2007). Biogas has the potentiality to be used in various sectors as listed below.

2.4.1 Electricity generation

Electricity can be generated from biogas by using an Internal Combustion Engine (ICE) or microgas turbines. The use of microgas turbine for the generation of electricity can substantially reduce NOx formation and a wide range of flexibility to meet the various load (Scarlat et al. 2018). Electricity generation from biogas may significantly reduce the dependency on fossil fuels as well as greenhouse gas emissions and reduce the threat to environmental pollution. The generated electricity can be used to run WWTPs or electric vehicles.

2.4.2 Heat generation

Biogas can be directly combusted in a boiler to generate heat energy. The generated heat can be used for district heating, drying of agricultural products, heating the digestate, and aquifer systems.

2.4.3 Combined heat and power generation

Biogas can be used as a fuel for cogeneration systems to generate heat and electricity. The electrical efficiency of an ICE fueled by biogas is in the range of 20–40% and a major fraction of energy is wasted as heat due to the high-temperature stream exit from the ICE system. However, the cogeneration efficiency can reach up to 90% of which 35% electrical and 65% thermal (Shipley et al. 2009). The application field of generated heat is the same as discussed in Section 2.4.2.

2.4.4 *Generation of biomethane*

Biomethane can be produced from biogas through cleaning and upgrading. Cleaning of biogas aims at removing moisture, hydrogen sulfide, oxygen, ammonia, siloxanes, carbon dioxide, carbon monoxide, hydrocarbons, and nitrogen. The cleaned biogas is converted to biomethane by physical and chemical treatment of adsorption, absorption, cryogenic and membrane separations, gas separation membranes as well as biological technologies (Kapoor et al. 2019). Biological techniques are not widely used in industry for the conversion of biogas to biomethane due to the longer processing time compared to the physicochemical method (Scarlat et al. 2018). Generated biomethane from biogas contains 95–97% of methane which is similar to natural gas and can be used for heating purposes, electricity generation, or the production of Compressed Natural Gas (CNG) for transportation (Ryckebosch et al. 2011).

Generated biomethane from biogas can replace natural gas to power vehicles with a significant reduction of the use of fossil fuel and the related greenhouse gas emission. Engine performance and efficiency are considerably higher when biomethane is used in hybrid or fuel cell vehicles compared to biodiesel or ethanol-fueled vehicle (Abanades et al. 2021, Faaij 2006). Biogas used for the transportation sector is normally termed bio-CNG and it is possible to store them for long-term as Liquified Biogas (LBG) for future use. Other forms of fuel such as hydrogen, gasoline, methanol, ethanol, and higher alcohol can be generated from LBG (Yang et al. 2014).

2.4.5 *Hydrogen production*

Hydrogen can be generated from biogas through the steam reforming process in presence of a catalyst. Hydrogen can be used to generate electricity, heat, food processing, ammonia, methanol, ethanol, and higher alcohol (Armor 1999).

2.4.6 *Fuel cells*

Fuel cells can generate electricity from biogas for microelectronic equipment, power distribution generators, power for electric cars, buildings with high efficiency, with low emission of CO_2 and NOx which reduce global warming threat (Alves et al. 2013). Fuel cells have the highest electrical and thermal efficiency of 60% and 40%, respectively, and can be integrated with other power systems like gas or microgas turbines (Pöschl et al. 2010).

3. Conclusion and Future Recommendation

Conventional biomass (wood, lignocellulose, algae, municipal solid waste, etc.), solar, wind, geothermal, tide and ocean are considered renewable energy sources in the world. The contribution of renewable sources was 14% to the world's primary energy supply in 2016. Till now SS is not considered a renewable energy source all over the world especially in developing countries in South Asia and Africa due to the leaking of updated data and conventional processing systems. On-site biogas generation from SS through anaerobic digestion could contribute to the energy required to run WWTPs as well as to the world's primary energy supply.

SS generation rate and composition are variable in different countries or different regions in the same country due to the differences of sewage systems as well wastewater treatment directives. Management of SS through agricultural reuse either directly or after conversion to fertilizer, landfilling, and incineration contributes to environmental pollution. Energy recovery as green fuels from SS through anaerobic digestion may be more advantageous compared to thermal treatments, such as gasification or pyrolysis, due to the mild operating conditions, complete removal of pollution risk to the environment and public health, and the higher energy content of the final product (biogas) compared to syngas obtained after thermal treatment (gasification or pyrolysis). Biogas generation rate and composition highly depend on the composition of feed materials and the digester configuration. Biogas has the potentiality to be used for the generation of electricity, heat, combined heat and power, biomethane, hydrogen, or to run a fuel cell.

Anaerobic digestion requires four steps to generate biogas. Hydrolysis and methanogenesis are the rate-limiting steps due to the slow growth rate of *microorganisms* involved in these steps. An additional pretreatment unit should be integrated with an anaerobic digester to add some useful *hydrolytic* and *fermentative bacteria* to reduce the processing time in the first steps as well as to add some *archaea* in methanogenesis steps to increase methane content in biogas and increase the conversion efficiency. In addition to this, inorganic (micro) nutrients, such as Fe, Ni, Co, and Cu, should be added to the digester to boost up the activities of the microbial community to improve the SS digestion performances.

Abbreviations

AD	Anaerobic Digestion
CHP	Combined Heat and Power
CNG	Compressed Natural Gas
DS	Dry Solid
EU	European Union
HIX	Humification Index
HRT	Hydraulic Retention Time
ICE	Internal Combustion Engine
IPCC	Intergovernmental Panel on Climate Change
LBG	Liquified Biogas
LHV	Lower Heating Value
OLR	Organic Loading Rate
PAHs	Polycyclic Aromatic Hydrocarbons
PCBs	Poly Chlorinated Biphenyls
SS	Sewage Sludge
VFAs	Volatile Fatty Acids
VS	Volatile Substances
WWTPs	WasteWater Treatment Plants

References

Abanades, S., Abbaspour, H., Ahmadi, A., Das, B., Ehyaei, M.A., Esmaeilion, F. et al. 2021. A critical review of biogas production and usage with legislations framework across the globe. Int. J. Environ. Sci. Technol. 1–24.

Abdelrahim, A., Brachi, P., Ruoppolo, G., Fraia, S.D. and Vanoli, L. 2020. Experimental and numerical investigation of biosolid gasification: Equilibrium-based modeling with emphasis on the effects of different pretreatment methods. Ind. Eng. Chem. Res. 59(1): 299–307.

Agrafioti, E., Bouras, G., Kalderis, D. and Diamadopoulos, E. 2013. Biochar production by sewage sludge pyrolysis. J. Anal. Appl. Pyrolysis. 101: 72–78.

Al-Sudani, H.I.Z. 2019. A review on groundwater pollution. Int. J. Eng. Sci. 6(5): 14–22.

Alvarez, J., Amutio, M., Lopez, G., Barbarias, I., Bilbao, J. and Olazar, M. 2015. Sewage sludge valorization by flash pyrolysis in a conical spouted bed reactor. Chem. Eng. J. 273: 173–183.

Alvarez, J., Lopez, G., Amutio, M., Artetxe, M., Barbarias, I., Arregi, A. et al. 2016. Characterization of the bio-oil obtained by fast pyrolysis of sewage sludge in a conical spouted bed reactor. Fuel Process. Technol. 149: 169–175.

Alves, H.J., Junior, C.B., Niklevicz, R.R., Frigo, E.P., Frigo, M.S. and Coimbra-Araújo, C.H. 2013. Overview of hydrogen production technologies from biogas and the applications in fuel cells. Int. J. Hydrog. Energy 38(13): 5215–5225.

Amin, F.R., Khalid, H., El-Mashad, H.M., Chen, C., Liu, G. and Zhang, R. 2021. Functions of bacteria and archaea participating in the bioconversion of organic waste for methane production, Sci. Total Environ. 763: 143007.

Andreoli, C.V., von Sperling, M., Fernandes, F. and Ronteltap, M. 2007. Biological Wastewater Treatment Series, Sludge Treatment and Disposal. IWA Publishing, London.

AQUASTAT. 2014. FAO global information system on water and agriculture. Wastewater section. http://www.fao.org/nr/water/aquastat/wastewater/index.stm. Accessed Date June 12, 2021.

Arjharn, W., Hinsui, T., Liplap, P. and Raghavan, G.S.V.. 2013. Evaluation of an energy production system from sewage sludge using a pilot-scale downdraft gasifier. Energy Fuels 27: 229–236.

Armor, J.N. 1999. The multiple roles for catalysis in the production of H_2. Appl Catal A. 176(2): 159–176.

Bache, R. and Pfennig, N. 1981. Selective isolation of Acetobacterium woodii on methoxylated aromatic acids and determination of growth yields. Arch. Microbiol. 130: 255–261.

Bludowsky, T. and Agar, D.W. 2009. Thermally integrated bio-syngas-production for biorefineries. Chem. Eng. Res. Des. 87(9): 1328–1339.

Bosshard, P.P., Zbinden, R. and Altwegg, M. 2002. Turicibacter sanguinis gen. nov., sp. nov., a novel anaerobic, Gram-positive bacterium. Int. J. Syst. Evol. Microbiol. 52: 1263–1266.

Braun, M. and Gottschalk, G. 1982. Acetobacterium wieringae sp. nov., a new species producing acetic acid from molecular hydrogen and carbon dioxide. Zentralblatt für Bakteriol. Mikrobiol. und Hyg. I. Abt. Orig. C Allg. Angew. und ökologische Mikrobiol. 3: 368–376.

Braz, G.H.R., Fernandez-Gonzalez, N., Lema, J.M. and Carballa, M. 2018. The time response of anaerobic digestion microbiome during an organic loading rate shock. Appl. Microbiol. Biotechnol. 102(23): 10285–10297.

Bunsow, W. and Dobberstein, J. 1987. Refuse-derived fuel: composition and emissions from combustion. Resour. Conserv. 14: 249–256.

Canziani, R. and Spinosa, L. 2019. Sludge from wastewater treatment plants. Industrial and Municipal Sludge, Butterworth-Heinemann Elsevier, 3–30.

Chakraborty, A. and Picardal, F. 2013. Neutrophilic, nitrate-dependent, Fe (II) oxidation by a Dechloromonas species. World J. Microbiol. Biotechnol. 29: 617–623.

Chen, C., Guo, W., Ngo, H.H., Lee, D-J., Tung, K-L., Jin, P. et al. 2016. Challenges in biogas production from anaerobic membrane bioreactors. Renew. Energy 98: 120–134.

Chen, J.L., Ortiz, R., Steele, T.W.J. and Stuckey, D.C. 2014. Toxicants inhibiting anaerobic digestion: A review. Biotechnol. Adv. 32: 1523–1534.

Chen, S., Niu, L. and Zhang, Y. 2010. Saccharofermentans acetigenes gen. nov., sp. nov., an anaerobic bacterium isolated from sludge treating brewery wastewater. Int. J. Syst. Evol. Microbiol. 60: 2735–2738.

Chen, Y., Cheng, J.J. and Creamer, K.S. 2008. Inhibition of anaerobic digestion process: A review. Bioresour. Technol. 99: 4044–4064.

Choi, Y-K., Ko, J-H. and Kim, J-S. 2017. A new type three-stage gasification of dried sewage sludge: Effects of equivalence ratio, weight ratio of activated carbon to feed, and feed rate on gas composition and tar, NH_3, and H_2S removal and results of approximately 5 h gasification. Energy 118: 139–146.

Chojnacka, A., Szczęsny, P., Błaszczyk, M.K., Zielenkiewicz, U., Detman, A., Salamon, A. et al. 2015. Noteworthy facts about a methane-producing microbial community processing acidic effluent from sugar beet molasses fermentation. PLoS One 10: 1–23.

Christodoulou, A. and Stamatelatou, K. 2016. Overview of legislation on sewage sludge management in developed countries worldwide. Water Sci. Technol. 73(3): 453–462.

Colón, J., Alarcón, M., Healy, M., Namli, A., Sanin, D., Taya, C. et al. 2017. Producing sludge for agricultural applications. IWA Publishing, 292–314.

Czatzkowska, M., Harnisz, M., Korzeniewska, E. and Koniuszewska, I. 2020. Inhibitors of the methane fermentation process with particular emphasis on the microbiological aspect: A review. Energy Sci. Eng. 8: 1880–1897.

Czatzkowska, M., Harnisz, M., Korzeniewska, E., Rusanowska, P., Bajkacz, S., Felis, E. et al. 2021. The impact of antimicrobials on the efficiency of methane fermentation of sewage sludge, changes in microbial biodiversity and the spread of antibiotic resistance. J. Hazard. Mater. 416: 125773.

Dahle, H. and Birkeland, N.K. 2006. Thermovirga lienii gen. nov., sp. nov., a novel moderately thermophilic, anaerobic, amino-acid-degrading bacterium isolated from a North Sea oil well. Int. J. Syst. Evol. Microbiol. 56: 1539–1545.

de Andrés, J.M., Narros, A. and Rodríguez, M.E. 2011. Behaviour of dolomite, olivine and alumina as primary catalysts in air-steam gasification of sewage sludge. Fuel 90: 521–527.

de Andrés, J.M., Vedrenne, M., Brambilla, M. and Rodríguez, E. 2019. Modeling and model performance evaluation of sewage sludge gasification in fluidized-bed gasifiers using Aspen Plus. J. Air Waste Manage. 69(1): 23–33.

Delibacak, S., Voronina, L., Morachevskaya, E. and Ongun, A.R. 2020. Use of sewage sludge in agricultural soils: Useful or harmful. Eurasian J. Soil Sci. 9(2): 126–139.

Demirbas, A., Taylan, O. and Kaya, D. 2016. Biogas production from municipal sewage sludge (MSS). Energy Source Part A 38(20): 3027–3033.

Demirer, G.N. and Chen, S. 2004. Effect of retention time and organic loading rate on anaerobic acidification and biogasification of dairy manure. J. Chem. Technol. Biotechnol. 79: 1381–1387.

Dichtl, N., Rogge, S. and Bauerfeld, K. 2007. Novel strategies in sewage sludge treatment. Clean-Soil Air Water 35: 473–479.

Dominguez, A., Menendez, J.A., Inguanzo, M. and Pis, J.J. 2006. Production of bio-fuels by high temperature pyrolysis of sewage sludge using conventional and microwave heating. Bioresour. Technol. 97: 1185–1193.

Elias, X. 2005. Tratamiento y valorización energética de residuos,ed Díaz de Santos.

European Commission. 2009. Environmental, economic and social impacts of the use of sewage sludge on land. Consultation Report on Options and Impacts, Report by RPA, Milieu Ltd and WRc for the European Commission, DG Environment under Study Contract DG ENV.G.4/ETU/2008/0076r.

Eurostat 2021. (https://ec.europa.eu/eurostat/statistics-explained/index.php?title=Water_statistics). Access Date June 17, 2021.

Eurostat 2021. (https://ec.europa.eu/eurostat/databrowser/view/ten00020/default/table?lang=en). Access Date June 7, 2021.

Eurostat 2021. (https://appsso.eurostat.ec.europa.eu/nui/submitViewTableAction.do). Access Date June 2, 2021.

Faaij, A. 2006. Modern biomass conversion technologies. Mitig. Adapt. Strat. Glob. Change 11(2): 343–375.

Fan, H. and He, K. 2016. Fast pyrolysis of sewage sludge in a curie-point pyrolyzer: The case of sludge in the city of Shanghai, China. Energy Fuels 30: 1020−1026.

Feng, Y., Zhang, Y., Quan, X. and Chen, S. 2014. Enhanced anaerobic digestion of waste activated sludge digestion by the addition of zero valent iron. Water Res. 52: 242–250.

Fericelli, P.D. 2011. Comparison of Sludge Treatment by Gasification vs. Incineration. LACCEI. WE1-10. Medellín, Colombia.

Gonzalez, A., Hendriks, A.T.W.M., van Lier, J.B. and de Kreuk, M. 2018. Pre-treatments to enhance the biodegradability of waste activated sludge: Elucidating the rate limiting step. Biotechnol. Adv. 36: 1434–1469.

Grégoire, P., Fardeau, M., Joseph, M., Guasco, S., Hamaide, F., Biasutti, S. et al. 2011. Isolation and characterization of Thermanaerothrix daxensis gen. nov., sp. nov., a thermophilic anaerobic bacterium pertaining to the phylum "Chloroflexi", isolated from a deep hot aquifer in the Aquitaine Basin. Syst. Appl. Microbiol. 34: 494–497.

Grobelak, A., Czerwinska, K. and Murtas, A. 2019. 7—General considerations on sludge disposal, industrial and municipal sludge. pp. 135–153. *In*: Prasad, M.N.V., de Campos Favas, P.J., Vithanage, M. and Mohan, S.V. (eds.). Industrial and Municipal Sludge. Butterworth-Heinemann.

Guo, Z., Zhou, A., Yang, C., Liang, B., Sangeetha, T., He, Z. et al. 2015. Enhanced short chain fatty acids production from waste activated sludge conditioning with typical agricultural residues: carbon source composition regulates community functions. Biotechnol. Biofuels 192: 1–14.

Hanif, M.U., Zwawi, M., Capareda, S.C., Iqbal, H., Algarni, M., Felemban, B.F. et al. 2020. Influence of pyrolysis temperature on product distribution and characteristics of anaerobic sludge. Energies 13: 79.

Harrison, E.Z., Oakes, S.R., Hysell, M. and Hay, A. 2006. Organic chemicals in sewage sludges. Sci. Total Environ. 367: 481–497.

Hendriks, A.T.W.M., van Lier, J.B. and de Kreuk, M.K. 2018. Growth media in anaerobic fermentative processes: The underestimated potential of thermophilic fermentation and anaerobic digestion. Biotechnol. Adv. 36: 1–13.

Horn, M.A., Ihssen, J., Matthies, C., Schramm, A., Acker, G., Drake, H.L. et al. 2005. Dechloromonas denitrificans sp. nov., Flavobacterium denitrificans sp. nov., Paenibacillus anaericanus sp. nov. and Paenibacillus terrae strain MH72, N_2O producing bacteria isolated from the gut of the earthworm Aporrectodea caliginosa. Int. J. Syst. Evol. Microbiol. 55: 1255–1265.

Huang, Y.-F., Shih, C.-H., Chiueh, P.-T. and Lo, S.-L. 2015. Microwave co-pyrolysis of sewage sludge and rice straw. Energy 87: 638–644.

Hupfauf, S., Plattner, P., Wagner, A.O., Kaufmann, R., Insam, H. and Podmirseg, S.M. 2018. Temperature shapes the microbiota in anaerobic digestion and drives efficiency to a maximum at 45°C. Bioresour. Technol. 269: 309–318.

IEA Bioenergy Countries' Report—Update 2018. Bioenergy policies and status of implementation. https://www.ieabioenergy.com/wp-content/uploads/2018/10/IEA-Bioenergy-Countries-Report-Update-2018-Bioenergy-policies-and-status-of-implementation.pdf.

IPCC Report 2014. (AR5 values: https://www.ipcc.ch/pdf/assessmentreport/ar5/wg1/WG1AR5_Chapter08_FINAL. pdf (pp. 73- 79)).

Jetten, M.S.M., Stams, A.J.M. and Zehnder, A.J.B. 1992. Methanogenesis from acetate: A comparison of the acetate metabolism in Methanothrix soehngenii and Methanosarcina spp. FEMS Microbiol. Rev. 88: 181–198.

Kadam, P.C. and Boone, D.R. 1996. Influence of pH on ammonia accumulation and toxicity in halophilic, methylotrophic methanogens. Appl. Environ. Microbiol. 62: 4486–4492.

Kadier, A., Kalil, M.S., Chandrasekhar, K., Mohanakrishna, G., Saratale, G.D., Saratale, R. et al. 2018. Surpassing the current limitations of high purity H_2 production in microbial electrolysis cell (MECs): Strategies for inhibiting growth of methanogens. Bioelectrochemistry 119: 211–219.

Kapoor, R., Ghosh, P., Kumar, M. and Vijay, V.K. 2019. Evaluation of biogas upgrading technologies and future perspectives: a review. Environ. Sci. Pollut. Res. 26(12): 11631–11661.

Kelessidis, A. and Stasinakis, A.S. 2012. Comparative study of the methods used for treatment and final disposal of sewage sludge in European countries. Waste Manage. 32: 1186–1195.

Kiselev, A., Magaril, E., Magaril, R., Panepinto, D., Ravina, M. and Zanetti, M.C. 2019. Towards circular economy: Evaluation of sewage sludge biogas solutions. Resour. 8: 91.

Kulling, D., Stadelmann, F. and Herter, U. 2001. Sewage Sludge—Fertilizer or Waste? UKWIR Conference, Brussels.

Kumar, K., Gupta, S.C., Baidoo S.K., Chander, Y. and Rosen, C.J. 2005. Antibiotic uptake by plants from soil fertilized with animal manure. J. Environ. Qual. 34: 2082–2085.

Kwarciak-Kozlowska, A., Krzywicka, A. and Galwa-Widera, M. 2016. The use of ozonation process in coke wastewater treatment. Rocz. Ochr. Srodowiska. 18(2): 61–73.

LeBlanc, R.J., Matthews, P. and Richard, R.P. 2009. Global atlas of excreta, wastewater sludge, and biosolids management: Moving forward the sustainable and welcome uses of a global resource: Un-habitat.

Leigh, J.A., Mayer, F. and Wolfe, R.S. 1981. Acetogenium kivui, a new thermophilic hydrogenoxidizing, acetogenic bacterium. Arch. Microbiol. 129: 275–280.

Li, W., Siddhu, M.A.H., Amin, F.R., He, Y., Zhang, R., Liu, G. et al. 2018. Methane production through anaerobic co-digestion of sheep dung and wastepaper. Energy Convers. Manag. 156: 279–287.

Li, Y., Chen, Y. and Wu, J. 2019. Enhancement of methane production in anaerobic digestion process: A review. Appl. Energy 240: 120–137.

Liu, C.M., Wachemo, A.C., Tong, H., Shi, S.H., Zhang, L., Yuan, H.R. et al. 2018. Biogas production and microbial community properties during anaerobic digestion of corn stover at different temperatures. Bioresour. Technol. 261: 93–103.

Liu, H., Yi, L., Zhang, Q., Hu, H., Lu, G., Li, A. et al. 2016. Co-production of clean syngas and ash adsorbent during sewage sludge gasification: synergistic effect of Fenton peroxidation and CaO conditioning. Appl. Energy 179: 1062–1068.

Liu, Y., Nilsen, P.J. and Maulidiany, N.D. 2019. Thermal pretreatment to enhance biogas production of waste aerobic granular sludge with and without calcium phosphate precipitates. Chemosphere 234: 725–732.

Liu, Y., Qiao, J., Yuan, X., Guo, R. and Qiu, Y. 2014. Hydrogenispora ethanolica gen. nov., sp. nov., an anaerobic carbohydrate-fermenting bacterium from anaerobic sludge. Int. J. Syst. Evol. Microbiol. 64: 1756–1762.

Looft, T., Levine, U.Y. and Stanton, T.B. 2013. Cloacibacillus porcorum sp. nov., a mucindegrading bacterium from the swine intestinal tract and emended description of the genus Cloacibacillus. Int. J. Syst. Evol. Microbiol. 63: 1960–1966.

Ma, K., Liu, X. and Dong, X. 2006. Methanosaeta harundinacea sp. nov., a novel acetate scavenging methanogen isolated from a UASB reactor. Int. J. Syst. Evol. Microbiol. 56: 127–131.

Magdziarz, A., Dalai, A.K. and Koziński, J.A. 2016. Chemical composition, character and reactivity of renewable fuel ashes. Fuel 176: 135–145.

Manara, P. and Zabaniotou, A. 2012. Towards sewage sludge based biofuels via thermochemical conversion—a review. Renew. Sust. Energ. Rev. 16: 2566–2582.

Margat, J. and Gun, J.V.D. 2013. Groundwater Around the World: A Geographic Synopsis (1st ed.). CRC Press.

Mateo-Sagasta, J., Raschid-Sally, L. and Thebo, A. 2015. Global wastewater and sludge production, treatment and use. Wastewater: Economic Asset in an Urbanizing World. Springer Dordrecht Heidelberg London, 15–38.

Matsuoka, M., Park, S., An, S., Miyahara, M., Kim, S., Kamino, K. et al. 2012. Advenella faeciporci sp. nov., a nitrite-denitrifying bacterium isolated from nitrifying–denitrifying activated sludge collected from a laboratory-scale bioreactor treating piggery wastewater. Int. J. Syst. Evol. Microbiol. 62: 2986–2990.

McCarty, P.L. and Smith, D.P. 1986. Anaerobic wastewater treatment. Environ. Sci. Technol. 20(12): 1200–1206.

Metcalf, K. and Eddy, A. 1991. Wastewater Engineering: Treatment, Disposal and Reuse. McGraw-Hill, New York.

Migliaccio, R., Brachi, P., Montagnaro, F., Papa, S., Tavano, A., Montesarchio, P. et al. 2021. Sewage sludge gasification in a fluidized bed: Experimental investigation and modeling. Ind. Eng. Chem. Res. 60: 5034−5047.

Mininni, G., Mauro, E., Piccioli, B., Colarullo, G., Brandolini, F. and Giacomelli, P. 2019. Production and characteristics of sewage sludge in Italy. Water Sci. Technol. 79(4): 619–626.

Nguyen, V-T. and Chiang, K-Y. 2021. Sewage and textile sludge thermal degradation kinetic study using multistep approach. Thermochimica. Acta 698: 178871.

Nguyen, V.K., Chaudhary, D.K., Dahal, R.H., Trinh, N.H., Kim, J., Chang, S.W. et al. 2021. Review on pretreatment techniques to improve anaerobic digestion of sewage sludge. Fuel 285: 119105.

Nguyen, D. and Khanal, S.K. 2018. A little breath of fresh air into an anaerobic system: How microaeration facilitates anaerobic digestion process. Biotechnol. Adv. 36: 1971–1983.

NGWA. 2021. (https://www.ngwa.org/what-is-groundwater/About-groundwater). Access Date: June 21, 2021.

Nielson, G.H., Hogue, E.J., Nielson, D. and Zebarth, B.J. 1998. Evaluation of organic wastes as soil amendments for cultivation of carrot and chard on irrigated sandy soils. Can. J. Soil Sci. 78: 217–225.

Oren, A. 2014. The family Methanotrichaceae. *In*: Rosenberg, E., DeLong, E.F., Lory, S., Stackebrandt, E. and Thompson, F. (eds.). The Prokaryotes. Springer, Berlin, Heidelberg.

Panchasara, H. and Ashwath, N. 2021. Effects of pyrolysis bio-oils on fuel atomisation—A review. Energies 14: 794.

Patel, G.B. and Sprott, G.D. 1990. Methanosaeta concilii gen. nov. sp. nov. ("Methanothrix concilii") and Methanosaeta thermoacetophila nom. rev., comb. nov. Int. J. Syst. Bacteriol. 40: 79–82.

Peces, M., Astals, S., Jensen, P.D. and Clarke, W.P. 2018. Deterministic mechanisms define the long-term anaerobic digestion microbiome and its functionality regardless of the initial microbial community. Water Res. 141: 366–376.

Pöschl, M., Ward, S. and Owende, P. 2010. Evaluation of energy efficiency of various biogas production and utilization pathways. Appl Energy 87(11): 3305–3321

Rehman, M.L.U., Iqbal, A., Chang, C-C., Li, W. and Ju, M. 2019. Anaerobic digestion. Water Environ. Res. 91: 1253–1271.

Ren, N., Wang, B. and Huang, J.C. 1997. Ethanol-type fermentation from carbohydrate in high rate acidogenic reactor. Biotechnol. Bioeng. 54: 428–433.

Roche, E., de Andrés, J.M., Narros, A. and Rodríguez, M.E. 2014. Air and air-steam gasification of sewage sludge. The influence of dolomite and throughput in tar production and composition. Fuel 115: 54–61.

Ruiz-Gómez, N., Quispe, V., Ábrego, J., Atienza-Martínez, M., Murillo, M.B. and Gea, G. 2017. Co-pyrolysis of sewage sludge and manure. Waste Manage. 59: 211–221.

Rulkens, W. 2008. Sewage sludge as a biomass resource for the production of energy: overview and assessment of the various options. Energy Fuels 22: 9–15.

Ryckebosch, E., Drouillon, M. and Vervaeren, H. 2011. Techniques for transformation of biogas to biomethane. Biomass Bioenerg. 35(5): 1633–1645.

Sato, T., Qadir, M., Yamamoto, S., Endo, T. and Zahoor, A. 2013. Global, regional, and country level need for data on wastewater generation, treatment, and reuse. Agric. Water Manag. 130: 1–13.

Scarlat, N., Dallemand, J-F. and Fahl, F. 2018. Biogas: Developments and perspectives in Europe. Renew. Energy 129: 457–472.

Schmidt, T., McCabe, B.K., Harris, P.W. and Lee, S. 2018. Effect of trace element addition and increasing organic loading rates on the anaerobic digestion of cattle slaughterhouse wastewater. Bioresour. Technol. 264: 51–57.

Seggiani, M., Vitolo, S., Puccini, M. and Bellini, A. 2012. Cogasification of sewage sludge in an updraft gasifier. Fuel 93: 486–491.

Shafiee, S. and Topal, E. 2009. When will fossil fuel reserves be diminished? Energy Policy 37(1): 181–189.

Shao, L., Wang, T., Li, T., Lü, F. and He, P. 2013. Comparison of sludge digestion under aerobic and anaerobic conditions with a focus on the degradation of proteins at mesophilic temperature. Bioresour. Technol. 140: 131–137.

Shcherbakova, V., Rivkina, E., Pecheritsyna, S., Laurinavichius, K., Suzina, N. and Gilichinsky, D. 2011. Methanobacterium arcticum sp. nov., a methanogenic archaeon from Holocene Arctic permafrost. Int. J. Syst. Evol. Microbiol. 61: 144–147.

Shimizu, S., Ueno, A., Tamamura, S., Naganuma, T. and Kaneko, K. 2013. Methanoculleus horonobensis sp. nov., a methanogenic archaeon isolated from a deep diatomaceous shale formation. Int. J. Syst. Evol. Microbiol. 63: 4320–4323.

Shipley, A., Hampson, A., Hedman, B., Garland, P. and Bautista, P. 2009. DOE report: Combined heat and power: Effective energy solutions for a sustainable future.

Singh, B.R. and Singh, O. 2012. Global trends of fossil fuel reserves and climate change in the 21st century, 167–192.

Spinosa, L. 2007. Wastewater Sludge: A Global Overview of the Current Status and Future Prospects. IWA Publishing, 6.

Spliethoff, H., Scheurer, W. and Hein, K.R.G. 2000. Effect of co-combustion of sewage sludge and biomass on emissions and heavy metals behaviour. Process Saf. Environ. Prot. 78(1): 33–39.

Strauber, H., Buhligen, F., Kleinsteuber, S. and Dittrich-Zechendorf, M. 2018. Carboxylic acid production from ensiled crops in anaerobic solid-state fermentation-trace elements as pH controlling agents support microbial chain elongation with lactic acid. Eng. Life Sci. 18(7): 447–458.

Thanh, P.M., Ketheesan, B., Yan, Z. and Stuckey, D. 2016. Trace metal speciation and bioavailability in anaerobic digestion: A review. Biotechnol. Adv. 34: 122–136.

Ulusoy, B., Anicic, B., Lin, W., Lu, B., Wang, W., Dam-Johansen, K. et al. 2021. Interactions in NOx chemistry during fluidized bed co-combustion of residual biomass and sewage sludge. Fuel 294: 120431.

Vukasinovic, M., Zdravkovic, V., Lutovac, M. and Zdravkovic, N. 2017. The effects of polychlorinated biphenyls on human health and the environment. Glob. J. Pathol. Microbiol. 5: 8–14.

Wang, M., Li, R. and Zhao, Q. 2019. Distribution and removal of antibiotic resistance genes during anaerobic sludge digestion with alkaline, thermal hydrolysis and ultrasonic pretreatments. Front. Environ. Sci. Eng. 13: 43.

Wang, P., Wang, H., Qiu, Y., Ren, L. and Jiang, B. 2018. Microbial characteristics in anaerobic digestion process of food waste for methane production—A review. Bioresour. Technol. 248: 29–36.

Wang, S., Jena, U. and Das, K.C. 2018. Biomethane production potential of slaughterhouse waste in the United States. Energy Convers. Manag. 173: 143–157.

Wang, K., Zheng, Y., Zhu, X., Brewer, C.E. and Brown, R.C. 2017. *Ex-situ* catalytic pyrolysis of wastewater sewage sludge—A micropyrolysis study. Bioresour. Technol. 232: 229–234.

WBA Global Bioenergy Statistics. 2018. http://www.worldbioenergy.org/uploads/181203%20WBA%20GBS%20 2018_hq.pdf.

Wei, L., Zhu, F., Li, Q., Xue, C., Xia, X., Yu, H. et al. 2020. Development, current state and future trends of sludge management in China: Based on exploratory data and CO_2-equivalent emissions analysis. Environ. Int. 144: 106093.

Weiland, P. 2010. Biogas production: Current state and perspectives. Appl. Microbiol. Biotechnol. 85: 849–860.

Werle, S. 2015. Gasification of a dried sewage sludge in a laboratory scale fixed bed reactor. Energies 8: 8562–8572.

Wiegel, J. 1990. Temperature spans for growth: hypothesis and discussion. FEMS Microbiol. Rev. 75: 155–170.

Wu, S. and Lai, M. 2011. Methanogenic archaea isolated from Taiwan's Chelungpu Fault. Appl. Environ. Microbiol. 77: 830–838.

Wyrwicka, A., Steffani, S. and Urbaniak, M. 2014. The effect of PCB-contaminated sewage sludge and sediment on metabolism of cucumber plants (Cucumis sativus L.). Ecohydrol. Hydrobiol. 14(1): 75–82.

Xenofontos, E., Tanase, A., Stoica, I. and Vyrides, I. 2016. Newly isolated alkalophilic Advenella species bioaugmented in activated sludge for high p-cresol removal. New Biotechnol. 33: 305–310.

Xie, Q., Peng, P., Liu, S., Min, M., Cheng, Y. and Wan, Y. 2014. Fast microwave-assisted catalytic pyrolysis of sewage sludge for bio-oil Production. Bioresour. Technol. 172: 162–168.

Xue, Y., Zhou, Y., Liu, J., Xiao, Y. and Wang, T. 2021. Comparative analysis for pyrolysis of sewage sludge in tube reactor heated by electromagnetic induction and electrical resistance furnace. Waste Manage. 120: 513–521.

Yang, L., Ge, X., Wan, C., Yu, F. and Li, Y. 2014. Progress and perspectives in converting biogas to transportation fuels. Renew Sustain Energy Rev. 40: 1133–1152.

Zhai, J., Tian, W. and Liu, K. 2011. Quantitative assessment of polycyclic aromatic hydrocarbons in sewage sludge from wastewater treatment plants in Qingdao, China. Environ. Monit. Assess. 180(1-4): 303–311.

Zhang, Z., Gao, P., Cheng, J., Liu, G., Zhang, X. and Feng, Y. 2018. Enhancing anaerobic digestion and methane production of tetracycline wastewater in EGSB reactor with GAC/NZVI mediator. Water Res. 136: 54–63.

Zhao, Z., Zhang, Y., Woodard, T.L., Nevin, K.P. and Lovely, D.R. 2015. Enhancing syntrophic metabolism in up-flow anaerobic sludge blanket reactors with conductive carbon materials. Bioresour. Technol. 191: 140–145.

Zuo, W., Jin, B., Huang, Y. and Sun, Y. 2014. Characterization of top phase oil obtained from co-pyrolysis of sewage sludge and poplar sawdust. Environ. Sci. Pollut. Res. 21: 9717–9726.

CHAPTER 15

Techno-economic Study of Microbial Green Fuels vs Plant Fuels

Jorge Aburto

1. Introduction

According to the International Energy Agency (IEA 2015), the world energy consumption continues to raise due to economic and population growth. The transport and residential energy have maintained a constant increase since the 70s, whilst industry has diminished its energy consumption due to technology development and measures of energy efficiency. Hence, the industrial branch of transport contributes with 17% CO_2 emissions, but residential, agriculture activities and unsustainable forest activities contribute with 45% of human activities (IATA 2015). In order to assume a reduction of greenhouse gases (GHG), an international commitment was reached at the 21th Conference of the Parties (COP21) in Paris, to disrupt economic growth from gas emission generation, mainly CO_2 and CH_4, through the Intended National Determined Contributions (INDC). Energy source comes primarily from fossil fuels with a contribution of 79.5%, followed by 10.4% of modern renewables, 7.8% corresponds to the conventional use of firewood and 2% of nuclear energy as stated by the renewable report (REN21 2018; Fig. 1). Biofuels contribute with a 0.9% to such energy matrix.

Petroleum, its petrolifers and petrochemicals still dominate such energy matrix, but the actual energy transition promotes a mixed matrix for electricity generation and also for fuel production. Here, bioenergy which refers to the use of biomass, i.e., organic matter of recent generation, might be an important primary and secondary source of energy, biofuels, bio-based chemicals and materials in a biorefinery. The latter should be understood as a holistic system for the sustainable processing of biomass through the integration of physical, chemical, biochemical and thermochemical processes.

In the transport sector, liquid biofuels are more common and are blended in different proportions with fossil fuels. Such biofuels may be used as oxygenate agents like bioethanol in gasolines in Mexico with a 5.8% content of bioethanol, or like gasoline components with a major content as 15 or 27% in USA or Brazil, respectively. In the case of biodiesel, it is blended till 20% with diesel. Biogas, obtained from the anaerobic digestion of manure, biomass and the organic fraction of urban

Gerencia de Transformación de Biomasa, Instituto Mexicano del Petróleo, Eje Central Lázaro Cárdenas Norte 152, Col. San Bartolo Atepehuacan, 07730, Alcaldía Gustavo A. Madero, Mexico.
Email: jaburto@imp.mx

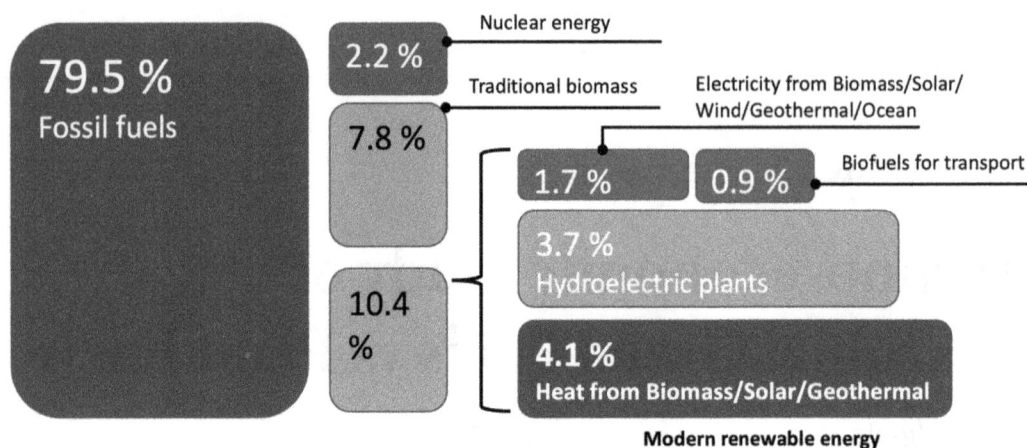

Fig. 1. Energy matrix (REN21 2018).

wastes, is mainly used for electric and thermal power generation, but some countries have introduced biogas to power public and freight transport (Roguslka et al. 2018). Concerning the aviation sector, biojet fuel has been recently introduced and it may be blended with jet fuel till a content of 30%. Ethanol is economically produced from biomass containing starch or sugar as saccharose, whilst biodiesel from lipids (1st generation biofuels), and the actual sources are starchy or sugar-rich plants (Drapcho et al. 2008) or seeds from oleaginous plants (Zhou et al. 2020). Nevertheless, this approach has caused some concerns about the food-water-energy nexus and the following biofuel generations try to overcome the limitations and challenges. 2nd generation biofuels deal with the valorization of cellulose, hemicellulose and lignin into liquid biofuels, which are biomass residues with the advantage that they do no compete with food production. Nevertheless, the associated technologies still require agricultural or forest land for biomass supply and some of them are not still economically competitive. Therefore, microbial based biofuels research and technology developments emerged as an attractive solution to the need of intensive agricultural and forest activities and an effort to reduce dependence on food and feed related sources as well as environmental impacts.

A systematic approach to determine and evolve the scientific and technical issues is through the techno-economic analysis (TEA), where the technical data are analyzed, customary in an engineering process scheme and the mass and energy balances are done according to the involved chemical reactions, kinetic and thermodynamic data. The technical performance achieved by this approach helps to guide further science and technology development, but also, to get an understanding of how the latter affects the economic and financial dimensions in order to move forward into the market. The present work deals with the techno-economic analysis of microbial green fuels and its comparison with plant-based biofuels. We considered the definitions of biofuel generations, their advantages and challenges, as well as the technology, economic and financial issues. Finally, the TEA of production of farnesene, microbial and yeast oils and plant-based oils are discussed.

2. Microbial and Plant-based Biofuels

Biofuels or its precursors may be obtained from heterotrophic and autotrophic microorganisms like bacteria, cyanobacteria, fungus, yeast and microalgae, but also from plant-based molecules as polysaccharides, simple sugars, lipids, terpene and polyphenols known as lignin (Bhatia et al. 2018, Shariat Panahi et al. 2019, Patel et al. 2020). Autotrophic microorganisms can use CO_2 as carbon source and solar light (photoautotrophs), CO_2 and electron donors (chemolithoautotrophs), and CO_2, organic substrate and electron donors (mixotrophs) to produce biomass, oils and polysaccharides that may be transformed into fuels and chemicals (Claassens et al. 2016). Meanwhile, heterotrophic microorganisms use organic substances to transform them to biomass, energy and secondary metabolites (Spagnuolo et al. 2019). In the case of plant-based biofuels, agricultural plants have to be sown, harvested and processed into major macromolecules like oils, polysaccharides, sugars and lignin, which need further treatment through conventional or advanced technologies like chemical, physicochemical, thermochemical, microbial, enzymatic or hybrid approaches to biofuels (Naz 2020).

Such microbial and plant-based biofuels or precursors have been widely classified as 1st, 2nd, 3rd and 4th generation based on biomass source (1st and 2nd) and microbial source or advanced processes (3d and 4th; Table 1; Abdullah et al. 2019, Alalwan et al. 2019, Pandeeti et al. 2019, Magda et al. 2020, Walls and Ríos-Solis 2020). First generation biofuels refer to the use of human food crops as sugarcane, sugar beet, starch from several sources; oleaginous plants as soy, sunflower, *Camelina sativa* L., *Jatropha curcas*, castor bean plant (*Ricinus communis* L.), firewood, among others. Conventional technologies as saccharification, fermentation with native yeast as *Saccharomyces cerevisiae* and distillation are used to produce ethanol; transesterification and esterification of oils and fats are employed to obtain biodiesel and glycerol; while residential heating and cooking as well as industrial power and steam generation are obtained from traditional burning of firewood and biomass residues (Amezcua-Allieri et al. 2019, Martínez-Hernández 2019a).

Second generation fuels refer to lignocellulosic materials, non-food crops, and dedicated crops where polysaccharides like cellulose and hemicelluloses are hydrolized, fermented with native but versatile microorganisms into ethanol (Sadhukhan et al. 2019, Amezcua-Allieri et al. 2017); or processed through hydrodeoxigenation (HDO) and isomerization into green diesel, biojet fuel,

Table 1. Common classification of biofuels.
(List non-exclusive; Pandeeti et al. 2019, Magda et al. 2020).

Generation	Biomass Source	Technology
First	• Sugar, starch.	• Starch enzymatic hydrolysis, sugar fermentation to ethanol.
	• Oil from oleaginous plants, animal fat and waste cooking oil.	• Transesterification of oil and fat into biodiesel.
	• Firewood; forest, agricultural and agroindustrial residues.	• Combustion for power and steam.
Second	• Lignocellulosic materials from agricultural, forestry, agro-industrial industries.	• Cellulosic ethanol.
	• Non-food crops.	• Hydrogenation of oils and fats to green diesel.
	• Dedicated crops as switchgrass, alfalfa, Miscanthus, among others (Yadav et al. 2019).	• Hydrotreatment and isomerization of vegetable oils and animal fats to biojet fuel.
		• Pyrolysis, gasification, hydrothermal processing of biomass to bio-oil, syngas, and biochar and their up-grading for biofuels, power and steam generation and chemicals.
	• Organic wastes and manure.	• Anaerobic digestion to biogas and further purification to biomethane.
Third	• Native microorganisms like bacteria, microalgae, oleaginous yeast and fungus (Rodrigues-Reis et al. 2020).	• Extraction of oil, polysaccharides and processing with conventional approaches.
		• Production of hydrogen.
Fourth	• Genetically modified organisms (GMO) as microalgae, yeast and bacteria (Abdullah et al. 2019).	• Separation of dedicated molecules (oil, polysaccharides, sugars).
		• Processing through conventional or advanced technologies.

green propane and naphtha (Valencia et al. 2018); or processed by thermochemical technologies like pyrolysis, gasification and hydrothermolysis into bio-oil, syngas, biochar that need certain up-grading in order to become biofuels, power and steam generation or bio-based chemicals (Table 1; Amezcua-Allieri and Aburto 2018). Organic residues and manure may also be considered here and through anaerobic digestion to produce biogas that requires further purification to separate the biomethane.

The use of native and unmodified microorganisms as oleaginous yeast and phototropic microalgae correspond to the known third biofuel generation. These microorganisms produce oils through heterotrophic fermentation or autotrophic CO_2 assimilation, respectively. Such oils can be further converted by conventional or advanced processing into biofuels. Microalgae are very versatile since they can uptake CO_2 and also utilize inorganic and organic compounds to store starch, lipid granules and proteins that may be converted into ethanol, oil-derived fuels like biodiesel, biogas, syngas and power and steam (Bhushan et al. 2020, Anto et al. 2020). Oleaginous yeasts have been widely studied because of their capability to use simple sugars like glucose, but mostly for their avidity of pentose like xylose and arabinose, monoaromatics from lignin hydrolysates as well as glycerol as carbon source (Sitepu et al. 2014, Spagnuolo et al. 2019).

Since native microorganisms have several constrains concerning substrate utilization, inhibitor sensibility, low growth rate, low titer grade of interesting metabolites such as starch and lipids, among others, latest research is focused on the development of genetically modified organisms (GMO), known as fourth generation (Shokravi et al. 2019, Abdullah et al. 2019). One of the latest and important contribution refers to the microbial production of isoprenoids comprising genetically modified organisms like *E. coli*, *S. cerevisiae* and *Y. lipolytica* (Walls and Rios-Solis 2020), and also, terpene synthesis through bacteria of genus *Clostridium* and *Moorella* (Koepke 2021).

Today, the more used biofuels are ethanol, biodiesel and biogas; all of them form 1st and 2nd generation technologies (IEA 2019a, WBA 2019). Ethanol is the most oxygenated agent blended with gasolines in spark ignition engines and it is obtained from fermentation of direct saccharose extracted from sugarcane or sugar beet, or from hydrolyzed starch from cereals, corn or tuber sources. USA is the first ethanol producer followed by Brazil with 60 and 32 billion liters in 2018, respectively (IEA 2019a). Diesel compression-ignition engines require a fuel with a cetane number larger than 40 (ASTM International 2016); hence, the transesterification product from fatty acid triacyl glycerides from plants and animals resulted in first generation biodiesel or its hydrotreatment produces 2nd generation hydrotreated vegetable oil (HVO). Current production of biodiesel and HVO reaches 14.9, 7.8 and 5.2 billion liters in the European Union, USA and Brazil, respectively (IEA 2019a). In the case of biogas production, the European Union leads with ca. 18 Mtoe, followed by China and USA with ca. 7 and 4 Mtoe, respectively (IEA 2020).

All these 1st generation biofuels require a biomass feedstock competing with food and are processed through conventional technologies. Ethanol and biogas production need microorganisms, mainly sugar fermentative or anaerobic heterotrophs ones, while biodiesel and glycerol are produced by a catalytically chemical conversion with plant oil, animal fat or waste cooking oils with methanol. Microbial biofuel production in 1st and 2nd generations require the feed of biomass components as sugars, starch, polysaccharides, lipids, proteins, and lignin. Therefore, such 1st and 2nd generation technologies require agricultural, forestry lands and their management, food sources and cattle feed crops, organic matter and water (WBA 2019). They may be considered as an initial step in energy transition to more sustainable production ways of biofuels. Pro and cons of such approaches vs. 3rd and 4th latest developments are summarized in Table 2.

Microbial biofuel production presents various advantages when compared to plant-based ones (Table 2; Nigam and Singh 2011, Abdullah et al. 2019, Shokravi et al. 2019); they may require small

Table 2. Advantages and challenges of microbial and plant-based biofuels. Technology readiness level (TRL; NASA 2012). (Nigam and Singh 2011, Abdullah et al. 2019, Shokravi et al. 2019).

Generation Technologies	Advantages	Challenges
First and second	• Energy security • Energy access to people • Locally distributed • Rural development • High biomass availability, especially lignocellulosic • Creation of new value chains • Competitive production costs in some economies • TRL = 9, i.e., in the market • Commercial production • Existing policies and incentives • Reduce GHGs emissions • Mature technologies on 1st generation biofuels • Co-products for other industries	• High land use • Land use change • Deforestation concerns • Fertilizer and pesticide use • Mono cultures • Highly intensive agriculture • Microbial dependence on biomass molecules • Food-water-energy nexus • Several steps in production and refining • Techno-economic risk with 2nd generation • Intensive capital investment and high operation costs in 2nd generation
Third and fourth	• Potential for reduced land use • Valorization of CO_2 emissions • In some cases, high target molecule content (starch, oil, protein) • Microbial accelerated growth rate compared to plants • Lower water footprint • Reduce GHGs emissions • CO_2 sequestering and conversion • Optimized biofuel production and separation • Minimum or non-dependence of food biomass sources	• Environmental, ecological and health risks for the use of GMO • Stability and diminution of interspecies gene transfer • Need for high concentration and flow of CO_2 current • Contamination risk of cultures • Techno-economic feasibility • TRL < 3, i.e., proven at lab scale • Intensive capital investment and high operation costs • Valorization of co-products • Poor policies and regulations • No commercial production

area for industrial facilities, hexoses fermentation for ethanol and anaerobic digestion for biogas and biomethane are mature technologies and already implemented worldwide in the market. Such biofuels have an established policy framework, an accelerated growth, low water footprint may be achieved through water recycling and produce valuable coproducts as dried distiller's grains, bagasse, microbial biomass, protein meal, fertilizers from vinasses, activated carbon and bio-based chemicals, among others (Vivekanandhan et al. 2013, Rosales-Calderon 2019). Nevertheless, these biofuels still employ organic carbon sources and then large land areas for their production, which is the same challenge faced by plant-based biofuels. The techno-economic analysis of biomass sowing and harvesting must also be considered but this subject is out of the scope of present work. Also, they require heterotrophic microorganisms such as *S. cerevisae*, *Clostridium* spp. and other methanogenic and sulfate-reducing microorganisms present in manure (Kushkevych et al. 2017).

An enormous breakthrough in biofuel production might be the use of 3rd and 4th generation biofuels where microorganisms use CO_2 emissions and waste streams. The dependence on organic based carbon sources from agriculture, agro-industries, and forestry residues might be diminished or avoided. Such approach uses native (3rd g) and genetically modified organisms (4th g) with a preferred mixotrophic metabolism over an autotrophic and heterotrophic one due to an enhanced growth and accumulation of macromolecules like lipids and starch (Shokravi et al. 2019), as well as a reduced food-energy-water conflict because of the absence of food sources and recyclability of water (Abdullah et al. 2019). The main concern here is the ecological and health concerns due to gene transfer risk between species and release of such GMOs to the environment that have been reviewed elsewhere (Henley et al. 2013, Szyika et al. 2017, Hilbeck et al. 2020), and that limit commercial application nowadays. In order to assess the feasibility of microbial and plan-based fuels, techno-economic analysis is usually undertaken and will be further discussed.

3. Techno-economic Analysis of Microbial and Plant-based Biofuels

During the different phases of fundamental and applied research, techno-economic analysis might give insights to address the scientific and technical questions (1<TRL<5) for early-medium developments of any conversion process including biofuels as well as economic and financial parameters to drive decisions for near-to-market technology developments (5<TRL<9; NREL 2011, EIA 2015, NASA 2012, Kargbo et al. 2021). In order to address such studies, we will present the fundamentals of techno-economic analysis and application to literature-based cases representing microbial vs. plant-based biofuels.

Techno-economic analysis may be explained in terms of (1) the technical performance of an identified prototype or technological development where all critical scientific and technical data that define energy and mass balances of a specific conversion pathway using process modelling are evaluated; and (2) the economic and financial behavior of such conversion pathway, including operational and capital costs, cash flow as well as financial parameters as internal rate of return (IRR), payback time (PBt), net present value (NPV), among others. An interesting set of issues to better approach the techno-economic analysis of a technology development with 1<TRL<7 is shown in Table 3. Such TRL 7 established that a prototype has been demonstrated under a real environment, whilst a TRL of 8 and 9 requires further robust analysis as Front-End Loading that can be found elsewhere (Saputelli et al. 2013, Midyette 2020, Newman et al. 2020). Moreover, the

Table 3. Set of technology, economic and financial issues to be addressed through techno-economic analysis (Adapted from IEA 2015, Aburto and Martínez-Hernández 2021).
Inside battery limits (ISBL), Outside battery limits (OSBL).

Technology Issues	Economic Issues	Financial Issues
• Definition of the product, service, technology (ISBL, OSBL)	• Costs of components (operational and capital) and their uncertainty	• Minimum acceptable rate of return
• Performance of the technology	• Cost of identified risks and externalities (social, health, environment, climate change)	• Market conditions Price
• Risks of the technology	• Estimated total cost of the technology configuration	Demand Economic Growth Competition Regulations Policy commitment
• Existing and near future competitive technologies	• Method of financing (debt, equity, financial sources)	• Sensitivity of rate of return
• Comparison of risks between own and competitive technologies	• Minimum unit production cost	• Payback time (PBt)
• Technology advancement in terms of costs, performance and risks	• Minimum selling price	• Gross Netback
• Identification of an optimized technology configuration	• Dependence of technology cost to external costs (energy, taxes, etc.)	• Minimum acceptable plant capacity
• Technology improvements to base case	• Fixed and variable cost	• Sensitivity of cost and performance to market and regulations conditions
	• Product cost and its price, production rate	

techno-economic analysis of biomass supply chain must also be considered but this subject is out of the scope of present work (Mungodla et al. 2019, Lo et al. 2021).

With respect to technological issues (Table 3), the techno-economic analysis must give insights concerning the conceptual engineering, including scheme of all main unit operations and their input and output mass flows, chemical reactions, engineering calculation involving equipment, material properties and operation conditions in every unit operation and energy requirements as steam and electricity. The use of process simulators to design conceptual engineering and mass-balance calculations is common today, such as Superpro Designer® from Intelligen Inc., Batch Plus® from Aspen Technology, Inc., (Petrides et al. 1989, Shanklin et al. 2001, Petrides and Siletti 2004), IMP Bio2Energy® from the Mexican Petroleum Institute (Cluster BCS 2021), BEFS Cogeneration tool developed (FAO 2021), among others. In the case of biotechnological processes, they are still many lacks on information concerning physicochemical properties of biomolecules; carbon-based simple formulae of microorganisms; kinetics, thermodynamics and stoichiometry of biotechnological and chemical reactions; as well as engineering data of bioprocesses, which must be approached through experimental data validation.

Economic and financial analysis must consider the price of all raw materials, operational and capital costs, and more importantly the minimum production cost and selling price as well as capacity plant. Sensitivity analysis of financial parameters is a main activity where minimum acceptable plant capacity and internal rate of return may be identified.

3.1 Techno-economic Analysis of Farnesene, Microbial and Yeast Oil, and Plant-based Oils

Microbial oil or intermediate compound production needs the implementation of a series of bioreactors for microbial seed production with focus on biomass generation, the fermentation step where metabolites as oil and intermediates compounds are produced and stored into the cell, and in some cases released to the medium. This has significant effect on processing, since the first requires the cell rupture and oil or intermediate compound separation, whilst the latter needs just oil or intermediate compound separation through centrifugation of solvent extraction process and its distillation. A resumed scheme process for the production of microbial oil, monoterpenes or sesquiterpenes (intermediate compounds) is shown in Fig. 2.

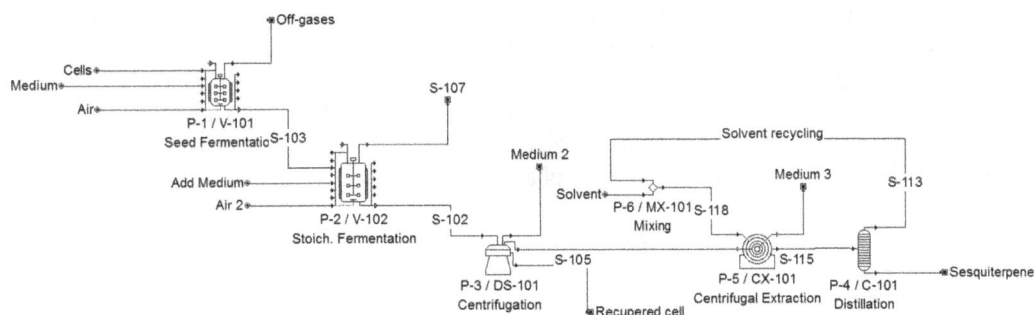

Fig. 2. Microbial oil, monoterpenes or sesquiterpenes production process (Pedraza-de la Cuesta et al. 2018).

Oleaginous microorganisms synthesize oil through a multienzymatic complex that can be found elsewhere (Athenaki et al. 2018) and summarized as:

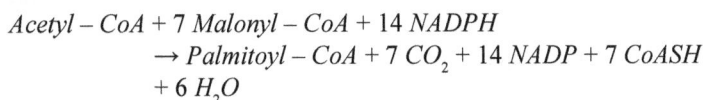

$$Acetyl - CoA + 7\ Malonyl - CoA + 14\ NADPH$$
$$\rightarrow Palmitoyl - CoA + 7\ CO_2 + 14\ NADP + 7\ CoASH$$
$$+ 6\ H_2O$$

Such summarized reaction or a more robust and complex series of reactions may be used to simulate oil production, but a simpler way is to refer to an oil microbial content as seen below in Table 4.

Isoprenoid derivatives have attracted attention as fuel replacements due to their physicochemical properties that allow them, after hydrogenation, to be used into fossil diesel blends. Monoterpenes (C10) and sesquiterpenes (C15) are produced via the isopentenyl diphosphate (IPP) through the mevalonate (MVA) or methylerythritol (MEP) pathways in plants or in native and genetically modified microorganisms (Niu et al. 2017, Jiang et al. 2018, Walls and Rios-Solis 2020) as shown:

MVA pathway to IPP:

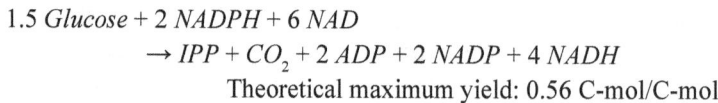

$$1.5 \; Glucose + 2 \; NADPH + 6 \; NAD$$
$$\rightarrow IPP + CO_2 + 2 \; ADP + 2 \; NADP + 4 \; NADH$$

Theoretical maximum yield: 0.56 C-mol/C-mol

MEP pathway to IPP:

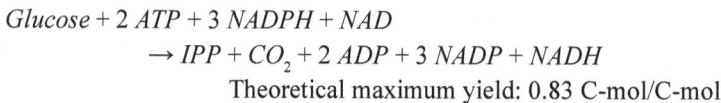

$$Glucose + 2 \; ATP + 3 \; NADPH + NAD$$
$$\rightarrow IPP + CO_2 + 2 \; ADP + 3 \; NADP + NADH$$

Theoretical maximum yield: 0.83 C-mol/C-mol

IPP yield from glucose is then higher in the MEP pathway when compared to the MVA one but more reducing agent (NADPH) is required (Niu et al. 2017). Another stoichiometric reaction for production of farnesene was studied by Gama-Ferreira and Petrides (2020). These two pathways are present in plants but their economic production is highly limited by their slow growth and low titer of isoprenoids. Then, the development of GMOs has been studied during last years to increment such titer (Wu et al. 2021) as well as their simulation process (Pedraza-de la Cuesta et al. 2018, Gama-Ferreira and Petrides 2020). Such stoichiometric reactions are important to proceed to process simulation but a more robust study might consider the kinetic and thermodynamic parameters that must be still developed.

On the other side, plants produce glucose through photosynthesis which, after a very multi-enzymatic complex, transformed and accumulated into their seeds as starch, cellulose and lipids (Baud 2018). Then, plant-based oils must be extracted from oleaginous plants as palm oil, soybean, sunflower, canola, camelina, *Jatropha curcas*, castor, among others (O'Brien 1998a, Beaudoin et al. 2014, Alherbawi et al. 2021), or waste cooking oils may be valorized, but they must be first filtered and refined (Sahar et al. 2018). Then, a common extraction process considers the reception of oleaginous plant or fruits into the facility, threshing for separation of oily nuts from bunches or shells, separation of oil by pressing and/or solvent extraction, clarification and refining to obtain exclusively the fatty acid triacyl glycerides phase (Fig. 3; based on Garcia-Nunez et al. 2015),

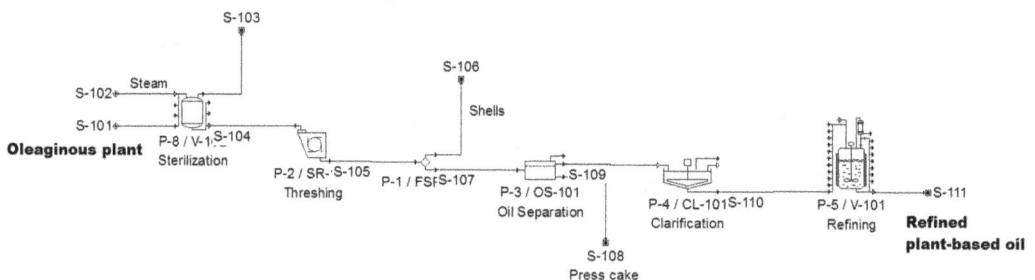

Fig. 3. Plant-based oil extraction process (Based on Garcia-Nunez et al. 2015).

without any other molecule like polysaccharides, volatiles, pigments, waxes and lecithin (Cheng 2017). Common refining procedures involve bleaching, degumming, winterization, dewaxing and deodorization as referred elsewhere (O'Brien 2018b). Special attention must be done to the salt and solid content of waste cooking oils as well as their free fatty acid and bromine index indicating the degradation of such oils.

Once the microbial, yeast or plant-based oils or intermediate compounds like farnesene are extracted from biomass, they need to be processed in order to obtain a tailored fuel capable to be used in compression-ignition engines or turbines. One way to produce green diesel (paraffins) or biojet fuel (mixtures of paraffins and iso-paraffins) is through the hydrotreatment of ester fatty acids (HEFA; Fig. 5), where the oily feedstock is heated into a first reactor with hydrogen and a catalyst in order to undertake the following reactions (Martínez-Hernández et al. 2019b):

Depropanation:

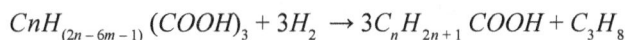

$$CnH_{(2n-6m-1)}(COOH)_3 + 3H_2 \rightarrow 3C_nH_{2n+1}COOH + C_3H_8$$

Hydrodeoxygenation reaction (DO):

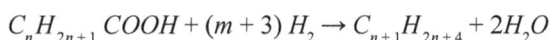

$$C_nH_{2n+1}COOH + (m+3)H_2 \rightarrow C_{n+1}H_{2n+4} + 2H_2O$$

Decarboxylation reaction (DCx):

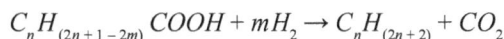

$$C_nH_{(2n+1-2m)}COOH + mH_2 \rightarrow C_nH_{(2n+2)} + CO_2$$

Decarbonylation reaction (DC):

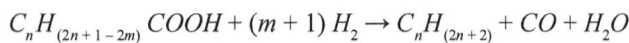

$$C_nH_{(2n+1-2m)}COOH + (m+1)H_2 \rightarrow C_nH_{(2n+2)} + CO + H_2O$$

where n is the number of carbon atoms and m is the number of insaturations in the aliphatic chains of triacyl glycerides. These reactions occur simultaneously in order to produce pair and odd paraffins which are often called as green diesel as well as green propane. Regarding carbon economy of such reactions, it comes out that depropanation and hydrodeoxygenation reactions must prevail over decarboxylation and decarbonylation ones, where CO_2 and CO are produced and affect the fuel's yield. The processing advantage of farnesene production is that it only requires a hydrogenation section of farnesene's insaturations to produce farnesane (2,6,10-trimethyldodecane; Fig. 4), but there are neither co-products nor emissions as CO_2 and CO. Such farnesane can be used in blends with jet fuel (ASTM 2020):

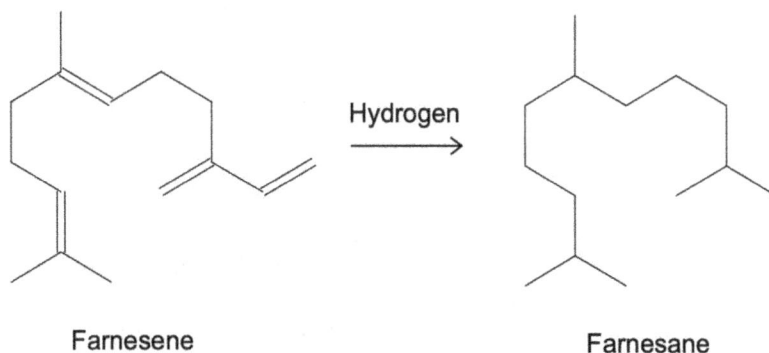

Farnesene   Farnesane

Fig. 4. Hydrogenation of farnesene to farnesane.

Now, such paraffins blend must undergo further reactions in a 2nd reactor to obtain the blend of paraffins and isoparaffins and is called as biojet fuel:

Hydrocracking of alkanes:

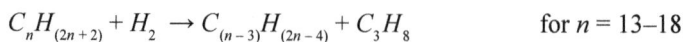

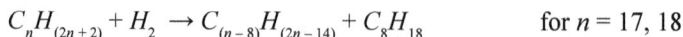

$$C_nH_{(2n+2)} + H_2 \rightarrow C_{(n-3)}H_{(2n-4)} + C_3H_8 \qquad \text{for } n = 13\text{--}18$$

$$C_nH_{(2n+2)} + H_2 \rightarrow C_{(n-8)}H_{(2n-14)} + C_8H_{18} \qquad \text{for } n = 17, 18$$

Isomerization:

$$C_nH_{(2n+2)} \rightarrow iso - C_nH_{(2n+2)} \qquad \text{for } n = 6\text{--}18$$

Further purification through distillation gives the main product (green diesel or biojet fuel) and co-products (propane, naphta; Fig. 5).

Technical assessment of biofuels' production from microbial and yeast oils, intermediates as monoterpenes and sesquiterpenes or plant-based oils requires to define the issues in Table 3, which are case specific and presented in Table 4. Process simulation, stoichiometric reactions or molecule objective yield must be first defined. Then, all streams as feedstocks, raw materials, products as well as energy requirements must be added in order to calculate the corresponding mass and energy balances to define the facility production capacity and yields of products and possible co-products. Most advanced developments (5<TRL<9) are based on the use of organic matter coming from agricultural, agro-industrial, and forest activities. Nevertheless, the costs associated with the production, processing and transport (Lo et al. 2021) of such feedstocks must be considered. Definitely, microbial, yeast and plant-based oils have the higher titer when compared to the monoterpene and sesquiterpene production (Table 4). This is a serious issue with respect to the technical feasibility to produce such intermediate compounds for biofuels, since fuel market is a huge volume and low-cost industry, but their use in high value fine chemicals is a more promising application of such technology in the near term. Nevertheless, intermediate compounds produced by native and genetically modified microorganisms are very interesting since, thanks to their autotrophic or mixotrophic metabolism, there is a less demand for organic matter and gives value to waste streams and emissions like CO_2 and methane. Indeed, methanotrophic metabolism allows native and GMO to use CH_4 as carbon source and transform it into lipids which can be converted into biodiesel (Fei et al. 2018). This could allow to disrupt or minimize the production of fuels from agricultural, agro-industrial and forest activities, but also to contribute to the mitigation of greenhouse gases (GHG) as well as climate change. Plant-based oils are currently the more feasible source for biofuel production but attention is needed with respect to the food-water-energy nexus, the land requirements, use of fertilizers, pesticides and herbicides that have an important environmental impact (Amezcua-Allieri et al. 2019, Martínez-Hernández et al. 2019a). Even if microbial oil may be obtained at high yield and comparable to plant-based oils, there are some issues related to scaling-up, culture contamination, inhibition and heterotrophic metabolism that requires the use of organic matter, which may increase associated costs.

With respect to economic analysis, capital and operational costs, energy density (MJ/kg, MJ/L, US\$/MJ; Mazloomi and Gomes 2012), unit production cost (UPC) as well as unit production revenue (UPR) must be identified with respect to plant capacity and specific product and co-products (Fig. 6). This information allows the evaluation of the financial parameters as CAPEX, OPEX, IRR, payback time (PBt; Fig. 7). We have selected five case studies concerning the production of microbial b-farnesene by a GMO (Gama-Ferreira and Petrides 2020), a microbial oil (Karamerou et al. 2021), a yeast oil (Jena et al. 2015), a palm oil converted into biojet fuel

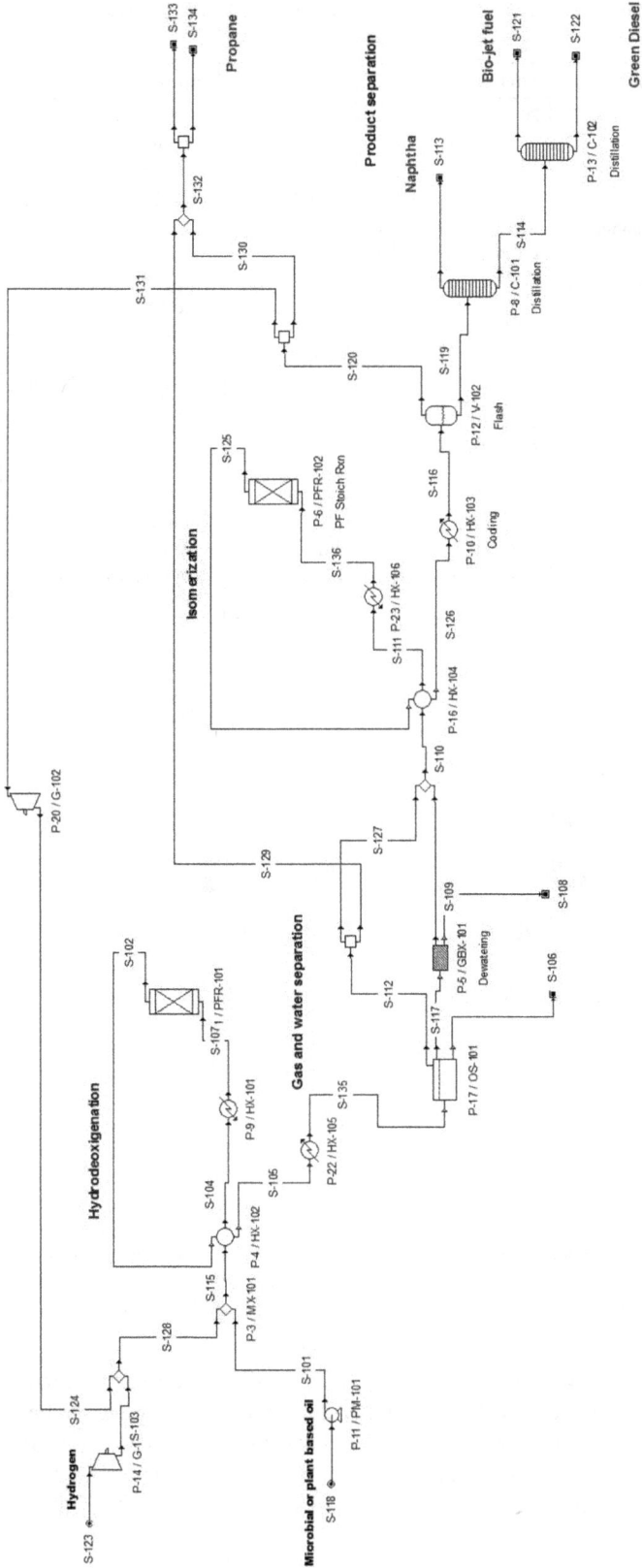

Fig. 5. Scheme for the production of advanced biofuels from microbial, plant-based oils or intermediates (Martínez-Hernández et al. 2019b).

Table 4. Technical characteristics of microbial, monoterpene and sesquiterpene and plant-based oil ([1]Athenaki et al. 2018, Patel et al. 2020, [2]Rolf et al. 2020, [3]Meadows et al. 2016, [4]Zhao et al. 2020, [5]Zhou et al. 2020).

Technology	Feedstock	Product	Titer	Risks	Improvements
Microbial oil production	Sugars, biomass hydrolysates, CO_2, CH_4, glycerol, wastes	Microbial oils	20–89%[1]	• Low TRL • Few pilot plants in operation • Contamination	• Nutraceutical production • High oil titer
Monoterpenes and sesquiterpenes	• Pure glycerol • Unrefined cane syrup • Waste cooking oil/CO_2	• Limonene • Farnesene • Bisabolane	• 3,600 mg/L[2] • 130,000 mg/L[3] • 973/22 mgL[4]	• Low titer • Low growth • Need for sugars • Low TRL • GMO management • Contamination	• Use of wastes as feedstocks • Improve mixotropic metabolism • Improve production scale • Lower dependence on organic matter
Plant-based oil	• Soybean • Sunflower • Corn • Oil palm • Coconut • Pine • Olive	• Plant-based oil[5]	• 18–24% • 46–50% • 4.5–4.8% • 50–55% • 66–74% • 58–69% • 31v56%	• Food-water-energy nexus • High land use • Need for fertilizers, herbicides and pesticides	• Production of biofuels, bio-based chemicals and electric and steam power • Increment of commercial production and deployment

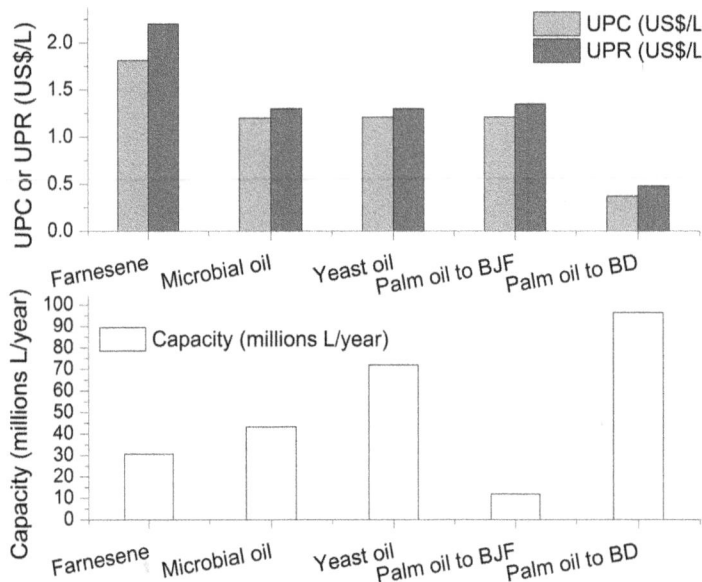

Fig. 6. Plant capacity, unit production cost (UPC) and unit production revenue (UPR) of techno-economic studies for the production of farnese, microbial oil, yeast oil, palm oil to biojet fuel (BJF) and palm oil to biodiesel (BD; Jena et al. 2015, Martínez-Hernández et al. 2019b, Gama-Ferreira and Petrides 2020, van Dyk and Saddler 2021, Karamerou et al. 2021).

(BJF; Martínez-Hernández et al. 2019b, van Dyk and Saddler 2021) and palm oil transformed in biodiesel (BD; Kermani et al. 2017). We observe that conventional conversion of palm oil to biodiesel permits a high plant capacity of 96 million liters per year with the lowest values for UPC and UPR. Such good economics are due to the use of a mature technology to produce biodiesel, high oil titer, an intensive palm oil industry and a large processing capacity with several co-products (red palm

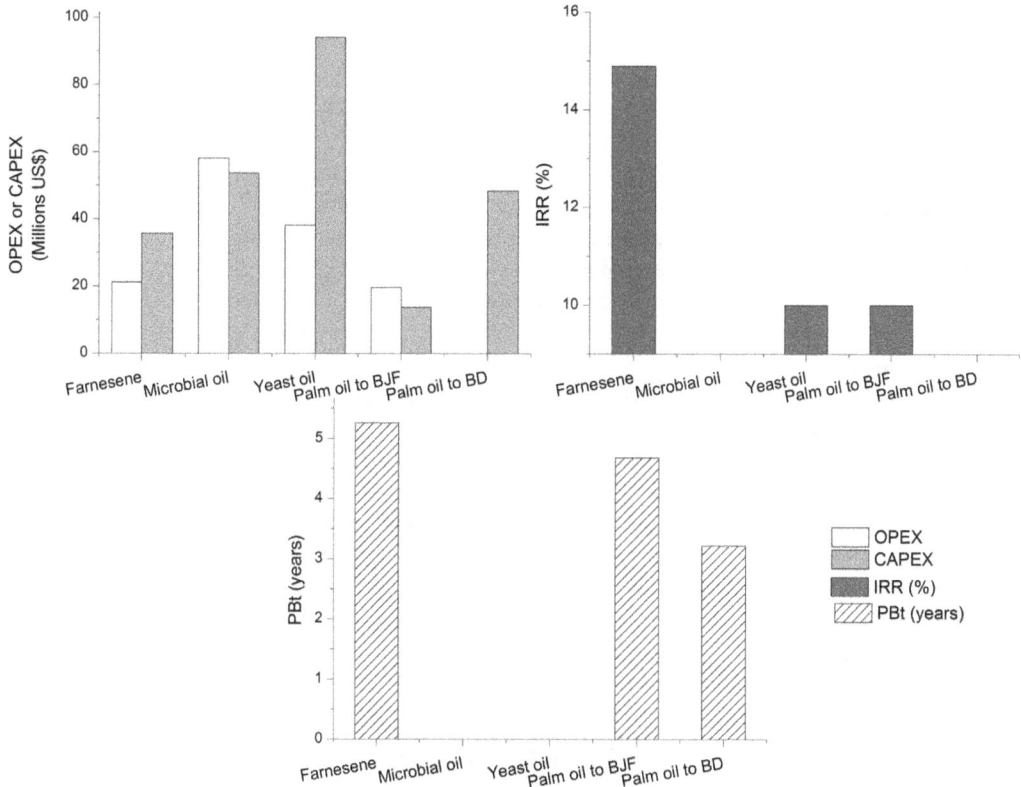

Fig. 7. Financial parameters estimated from techno-economic studies for the production of farnesene, microbial oil, yeast oil, palm oil to biojet fuel (BJF) and palm oil to biodiesel (BD; Jena et al. 2015, Martínez-Hernández et al. 2019b, Gama-Ferreira and Petrides 2020, van Dyk and Saddler 2021, Karamerou et al. 2021).

oil, refined palm oil, lecithin, empty fruit bunches for energy or ethanol) that enhanced financial parameters including the lowest PBt among these studies (Fig. 7). Nevertheless, special attention refers to sustainability issues associated with such intensive palm oil production (Jamaludin et al. 2018) that makes it a high-risk investment.

The latter also concerns the production of biojet fuel from palm oil, but since it needs an alternative technology based on hydrotreatment and isomerization of triacyl glycerides to convert them in paraffins and iso-paraffins, respectively, they have relative higher OPEX and CAPEX with a smaller reported plant capacity when compared to biodiesel production. Here, palm oil processing requires hydrogen usually obtained from the reforming of natural gas, coal or liquified petroleum gas (LPG) with an associated cost of 1–2 US$/kg H_2, which does not support the sustainability of such biofuel since 10-ton CO_2/ton H_2 are produced, while the production of renewable hydrogen from electrolysis rounds about 8–10 US$/kg H_2 and its CO_2 intensity depends on electricity source (IEA 2019b).

Plant capacities for the production of farnesene, microbial and yeast oils vary from 30 to 70 million liters per year and show the higher values for UPC and UPR (Fig. 6). This is attributable to the low TRL (less than 5) and significant need for further research and technology development that may help to reduce economic and financial uncertainties (Fig. 7). Furthermore, these oils or intermediates still need to be processed in order to obtain an actual usable fuel in current engines or turbines, or "drop-in" fuel, which certainly will cause higher costs (OPEX, CAPEX and UPC), and making these approaches less competitive with respect to fossil fuels. Nevertheless, these developments must diminish land area for crops cultivation that allows the grow of microorganisms that uses organic carbon. More importantly, the fixation of CO_2 by microrganisms may facilitate the decoupling of energy production from food and feed production.

4. Conclusion

Techno-economic analysis is a useful tool for the assessment of scientific, technical, economic and financial status of a specific technology development. It might raise the corresponding hypothesis and questions that drive the science and technology of oleaginous microorganisms and plants for the production of sustainable value chains as a whole, from (1) the production and valorization of biomass, organic carbon, CO_2 emissions and waste streams, to (2) the green conversion of oils and intermediates that allow to identify limitations and opportunities of physicochemical and biotechnology processing to obtain biofuels, electric and heat power, bio-based chemicals and materials; and (3) the supply logistics of raw materials, intermediates, products and co-products to satisfy the needs of energy, food, feed, water, health of population with respect to environment as well as climate change mitigation and adaptation challenges. There is a need for physicochemical, stoichiometric, kinetic and thermodynamic data of biomass, microbial and plant-based biofuels in order to obtain robust process simulations and TEA results. Techno-economic analysis of production of microbial and plants oils must be complemented with life cycle and sustainability analysis based on population needs. 1st and 2nd generation biofuels may still be enhanced to achieve better performance, and 3rd and 4th generation biofuels still need a deeper research and technology development in order to become technically and economically accessible to the population.

Acknowledgements

The Author acknowledges the financial support from British Council, Newton Fund Impact Scheme through project NFIS-540821111 as well as Instituto Mexicano del Petróleo co-financing through project Y.62001 "A decision support platform for bioenergy technology deployment and policy making in Mexico".

References

Abdullah, B., Muhammad, S.A.F.S., Shokravi, Z., Ismail, S., Kassim, K.A., Mahmood, A.N. and Aziz, M.M.A. 2019. Fourth generation biofuel: A review on risks and mitigation strategies. Renew. Sust. Energ. Rev. 107: 37–50.

Aburto, J. and Martínez-Hernández, E. 2021. Is sugarcane a convenient feedstock to provide ethanol to oxygenate gasolines in mexico? A process simulation and techno-economic-based analysis. Front. Energy Res. 8: 612647.

Alalwan, H.A, Alminshid, A.H. and Aljaafari, H.A.S. 2019. Promising evolution of biofuel generations. Subject review. Renew. Energ. Focus. 28: 127–139.

Alherbawi, M., McKay, G., Mackey, H.R. and Al-Ansari, T. 2021. Jatropha curcas for jet biofuel production: Current status and future prospects. Renew. Sust. Energ. Rev. 135: 110396.

Amezcua-Allieri, M.A., Martinez-Hernandez, E., Anaya, O., Magdaleno-Molina, M., Melgarejo-Flores, L.A., Palmerín-Ruiz, M.E., Zermeño-Eguia-Lis, J.A., Rosa-Molina, A., Enríquez-Poy, M. and Aburto, J. 2019. Techno-economic analysis and life cycle assessment for energy generation from sugarcane bagasse: Case study for a sugar mill in Mexico. Food and Bioproduct Processing 118: 281–292.

Amezcua-Allieri, M.A. and Aburto, J. 2018. Chapter 6. Conversion of lignin to heat and power, chemicals or fuels into the transition energy strategy. pp. 145–160. *In*: Poletto, M. (ed.). Lignin: Trends and Applications. IntechOpen, London, United Kingdom.

Amezcua-Allieri, M.A., Sánchez-Durán, T. and Aburto, J. 2017. Study of chemical and enzymatic hydrolysis of cellulosic material to obtain fermentable sugars. Journal of Chemistry 2017: 5680105.

Anto, S., Mukherjee, S.S., Muthappa, R., Mathimani, T., Deviram, G., Kumar, S.S., Verma, T.N. and Pugazhendhi, A. 2020. Algae as green energy reserve: Technological outlook on biofuel Production. Chemosphere 242: 125079.

ASTM International (ASTM). 2020. ASTM D7566-20c: Standard Specification for Aviation Turbine Fuel Containing Synthesized Hydrocarbons.

ASTM International. 2016. D975-16: Standard Specification for Diesel Fuel Oils. Consulted Online on April 2014.

Athenaki, M., Gardeli, C., Diamantopoulou, P., Tchakouteu, S.S., Sarris, D., Philippoussis, A. and Papanikolaou, S. 2018. Lipids from yeasts and fungi: physiology, production and analytical considerations. Journal of Applied Microbiology 124: 336–367.

Baud, S. 2018. Seed as oil factories. Plant Reproduction 312: 213–235.

Beaudoin, F., Sayanova, O., Haslam, R.P., Bancroft, I. and Napie, J.A. 2014. Oleaginous crops as integrated production platforms for food, feed, fuel and renewable industrial feedstock. OCL 21: D606.

Bhatia, S.K., Joo, H.S. and Yang, Y.H. 2018. Biowaste-to-bioenergy using biological methods—A mini-review. Energy Conversion and Management 177: 640–660.

Bhusha, S., Kalrac, A., Simsek, H., Kumard, G. and Prajapatia, S.K. 2020. Current trends and prospects in microalgae-based bioenergy production. J. Environ. Chem. Engin. 8: 104025.

Cheng, M.H. 2017. Sustainability analysis of soybean refinery: soybean oil extraction process. Graduate Theses and Dissertations. 15277. https://lib.dr.iastate.edu/etd/15277.

Claassens, N.J., Sousa, D.Z., Martins dos Santos, V.A.P., de Vos, W.M. and van der Oost, J. 2016. Harnessing the power of microbial autotrophy. Nature Reviews Microbiology 14: 692–706.

Cluster, B.C.S. 2021. Consulted on June 2021: https://www.wegp.unam.mx/Bio2Energy.

Drapcho, C.E., Nhuan, N.P. and Walker, T.H. 2008. Biofuel feedstocks. *In*: Hager, L.S. (ed.). Biofuels Engineering Process Technology. MacGraw-Hill, USA.

Fei, Q., Puri, A.W., Smith, H., Dowe, N. and Pienkos, P.T. 2018. Enhanced biological fixation of methane for microbial lipid production by recombinant *Methylomicrobium buryatense*. Biotechnol. Biofuels 11: 129.

Food and Agriculture Organization (FAO). 2021. BEFS Cogeneration tool. Consulted in June 2021: http://www.fao.org/energy/bioenergy/bioenergy-and-food-security/assessment/befs-ra/energy-end-use/en/.

Gama-Ferreira, R. and Petrides, D. 2020. Production of β-Farnesene: Modeling and Evaluation with SuperPro Designer®. Consulted online on june 2021: https://www.researchgate.net/publication/341103805_Production_of_Farnesene_a_Terpene_via_Fermentation_-_Process_Modeling_and_Techno-Economic_Assessment_TEA_using_SuperPro_Designer.

Garcia-Nunez, J.A., Garcia-Perez, M., Rodriguez, D.T., Ramirez, N.E., Fontanilla, C., Stockle, C., Amonette, J., Frear, C. and Silva, E. 2015. Evolution of palm oil mills into biorefineries: Technical, and environmental assessment of six biorefinery options. *In*: Abatzoglou, N. and Meier, D. (eds.). Biorefinery I: Chemicals and Materials From Thermo-Chemical Biomass Conversion and Related Processes. ECI Symposium Series. Consulted on June 2021: http://dc.engconfintl.org/biorefinery_I/9.

Henley, W.J., Litaker, R.W., Novoveská, L., Duke, C.S., Quemada, H.D. and Sayre, R.T. 2013. Initial risk assessment of genetically modified (GM) microalgae for commodity-scale biofuel cultivation. Algal Res. 2: 66–77.

Hilbeck, A., Meyer, H., Wynne, B. and Millstone, E. 2020. GMO regulations and their interpretation: How EFSA's guidance on risk assessments of GMOs is bound to fail. Environ. Sci. Eur. 32: 54–69.

International Energy Agency (IEA). 2015. Energy and Climate Change. World Energy Outlook Special Report. Consulted online in June 2021.

International Energy Agency (IEA). 2019a. Renewables 2019: Analysis and forecast to 2024. Consulted on June 2021: https://iea.blob.core.windows.net/assets/a846e5cf-ca7d-4a1f-a81b-ba1499f2cc07/Renewables_2019.pdf.

International Energy Agency (IEA). 2019b. The future of hydrogen. Consulted online on June 2021: https://iea.blob.core.windows.net/assets/9e3a3493-b9a6-4b7d-b499-7ca48e357561/The_Future_of_Hydrogen.pdf.

International Energy Agency (IEA). 2020. Outlook for biogas and biomethane: Prospects for organic growth. Consulted on June 2021: https://iea.blob.core.windows.net/assets/03aeb10c-c38c-4d10-bcecde92e9ab815f/Outlook_for_biogas_and_biomethane.pdf.

International Air Transport Association (IATA). 2015. Consulted Online in June 2021: http://www.iata.org/Pages/default.aspx.

Jamaludina, N.F., Hashima, H., Muis, Z.A., Zakaria, Z.Y., Jusoh, M., Yunus, A. and Murad, S.M.A. 2018. A sustainability performance assessment framework for palm oil mills. Journal of Cleaner Production 174: 1679–1693.

Jena, U., McCurdy, A.T., Warren, A., Summers, H., Ledbetter, R.N., Hoekman, S.K., Seefeldt, L.C. and Quinn, J.C. 2015. Oleaginous yeast platform for producing biofuels via co-solvent hydrothermal liquefaction. Biotechnol. Biofuels 8: 167–186.

Jiang, W., Gu, P. and Zhang, F. 2018. Steps towards 'drop-in' biofuels: Focusing on metabolic pathways. Current Opinion in Biotechnlogy 53: 26–32.

Karamerou, E.E., Parsons, S., McManus, M.C. and Chuck, C.J. 2021. Using techno-economic modelling to determine the minimum cost possible for a microbial palm oil substitute. Biotechnol. Biofuels 14: 57–76.

Kargbo, H., Harris, J.S. and Phan, A.N. 2021. "Drop-in" fuel production from biomass: Critical review on techno-economic feasibility and sustainability. Renewable and Sustainable Energy Reviews 135: 110168.

Kermania, M., Celebi, A.D., Wallerand, A.S., Ensinas, A., Kantor, I.D. and Maréchal, F. 2017. Techno-economic and environmental optimization of palm-based biorefineries in the Brazilian context. *In*: Espuña, A., Graells, M. and Puigjaner, L. (eds.). Proceedings of the 27th European Symposium on Computer Aided Process Engineering—ESCAPE 27 October 1st–5th, 2017, Barcelona, Spain © 2017 Elsevier B.V.

Koepke, M. 2021. Microbial fermentation for the production of terpenes. US patent 10,913,958.

Kushkevych, I., Vítězová, M., Vítěz, T. and Bartoš, M. 2017. Production of biogas: Relationship between methanogenic and sulfate-reducing microorganisms. Open Life Science 12: 82–91.

Lo, S.L.Y., How, B.S., Leong, W.D., Teng, S.Y., Rhamdhani, M.A. and Sunarso, J. 2021. Techno-economic analysis for biomass supply chain: A state-of-the-art review. Renewable and Sustainable Energy Reviews 135: 110164.

Magda, R., Szlovák, S. and Tóth, J. 2020. Chapter 7—The role of using bioalcohol fuels in sustainable development. pp. 133–146. *In*: Bochtis, D., Banias, G., Achillas, C. and Lampridi, M. (eds.). Bio-Economy and Agri-Production: Concepts and Evidence. Academic Press, London, United Kingdom.

Martinez-Hernandez, E., Magdaleno-Molina, M., Melgarejo-Flores, L.A., Palmerín-Ruiz, M.E., Zermeño-Eguia-Lis, J.A., Rosas-Molina, A., Aburto, J. and Amezcua-Allieri, M.A. 2019a. Energy-water nexus strategies for the energetic valorization of orange peels based on techno-economic and environmental impact assessment. Food and Bioproduct Processing 117: 380–387.

Martínez Hernández, E., Ramírez Verduzco, L.F., Amezcua-Allieri, M.A. and Aburto, J. 2019b. Process simulation and techno-economic analysis of bio-jet fuel and Green diesel production—Minimum selling prices. Chem. Eng. Research Design 146: 60–70.

Mazloomi, K. and Gomes, Ch. 2012. Hydrogen as an energy carrier: Prospects and challenges. Renew. Sustain. Energ. Rev. 16: 3024–3033.

Midyette, D. 2020. FEL1 and 2—The beginnings of successful paper machine conversions. PEERS/IBBC Virtual Conference, pp. 161–172.

Mungodla, S.G., Linganiso, L.Z. Mlambo, S. and Motaung, T. 2019. Economic and technical feasibilities studies: Technologies for second generation biofuels. Journal of Engineering, Design and Technology 17: 670–704.

NASA. 2012. Technology Readiness Level. Consulted on June 2021: https://www.nasa.gov/directorates/heo/scan/engineering/technology/technology_readiness_level.

National Renewable Energy Laboratory (NREL). 2011. Process Design and Economics for Biochemical Conversion of Lignocellulosic Biomass to Ethanol Dilute-Acid Pretreatment and Enzymatic Hydrolysis of Corn Stover. Task No. BB07.2410.

Naz, T., Nazir, Y., Fazili, A.B.A., Mustafa, K., Bai, X. and Song, Y. 2020. Transformation of Lignocellulosic Biomass into Sustainable Biofuels: Major Challenges and Bioprocessing Technologies.

Newman, D., Begg, S. and Welsh, M. 2020. Society of Petroleum Engineers—SPE Asia Pacific Oil and Gas Conference and Exhibition 2020, APOG 2020.

Nigam, P.S. and Singh, A. 2011. Production of liquid biofuels from renewable resources. Progress in Energy and Combustion Science 37: 52–68.

Niu, F.X., Lu, Q., Bu, Y.F. and Liu, J.Z. 2017. Metabolic engineering for the microbial production of isoprenoids: Carotenoids and isoprenoid-based biofuels. Synthetic and Systems Biotechnology 2: 167–175.

O'Brien, R.D. 1998a. Raw materials. pp. 1–45. *In*: Fats and Oils: Formulating and Processing for Applications. Technomic publishing Company Inc., Basel, Switzerland, 1998.

O'Brien, R.D. 1998b. Fats and oil processing. pp. 47–180. *In*: Fats and Oils: Formulating and Processing for Applications. Technomic publishing Company Inc., Basel, Switzerland, 1998.

Pandeeti, E.V.P., Sangeetha, V. and Deepika, R.G. 2019. Chapter 9—Emerging trends in the industrial production of chemical products by microorganisms. pp. 107–125. *In*: Buddolla, V. (ed.). Recent Developments in Applied Microbiology and Biochemistry. Academic Press, 2019, London, United Kingdom.

Patel, A., Karageorgou, D., Rova, E., Katapodis, P., Rova, U., Christakopoulos, P. and Matsakas, L. 2020. An overview of potential oleaginous microorganisms and their role in biodiesel and omega-3 fatty acid-based industries. Microorganisms 8: 434–474.

Pedraza-de la Cuesta, S., Knopper, L., van der Wielen, L.A.M. and Cuellar, M.C. 2018. Techno-economic assessment of the use of solvents in the scale-up of microbial sesquiterpene production for fuels and fine chemicals. Biofuels, Bioproducts & Biorefining. DOI: 10.1002/bbb.1949.

Petrides, D.P. and Siletti, C.A. 2004. The role of process simulation and scheduling tools in the development and manufacturing of biopharmaceuticals. *In*: Ingalls, R.G., Rossetti, M.D., Smith, J.S. and Peters, B.A. (eds.). Proceedings of the 2004 Winter Simulation Conference. Consulted on June 2021: https://www.researchgate.net/publication/4111967_The_role_of_process_simulation_and_scheduling_tools_in_the_development_and_manufacturing_of_biopharmaceuticals.

Petrides, C., Cooney, C.L., Evans, L.B., Field, R.P. and Snoswell, M. 1989. Bioprocess simulation: An integrated approach to process development. Computer & Chemical Engineering 13: 553–561.

Renewable Energy Policy Network for the 21th Century. REN21. 2018. Renewables 2018 Global Status Report. Consulted online in June 2021: https://www.ren21.net/wp-content/uploads/2019/08/Full-Report-2018.pdf.

Rodrigues Reis, C.E., Bento, H.B.S., Carvalho, A.K.F., Rajendran, A., Hu, B. and De Castro, H.F. 2020. Critical applications of Mucor circinelloides within a biorefinery context. Critical Reviews in Biotechnology 39: 555–570.

Rogulska, P. Bukrejewski and Krasuska, E. 2018. Biomethane as transport fuel. *In*: Biernat, K. (ed.). Biofuels—State of Development. IntechOpen, DOI: 10.5772/intechopen.75173. Consulted online in June 2021: https://www.intechopen.com/books/biofuels-state-of-development/biomethane-as-transport-fuel.

RosalesCalderon, O. and Arantes, V. 2019. A review on commercial scale high value products that can be produced alongside cellulosic ethanol. Biotechnology for Biofuels 12: 240–298.

Sadhukhan, J., Martinez-Hernandez, E., Amezcua-Allieri, M.A., Aburto, J. and Honorato, J.A. 2019. Economic and environmental impact evaluation of various biomass feedstock for bioethanol production and correlations to lignocellulosic composition. Bioresource Technology Reports 7: 100230.

Sahar, S.S., Iqbal, J., Ullah, I., Bhatti, H.N., Nouren, S., Rehman, H.U., Nisar, J. and Iqbal, M. 2018. Biodiesel production from waste cooking oil: An efficient technique to convert waste into biodiesel. Sustainable Cities and Society 41: 220–226.

Saputelli, L., Black, A., Passalacqua, H. and Barry, K. 2013. Front-End-Loading (FEL) process supporting optimum field development decision making. Presented at the Society of Petroleum Engineers (SPE) Kuwait Oil and Gas Show and Conference held in Mishref, Kuwait, 7–10 October 2013.

Shanklin, T., Roper, K., Yegneswaran, P.K. and Marten, M.R. 2001. Selection of bioprocess simulation software for industrial applications. Biotechnol. Bioeng. 72: 483–9.

Shariat Panahi, H.K., Dehhaghi, M., Kinder, J.E. and Chukwuemeka Ezeji, Th. 2019. A review on green liquid fuels for the transportation sector: A prospect of microbial solutions to climate change. Biofuel Research Journal 23: 995–1024.

Shokravi, Z., Shokravi, H., Aziz, M.A. and Shokravi, H. 2019. The fourth-generation biofuel: A systematic review on nearly two decades of research from 2008 to 2019. pp. 213–251. *In*: Aziz, M.A., Kassim, K.A., Bakar, W.A.W.A., Marto, A. and Muhammad, S.A.F.S. (eds.). Fossil Free Fuels. CRC Press, Boca Raton, FA, USA.

Sitepu, I.R., Garay, L.A., Sestric, R., Levin, D., Block, D.E., German, J.B. and Boundy-Mills, K.L. 2014. Oleaginous yeasts for biodiesel: Current and future trends in biology and production. Biotechnology Advances 32: 1336–1360.

Spagnuolo, M., Yaguchi, A. and Blenner, M. 2019. Oleaginous yeast for biofuel and oleochemical production. Current Opinion in Biotechnology 57: 73–81.

Szyjka, S.J., Mandal, S., Schoepp, N.G., Tyler, B.M., Yohn, C.B., Poon, Y.S., Villareal, S., Burkarta, M.D., Shurin, J.B. and Mayfield, S.P. 2017. Evaluation of phenotype stability and ecological risk of a genetically engineered alga in open pond production. Algal Res. 24: 378–386.

U.S. Energy Information Administration (EIA). 2015. Technical Economic Analysis Guide—DRAFT. U.S. Department of Energy. Consulted online in June 2021: https://www.eia.gov/outlooks/documentation/workshops/pdf/tea_guide_071015_draft.pdf.

Valencia, D., Garcia-Cruz, I., Uc, V.H., Ramirez-Verduzco, L.F., Amezcua-Allieri, M.A. and Aburto, J. 2018. Unravelling the chemical reactions of fatty acids and triacylglycerides under hydrodeoxygenation conditions based on a comprehensive thermodynamic analysis. Biomass and Bioenergy 112: 37–44.

Van Dyk, S. and Saddler, J. 2021. Progress in commercialization of biojet fuel/sustainable aviation fuels: Technologies, potential and challenges. IEA Bioenergy. Consulted oline on June 2021: https://www.ieabioenergy.com/wp-content/uploads/2021/06/Task-39-Progress-in-the-commercialisation-of-biojet-fuels-FINAL-May-2021.pdf?utm_campaign=IEA%20Bioenergy%20Press%20Release%2029%20June&utm_medium=email&utm_source=EOACLK.

Vivekanandhan, S., Zarrinbakhsh, N., Misra, M. and Mohanty, A.K. 2013. Coproducts of biofuel industries in value-added biomaterials uses: A move towards a sustainable bioeconomy. *In*: Fang, Z. (ed.). Liquid, Gaseous and Solid Biofuels. IntechOpen, London, United Kingdom.

Walls, L.E. and Rios-Solis, L. 2020. Sustainable production of microbial isoprenoid derived advanced biojet fuels using different generation feedstocks: A review. Frontiers in Bioengineering and Biotecnology 8: 599560.

World Bioenergy Association (WBA). Global bioenergy statistics 2019. 2019. Consulted on June 2021: http://www.worldbioenergy.org/uploads/191129%20WBA%20GBS%202019_HQ.pdf.

Wu, W., Gladden, J.M., Wu, B.C.P. and Davis, R.W. 2021. Terpene synthases for biofuel production and methods thereof. US patent 10,947,563.

Zhou, Y., Zhao, W., Lai, Y., Zhang, B. and Zhang, D. 2020. Edible plant oil: Global status, health issues, and perspectives. Front. Plant Sci. 11: 1315.

Index

For Product Safety Concerns and Information please contact our EU
representative GPSR@taylorandfrancis.com
Taylor & Francis Verlag GmbH, Kaufingerstraße 24, 80331 München, Germany

www.ingramcontent.com/pod-product-compliance
Lightning Source LLC
Chambersburg PA
CBHW080924220326
41598CB00034B/5667